Lecture Notes in Artificial Intelligence

Subseries of Lecture Notes in Computer Science
Edited by J. G. Carbonell and J. Siekmann

Lecture Notes in Computer Science
Edited by G. Goos, J. Hartmanis and J. van Leeuwen

Springer
Berlin
Heidelberg
New York
Barcelona
Budapest
Hong Kong
London
Milan
Paris
Singapore
Tokyo

Ian Smith (Ed.)

Artificial Intelligence in Structural Engineering

Information Technology for Design, Collaboration, Maintenance, and Monitoring

Series Editors
Jaime G. Carbonell, Carnegie Mellon University, Pittsburgh, PA, USA
Jörg Siekmann, University of Saarland, Saarbrücken, Germany

Volume Editor

Ian Smith
Structural Engineering and Mechanics (IMAG-DGC)
EPFL–Federal Institute of Technology
CH-1015 Lausanne, Switzerland
E-mail: Ian.Smith@epfl.ch

Cataloging-in-Publication Data applied for

Die Deutsche Bibliothek - CIP-Einheitsaufnahme

Artificial intelligence in structural engineering : information technology for design, collaboration, maintenance, and monitoring / Ian Smith (ed.). - Berlin ; Heidelberg ; New York ; Barcelona ; Budapest ; Hong Kong ; London ; Milan ; Paris ; Singapore ; Tokyo : Springer, 1998
(Lecture notes in computer science ; 1454 : Lecture notes in artificial intelligence)
ISBN 3-540-64806-2

CR Subject Classification (1991): I.2, D.2, J.2, J.6, H.5

ISBN 3-540-64806-2 Springer-Verlag Berlin Heidelberg New York

Printed in Germany

Typesetting: Camera ready by author
SPIN 10638211 06/3142 – 5 4 3 2 1 0 Printed on acid-free paper

Preface

Information technology applications in structural engineering are hindered by factors such as holistic knowledge, interdependent tasks, incomplete information and constantly changing contexts. Lower hardware costs, Internet communication, advances in product and process modelling, improved human-computer interfaces, faster computation and other advances in the information sciences have now created favourable conditions for increasing the number of useful applications. However, this is not enough for many important tasks.

Deductive tasks such as structural analyses have been successfully supported for several decades. However, applications to tasks such as design and diagnosis have not provided practising engineers with equivalent levels of assistance. Nevertheless, when engineers perform design and diagnosis tasks, they often add much more value to a project than when structural analysis tasks are carried out. Part of the difficulty originates in the nature of the task. Design and diagnosis tasks require abductive inference; engineers must reverse cause-effect and structure-behaviour relationships to satisfy multiple goals. To be reliable, black-box abductive inference requires complete domain models and this is impossible in structural engineering. Contextual parameters such as politics, economics and local conditions are nearly always important.

In addition to design and diagnosis, computer support for collaboration between multiple actors has much potential for improving the effectiveness of structural engineers. Project delays, excessive costs and even accidents are almost invariably linked to bad collaboration, especially when many changes have been necessary. Since a primary function of computing is to store and transmit data, support for collaboration seems easy. However, this is not the case in structural engineering. Structural engineers must work in parallel with architects, tradespeople, contractors and fabricators. Nearly every actor views construction projects differently and there are many competing goals. As a result, support for collaboration has required much more research than originally thought necessary.

One issue is common to all of these tasks. Representation, manipulation and use of structural engineering knowledge is inevitably a major challenge for each and every application. Since such aspects of knowledge have been the subject of study in artificial intelligence (AI) research since the late 1950s, it is understandable that researchers are looking to this field for inspiration. However, simple reuse of AI algorithms and representational frameworks is not possible. Most structural engineering tasks are complex, changing, long and poorly defined. Structures must be safe, serviceable, environmentally correct, useful, aesthetically pleasing, easy to build and long lasting. Therefore, structural engineering is one of AI's most difficult application fields.

The papers in this volume present the state of the art of AI in structural engineering. Not all contributions deal directly with structural engineering issues since much excellent and applicable research is underway in other areas. This work describes key advances that are becoming catalysts for fully exploiting the current favourable conditions which were mentioned above for increasing the number of useful applications.

I would like to thank all of the authors and particularly, every member of the IAB (International Advisory Board – see list overleaf) for contributing papers to this volume. The quality of the papers is a direct result of the work of reviewers who unselfishly provided constructive criticism so that authors could improve their papers. I would thus like to thank the members of the Programme Committee, the Organising Committee and the IAB for ensuring that each paper received at least three reviews. I am also grateful to K. Shea, B. Raphael and R. Stalker for additionally helping with review organisation, paper management and many administrative tasks. Finally, H. Flühler and C. Stamm of the Centro Stefano Franscini and ETHZ provided much advice and support.

Lausanne, May 1998 Ian F.C. Smith

Organisation

These papers were presented at a conference organised by the European Group for Structural Engineering Applications of Artificial Intelligence (EG-SEA-AI).
http://www.strath.ac.uk/Departments/Civeng/egseaai/
The conference was made possible through a generous grant from the Centro Stefano Franscini, Monte Verità, Ascona, Switzerland.

Conference Chair

I.F.C. Smith, EPFL, Switzerland

Program Committee

B. Kumar University of Strathclyde, UK, Chair
C. Moore University of Wales, Cardiff, UK
J. Bento IST, Lisbon, Portugal
P. Salvaneschi ISMES, Bergamo, Italy

Organising Committee

I.F.C. Smith, EPFL
S. Boulanger, EPFL
B. Raphael, EPFL
K. Shea, EPFL
R. Stalker, EPFL

International Advisory Board

M. Abe, University of Tokyo, Japan
J.W. Baugh, University of North Carolina, USA
S.J. Fenves, Carnegie Mellon University, USA
G. Fischer, University of Colorado, USA
M. Fischer, Stanford University, USA
R. Fruchter, Stanford University, USA
J.H. Garrett, Carnegie Mellon University, USA
J.S. Gero, University of Sydney, Australia
J.C. Kunz, Stanford University, USA
K.H. Law, Stanford University, USA
M.L. Maher, University of Sydney, Australia
W.J. Mitchell, MIT, USA
D.T. Ndumu, BT Labs, UK
F. Peña-Mora, MIT, USA
K. Roddis, University of Kansas, USA
R.J. Scherer, TU Dresden, Germany
G. Schmitt, ETHZ, Switzerland

Table of Contents

Long Papers

Short Papers

Structural Monitoring of Civil Structures Using Vibration Measurement – Current Practice and Future –

Masato Abé

Department of Civil Engineering, The University of Tokyo, Tokyo 113-8656, Japan
masato@kyouryou.t.u-tokyo.ac.jp
http://kyouryou.t.u-tokyo.ac.jp/~masato/index.html

Abstract. Structural monitoring is an attractive solution for the maintenance of civil infrastructure to assure the reliability of the entire network of infrastructure system with reasonable level of investment. Structural vibration, which can relatively readily measured without damaging structures, reflects the integrity and loading conditions of structures. In this paper, state of the art of application of vibration measurement to structural monitoring is reviewed with variety of examples in engineering practices. Potential future development with application of artificial intelligence technology is also discussed.

1. Introduction

Health monitoring of civil structures has been attracting much attention from both research community and practicing engineers as more and more structures start degradation by aging in especially many developed countries. By monitoring actual load and actual strength of structures, the uncertainty inherent in design calculation, which is represented as factor of safety, can be reduced. Hence, remaining life of existing structures can be accurately evaluated and, in most cases, extended. At the same time, appropriate repair or strengthening of structures based on the monitored state would further lengthen the life of structure. Because the cost for monitoring and repair is much lower than the cost for reconstruction of new structures, monitoring is the economical solution to maintain the vast stock of civil infrastructure. At the developed countries where the number of existing structures is enormous, prioritization of repair investment is also necessary to maintain the reliability of the entire infrastructure system with reasonable amount of investment.

Structural vibration, which can relatively readily measured without damaging structures, reflects the integrity and loading conditions of structures. It is often used to monitor the status of existing structures, which is usually not tractable by analytical methods or laboratory testing. For example, it has been used to evaluate the effect of dynamic loadings, such as wind, earthquake, or traffic loadings, and also to verify the performance of passively/actively controlled structures. Application of vibration measurement for health monitoring of civil structures has also been actively studied in recent years.

The purpose of this paper is to provide overview on the state of the art of structural monitoring using vibration measurement, and discuss the potential future development with possible application of artificial intelligence technology. The review in this paper is intended to be rather problem specific than extensive, to provide practical viewpoints on vibration monitoring. For this purpose, all the examples are taken from the application practices which the author has participated in or is familiar with, and hence mostly taken from Japanese practice. The focus of the discussion is on highway and railway bridges, which represent the essential and typical elements of civil infrastructure.

At the beginning of the paper, current health monitoring practice is explained using the example of inspection strategy proposed by Railway Technical Research Institute (RTRI) of Japan. Following this introduction, application of vibration measurement to structural monitoring is discussed. First, various possible applications of vibration measurement and supporting technology are summarized with relation to typical life-cycle of bridges. Then, the basic framework of vibration monitoring is presented, and its difficulty and potential development are discussed. These remarks show the direction to successful monitoring, where potential improvement is expected by application of artificial intelligence. At the last part of the paper, several examples of monitoring using vibration measurement are introduced along with on-going research project at Bridge & Structure Laboratory of the University of Tokyo, to substantiate the general remarks in the first half of the paper.

2. Health Monitoring of Civil Structures

To clarify the monitoring technology in current practice, inspection and repairing strategy of railway bridges implemented by RTRI of Japan, shown in Fig.1, is explained as an example [1]. Due to fatigue and other environmental effects such as corrosion, structural performance tends to degrade with time from initial performance level (level 1) as shown in the solid line in the figure. Not only structural degradation but also change in loading condition, such as increase of speed or weight of train, causes similar effect. When the remaining structural performance reaches at the maintenance limit (level 3), the structure need to be repaired to maintain serviceability. Usually, investment starts earlier than this maintenance limit (level 2) so that the reliability level of the entire network of railway system be kept at certain level. In addition to this gradual degradation process, rare event such as earthquake or ship crash may happen which can lower the performance level of the structure considerably (point *c*).

The inspection procedures are shown Fig.2 and 3. Detailed inspection of Fig.3 is conducted if damage is found at the regular inspection of shown in Fig.2. These inspections mainly rely upon visual and qualitative information and judgment of experienced experts. In the inspection, the structural state is classified into the ranks *AA, A1, A2*, .. etc, in Fig.1. These ranks are assigned by expert judgment based on reference values of measured stress and allowable stress as defined in Table 1. The following stress ratio,

$$SR=\sigma_m/\sigma \tag{1}$$

is used for classification, where σ: induced stress when the bridge is loaded by train with the allowable maximum speed and σ_m: allowable stress which reflects the effect of aging and fatigue. Because measurement of these values is extremely difficult especially at service conditions, knowledge and judgment of expert are crucial at this decision.

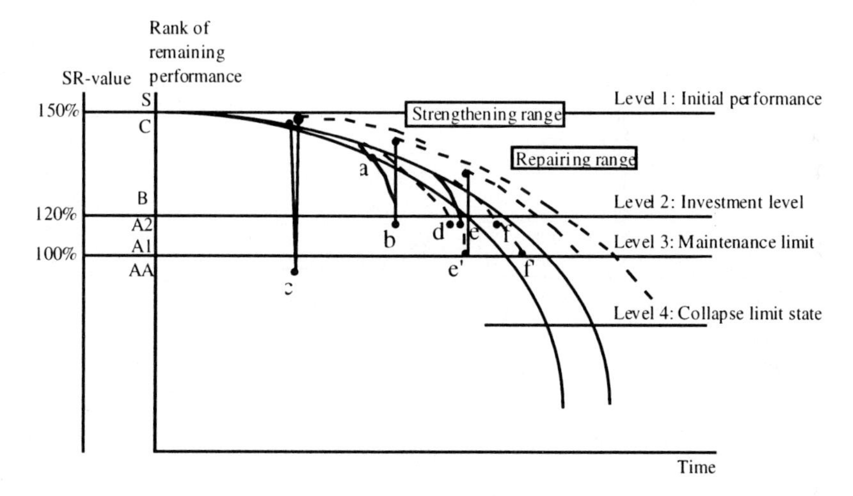

Fig. 1. Repairing strategy and performance degradation process.

a: Degradation of rank *C*, precaution is required
b: Rank *A2*, will be repaired with appropriate scheduling
c: Emergent action will be required due to accident or major natural disaster
d: Same level of degradation of *b*, but less investment for repair would be required.
e : Rapid degradation is expected
e' : Due to rapid degradation, it is expected to reach rank *AA* before the next regular inspection
f: Slower degradation
f' : Since the degradation is slower, there will be enough time for detailed inspection before reaching *AA*.
Although *e* and *f* are at the same level of performance, priority of repair will be placed to *e* because the degradation is faster.

Table 1. Classification of structural status in Fig. 1.

Rank	Actual remaining stress ratio (SR-value)
AA	SR≤100%
A1/A2	100%<SR≤120%
B	120%<SR≤150%
C/S	SR>150%

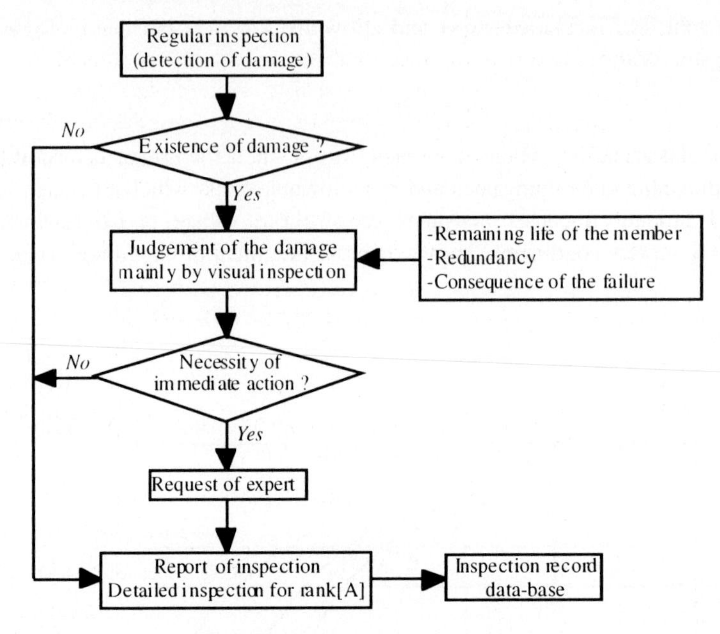

Fig.2. Regular inspection procedure.

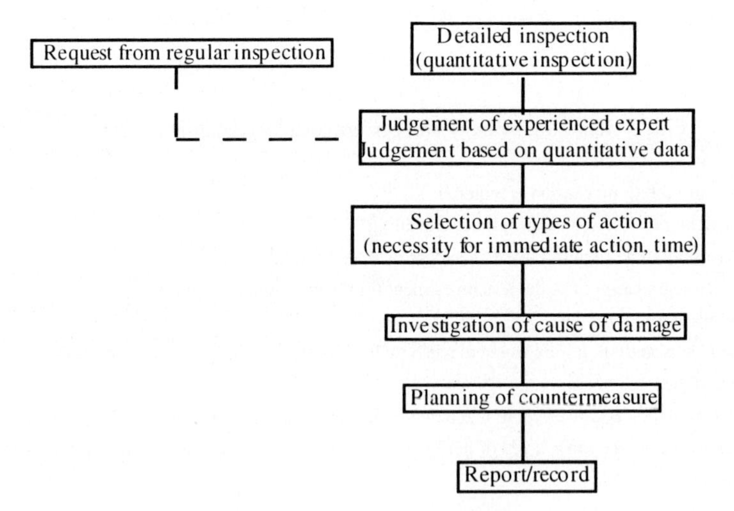

Fig.3. Detailed inspection procedure.

By automating this inspection procedure using advanced monitoring technology and obtaining quantitative information, the reliability of the maintenanced structure is expected to be improved while the cost of inspection would reduce considerably. In this paper, this automation which includes judgment is referred to as *structural monitoring*. Due to the scale, the accessibility and the surrounding noisy environment of civil structures, monitoring technology which has the ability to grasp the structural state with relatively limited number of measurement points, and the robustness to sustain high level of environmental uncertainty is required. These fundamental requirements make application of vibration measurement attractive over other conventional quantitative non destructive evaluation (NDE) methods.

3. Application of Vibration Measurement to Structural Monitoring

3.1. Life-Cycle of Structure and Vibration Monitoring

Structural vibration reflects information on loading, and structural parameters such as mass, stiffness and damping, and has been applied to monitor structural and loading status at various situations. Fig.4 shows the life cycle of highway bridge structure with possible application of monitoring by vibration measurement [2]. In certain cases, such as vibration control or evaluation of serviceability/environmental effects, information of vibration itself is of interest, while in other cases such as health monitoring or load estimation, vibration measurement is used as a mean to obtain structural or loading information.

The technology which supports monitoring using vibration measurement is summarized in Table 2. Measurement, data processing, and modeling are the essential and fundamental technology to support vibration measurement. Items categorized in advanced technology are the necessary technological advancement to develop the intelligent structural monitoring system Advanced monitoring technology is expected to provide more reliable and spatially distributed information with high resolutions, while the decision making technology would automate the expert judgment procedures such as those given in Figs.2 and 3.

3.2. Fundamental Scheme of Monitoring Using Vibration Measurement

As noted previously, monitoring using vibration measurement can be classified into the following two categories: (i) monitoring of vibration itself, such as evaluation of comfort or control of structures; and (ii) monitoring of structural/loading state via vibration measurement. Monitoring in category (i) is generally a straight forward procedure, because the quantity of interest is directly measured. On the other hand, problems in category (ii), which need extraction of information from vibration measurement, are more challenging, because the system must be solved inversely to

LIfe-Cycle of Highway Bridges	Purpose of Monitoring	Required Monitoring
Planning/Design	-Prediction of structural performance and loading conditions	-Model experiment -Measurement at existing structures
Construction	-Control of Precision of Shape -Improvement of Construction Method -Working Condition	-Monitoring of static deformation, temperature, etc. - Measurement of dynamic properties
Completion	-Performance Evaluation	-Monitoring of dynamic properties -Comparison with structural design
Serivice Maintenance	-Safety evaluation -Prediction of remaining life -Serviceability evaluation	-Monitoring for serviceability/health conditions in service/extreme loading (disaster) conditions -Monitoring for fatigue/damage status -Monitoring for actual stress/loading
	-Evaluation of riding comfort	-Monitoring for road roughness
	-Environmental requirement	-Monitoring of noise and vibration at structure and surrounding environment
Retrofit/Strengthning	Selection of retrofit method Evaluation of retrofit effect	-Monitoring by vibration/acoustic measurement or NDE
Demolish/recycle		

Fig.4. Application of vibration monitoring in life-cycle of highway bridges.

Table 2. Supporting technology for vibration monitoring

Fundamental technology	
Fundamental tools for measurement	-Sensors -Data acquisition/processing systems -Loading method for testing
data processing	-digital filters -Analysis and identification methods (time/frequency domain) -Estimation of damping
Modeling technique	-Modeling of structure (girder, slab, bearing, connections, etc) -Fatigue, fracture/damage model -Structural analysis method -Modeling of surrounding environment (traffic/wind)
Advanced technology	
Advanced monitoring systems	-Optical fibers -Laser/long range measurement system -Smart/intelligent structures -Real-time monitoring systems -Innovative sensor technology and measurement systems
Decision making	-Methods to connect measurement and conditions of structure/ loading -Identification of damage and fatigue status -Cost and reliability evaluation -Prediction of future status

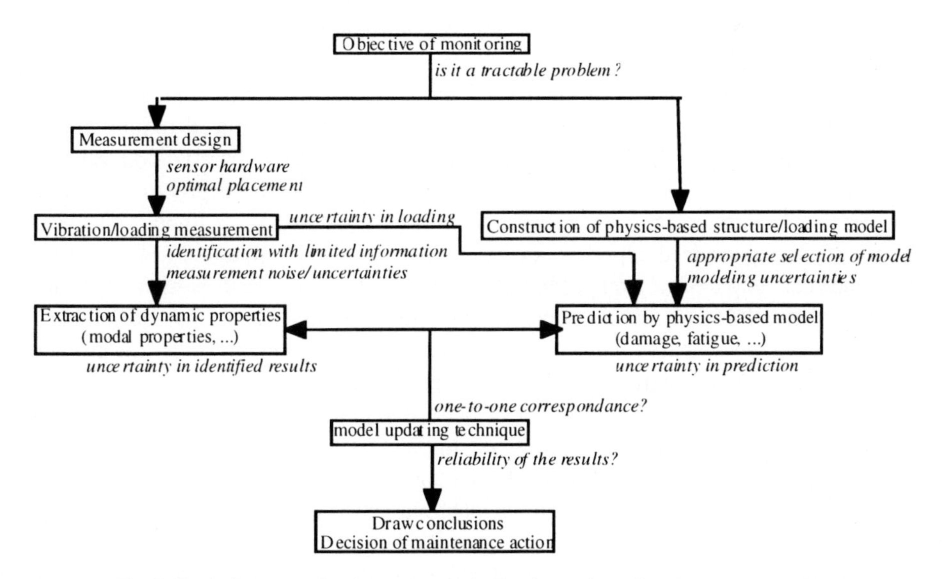

Fig.5. Basic framework of structural monitoring using vibration measurement.

obtain structural or loading information, and this inverse problem can easily be ill-posed as will be explained in the examples in the next section.

Fig.5 shows the basic framework of monitoring of category (ii). First, objective of monitoring should be appropriately established so that the problem is tractable by available technical and investment resources. The left hand side branch shows the procedure of measurement. Measurement system needs to be carefully designed with appropriate selection of hardware which has sufficient resolution to satisfy the objective, and the sensors should be allocated to avoid any unobservability. However, even with high level of caution, there always exists uncertainty in measured result introduced natural or human causes. This uncertainty can be magnified by data processing, and identified dynamic properties or loading are expected to contain certain amount of uncertainty.

To extract information on structural/loading state, physics-based model is also constructed as shown in the right hand side of the figure. Because of inherent uncertainty of modeling, i.e. unmodeled dynamics, the prediction based on physics-based model is also with uncertainties, which can even be larger than uncertainty in measurement.

These identified results from measurement and prediction by physics are compared, and the structural status is evaluated by updating the model according to measured results. In most cases, civil structures have a large number of degrees of freedom, and the measurement resources are far from sufficient to verify all the degrees. Therefore, there generally exist multiple possible realization of physics-based model to recreate the same measurement, which makes application of analytical approach almost impossible. The problem becomes even harder due to the above noted measurement and modeling uncertainties. Hence, heuristic judgment of experienced expert is usually introduced at this stage to eliminate several theoretically-possible but practically-unlikely conclusions, and to extract meaningful information on structural and loading states.

The items in italic letters in Fig.5 show the bottlenecks in this framework which cause these difficulties, where expert knowledge is usually introduced to interpret the results. Among these bottlenecks, major problems in vibration based monitoring can be summarized as: (a) insufficient measurement to verify all the degrees of freedom; and (b) uncertainties in both measurement and modeling. Although heuristic reasoning of experienced experts using experiences of similar situations and qualitative inference can help extract limitted but useful information even at these difficulties, it is almost impossible to obtain quantitative information. On the other hand, mathematics/physics-based reasoning gives quantitative information when the problem is well-posed, which is hardly the case in this problem.

Therefore, appropriate combination of these two reasoning methods is expected to constitute the breakthrough to successful implementation of structural monitoring. Application of artificial intelligence technology, such as knowledge based expert system, inference with fuzzy logic, or artificial neural networks, where heuristic knowledge and experiences can be quantified and connected to mathematical treatment, appears to be the most natural and attractive way to tackle this challenging problem.

4. Examples of Structural Monitoring Using Vibration Measurement

4.1. Monitoring for Vibration and Control

This type of monitoring is categorized as "category (i)" defined in the previous section, where vibration measurement itself is of main interest. For example, vibration testing after completion of long-span bridges is a standard procedure to verify the dynamic properties predicted by design, because these properties are closely related to aerodynamic stability of the structure. One of the recent example is the testing of Kap Shui Mun Bridge in Hong Kong where standard spectral analysis using ambient vibration measurement is applied to verify its performance[3].

Vibration measurement is also essential for application of structural control. To suppress wind-induced vibration so as to improve comfort and working conditions, active control has been implemented in several civil structures since late 1980's. For instance, active control is often used to ensure the working conditions during construction of cable stayed bridges or suspensions bridges because their towers are extremely flexible at the construction stage where they need to stand alone without cables. In Hakucho Bridge, active mass damper (AMD) is installed during construction as shown in Fig.6 [4]. Table 3 gives the design criteria for the control system at various construction stages of the tower. To observe the contribution of multiple modes, accelerometers are distributed relatively densely, and anemometer is also used to measure wind velocity. Because the existence of modeling and measurement uncertainties in the real structure is inevitable, direct velocity feedback control is employed to guarantee the absolute stability of the structure.

As one can observe at these examples, the monitoring of category (i) is at the state of certain level of maturity in practical applications.

Table 3. Design criteria for control system

Construction stage of tower	Objective natural frequency [Hz]	Target logarithmic damping ratio
After completion of middle lateral connection	0.40-0.52	0.570
Before completion of top lateral connection	0.22-0.29	0.250
At the completion	0.21-0.27	0.210
Stand-alone	0.20-0.26	0.180

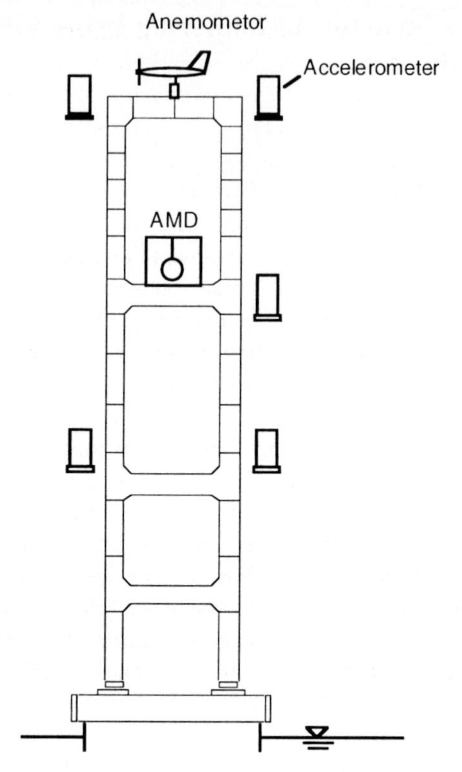

Fig.6. Control system installation at Hakucho Bridge.

4.2. Monitoring for Performance Evaluation

Vibration data is also used to evaluate the performance of structures at actual loading. In 1995 Kobe earthquake, seismic records of main shock and aftershocks are obtained at Matsunohama district of Hanshin Expressway (Fig.7), which is approximately 35[km] east of epicenter. The instrumentation is shown in Fig.8, and the observed maximum ground acceleration was 144[cm/s^2]. Because this bridge is one of the first base-isolated bridges using lead rubber bearings (LRB) in Japan, its performance is evaluated using this observed record by the author and his colleagues [5]. In this investigation, the authors tried to quantify the uncertainty in measurement and modeling.

The measured data contained unnaturally large amount of high frequency component, which was, at first, suspected to be the noise introduced by the sensors' performance. Finding this noise has some distinct spectral characteristics, the authors have examined the attachment details of the accelerometer and identified the cause as the local oscillation of lateral stringer. Then, digital filter is designed using the transfer function of actual measurement of stringer vibration with respect to the main girder vibration in order to eliminate this noise. Because of this filtering treatment, the uncertainty of measurement data is reduced considerably. Then, damping and stiffness of lead rubber bearings are identified using spectral method. During identification, Bootstrap analysis [6] is applied to evaluate the amount of uncertainty in identified values. Confidence interval with 99% is found to be less than 0.1% of the averaged value, which implies that uncertainty in measurement and hence in identified values are relatively small.

The main source of uncertainty for modeling is identified as friction of bearing supports at both ends of the girder. Considering the variation of friction with respect to aging, corrosion, humidity, etc, the authors have quantified the possible range of friction coefficient based on available test results of the bearings. Fig. 9 shows the comparison of the identified stiffness value and the predicted value of physics-based model. The effect of uncertainty bound of modeling is observed to be extremely large especially at low level seismic input, due to the uncertainty in modeling of friction. The effect of uncertainty in friction reduces as response amplitude increases, because the magnitude of Coulomb friction force is constant regardless of the amplitude level.

In this investigation, uncertainty in identified values is small, but modeling uncertainty is found very large in case of small seismic events. If the monitoring objective is to verify performance using only small seismic events, it was not an appropriate tractable problem and the framework of Fig.5 fails from the very beginning. In addition, appropriate modeling of attachment details of accelerometer and friction force required engineering judgment and insight of experienced experts.

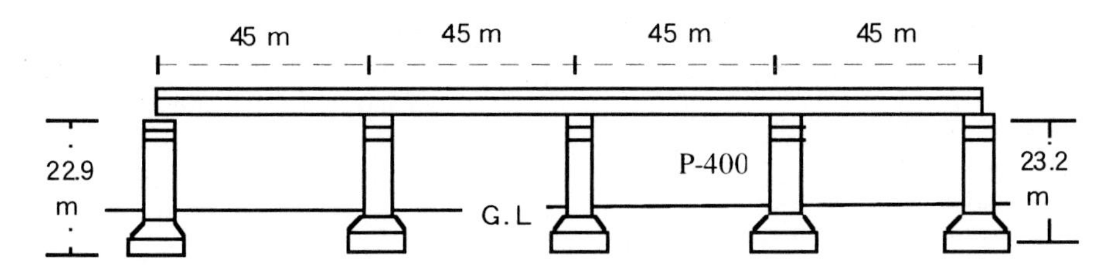

Fig.7. Matsunohama base-isolated bridge.

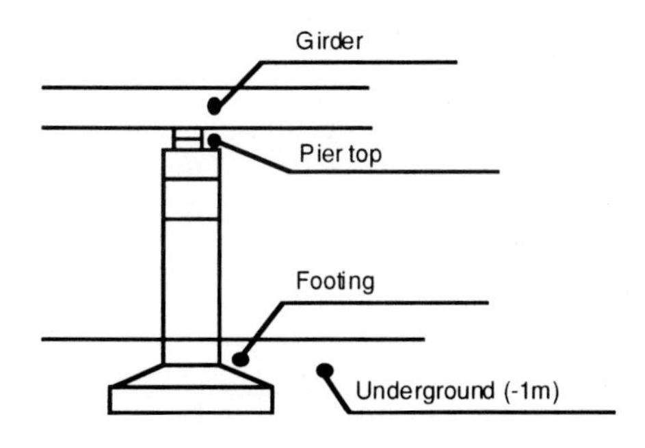

Fig.8. Location of measurement.

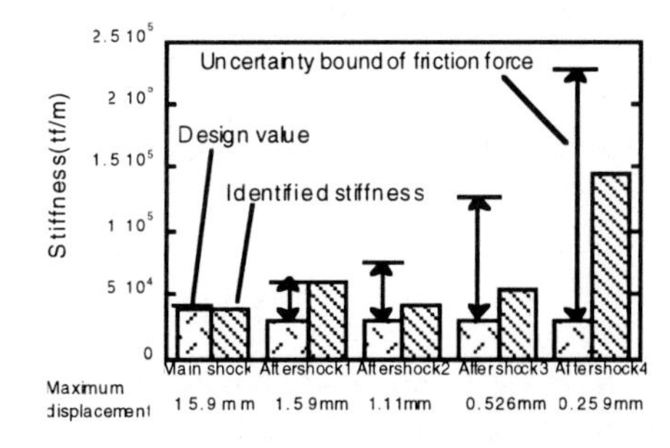

Fig. 9. Identified stiffness of LRB and prediction by design values.

4.3. Monitoring for Damage Detection

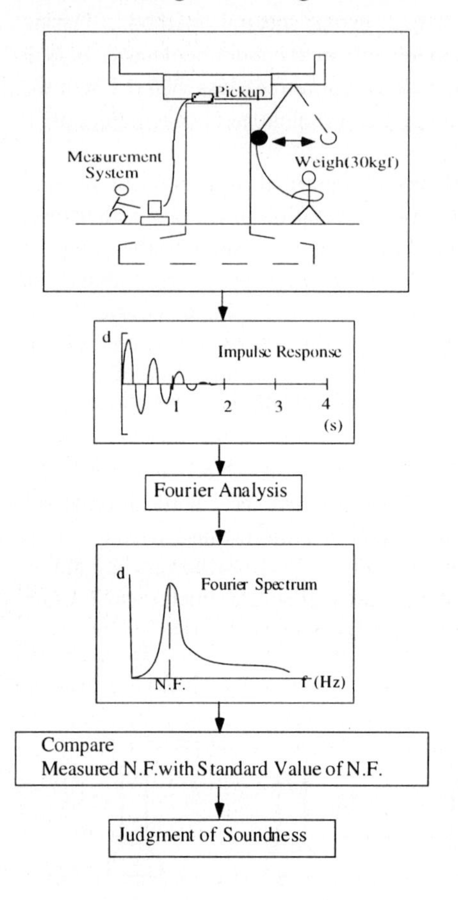

Fig.10. Procedure of impact vibration test to evaluate soundness.

Vibration measurement has also been used to verify the performance of railway bridges by RTRI [7]. The procedure is shown in Fig.10. Portable measurement system is employed rather than permanent installation of sensors. The fundamental natural frequency of the structure is identified using impact vibration test and spectral analysis. The identified natural frequency is compared with standard value, and the performance of the bridge is classified by the soundness index of Table 4. The standard value of natural frequency is derived from statistical regression analysis of existing railway bridges of the same type which is listed in Table 5. This procedure is especially suitable to the structures where standard design is applied, and a large number of natural frequency measurement can be obtained for the same type of structures.

The definition of soundness index and standard value come from experiences of actual damaged piers and is not supported by physics-based modeling. Hence, the information obtained in this procedure is basically qualitative and quantitative estimation of reliability for the soundness index in Table 4 is unknown. It should also be noted that introduction of certain uncertainty is incvitable to obtain regression equations in Table 5.

Table 4. Classification of soundness.

Soundness index	Soundness classification	Recommended action
<0.70	A1	Immediate restoration
<0.85	A2	Precaution
>0.86	B	No need for repair

Table 5. Standard values of natural frequency for rigid framed viaduct in transverse direction

Foundation type	Standard value for natural frequency
Spread foundation	$N.F.=4.03N^{0.14}/Hd^{0.37}$
Pile foundation	$N.F.=5.12(nD^3N^{0.25})^{0.01}/Hd^{0.34}$

N: Weighted average of standard penetration test results
Hd: Height of pier from ground level [m]
D: Diameter of pile [m]
n: Number of piles

4.4. Development for Quantitative Evaluation of Structural Status

The previous example of RTRI demonstrates practical applicability of vibration measurement for structural monitoring. To obtain quantitative information on structural status, applicability of long range measurement system and model updating technique are currently investigated at the Bridge & Structure Laboratory of the University of Tokyo.

One of the fundamental difficulties in monitoring using vibration measurement is the restriction of number of measurement points which reduces reliability of model updating. Number of measurement positions is limited mainly due to difficulty of installation especially at high positions, such as bridge girders. For example, in the installation of Fig.10, measurement point is only a few. To obtain quantitative information, large number of spatially distributed measurements with high resolutions are essential. Among currently available hardware, this goal can be achieved either by fiber optic sensors, or by long range optical/acoustical sensors, such as camera, laser or ultrasonic applications. In our development, long range optical measurement systems are employed to make the system portable.

At our first trial, CCD camera is used to measure the impact vibration test of Gimyo-Gawa Bridge shown in Fig.11. Fig.12 shows the installation of measurement system. Fig.13 shows the typical time history obtained by the measurement. Although the measurement result is found to contain first mode oscillation of the girder as distinct spectral content, it is also observed to include large amount of noise content. By comparing with accelerometer reading at the same position, the authors try to investigate the cause and to quantify the uncertainty in the measurement. By several trials, this uncertainty is identified as the effect of thermal fluctuation of air. Because thermal fluctuation is influenced by other uncertain causes such as temperature at air/road surface and wind velocity, the authors have not yet been successfully modeled these uncertainties to extract meaningful information from measurement.

Laser Doppler vibrometer is then tested as another hardware configuration. Because this device uses Doppler effect of the laser light to identify velocity of reflection surface, it is expected to be less sensitive to thermal fluctuation. Ambient vibration measurement using two story building model structure (Fig.14, Fig.15) shows high potential of this method. Although the maximum displacement response of the structure is approximately 0.004[mm] in this example, spectral contents can be identified clearly as shown in Fig.15.

Using these measured results, parameter variation of the structure is identified using sensitivity based modal updating technique [8]. It can easily be shown that the necessary condition of existence of solution is that,

$$p \leq q_f + q_s(s - 1) \tag{2}$$

where, p: number of elements whose properties may possibly be changed; q_f: number of natural frequencies identified by measurement; q_s: number of mode shape ratios identified; and s: number of measurement point [9]. Here, mode shape ratio is the ratio of Fourier spectra of each measurement point at the specified modal frequency. This mode shape ratio is employed to eliminate the effect of input force, which is unknown in ambient vibration measurement.

In this testing, 20% of story mass is added to the second floor of the model structure of Fig.14 to simulate the change in structural status. Because the model is basically a two degrees of freedom structure, p in Eq. (2) is simply equal to 2. Number of measurement point s is also taken to be 2 by measuring acceleration of both floors. Fig.16(a) shows the identified mass increase with q_f =2 and q_s =0, and Fig.16(b) is for q_f =1 and q_s =1. Although both cases satisfies Eq.(2) and mathematical observability, considerable error in estimation is introduced in the second case. This error is probably caused by larger uncertainty in identification of mode shape than in natural frequencies. This kind of reasoning of the cause of error can easily be done by expert, although mathematical quantification of uncertainty is difficult. Therefore ,model updating technique, which can also accommodate the effect of uncertainty in data, is necessary to make the conclusion more reliable. Application of artificial intelligence is an appropriate method to implement this heuristic reasoning of expert into model updating procedure, which would constitute one of the key technology for implementation of automated intelligent monitoring. Currently, construction of rule based inference system to combine the expert knowledge with mathematical results is being studied.

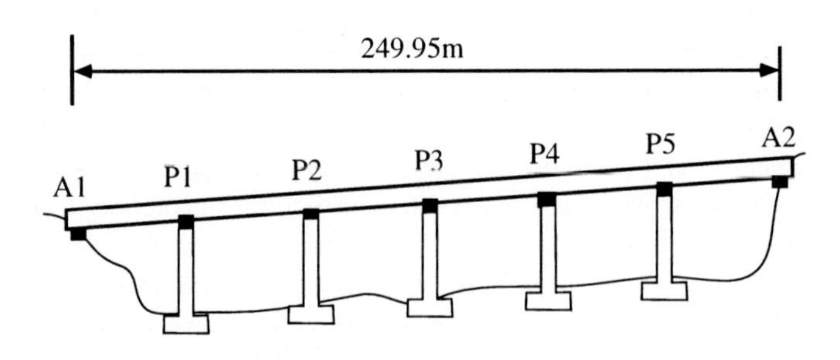

Fig.11. Gimyo-Gawa Bridge

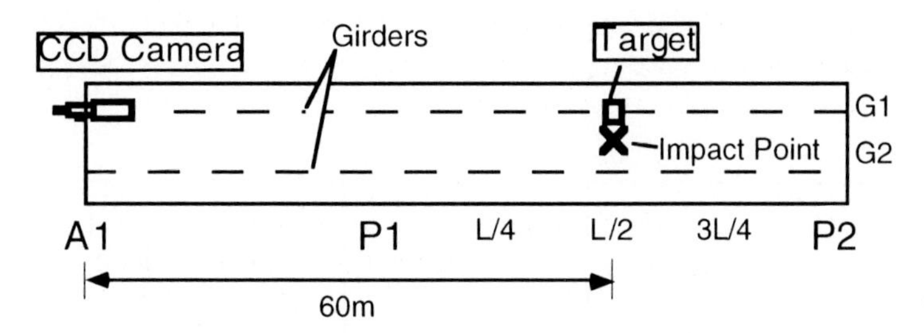

Fig.12. Configuration of impact vibration test.

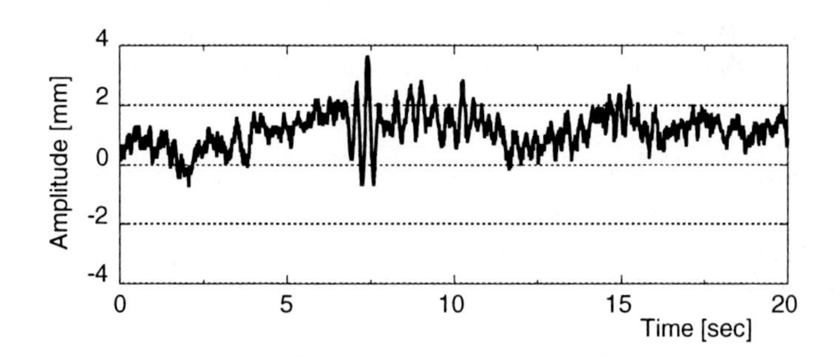

Fig.13. Measurement by CCD camera.

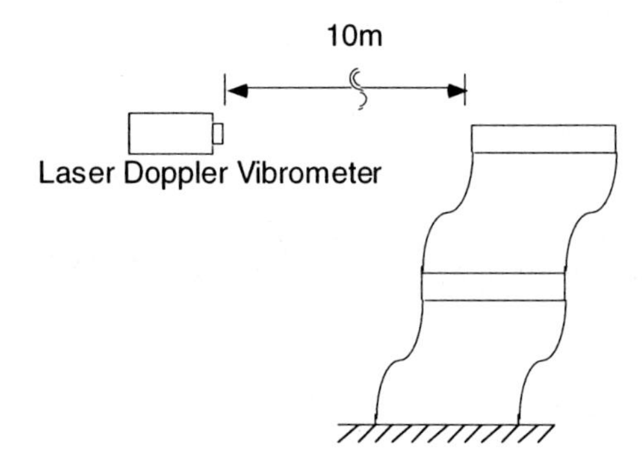

Fig.14. Test configuration using laser Doppler vibrometer.

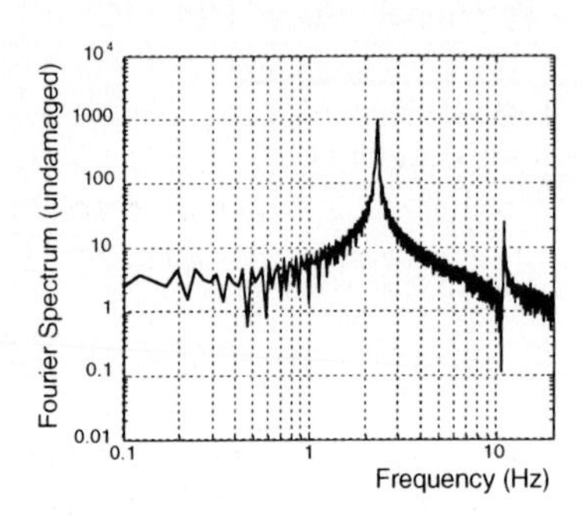

(a) Original state.

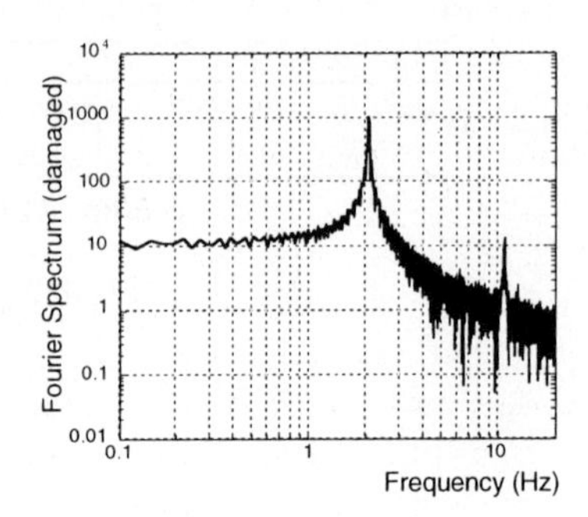

(b) With mass increase of 20% to the second floor.

Fig.15. Measured power spectrum of ambient vibration.

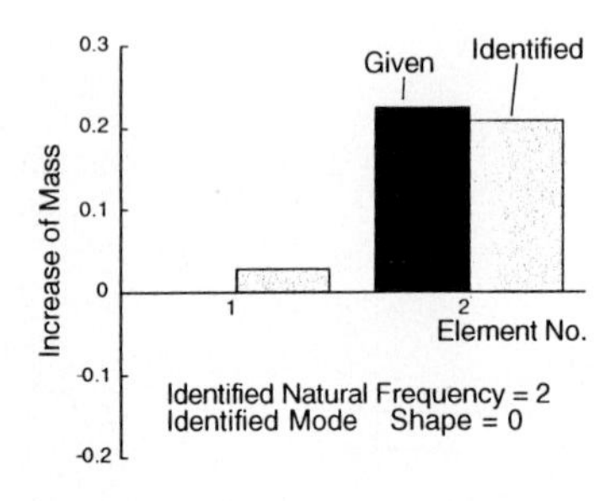

(a) Using two natural frequencies (q_f =2 and q_s =0).

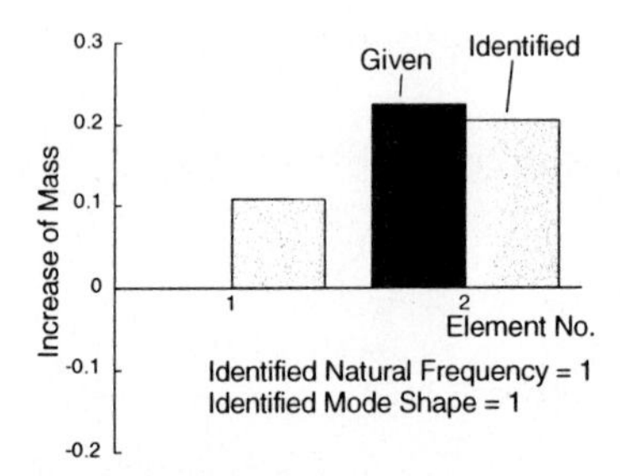

(b) Using a natural frequency and a mode shape ratio (q_f =1 and q_s =1).

Fig.16. Identification of increase of mass by sensitivity analysis.

6. Summary

The paper reports brief overview on structural monitoring using vibration measurement with emphasis on practical implementation. Here, vibration measurement is categorized into two groups:
(i) monitoring of vibration itself, such as evaluation of comfort or control of structures;
(ii) monitoring of structural/loading state via vibration measurement.

Category (i) is a straight-forward application of vibration measurement, and there exist a large number of successful examples. However, measurement in category (ii) involves inverse analysis with major difficulties in:
(a) insufficient measurement to verify all the degrees of freedom;
(b) uncertainties in both measurement and modeling.

Although structural monitoring using vibration measurement has been producing some promising results in various examples, it still remains in extraction of qualitative information, because of these difficulties. Insights and experience of human experts can extract useful information even with these difficulties, which needs to be incorporated into future automated monitoring system. Appropriate combination of these two reasoning methods would bring the breakthrough to successful implementation of structural monitoring. Application of artificial intelligence technology, where heuristic knowledge and human-like reasoning is quantified and connected to mathematical results, appears to be the most natural and attractive way to tackle this challenging problem.

Acknowledgment

The author would like to thank Dr. A. Yamashita of RTRI for providing valuable information and advice.

References

1. Sugidate, M., Ichikawa, A., Koshiba, A., Abe, M.: The evaluation method of serviceability of steel railway bridges in bridge maintenance system (BMC), *Railway Technical Research Institute Report*, Vol.8, No.8 (1994) 31-36 (in Japanese)
2. Bridge Vibration Monitoring Subcommittee: *Guideline for Monitoring of Bridges using Vibration Measurement*, Structural Engineering Committee, Japan Society of Civil Engineers (under preparation)
3. Chang, P. T. Y., Chang, C. C., Zhu, L. D.: A streamlined ambient vibration analysis program for long-span bridges, *Proceedings of the 6th East Asia-Pacific Conference on Structural Engineering & Construction*, Taipei, Taiwan (1998) 375-380
4. Fujino, Y., Yamaguchi, H.: Application of vibration control in civil structures with emphasis on active control, *J. Japanese Society of Steel Construction* (1994) 16-22 (in Japanese)

5. Abé, M., Fujino, Y., Yoshida, J.: Seismic response analysis of base-isolated bridges during 1995 Kobe earthquake based on observed records, *Proc. Japan Society of Civil Engineers* (submitted for publication, in Japanese).
6. Hunter, N.F., Paez, T. L.: Application of the Bootstrap to the analysis of vibration test data, *Proc. 66th Shook and Vibration Symposium*, Biloxi, Mississippi (1996) 99-108
7. Nishimura, A., Tanamura, S.: A study on integrity assessment of railway bridge foundation, *Railway Technical Research Institute Report*, Vol.3, No.8, (1989) 41-49 (in Japanese)
8. Doebling, S. W., Farrar, C. R., Prime, M. B., Shevitz, D. W.: Damage identification and health monitoring of structural and mechanical systems from changes in their vibration characteristics: A literature review, *LA-13070-MS*, Los Alamos National Laboratory (1996)
9. Abé, M.: Structural damage detection by natural frequencies, *37th AIAA/ASME/ASCE/AHS /ASC Structures, Structural Dynamics, and Materials Conference*, Salt Lake City, Utah (1996) 1064-1069

Object-Oriented Software Patterns for Engineering Design Standards Processing

Mustafa Kamal Badrah [1] , Iain MacLeod [2] and Bimal Kumar [3]

[1] PhD Candidate, Department of Civil Engineering, University of Strathclyde,
107 Rottenrow, Glasgow G4 0NG, Scotland
m.k.badrah@strath.ac.uk

[2] Professor, Department of Civil Engineering, University of Strathclyde,
107 Rottenrow, Glasgow G4 0NG, Scotland
i.a.macleod@strath.ac.uk

[3] Senior Lecturer, Department of Civil Engineering, University of Strathclyde,
107 Rottenrow, Glasgow G4 0NG, Scotland
b.kumar@strath.ac.uk

Abstract. An object-oriented software pattern identifies the framework of the participating objects, their roles and collaborations for a specific problem context. Three patterns for engineering design standards processing are described in this paper. These relate to design standards, design cases and design product data. They employ the reusability characteristic of object-oriented software technology in its wider prospect. In addition to providing object frameworks that can be used/adapted frequently in the development of standards processing systems, the reusability is also employed, by using the delegation mechanism, for modelling standards cross-references and for the integration of the standards and design cases object models.

1 Introduction

The roots of object-oriented software patterns go back to Christopher Alexander's [1] philosophy about patterns in architectural design [9]. Alexander's pattern "describes a problem which occurs over and over and then describes the core of the solution to that problem in such a way that you can use this solution a million times over without ever doing it the same way twice" [9]. In analogy to Alexander's patterns, an object-oriented software pattern identifies the participating classes and instances, their roles and collaborations, and the distribution of responsibilities for a specific problem context [9], [17].

Gamma, et. al [9] developed object-oriented software patterns which are for general purpose and whose main objective is the development of reusable object-oriented software. Software reusability has two main principal levels which are reuse of software components and reuse of software architecture; where the concept of architecture encompasses several components and the appropriate glue between them [17].

In this research, some of Gamma's patterns are adapted for engineering design standards processing. Section 2 reviews research in engineering design standards processing, Section 3 describes object-oriented software patterns adapted for standards processing and Section 4 discusses the conclusions.

2 Engineering Design Standards Processing

Design standards play a significant role in all engineering fields. Design standards state performance requirements, describe design processes and give design guidelines for engineering entities. Different approaches for design standards representation and processing have had developed over the last four decades. These can be classified into *object-based* approaches (i.e. based on object-oriented software technology) and *non-object-based* approaches.

Non-object-based approaches include: Using decision tables [6], [7], [8], [10], using first-order predicates [15], [19] and using rules [16], [18]. Garrett [11] argues that although these approaches adopt declarative representation of the requirements of the design standards, they do not provide enough representational flexibility and thus lead to unnecessary complexity. The object-oriented approach, by Garrett [11], overcomes the shortcomings of previous approaches by using objects to represent requirements and data items in the standards; abstract behavior limitations; and design systems or components. Other object-based approaches include: The object-logic model [21], the object-oriented model for computational verification systems [14], and the context-oriented model [5].

However, the previously mentioned approaches are based primarily on the design standards as the main knowledge source for component or system design. A design standard forms a major knowledge source for the design of a regular component/system (that is within the scope of the standard), e.g. an RC T beam designed to BS 8110 [4]. However, there are some situations that can not be handled either directly or completely by a design standard; these are non-standard designs, e.g. an RC channel beam designed to BS 8110 [4].

It is discussed in [2] that a case-base of non-standard designs can form a good knowledge source not only for existing non-standard designs but also for expected new design contexts. Here, a design case is a set of approved and detailed recommendations for the design processing of a conceptual entity. Based on experience gained by the prototype system SADA-II (Standards Automated Design Assistant II), it is concluded that the model of design cases should complement the model of the design standards to form two main knowledge sources for both regular and non-standard design. SADA-II is different from previous object-based standards processing approaches in that [2]:

1. It utilizes more than one main knowledge source (namely the design standards and design cases) in addition to other knowledge sources and;

2. It employs the reusability aspect of object-oriented software technology for multiple purposes e.g. modelling the standards cross-references and integration of the standards model and the design cases model.

The software patterns for standards processing described in this paper are explained in general terms without focusing on a specific design standard or programming language . These patterns provide off-the-shelf elements for a standard processing developer to use or adapt in order to build up an application according to the problem context. The application of these patterns are demonstrated in SADA-II for reinforced concrete framework design; for detailed information about implementation issues, see [2].

3 Object-Oriented Patterns for Standards Processing

Object-oriented decomposition is defined as the process of breaking a system into parts, each of which represents some class or object form the problem domain [3]. By examining an object-oriented decomposition for the design processing of both of regular and non-standard structural components, three general correlated groups of classes and objects can be identified. These relate to design standards, design cases and the design product data. The three patterns presented in this section describe the frameworks of the classes and objects of each of these groups providing their roles, collaborations and responsibilities.

Before introducing these patterns, some terms related to object-oriented software technology need to be defined. *Object composition* is the assembling of objects (aggregate parts) to get more complex functionality. *Delegation* is an implementation mechanism in which an object forwards or delegates a request to another object; the delegate carries out the request on behalf of the original object [9]. Object Modelling Technique (OMT) notations [20] are used in the class and object diagrams, see Table 1.

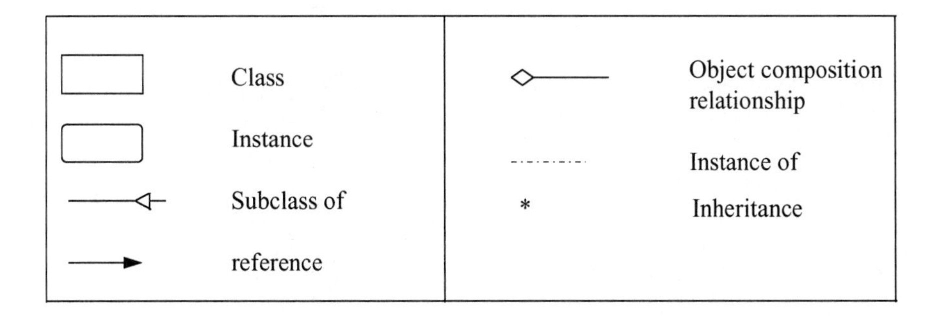

Table 1. Some OMT and Other Notations

3.1 Standards Design Issue Pattern (SDIP)

A design standard can be viewed as a collection of recommendations given for different *design issues* regarding different types of design entities, so a design standard is issue-oriented. Where a standards design issue is a topic in the design standards that involves specific requirements which need to be satisfied during the design synthesis or conformance checking processes and that concerns a grouping of design entities. As a standards design issue example, consider shear in an RC beam; BS 8110 [4] gives recommendations (in terms of rules and expressions) for different shapes the beam may take (e.g. rectangular beam, flanged beam, etc.) and for different situations (using links-system or both links and bent-up bar systems).

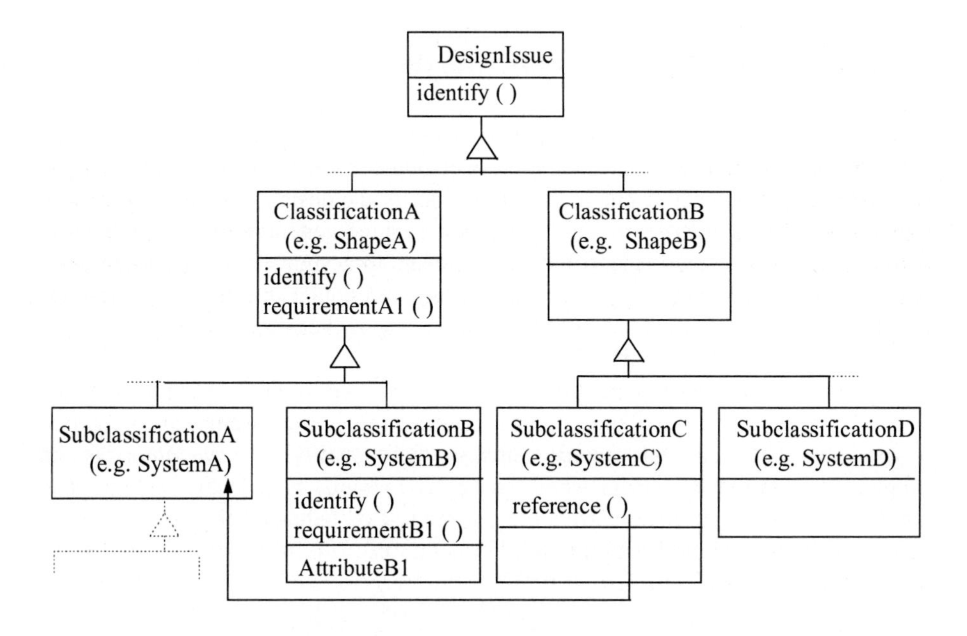

Fig. 1. Standards Design Issue Pattern

SDIP provides a framework to characterize a standards design issue in an object-oriented decomposition, see Fig. (1). The top node of SDIP is a class which may have multilevel of subclasses. Every class of this pattern has an 'identify' method (operation) which serves as an interface to this class. In a non-leaf class node, 'identify' retrieves the appropriate lower level subclasses according to the design entity properties. In a leaf class, 'identify' lists the relevant standards requirements coded in this class or inherited from a parent class(es). The attributes (instance variables) in the leaf classes include:

- The design entity properties which are desired to be as output data after this entity is being processed and;

- The provision numbers against which the design entity is checked or sized.

This pattern also applies the delegation mechanism to model standards cross-references, [2]. For example, a class representing the standards requirements for bending moment of a flanged beam having its neutral axis within the flange can reference the class representing a singly-reinforced rectangular beam; since both of these classes share the same requirements.

3.2 Design Case Pattern

The main difference between a design standard and a design case-base (proposed for non-standard design processing) regarding object-oriented decomposition is that a design standard is issue-oriented, see Section 3.1, while a design case-base is entity oriented (i.e. design recommendations are given only for each individual case). For example, the shear design recommendations of an inverted-T precast concrete beam (as a non-standard design entity) are intended only for this particular case.

DCP provides a framework to characterize a design case in an object-oriented decomposition. Fig. (2) shows a schematic diagram for DCP in which design cases are represented as instances of a three-level class hierarchy representing (Case, Design Issue Set and Design Issue) Where a design issue of a case is a design topic that involves specific requirements which need to be satisfied during the design synthesis or conformance checking of the corresponding design entity.

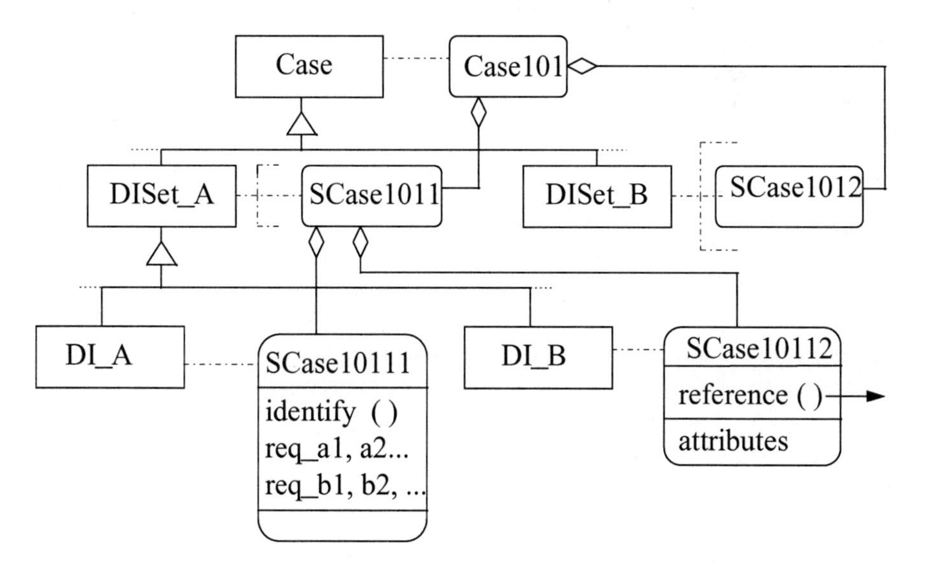

Fig. 2. Design Case Pattern. The (*arrow*) to the (*right*) denotes a reference to objects in the standards object model

Every case is a composition of different subcases; therefore, information such as design issue sets and lists of design issues in each set can be accessed at the case level. A leaf object represents a design issue and includes the same attributes as the leaf classes of SDIP except that a reference to the *design methodology document* is included rather than the clause numbers. The design methodology document contains the basis of the design recommendations upon which the requirements, coded in the Design-issue instances, are established. Each leaf object has also an 'identify' method which prompts to the user a message to select a design methodology if a design issue has more than one. For the selected design methodology, 'identify' lists the relevant set of requirements. This pattern also utilizes the delegation mechanism to reference other objects (or set of objects) in the design standard object model as appropriate [2].

For more complicated design situations, a design case may have design issues related to more than one part. For example, design issues of an RC channel beam, see Section 3.5, may be categorized to those related to the nib and those related to the whole beam excluding the nib design issue. In such situations, each design issue category can be represented in a case part and composition of these parts will characterize the composed case.

3.3 Design Data Pattern (DDP)

The need to integrate a standards processing application with a design product model is highlighted by Hakim [12], [13]. This has been considered in the early stage of the object-oriented decomposition for the standards processing application. In this application, the product model can be defined as a symbolic representation of the design data in an attribute-value format. The product model has twofold functionality which are:

1. An interface to exchange data between the standards processing application and other CAD applications and;
2. A repository from which design data required for design processing can be retrieved and in which output design data can be stored.

DDP provides a framework to characterize design product data in an object-oriented decomposition. Fig. (3) shows the schematic diagram of DDP in which the object 'aDesignEntity' is a composition of different parts (e.g., 'aDesignEntity_PartA', 'aDesignEntity_PartB', etc.,)'. DDP objects include mainly attributes representing design data where the common attributes are in higher levels and specific attributes are in lower levels.

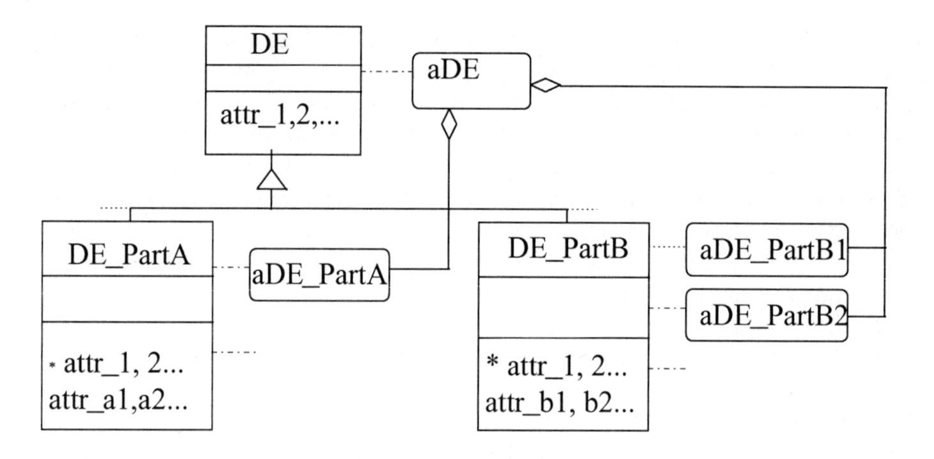

Fig. 3. Design Data Pattern

3.4 Patterns Collaboration

The evaluation process (e.g. the conformance checking or sizing of a structural component) starts by instantiating one of DDP classes and its parts. The design data can be obtained through the DDP interface by a user-query and/or data-exchange with other CAD applications. For regular components, the object representing the design entity will reference an SDIP class which will retrieve the design entity data to carry out the evaluation processes or delegate the responsibility to other class(es). Similarly, for a non-standard component the object representing the design entity will reference a DCP object which will retrieve the design data to carry out the evaluation process or delegate the responsibility to SDIP classes. In both cases, the output data will be stored back in DDP objects.

3.5 Example: RC Channel Beam Object Model

The purpose of this example is: 1) to illustrate the use of Design Case and Design Data Patterns and the collaboration among the different presented patterns. 2) to give a glimpse about the evaluation process, For more detailed information about the evaluation process, see [2].

In this example, an RC channel beam (a non-standard design component if designed according to BS 8110 [4]) is represented in 'Case401', see Fig. (4). For simplification, design issues related to the nib are not considered. An RC beam is usually designed to the Ultimate Limit State and the Serviceability Limit State which are represented as two design issue sets. The ULS set includes bending moment, shear and torsion design issues; and the SLS set includes deflection and flexural cracking design issues. The design of a channel beam for bending moment, shear, deflection and flexural cracking can be handled using BS 8110 requirements either directly or after making some assumptions. So that, objects representing these issues (SCase40111, SCase40112, SCase40114, SCase40115) will refer to the corresponding objects in the standards object model using the delegation mechanism. However, regarding torsion, four requirements are identified in 'SCase40113' which are:

1. Torsion resistance by core
2. Excessive cracking in the bottom flange
3. Excessive cracking in the top flange
4. Torsion-links arrangement

The 'Attributes' in 'SCase40113' includes the output data of requirement evaluation and a reference to the design methodology document upon which this requirements are established.

Design data of a channel beam is represented in 'ChannelBeam', see Fig. (4-b). This is composed of the following parts:

- Chn_WCom: This includes data regarding the whole component which are required for some design issues e.g. deflection.
- Chn_Secs: This includes data regarding the component sections that need to be sized or checked for some design issues (e.g. bending moment and shear).
- Chn_Nib: This includes data regarding the nib of the channel beam.

'B1' is a design entity that needs to be evaluated; so that, firstly design data need to be stored in 'B1' and its parts (B1_Sec1, B1_Sec2, B1_WComp, B1_Nib). Secondly, for requirement identification and evaluation, messages will be passed to 'SCase401' aggregate parts and may be delegated to the design standard objects if appropriate. And thirdly, output data will be stored in 'B1' aggregate parts.

4 Conclusions

Three design patterns and their collaboration for standards processing are presented. These relate to design standards, design cases and design product data. These patterns provide off-the-shelf elements for a standard processing developer to use or adapt in order to build up an application according to the problem context. The application of these patterns are demonstrated in the prototype system SADA-II for reinforced concrete framework design [2].

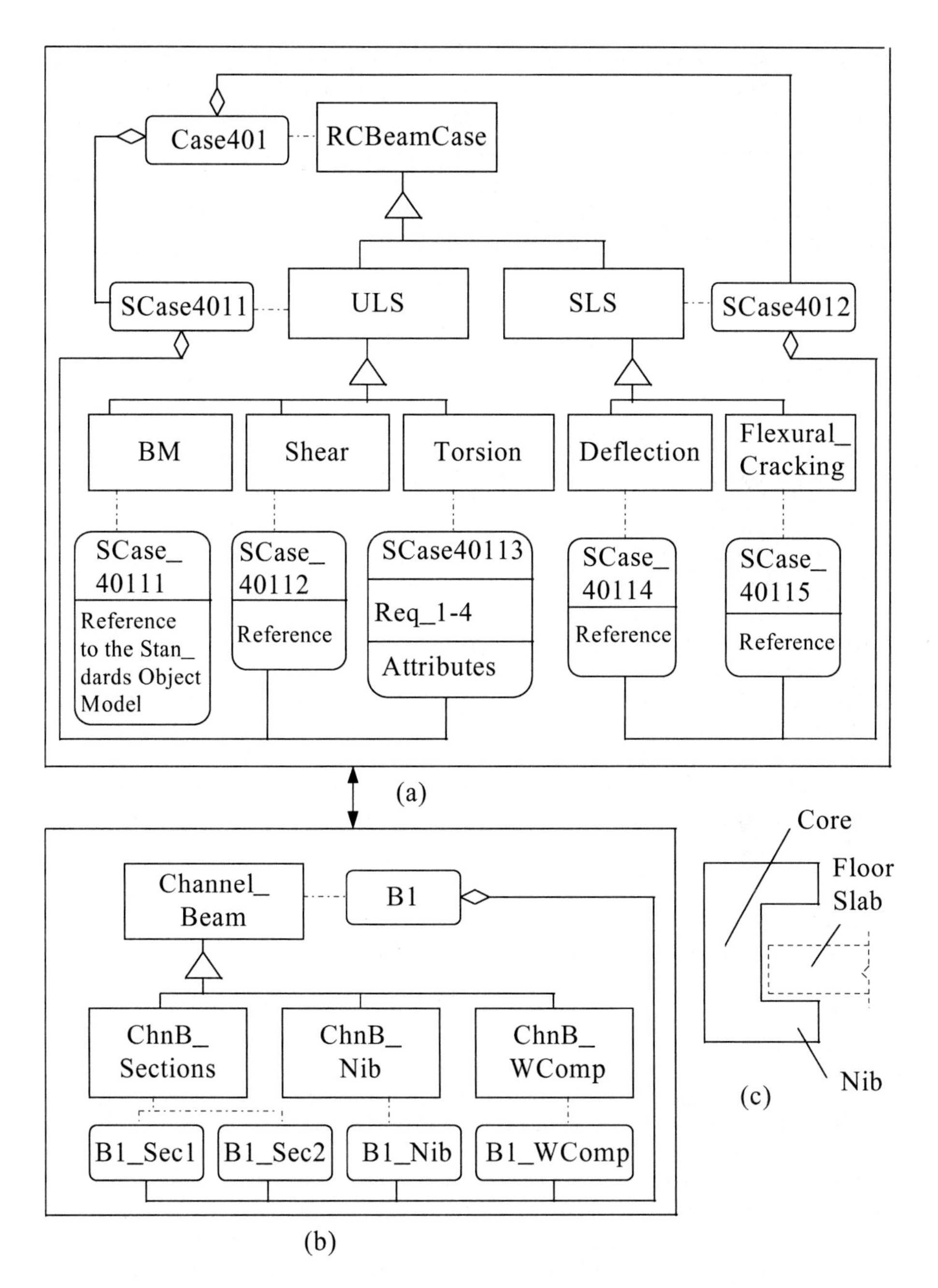

Fig. 4. Object Diagram of Example 1 (*a*) is the design case object diagram, (*b*) is the design data object diagram and (*c*) is a cross section in an RC channel beam

The presented patterns employ a most important feature of object- oriented software technology, viz. reusability. Reusability has been employed at the component and architecture levels. Reusability at the '*component*' level is achieved by using the delegation mechanism for reusing behavior embedded in objects that are already built. This is applied for modelling standards cross-references and integration of the standards and the design cases object models [2]. Reusability at the '*architecture*' level is achieved by providing object frameworks which would be used/adapted frequently in the development of standards processing systems.

References

1. Alexander, C., et. al: A Pattern Language. Oxford University Press, New York (1977)
2. Badrah, M. K., MacLeod, I.; and Kumar, B.: Using Object Communication for Design Standards Modelling. To appear in the July issue of the Journal of Computing in Civil Engineering, ASCE (1998)
3. Booch, G.: Object-Oriented Analysis and Design with Applications. The Benjamin-Cummings Series in Object-Oriented Software Engineering. The Benjamin-Cummings Publication Company, California (1994)
4. BS 8110: The Structural Usage of Concrete, Part 1-3. BSI, London (1997)
5. Condamin, E., Scherer, R. J., Garrett, J. H., Jr., and Kiliccote, H.: Context-Oriented Modelling of Eurocodes. In Scherer (ed.): "ECPPM'94: Product and Process Modelling in the Building Industry". A. A. Balkema, Rotterdam, The Netherlands (1995)
6. Elam, S. L., and Lopez, L. A.: Knowledge Based Approach to Checking Deigns for Conformance with Standards. Technical Report, Civil Engineering Systems Laboratory Research No. 9, University of Illinois at Urbana-Champaign, Urbana, Ill (1988)
7. Fenves, S. J.: Tubular Decision Logic for Structural Design. J. Struct. Engrg. Div., ASCE. 92(6) (1966) 473-490
8. Fenves, S. J., Wright, R. N., Stahl, F. I., and Reed, K. A.: "Introduction to SASE: Standards Analysis, Synthesis and Expression". Technical Report. NBSIR 873513, National Bureau of Standards, Washington, D. C. (1987)
9. Gamma, E., Helm, R., Johnson, R., and Vlissides, J.: Design Patterns: Elements of Reusable Object-Oriented Software. Addison-Wesely, USA (1995)
10. Garrett, J. H., Jr., and Fenves, S. J.: A Knowledge-Based Standard Processor for Structural Component Design. Engrg with Computers, 2(4) (1987) 219-238
11. Garrett, J. H., Jr., and Hakim, M. M.: Object-Oriented Model of Engineering Design Standards. J Computing in Civil Engrg, 6(3) (1992) 323-347
12. Hakim, M. M., and Garrett, J. H., Jr.: Issues in Modelling and Processing Design Standards. Computers and Building Standards Workshop, University of Montreal, Canada (1992)
13. Hakim, M. M. and Garrett, J. H., Jr.: A Description Logic Approach for representing Engineering Design Standards. Engrg with Computers, Vol. 9. Springer-Verlag London Limited (1993) 108:124
14. Holewik, P., and Hartmann, D.: An Object Oriented Model for the Representation of Knowledge to Develop Computational Verification Systems. In Kumar, B. (ed.): Information Processing in Civil and Structural Engineering Design. Civil-Comp Press, Edinburgh, UK (1996)

15. Jain, D., Law, K. H., and Krawinkler, H.: On Processing Standards with Predicate Calculus. In Will, K. (ed.): Proc., 6th Conf. on Computing in Civ. Engrg., ASCE, New York, N. Y. (1989)
16. Kumar, B.: Knowledge Processing for Structural Design. PhD Thesis, Dep. of Civil Engrg., University of Edinburgh, UK (1989)
17. Pree, W.: Design Patterns for Object-Oriented Software Development. Addison-Wesely, USA (1995)
18. Rasdorf, W. J., and Wang, T. E.: Generic Design Standards Processing in an Expert System Environment. J. Computing in Civil Engrg., ASCE, 2(1) (1988) 68-87
19. Rasdorf, W. J., and Lakmazaheri, S.: Logic-Based Approach for Modelling Organization of Design Standards. J. Computing in Civil Engrg., ASCE, 4(2) (1990) 102-123
20. Rumbaugh, J., et. al: Object-Oriented Modelling and Design. Prentice Hall, Englywood Cliffs, NJ, USA (1991)
21. Yabuki, N., and Law, K. H.: An Integrated Framework for Design Standards Processing. Technical Report No. 67, Stanford University, USA (1992)

Design and Verification of Real-Time Systems

John W. Baugh Jr.

Associate Professor, Department of Civil Engineering,
North Carolina State University, Raleigh, NC 27695-7908 USA
jwb@eos.ncsu.edu

Abstract. Advances in sensing, effecting, and computational technologies will change the way we design, construct, and monitor systems that interact with the physical world. Building structures will sense and respond to external loads, autonomous robots will occupy construction sites, and intelligent vehicles will monitor network flows to circumvent delays. As engineers, we must be prepared to work with the information technologies that underpin these coming systems. This paper addresses several of the prominent technical concerns in designing real-time systems that control some aspect of their environment. The view taken is that, by considering software systems to be an engineering artifact, we can begin to develop the kinds of quantitative approaches found in other areas of engineering design.

1 Introduction

Real-time applications are those in which responsiveness is critical to the system's operation. Computer systems for controlling robotic excavators, operating traffic monitoring beacons, and dissipating structural vibrations all have real-time requirements. For example, active structures contain an embedded computer system that monitors structural vibrations and controls actuators to dampen their effects. To function correctly, the computer system must operate within prescribed timing bounds; to function safely, it must be designed with redundant components, i.e., multiple processors and other subsystems.

How can embedded, multiprocessor computer systems be developed so that a reasonable degree of confidence in their performance is ensured? A common approach is to build a complete prototype of the control system and to experimentally verify its performance. However, in most cases it is difficult if not impossible to design a test that covers the range of possible behaviors, not the least of which are timing variations and device faults. As is well known, experimental testing alone can show only the presence, not the absence, of problems. This is particularly true with respect to computer systems, which do not behave linearly, i.e., small changes in input can cause vast differences in output and timing behavior.

An alternative is to view the development of real-time systems as an engineering activity supported by an underlying science. Can computer software be considered an engineering artifact, and can its design benefit from traditional engineering tools, such as mathematical specification and quantitative analysis? Theories of concurrency, timed processes, and state-space representation are beginning to provide a basis for quantitative design approaches. This paper discusses the role of quantitative techniques in designing multiprocessor systems for real-time control. Such techniques are intended to promote the correctness and fault tolerance of concurrent and real-time systems from the very early stages of design, i.e., before writing code.

2 The Business of "Sense and Communicate"

The current cost-performance ratios of computer hardware, notably of microprocessors and DRAMs, are moving the computer industry's centroid further away from mainframes and PCs and towards what has been referred to as "ubiquitous computing," wherein computers are are embedded throughout the physical environment. Ted Lewis in his September 1997 Binary Critic column for IEEE's *Computer* magazine writes

> "Various estimates place the non-PC, embedded-systems market at 1,000 times bigger than the huge PC industry. According to Tom Portante of Ernst and Young, over 2 billion chips and microcontrollers were sold worldwide in 1994 in contrast to 10 million personal computers.... Therefore, the business of 'sense and communicate' is already many times larger than 'compute and store.' This huge market is changing the computer industry."

And it is also changing our thinking. Modern society has come to expect automated solutions; it no longer accepts, for example, a delay at a traffic signal when there is no opposing traffic. "Just a little bit of intelligence," we think to ourselves; the notion of intelligent objects has crept into our thinking. Like transportation systems, other civil engineering artifacts are beginning to have the same expectations placed upon them.

2.1 Some Applications

In the twenty-ninth Shaw lecture at North Carolina State University, Professor Steve Fenves asks "Why do structural elements of a building have to be designed for a live load factor of 1.6 or greater?" The implication is that they *don't*, and he notes that, technically at least, active load indicators could be used to warn occupants when a structure is overloaded. Fenves also notes a number of other technical possibilities, including the "effortless workspace" with chips embedded in all sorts of tools, e.g., pneumatic wrenches that sense both torque and bolt elongation.

In a structural engineering context, the business of "sense and communicate" has enormous potential when it takes on the added dimension of "control," and numerous structural control studies have been performed since Yao [1] first introduced the idea (Spencer [2] gives an overview of the field). Of course, the intent is to enhance the ability of buildings, bridges, transmission towers and other structures to respond to natural or other disasters, such as strong earthquakes, severe wind, or even terrorist attacks. Such structures combine sensors, actuators, and computer-controlled feedback to improve structural performance and reliability. Active structural control might also have a role in "salvaging" damaged or underdesigned structures.

The potential of active control is not limited to structural *systems*, but is also seen at the *member* level. Studies at MIT by Andy Berlin [3] show that buckling loads can be increased by a factor of 5.6 by incorporating networks of strain gauges, controllers, and piezo-ceramic actuators in columns. Berlin notes that dynamically-stabilized members could be used for lightweight, deployable structures, or for structures that support large transient loads, such as airplane landing gear.

2.2 System Characteristics

A recent DARPA-ITO Workshop on Software-Enabled Control brought together researchers in computer science, control theory, and engineering to discuss the challenges and opportunities in developing control-oriented software. Military applications, including comprehensive, "theater-level" planning, deployment, and control of military resources were of obvious interest, as well as civilian ones, including active structures, intelligent transportation systems, and integrated governmental air-traffic management (GATM).

As described in our report [4], these systems represent diverse applications, but have several common characteristics.

- *Different loci of control*
  In contrast to flat, single-loop control systems, these applications may operate (concurrently) with several different levels of control abstraction, ranging from single members to groups of members to entire structures. Each level of abstraction has its own control needs, and coordination among levels is required to ensure that the actions of a single member do not compromise higher level control issues.

- *Different time scales*
  Deployable structures, which undergo large controlled geometric changes, may have timing requirements on system components that cover vastly different time scales. During operation, the dynamic characteristics of space cranes, for instance, may vary by orders of magnitude and require various degrees of control.

- *Decentralized control*
  For complex systems with a large array of sensors and interacting actuators, one cannot assume the availability of global state information. Rather, many individual controllers operate with local information to produce a desirable "emergent behavior."

- *Importance of fault-tolerance and exception-handling*
  These applications are characterized by spatially dispersed components and highly variable environments. In such contexts unexpected scenarios are certain to arise, and components will fail; run-time software support must be provided for identifying and mitigating major classes of errors, and for tolerating faults.

These and other characteristics distinguish the applications envisioned from those of single-loop control systems.

2.3 Technical Challenges

Models are simplified representations of reality used to provide insight, and they are fundamental to engineering design. While there are a number of technological obstacles to the implementation of practical systems for structural control, many of the challenges identified [4] are of a modeling nature:

1. Coordination of interacting processes
2. Integration of logical aspects of control with traditional control
3. Modeling and control of hybrid systems
4. Uncertainty, noise, and external disturbances
5. Unpredictability and design of robust controllers
6. Stochastic control of discrete event systems
7. Sensor fusion
8. Modularity

The ability to conceptualize a system and predict its behavior is a basis for the kind of iterative design scenarios found in traditional engineering applications. The difference here is that the artifacts to be modeled include *software systems,* which, unlike physical systems, are discrete, highly nonlinear, and complex, i.e., we cannot use continuity arguments to "fill in" the gaps. The addition of timing requirements further complicates the matter, since nondeterministic delays are sure to be found in any embedded computer system. The combination of discrete systems and nondeterministic delays means that conventional simulation techniques, e.g., Monte Carlo, are of very little use in predicting the behavior of real-time systems.

In the following discussion, I will attempt to articulate some of the prominent modeling concerns and capabilities in real-time software design, and to place them in an engineering design framework that ranges from system-level requirements to component-level specifications and implementation. Two key aspects of this design scenario, in the vernacular of the software engineering community, are verification and validation, where verification asks "Is the product right?" and validation asks "Is it the right product?" When viewed in the context of computer control systems, the latter of these, validation, can be thought of as what control theorists typically already do. An accepted practice of verification, on the hand, has yet to be established.

3 Defining Requirements for Real-Time Systems

The requirements imposed on a computer system by a particular control strategy are not often articulated, but are tested implicitly (if at all) by constructing a system and observing its behavior. While a control strategy may be formalized, computer systems do not behave as idealized by control laws, but have delays, timing variations, and worse, e.g., hardware, sensor, and actuator faults. If control laws are only abstractions of desired behavior, then how well must they be approximated by computer systems? The answer to this question lies at the interface between declarative mathematics (control theory) and an algorithmic implementation (computer systems). If computer system requirements can be stated precisely, then the roles of control engineers and computer scientists can be better defined and understood, to the benefit of both communities.

Because there has been little experience with defining the computational requirements for control systems, experimentation with practical prototypes is needed to validate any specification approach. What is the "right" form for such requirements? For example, hierarchical transition diagrams with timing bounds may allow some temporal requirements to be captured. However, time-invariant properties may also be required of the system, analogous to the frequency-domain view of dynamical systems. In addition, one must determine which behaviors are worthy of attention and specification, e.g., do sensor delays and timing variations have a significant impact on the overall capabilities of the system? These questions and many others can be assessed only by designing and modeling computer *implementations* of control systems.

3.1 A Framework

Control systems design requires that intentions be carefully defined, including desired functional behavior, physical capabilities and limitations of components, and expected disturbance signals. Once these are well understood, dynamic performance requirements can be articulated that ensure satisfactory behavior of the idealized control system.

A framework for designing and verifying control systems should begin here, with a control theorist's notion of requirements. That is, system requirements should be given in terms of the actual physical or process models, whether based on time- or frequency-response, or perhaps even pole-zero locations. For example, a controller for a mechanical system might be required to limit displacement time histories to 5", or the frequency of vibration to 20 Hz. Traditionally, these requirements are used to ensure that a control system *as idealized by declarative mathematics* is satisfactory. Such idealizations would include models of sensors, actuators, and controllers, for which many control theoretic packages are available.

By working from control system specifications, this approach results in a design scenario that encompasses the tools and techniques of control theory, e.g., tools such as MATLAB, MATRIXx, and CtrlC, that simulate dynamic response and support root-locus, frequency-response, and state-space design.

Getting down to the level of software implementations, and the requisite verification process, involves several levels of modeling that essentially constitute a refinement strategy. Intermediate forms between the extremes of control theoretic models and finite state systems could be based on various temporal logics and process algebras, as described below. General strategies are needed for generating *verification conditions* from control theoretic specifications, where the conditions may be phrased in terms of finite state *observer* processes, or perhaps in terms of a high level notation, such as a process algebra designed for control applications. These conditions represent proof obligations that must be discharged during the verification process, and that guarantee satisfactory performance. Eventually, abstract requirements become inter-event specifications in a finite state model of the control system, which can capture details not present in a control theoretic description, e.g., nonlinearities, signal noise, parameter variations, and actuator saturation.

3.2 Classes of Control Specifications

As matters of convenience and expressiveness, control specifications may be best articulated in one of various forms. Consider, for example, the application of control technology to structural systems. Active structures may have several dynamic performance requirements given in one or more of the following forms.

- *Time histories*
  Time-varying parameters, such as displacements, velocities, and accelerations, may have certain acceptable ranges of values. Control laws attempting satisfy such a specification are referred to as *bounded state control algorithms.*

- *Frequency response*
  In practice, time-domain response is rarely used in the design of structural systems. Rather, structures are often required to satisfy certain frequency-domain characteristics. For example, to avoid the likelihood of resonant excitations, a control system may be used to increase a building's fundamental natural frequency. In contrast, a flooring system may be designed to exhibit a low fundamental frequency so that occupants cannot perceive vibrations during normal usage of the floor.

- *Pole-zero locations*
  A transfer function between the input and output of a control system has poles, or, equivalently, characteristic roots, on a complex plane whose locations affect transient behavior, e.g., stability, settling time, and overshoot. For example, pole locations in the left half-plane ensure stability. To reduce building fatigue and damage, disturbances can be quickly damped by choosing pole locations that minimize transient settling time.

Because of their generality, these classes of specification can be applied in a number of different contexts. One can envision their application even in areas not traditionally viewed as classical control theory, such as intelligent transportation systems for incident management and traffic routing. For example, an ITS may consist of sensors, e.g., loop detectors, beacons, and probe vehicles, as well as "actuators" that include traffic signal plans and variable message signs. In that context, the following specifications can be stated as constraints on time histories:

- A time lag of x units must be maintained between two well-behaved vehicles entering an intersection from different approaches, where x is a design variable determined from the Manual on Uniform Traffic Control Devices.
- A vehicle waiting for a green signal at an approach should not wait for more than x units of time.

Likewise, constraints on frequency response may also be stated:

- Because of buffer capacity, a beacon pulse collector is required to transmit its data to a traffic management center every 5 seconds.

3.3 An Example: Pulse Control

Pulse control algorithms [5] limit the vibratory displacement of structures by applying an opposing pulse at a higher frequency to "break up" resonant forces. The major design variables are the pulse period, Δt_p, and the pulse duration, Δt, whose values are determined by the natural frequencies of the system, the expected forcing functions, and the desired level of displacement control. To determine the appropriate pulse magnitude to apply at a given interval, the state variables of the system (e.g., displacements, velocities, and/or accelerations) must be measured beforehand. Pulse magnitudes are calculated using Soong's algorithm [5], and these values are nonzero only when the predicted structural response during the following interval exceeds a predefined limit. Within the pulse intervals, the algorithm samples state variables, which are used to update the parameters of the structural model that may change over time, e.g., modal parameters such as natural frequency.

While Δt_p and Δt are useful abstractions of the timing properties of pulse control, they may give one the misleading impression that pulses can and must be applied at exact intervals, which is impractical given the potential timing variations of both thc hardware and software components of the system. The results of a parametric study on simulated pulse control with time delays [6] can be used to establish the following timing requirements. As shown in that study, acceptable values of ϵ and δ_{SP} are determined by a variety of problem characteristics.

- The period of pulse application should approximate Δt_p, i.e.,

$$\Delta t_p - \epsilon_l \leq \mathit{Pulse\ Period} \leq \Delta t_p + \epsilon_u \tag{1}$$

- The pulse duration should approximate Δt, i.e.,

$$\Delta t - \epsilon_l' \leq \mathit{Pulse\ Duration} \leq \Delta t + \epsilon_u' \tag{2}$$

– The delay between sampling and pulse application should not exceed an amount δ_{SP}.

The foregoing requirements have been stated as simple upper and lower bounds on inter-event delays in the pulse control algorithm. Depending on the approach used for verification, these requirements must be faithfully translated into an appropriate mathematical notation—the candidate hardware and software design will be required to satisfy them.

4 Concurrency

Control-oriented software exhibits a high degree of concurrency, whose coordination is a central activity of the system design process. While the expression of concurrency involves only a few linguistic features, these can be can be combined in such different and varied ways as to yield interesting and unusual concerns quite distinct from sequential programming. Indeed, the modeling and verification techniques subsequently described are an attempt to address these concerns.

4.1 Need for Concurrency

To mitigate the effects of hardware faults, practical real-time systems are designed with redundant components. Systems in which components are physically remote are generally termed *distributed*; other forms of concurrency in real-time systems include the following.

Reactive behavior. Control systems maintain a relationship with their environment and react and respond to external events. These *reactive* systems are control- or event-driven, and involve a complex interaction of components. They behave not by manipulating data in a functional way to produce a result, as do numerical algorithms, but rather by responding to a variety of events, signals, and interrupts in a timely manner. Such implementations can become complex and intricate, particularly in a highly distributed, networked environment. The concurrency issues for these applications are similar to those found in other reactive systems, e.g., graphical user interfaces.

Multiplexing. When a *read* statement is issued in a conventional programming language, execution of the process halts until the input is available. Similarly, in a distributed system, when the input is a message sent from another processor, the receiving process likewise blocks. To make progress while waiting on network input, a process must "fork" into two separate *threads* of control. Although one thread will block waiting on input, the other can continue to make progress. Such an approach enables a process to poll others, e.g., when reclaiming the analysis results of substructures assigned to physically remote processors.

Process monitoring. Without the notion of concurrency, it is difficult to make programs introspective in the sense that they monitor and adjust the performance of an algorithm. Such a capability is particularly useful in a multi-level control scenario. For example, a monitoring process might implement a voter scheme to detect and isolate faulty sensors. Monitoring processes can also ensure a level of responsiveness when results need to be obtained in a timely manner, e.g., by returning the best available result within a fixed amount of time.

4.2 Design Concerns

Programmers of sequential systems must ensure that their programs satisfy post-conditions (a safety concern) and that they terminate (a liveness concern). Satisfying both means the program is "correct." As might be expected, it is more difficult to manage computation on a network than on a single processor. In addition to the broader concerns of algorithmic granularity and load balancing are difficulties in managing concurrency, synchronization, and communication between processes. Because it is impossible to reproduce all possible timing variations that might occur in any process, testing concurrent systems is often inadequate to show determinacy, i.e., that its functional behavior is independent of timing issues.

Safety. To ensure safety in concurrent systems, objects must maintain a consistent state; e.g., nodal equilibrium may be an invariant of an iterative analysis algorithm. Safety arguments are complicated when multiple threads of control operate on the same data. Avoiding state changes, as in functional programming, can often simplify the analysis. If an object does undergo state changes, then access to that object must be managed, either through explicit synchronization or by structurally enforcing it.

Liveness. Like termination arguments, liveness in concurrent systems means that the computation is capable of making progress. Unfortunately, safety and liveness are often at odds. For safety, processes may be required to synchronize before updating an object, possibly resulting in *deadlock*: imagine two processes each requiring access to two objects to make progress. If the processes request exclusive access to the two objects in different orders, neither may be capable of progressing since each has something the other wants. A variety of other liveness problems are possible, including *starvation*, when a process is never allowed to run because others have taken over CPU resources.

4.3 Linguistic Issues

Real-time operating systems and languages provide basic facilities for expressing concurrency. These facilities include support for instantiating threads of control, or "lightweight" processes, that execute independently yet share resources and data. Synchronization among threads can be ensured by requiring that they obtain a "lock" on an object for exclusive access. Activities across threads are coordinated by allowing threads to block, or wait, for an object to change state, as well as the complementary behavior: threads may notify others that an object has changed state.

By offering linguistic support for data abstraction [7], object-oriented languages can reduce the amount of detail confronting real-time programmers. Support for data abstraction, combined with a well-integrated, object-oriented threading system, means that common patterns of concurrency can be implemented and used as a black box. These patterns, or models, are abstract descriptions of how objects are structured and used in a computation. The basic rationale here is to offer a toolbox of "metaphors" that help implementors structure concurrent programs in a meaningful way. Examples of patterns include dataflow models, task pools for balancing loads in a heterogeneous network, and recursive divide and conquer strategies.

5 Design Notations

While programming language solutions are typically the first sought for a given software development problem, they rarely seem to fulfill their "silver bullet" promise, to borrow a phrase from Fred Brooks. In fact, trying to develop languages that admit only safe and live programs is like trying to create beams and columns whose combination always produces stable structures.

Of course, engineers do not require that structural members somehow be incapable of being combined inappropriately, but rather rely on Newton's laws, theories of deformable bodies, and so on, to determine how those members *should be* combined. Analogously, a basic theory of concurrency can be built on top of simple models, e.g., communicating finite state machines, or using suitable logics, such as temporal or linear logic. Petri nets and Statecharts are based on the former, and may be used to reason about reactive systems. These tools model the states that components might be in, and the events that cause them to change state. Finite state systems capture this control while abstracting from the functional behavior of a component. Hybrid approaches, e.g., model checking, combine a basic theory of finite state systems with modal logics, allowing designers to prove that a concurrent system is in some sense equivalent to a simpler sequential system (the specification). The benefits of these approaches have been demonstrated in a number of domains other than control, such as construction process modeling [8].

In addition to concurrency models, various approaches have been proposed for formally describing the timing properties of real-time systems to enable quantitative analysis. Real-time logic (RTL) [9] is an expressive specification language, and many relations of interest are easy to describe within it. For example, the following expression describes timing bounds on a process that samples data:

$$\forall i \bullet @(\uparrow Sample, i) + L_S \leq @(\downarrow Sample, i) \leq @(\uparrow Sample, i) + U_S \tag{3}$$

where the start and stop events of a given action (e.g., *Sample*) are denoted by $\uparrow$ and $\downarrow$, respectively, and L_S and U_S denote lower and upper timing bounds. However, while the expressiveness of RTL is welcome for systems modeling, the price is its undecidability, meaning that verification may require ingenuity to make proofs go through, i.e., interaction with the designer. A complete description of a multiprocessor RTL model for active control can be found elsewhere [10].

5.1 Process Algebras

Process algebras [11–13] offer an approach to handling undecidability that is based on the use of compositionality (i.e., component-wise analysis) and semantic equivalence. Such formalisms typically contain a language consisting of a small number of constructs for building concurrent systems and a formally defined semantic equivalence that indicates when two processes "behave the same" and may be used interchangeably. Although many dialects have been proposed, the most commonly used and accepted is Milner's Calculus of Communicating Systems (CCS) [13].

Transition systems may be described in CCS by defining a set of computing agents, $\mathcal{P}$, and a set of actions, *Act*, that can be performed. For example, the notation

$$P = a.Q \tag{4}$$

denotes an agent P that first performs an action a and then behaves like the agent Q. For this transition system, $\mathcal{P} = \{P, Q\}$ and $Act = \{a\}$. More interesting behavior can be obtained through recursion. For example, the agent

$$R = a.b.c.R \tag{5}$$

performs the infinite sequence of actions a, b, c, a, b, c, ...

Two agents can act in parallel, e.g., $P \mid Q$; this is referred to as composite behavior. Parallel agents can communicate using an action and its complement, represented as a and $\overline{a}$, respectively. For example, the agents *Sensor* and *Actuator* below communicate synchronously over channel x

$$Sensor = sampleState.x.Sensor \tag{6}$$

$$Actuator = \overline{x}.activateMember.Actuator \tag{7}$$

resulting in the composite behavior *sampleState*, *activateMember*, *sampleState*, *activateMember*, ... When the two agents communicate they create what is referred to as an internal action, or τ. This internal action allows two systems to be compared based only on their external actions, and ignoring any internal communication. Other operations in CCS include nondeterministic choice (+), hiding, and relabeling.

The semantics are given operationally via inference rules that define the transitions available to CCS processes. Rules for each operator are given using the transition relation $\longrightarrow \subseteq Proc \times A \times Proc$, where *Proc* is the set of processes. For example, one rule for nondeterministic choice

$$\frac{p \xrightarrow{a} p'}{p + q \xrightarrow{a} p'}$$

states that if the process p can perform action a and become p', then the process $p + q$ can also perform an action a and become p'.

It is important to note the features present (and not present) in a process algebraic model. Actions such as *sampleState* and *activateMember* are merely names, albeit mnemonic ones, and are not intended to capture functional behavior. That is, the *Sensor* process does not tell us the magnitude of the state sampled, nor does it convey state information to the *Actuator* process. What *is* captured are the sequence of actions and coordination of communication events that are of interest to the modeler.

5.2 Temporal Process Algebras

To model timing behavior, process algebras can be defined with a temporal semantics, e.g., Temporal CCS [14]. In such a language, an action $d \in Nat$ represents a delay of d time units. Letting $L \subseteq Act$ be a set of channels, processes in the algebra may be built using the following grammar:

$$P ::= nil \mid a.P \mid P_1 + P_2 \mid P_1|P_2 \mid P\backslash L \tag{8}$$

Intuitively, these constructs may be understood in terms of the communication actions and units of delay (or idling) they may engage in. Non-delay actions are assumed to be duration-less; so time passes only when processes are capable of "idling." If $a \notin Nat$ then $a.P$ is a process that immediately engages in action a and then behaves like P; it is incapable of idling. If $a \in Nat$ then $a.P$ idles for a time units and then behaves like P. $P_1|P_2$ represents the parallel composition of P_1 and P_2. For the composite system to idle, both components must be capable of idling. As in other process algebras, the complete semantics are given operationally via inference rules that define the transitions available to processes.

Using a variant of TCCS, one can model the upper and lower timing bounds on a process that samples data as follows:

$$\uparrow Sample = L_S.delay(U_S - L_S, \downarrow Sample) \tag{9}$$

where $delay(n, P)$ is a nondeterministic delay between 0 and n time units:

$$delay(0, P) = P \tag{10}$$

$$delay(n, P) = \tau.P + \tau.1.delay(n-1, P) \tag{11}$$

In this definition, an internal action, τ, on both legs of the choice operator is used to defeat the *maximal progress property* found in some algebras: any process capable of a τ is incapable of idling.

5.3 Graphical Notations

Of several graphical notations for modeling real-time behavior, Modechart [15] supports the representation of systems as hierarchical state machines. The language includes two basic constructs: *modes*, which may be thought of as "structured states", and transitions between modes. Modes may be one of three types: *primitive*, *serial*, and *parallel*. Primitive modes have no internal structure and correspond precisely to states in a traditional finite-state machine. Modechart systems change modes by engaging in transitions, which are equipped with three components: a condition, a time bound of the form $[l, u]$, and an effect. A complete formal semantics of Modechart is given via a translation into the logic RTL [16].

6 Modeling and Verifying Real-Time Control Systems

It should be obvious that the tasks required of a real-time system for active structural control cannot be carried out instantaneously:

- Because practical control systems contain redundant components there must be communication between processors, and the latencies are often difficult to predict.
- Actuators are usually displacement controlled, so a lower-level control system must be implemented if forces are to be specified. Since control laws require that actuators apply different loads at different times, the ramp-up times for these loads will necessarily vary.
- Implementing force control, sampling and communicating data, and performing computations all take finite time.

These timing behaviors, as well as the logical aspects of control and communication, can be modeled using the techniques described. By adopting such an approach, one abstracts from various functional behaviors, including: physical, or structural models, network communication models, and hardware models, which get "converted" to inter-event delays. Of course, control theoretic tools are at the other end of the spectrum, and offer support for physical and structural modeling while abstracting from computer implementation of control laws.

6.1 Modeling Issues

To conceptualize a system, the timing properties of a given hardware and software configuration may be initially estimated by manufacturer's specifications, and subsequently verified by performance testing if the candidate architecture is selected. Hence, using the candidate system, timing bounds on components can be considered *requirements* for actual software routines and hardware components. If the actual components operate within the prescribed timing bounds the implementation is guaranteed to satisfy both safety and liveness properties.

With respect to control systems design, current techniques for modeling real-time implementations offer both advantages and disadvantages. Process algebraic models, for example, support several different approaches to verification, as described below, which enable one to guarantee both safety and liveness properties of a concurrent system. However, such an approach is conservative in the sense that the inter-event delays used to model components are static, i.e., PDEs cannot be "stuffed" into the model to capture timing properties with better fidelity. This means that timing bounds may become unacceptably loose in the course of modeling a given component.

As a modeling example, consider an experimental setup for active control reported by Soong [5] in which a structure is controlled by a diagonal, hydraulic member that is connected to a microprocessor via a digital-to-analog converter. Accelerations and displacements are measured using piezoelectric accelerometers and strain-gauge bridges that are assembled on the columns of the structure. These signals are sent to a digital data-acquisition system, which is connected to a spectrum analyzer for system identification.

With the timing values reported in that study, a multi-processor system can be conceptualized with a processor each for sampling the state of the structure, operating the hydraulic actuator, and controlling the overall system. A Modechart representation of the pulse control algorithm is given in Fig. 1, along with the corresponding process algebraic description. The control processor communicates with the sensor process through the action xs and with the actuator process through $'xp$. The upper and lower timing bounds of the state transitions are modeled as described above, and appear in units of tenths of a millisecond. A separate timer determines whether it is time to apply a pulse, or whether the sensor results should be used instead to update the parameters of the control model (i.e., for systems identification). The timer monitors the passage of the time until it reaches a value equal to the pulse application time (e.g., 135 in our example), then it performs a synchronization action, *clock*, causing the controller to move to $\uparrow$*Calculate*.

6.2 Verification

The implementation of a real-time system results in a collection of components with various timing properties. How can the designer of a real-time system predict the actual timing properties of the overall system, e.g., expected pulse intervals and durations? Of course, simulating the timing behavior of computer systems is easier than building actual prototypes, and it allows one to watch the computational process unfold in various ways. Because it is impossible to reproduce all possible timing variations, however, traditional simulation techniques suffer from some of the same problems that limit the usefulness of experimenting with prototypes, i.e., incompleteness of the test cases. The goal of verification is to answer questions such as "will property P *always* hold?", e.g., where P may be a timing property.

Answering such questions may require either "symbol pushing," i.e., formal proofs, or "model checking," which explores the complete state-space of a model.

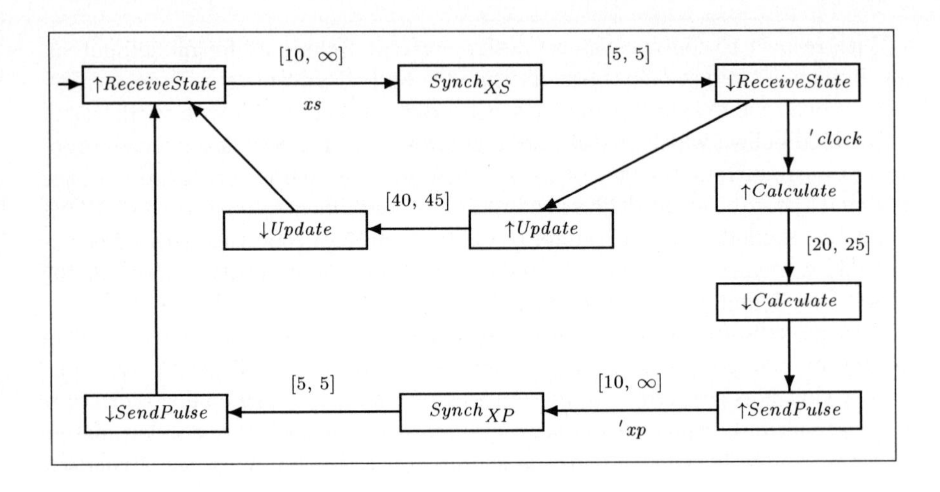


```
bi Control StartReceiveState
   where StartReceiveState = 10.StartReceiveState1
      and StartReceiveState1 = xs.5.StopReceiveState + 1.StartReceiveState1
      and StopReceiveState = 'clock.StartCalculatePulse + 1.StartUpdateModel
      and StartUpdateModel = 40.delay(5,StopUpdateModel)
      and StopUpdateModel = StartReceiveState
      and StartCalculatePulse = 20.delay(5,StopCalculatePulse)
      and StopCalculatePulse = StartSendPulse
      and StartSendPulse = 10.StartSendPulse1
      and StartSendPulse1 = 'xp.5.StopSendPulse + 1.StartSendPulse1
      and StopSendPulse = StartReceiveState
   end
```

Fig. 1. Pulse Control Process Model

The latter of these, model checking, has become the primary means of logical design validation used in the design of complex microprocessors, and numerous practical algorithms for verification have been developed around this application. Hence, the verification process can often be fully automated, and tools such as the NCSU Concurrency Workbench [17] are capable of automatically verifying (temporal) process algebraic models.

An overall design scenario, then, would consist of working down from requirements, working up from code and hardware, and meeting at a finite state representation that ultimately answers "yes" or "no." Thus, if requirements can be faithfully captured within the verification language, and if timing bounds on components are satisfied by the implementation, then one can be sure that system requirements have been met.

Several approaches may be used to get requirements down to the level of something that can be verified:

- *Equivalence checking* shows that two models engage in the same observable actions. Requirements, then, are expressed as a simple machine that exhibits desirable behaviors. Of course, such an approach demands that all behaviors of interest be captured in that machine.

- *Temporal logics* [18] allow assertions to be formulated about a system's behavior, e.g., safety and liveness properties, as it evolves over time. Because temporal logics support the expression of individual properties, designers can verify behaviors of interest without having to say anything about those not of interest.

- *Observer processes* [19] run in parallel with a real-time model and observe its behavior by attempting to synchronize with events of interest in the model. Conceptually, the model raises flags that are consumed by an observer, which moves to an error state if flags fail to appear when and where required.

For example, in the pulse control problem, pulse intervals can be verified by having the actuator process raise a flag when a pulse is applied to the structure, and by constructing an observer that times the occurrence of the flags. Such an approach resulted in in a 7819-state system which was shown to be deadlock-free and to satisfy the required timing bounds [19].

6.3 Fault Tolerance

In most real-time applications, some provision must be made to mitigate the effect of certain classes of errors at run time—such behavior is referred to as *fault tolerance.* Reliability of real-time systems may be enhanced through hardware and software redundancy. Interestingly, however, care must be taken since adding redundancy greatly increases the overall complexity of a real-time system (potentially leading to *less* reliable systems). Thus, the need for quantitative approaches is increased since not only must *correct* behavior in a *correct* environment be established, but also *dependable* behavior in an *arbitrarily incorrect* environment, i.e., in the presence of device faults, sensor errors, etc.

Redundant systems can be designed in various ways to behave reliably in the presence of faults. For example, consider the situation in which multiple components "vote" on the proper signals for actuator control—if a particular component consistently votes against others, that component may be suspected as faulty and taken offline without affecting performance, though reducing overall tolerance to subsequent faults. A "statically" redundant system, like the above, can be contrasted with dynamically redundant systems, where (duplicate) processors reconfigure themselves at run time to avoid interfering with each other. Both types of systems have been designed, modeled, and verified using process algebraic and model-checking techniques [20]. The number of states required to represent these systems (including unreachable states) is in excess of 10^{19}.

7 Conclusions

The demand for real-time computing solutions will undoubtedly multiply in all areas, and civil engineers will be increasingly called upon to provide and work with real-time products and processes. While a number of technological obstacles to such remain, many others are of a fundamental nature and concern the design and development process itself. The envisioned structures or processes being controlled are often *safety-critical* in that human lives and physical well-being depend on them; in such environments, ensuring that these systems behave correctly and reliably is of utmost importance.

The combination of "soft" and physical systems presents many challenges to the designer. In the area of requirements, there is a need to better understand how control system specifications can be captured and used in the verification process. Not only of concern is the form of these specifications, e.g., whether based on appropriate logics or finite state observers, but also the manner in which they can be obtained. While parametric studies might be used, it may be possible to reason from fundamental principles if a functional semantics is given to the actions of finite state models [8]. Intuitively, since both requirements and design specifications contain functional information, one might distill the functional behaviors from both, leaving only a set of temporal verification conditions to be discharged.

Challenges also exist in the areas of modeling and verification, and include the following:

- improving the fidelity with which timing bounds are modeled in real-time components
- reducing the state space of models using component-wise minimization approaches [20] or partial evaluation techniques
- developing better tools to support debugging finite-state systems, e.g., efficient algorithms for generating diagnostic traces [21]
- providing automatic translation of intuitive graphical languages, like Modechart, into verification languages
- relating functional specification approaches, e.g., equational logics, with real-time specification approaches, and both with realizations in conventional programming languages

More generally, improvements in predictive capabilities are needed to foster experimentation with different software architectures for real-time systems, leading to better understanding. Ultimately, these improvements may lead to a "packaging" of approaches for control in which subsystems can be readily composed with respect to the interacting domains of function, time, and fault tolerance.

Acknowledgments

Colleagues over the years have made various contributions to the views expressed in this paper. The author would particularly like to thank Jeannette Wing, Wael Elseaidy, Harpreet Chadha, Gopal Kakivaya, and Rance Cleaveland.

References

1. J. T. P. Yao. Concept of structural control. *ASCE J. Struct. Div.*, pages 1567–1574, 1972.
2. B. F. Spencer Jr. and M. K. Sain. Controlling buildings: A new frontier in feedback. Special Issue of the *IEEE Control Systems Magazine* on Emerging Technology, 17(6):19–35, December 1997.
3. A. A. Berlin. Active control of buckling using piezo-ceramic actuators. In *Smart Structures and Materials 1995: Industrial and Commercial Applications of Smart Structures Technologies (SPIE)*, volume 2447, pages 141–154, 1995.
4. DARPA-ITO Workshop on Software-Enabled Control. http://www.dyncorp-is.com/darpa/meetings/sw/software-enabled.html.
5. T. T. Soong. *Active Structural Control.* Longman Scientific, New York, 1990.
6. B. D. Rose and J. W. Baugh Jr. Parametric study of a pulse control algorithm with time delays. Technical Report CE-303-93, Department of Civil Engineering, North Carolina State University, Raleigh, NC, August 1993.
7. J. W. Baugh Jr. and D. R. Rehak. Data abstraction in engineering software development. *Journal of Computing in Civil Engineering*, 6(3):282–301, July 1992.
8. J. W. Baugh Jr. and H. S. Chadha. Semantic validation of product and process models. *Journal of Computing in Civil Engineering*, 11(1):26–36, 1997.
9. F. Jahanian and A. K. Mok. Safety analysis of timing properties in real-time systems. *IEEE Transactions On Software Engineering*, 12(9), September 1986.
10. J. W. Baugh Jr. and W. M. Elseaidy. Real-time software development with formal models. *Journal of Computing in Civil Engineering*, 9(1):73–86, 1995.
11. J. C. M. Baeten and W. P. Weijland. *Process Algebra*, volume 18 of *Cambridge Tracts in Theoretical Computer Science.* Cambridge University Press, Cambridge, England, 1990.
12. C. A. R. Hoare. *Communicating Sequential Processes.* Prentice-Hall, 1985.
13. R. Milner. *Communication and Concurrency.* Prentice-Hall, 1989.
14. A. Moller and C. Tofts. A temporal calculus of communicating systems. In *Proceedings of CONCUR'90*, pages 401–415. Lecture Notes in Computer Science 458, Springer-Verlag, 1990.
15. F. Jahanian and D. A. Stuart. A method for verifying properties of Modechart specifications. In *IEEE 9th Real-Time System Symposium*, pages 12–21. IEEE Computer Society Press, 1988.
16. F. Jahanian and A. K. Mok. Semantics of Modechart in Real Time Logic. In *21st Hawaii International Conference on System Science*, pages 479–489, 1988.
17. R. Cleaveland and S. Sims. The NCSU Concurrency Workbench. In R. Alur and T. Henzinger, editors, *Computer Aided Verification (CAV '96)*, pages 394–397. Lecture Notes in Computer Science 1102, Springer-Verlag, 1996.
18. E. A. Emerson. Temporal and modal logic. In J. van Leeuwen, editor, *Handbook of Theoretical Computer Science*, volume B, pages 995–1072. North-Holland, 1990.
19. W. M. Elseaidy, J. W. Baugh Jr., and R. Cleaveland. Verification of an active control system using temporal process algebra. *Engineering with Computers*, 12:46–61, 1996.
20. W. M. Elseaidy, R. Cleaveland, and J. W. Baugh Jr. Modeling and verifying active structural control systems. *Science of Computer Programming*, 29(1–2):99–122, July 1997.
21. G. R. Kakivaya and J. W. Baugh Jr. Distinguishing formulas for bisimulation inequivalence. Unpublished working paper.

Using Knowledge Nodes for Knowledge Discovery and Collaboration

Per Christiansson

Aalborg University, Prof. IT in Civil Engineering.
Sohngaardsholmsvej 57, 9000 Aalborg
pc@civil.auc.dk, http://www.civil.auc.dk/i6

Abstract. Today most of the information we produce is stored digitally. We are slowly forced to leave behind us thinking about information as something stored in physical containers as books, drawings etc. We make it possible to dynamically create logical containers of information on the fly. The paper focuses on how we in the future can aggregate, classify and generalize digitally stored information in order to make it more accessible and how we can define underlying knowledge container models to support knowledge discovery and collaboration. Examples are picked from ongoing research and the outcomes are generally valid and in particular for the structural engineering field.

1 Introduction

Today most of the information we produce is stored digitally. We are slowly forced to leave behind us thinking about information as something stored in physical containers as books, drawings etc. We make it possible to create logical containers of information on the fly. This requires high level integration of those intranets, extranets, and Internet to which the physical containers (hard discs etc.) are connected. We know that the information is there somewhere in the cyberspace but how can we reach it and assess what we get back in terms of completeness and other quality parameters?

At the same time huge steps are taken on the building up of a global 'operating system' where agents and objects thrive - RDF (Resource Description Framework) to describe and exchange metadata over the networks, XML (Extensible Markup Language) to create application specific metadata formats, CORBA (Common Object Request Broker Architecture) for handling distributed objects and intelligent agents communication in client/server environments, and multicast protocols for optimal flow of information from one source to many receivers.

The paper focuses on how we in the future can aggregate, classify and generalize digitally stored information in order to make it more accessible and how we can define supportive underlying meta level knowledge container. Examples are picked from ongoing research and the outcomes are generally valid and in particular for the structural engineering field.

2. Areas of Interest

As digital information will be easy accessible and flexibly packaged more focus will be on new tools for knowledge communication and competence collaboration as well as tools for knowledge experience capturing and storage for later use in projects and re-use in other projects. In parallel the knowledge discovery and data mining, KDD, tools will evolve.

A more or less conscious knowledge discovery process will take place in the project, global and even user digital domains. The increasing interest in the area is confirmed as you traverse the web; 'URL's for Data Mining' at http://www.galaxy.gmu.edu/stats//syllabi/DMLIST.html, 'Knowledge Discovery and Data Mining Web References' at http://www.cs.uah.edu/~infotech/mineproj.html, and 'Knowledge Discovery & Data Mining Web References' at http://www.kdd.org/.

We can thus distinguish some areas of particular interest for future research;

- to what detail will we classify digital information containers?
- what should the information container/wrapping granularity be to optimally support creation of digital knowledge containers?
- what knowledge representations will information containers support?
- who will mark information with subjective opinions (except from the authors)?
- how can information containers be associated with each other on different abstraction levels?
- how is bottom-up (meaning derived from content) and top-down (through classifiers and formalized structures) information search supported?
- how do intelligent agents navigate and find certain information patterns?
- how do we handle revised information?

The remainder of the paper will contribute to provide answers and general models to the above questions.

3. The Serfin and Merkurius Knowledge Nodes

A structural engineer is searching for information and possibly knowledgeable persons in the area of structural loadbearing capacity. He especially looks for high temperature steel properties in connection with repair of fire loaded paint protected beams. He contacts the Merkurius URL (Uniform Resource Locator) on the Internet. Merkurius, see figure 1, is a communication and information resource (demonstrator under development) through which knowledge produced at the Lund University is accessible. Information can be reached in three modes (a) through indexed free text search combined with search on documents similar to a found document, (b) by use of the public project and idea capture area where he can pose questions and look for potential project participants or (c) through establishment of a personal contact with a knowledgeable person at the university. In figure 1 it can be seen how the search domain may be restricted ('ange sökområde') to the local Merkurius knowledge container (concerning the knowledge communication process and information search itself), Lund University or the world.

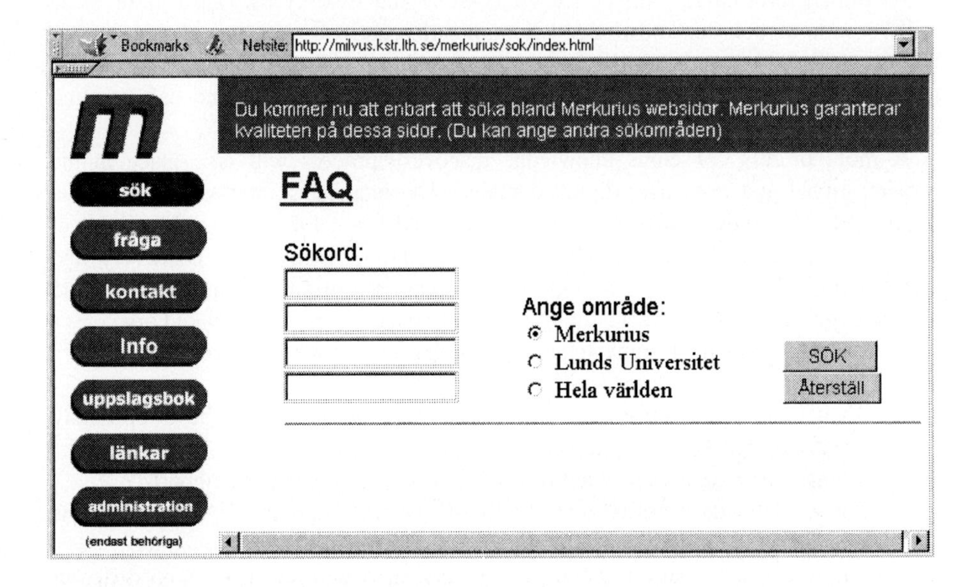

Fig. 1. The Lund University Industry Knowledge Node for access of knowledge produced at the university and for capturing and display of ideas for future projects.

The engineer finds a reference to another URL, Serfin, via a set of keywords already used. The Serfin knowledge node, [2], is a communication and information resource for handling technical building maintenance knowledge. Figure 2 shows how he can choose between a coarse top-down search using controlled vocabularies (with optional graphic support) for five knowledge domians or plain free text search.

Both systems embody mechanisms for capturing and quality marking of stored knowledge. In the Merkurius system this process already exist in the university research and teaching procedures.

The Merkurius and Serfin systems contains digital information packaged as documents. These documents may in its turn contain text, images, graphics, video, sound, encapsulated calculation routines (in objects), etc. Documents are to some extent 'classified' with regard to covered knowledge domain and detailing level. Below we will further discuss how structures, content and functionality can be improved through high level modeling.

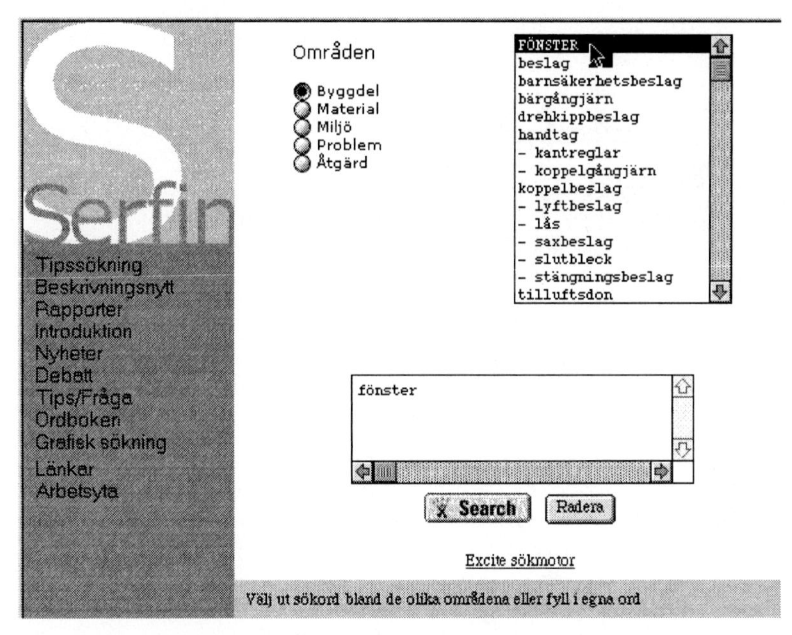

Fig. 2. Searching technical maintenance building information in the Serfin knowledge node. Choose 'Building Part' 'Window' ('Byggdel' 'Fönster') and eventually also Material, Environment, Problem type, Action - (Material, Miljö, Problem, Åtgärd). Add free text at your own wish. Relevance ordered feed-back is provided as well as search on similar documents. You can also send in a question or a tip of your own.

4. Logical Knowledge Containers and Knowledge Nodes

The personal competence and competencies co-operation will in the future as stated above be of central interest. Our personal information storage containers, today often stored in our personal portable computers, accommodate information with highly personal structure and semantics. When we exchange ideas and collaborate with other persons in *projects* we have to harmonize and to some extent formalize our common language.

Three overlapping levels of logical information repositories can be distinguished (1) the personal user dependent, (2) the project/cultural and (3) the global community dependent, see figure 3. On each level we will find long term rather well formalized containers in the form of databases and object stores, which are viewed and handled in project/cultural context through for example Structured Query Language, SQL, and web browser interfaces. The inter project/cultural linkages can be facilitated with RDF and dynamically adapted on the user levels through use of for example XSL, Extensible Style Language, to specify web document styles.

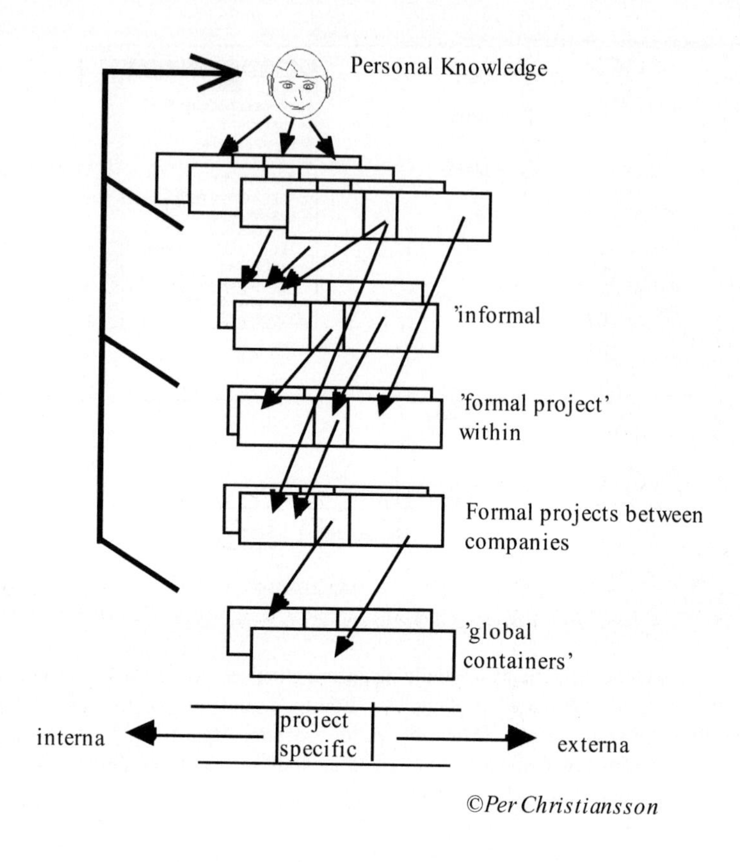

Fig. 3. Information gets more and more formal on its way to the long life global containers. Three levels containers can be distinguished; the personal, projects and the global level.

Persons and artifacts connect to the *Dynamic Knowledge Net*, DKN, [4]. The Internet and its services as World Wide Web today constitutes the DKN. DKN will evolve and perhaps (using metaphors) possess resemblance to the human brains dendrites and axons connecting what in artificial neural networks are called artificial neurons or Processing Elements, PE. [7].

A Knowledge Node is kind of high level processing unit and today equal to an URL, Uniform Resource Locator, on the Internet. A knowledge node, [3], has three main functions (a) dissemination of information on request or automatically channeled, (b) two way communication and feed-back capabilities through multimedia interfaces, and (c) access to a local knowledge bank and possibly meta knowledge about other knowledge nodes, see figure 4. The Merkurius and Serfin systems described above are example on Knowledge Nodes.

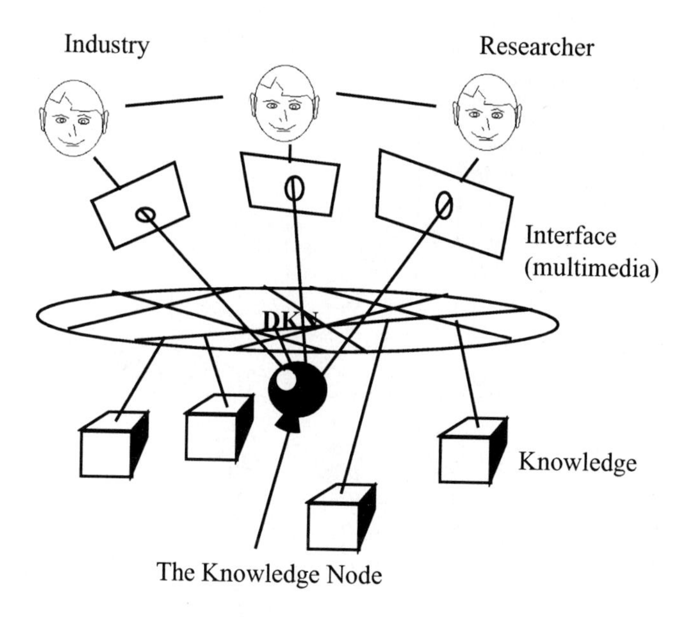

Fig. 4. The Knowledge Node can be regarded as a meta knowledge container and knowledge access control mechanism.

From [2] "The traditional physical information/knowledge containers as books, films, images, papers, etc. are at present in many cases also (or even only) stored in digital form in what we call logical ('virtual') knowledge containers. This latter containers have properties that from now on will completely change our view on how knowledge are structured and represented and interactively presented".

Figure 5 shows how the information access (line '1' in figure 5) to conventional physical knowledge containers as books and video tapes will change when most information is stored in digital format and packaged dynamically for different needs in non-physical (logical) containers. It is also shown how it is possible during collaboration to share information in a common workspace through multimedia interfaces ('3' in figure 5). . We talk about *logical containers* as contrast to physical when the physical wrapping is of importance (books, CDs, hard disks, video tapes, etc.).

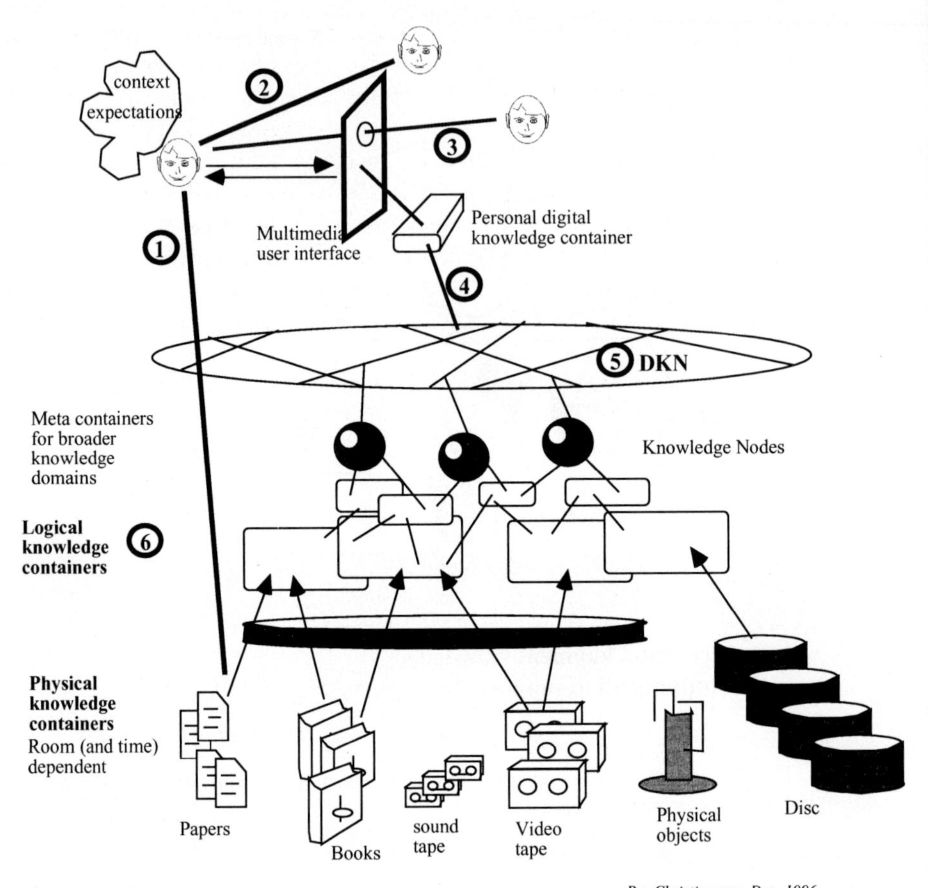

Fig. 5. We will, which is already a fact, communicate and handle digitally stored knowledge in new ways. (1) user searches and finds knowledge stored on paper. (2) Persons meet in real life or (3) use simple multimedia interfaces as telephone. The multimedia interface expands to incorporate more of our senses (Computer Supported Collaborative Work, CSCW, and Virtual Reality, VR, etc.). (4) Part of your personal computer stored knowledge may be connected to the (5) Dynamic Knowledge Net, DKN, see [4]. Logical information containers (6) can be created and dissolved with little effort.

5. Knowledge navigation and search

The human brain is very good at discovering (often unconsciously) subtle hidden patterns in information. With improved search and presentation IT-tools we get help in this process. But we also get some help for deeper analyses to uncover hidden knowledge. We need this help to save time.

We may use tools like WEBSOM, [9], to automatically cluster information and provide us with an ordered map where similar documents lie near each other on the map. In this case the method is based on an unsupervised learning algorithm for analyzing and visualizing high-dimensional statistical data. We can train a neural net through supervised learning for example by feeding it with trigrams (consecutive letters from a text, three at a time) thus finding typical patterns in the text, [12], or train an intelligent agent to help us filter found web-documents based on a user meta model, [8] . We can also use more straightforward navigation tools which provide us with different views for graphic navigation in an URL (for example the Mapucciono Java applet (http://www.ibm.com/java/education/mapuccino/java.map.html) from IBM.

Figure 6 provides a basic model with three facets to access information in a selected digital knowledge container.

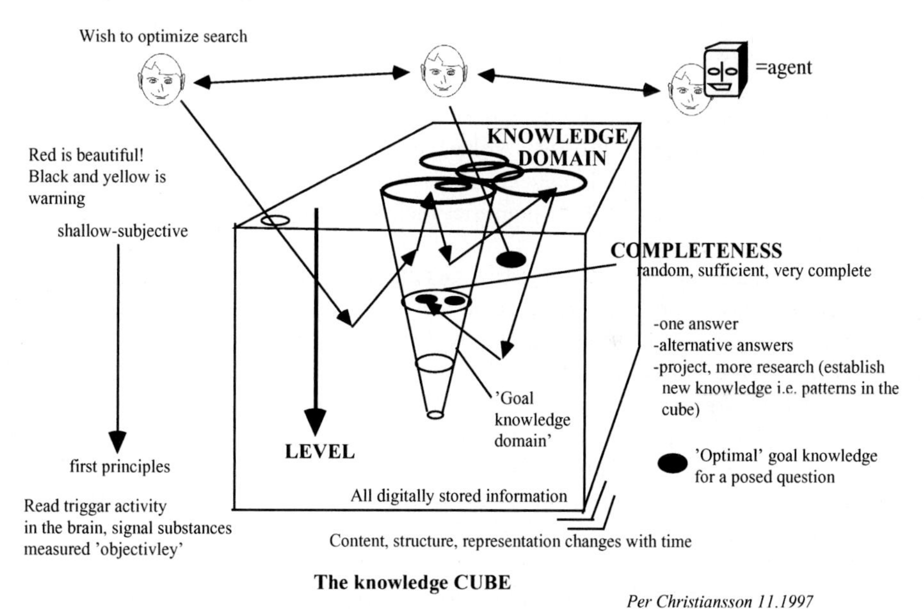

Fig. 6 Knowledge accessed from a node may be characterized according to level, completeness and domain. The one-way arrows denote the users search path towards a known or emerging goal knowledge domain. The result is stored in a logical knowledge container and is part of the 'goal knowledge domain'.

For example, a user wants to search the WWW for technical maintenance tips on removal of fire protection paint from wooden beams or frames. He will possibly be helped by an intelligent agent and start looking in metadata repositories for further links to information in the 'area' of technical maintenance, 'created' after 1990 in the Nordic countries. He may also do a discovery search world wide with no constraints on region or material (the right one-way upward arrow in figure 6 pointing to a new domain). After thus having narrowed in a potential goal domain he continues detailed indexed search in the 'description' parts of those web-objects. These analyses may well lead to references and a jump to another unexpected knowledge domain. The search ends with a collection of supposedly sufficiently good advises.

The requisites to develop IT-tools to make the scenario come through are present and under development today namely the RDF, Resource Description Framework, and XML, eXtensible Markup Language. See [13], [10].

From [13]: " RDF metadata can be used in a variety of application areas; for example: in *resource discovery* to provide better search engine capabilities; in *cataloging* for describing the content and content relationships available at a particular Web site, page, or digital library; by *intelligent software agents* to facilitate knowledge sharing and exchange; in *content rating*; in describing *collections* of pages that represent a single logical "document"; for describing *intellectual property rights* of Web pages, and in many others. RDF with *digital signatures* will be key to building the "Web of Trust" for electronic commerce, collaboration, and other applications."

RDF using the XML as its main carrier syntax allows us to handle name spaces for different knowledge domains and hopefully support web client mediation between databases.

The RDF data model can be represented as a set of triples {Property Type, Node/Resource, Node or Property Value} or serialized to a tagged text using the XML, eXtensible Markup Language. (This XML-file can be parsed to a tree-like object structure which in its turn simplifies meta level object handling in the Dynamic Knowledge Net, DKN).

XML (a subset of SGML, Standard Generalized Markup Language) extends the HTML, Hypertext MarkUp Language, in that it focuses on content only and leave the user views (part of the 'user models') to be defined in a separate XSL, Extensible Style Language. XML uses the same formalism as HTML i.e. documents are expressed as nested tagged expressions (<author> <first> nn </first> <last> mmm </last></author>). Mark-up languages based on XML are developed now for different areas, for example; Conceptual Markup Language, CKML, for handling conceptual spaces [5], and to support Electronic Data Interchange, EDI, [1]. See also [10].

6. Serfin and Merkurius meta level information

The MERKURIUS, figure 1, and Serfin system, figure 2, today do not contain meta-tags. Dublin Core meta-tags, [6], can be semi-automatically created using Reggie, a Dublin Core metadata Java Applet based editor, [10]. SubElements proposals are given from pull-down menus. There is also a Dublin Core Generator, DCdot, from University of Bath, which can generate metadata on existing html pages. See http://www.ukoln.ac.uk/metadata/dcdot.

Table 1. Dublin Core Metadata generated by Reggie, [10]

<META NAME="DC.Creator.PersonalName" CONTENT="(LANG=sv) Hans Nilsson">
<META NAME="DC.Subject" CONTENT="(LANG=en) hot work, hot air, open flame, window, paint removal">
<META NAME="DC.Description" CONTENT="(LANG=en) Removal of paint from tree frame">
<META NAME="DC.Publisher" CONTENT="(LANG=en) SERFIN Expert">

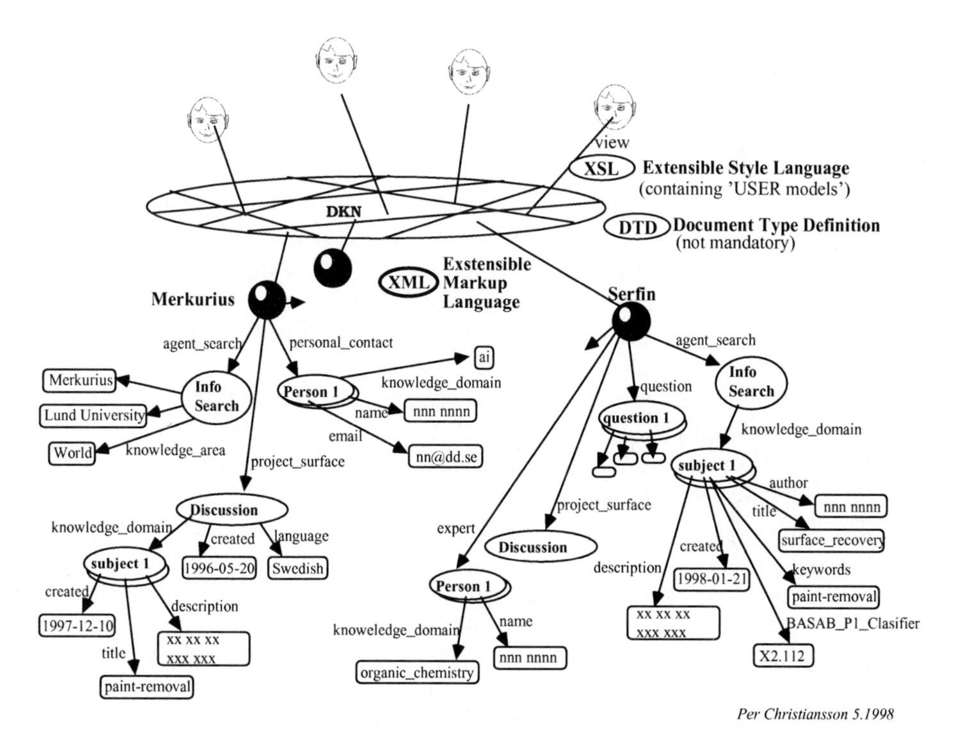

Fig. 7. Part of the top level contents of the Merkurius and Serfin knowledge nodes expressed as directed graphs according to the Resource Description Framework, RDF. The application areas for the XML, eXtensible Markup Language, XSL, Extensible Style Language, and Document Type Definition, DTD, (logical structure of document) are also shown.

The fifteen Dublin Core metadata tags contain: Title, Author or Creator, Subject and Keywords, Description, Publisher (of the electronic version), Other Contributor, Date, Resource Type (technical report, etc.), Format (html, pdf,...), Resource Identifier (retrieval identifier), Source (from the electronic version it was derived), Language, Relation (with other resources), Coverage (geographical or temporal), Rights Management (link to ownership information).

Figure 7 shows how the Knowledge Nodes Merkurius and Serfin attached to the Dynamic Knowledge Net, DKN, can be descried using directed graph notation according to the forthcoming Resource Description Framework, RDF. Such a description can be used in the conceptual modeling of the systems and later to facilitate high level couplings between the knowledge nodes. For example to discover pertinent competence persons and projects in other knowledge domains, for comparative analysis of different knowledge domains, and to harmonize application vocabulary development.

7. Conclusions

We can now see a clear break-point in the development of the future meta leveling of the globally stored information and the development of a knowledge node framework. Much work will be spent on compiling non-overlapping and comparable vocabularies and name spaces for different application areas.

The container descriptions (now 'A longer, textual, description of the resource in Dublin Core terminology) are mostly written by their authors. But other commentary and feed-back descriptions will also be written and associated with the same content. These will be very important when container content quality shall be estimated.

There are clear links between RDF and Entity-Relationship descriptions which will be helpful when WEB documents and objects are going to be generated from long term highly formalized relational database containers.

The abstraction process (aggregation, characterization, and generalization) will be even more interesting than before in connection with studying collaboration between different competencies (architects, engineers, clients, environmental planners,..) in order to capture, formalize and link 'equivalent' concepts.

The agent concept will be used extensively to wrap different kinds of complex and compound knowledge representations. The above related languages will support the definition of both the inter agent and agent human communication formalisms.

We now experience the beginning of a shift to a global totally digital information handling. It is only five years since we started publish on the web and we are already in a phase of re-engineering it. May be it is time to reconsider some of the pioneering works done by for example Ted Nelson (HomePage at http://www.sfc.keio.ac.jp/~ted/index.html.) regarding version handling and hypertext growth.

Acknowledgments

I want to thank my research colleagues Fredrik Stjernfeldt and Gustav Dahlström at the KBS-Media Lab, Lund University, for their collaboration in the MERKURIUS (The Foundation for Knowledge and Competence Development KKS-2343:I/95) and SERFIN projects (The Swedish Building Research Council, BFR-950549-0).

References

1. Bryan, M.,: Guidelines for Using XML for Electronic Data Interchange. Version 0.05, 25th January (1998). XML/EDI Group. http://www.geocities.com/WallStreet/Floor/5815/guide.htm
2. Christiansson, P.: Experiences from developing a Building Maintenance Knowledge Node. In CIB Proceedings Information Technology Support for Construction Process Re-Engineering, IT-CPR-97. (1997) 89-101. (http://delphi.kstr.lth.se/reports/cibw78cairns1997.html).
3. Christiansson, P.: Knowledge communication in the building industry. The Knowledge Node Concept. In Construction on the Information Highway. CIB Proceedings 198 (ed. Z. Turk) (1996) 121-132. (http://delphi.kstr.lth.se/reports/cibw78bled96.html)
4. Christiansson, P.: Dynamic Knowledge Nets in a changing building process. Automation in Construction, Vol 2, nb 2, Elsevier Science Publishers B.V. Amsterdam, (1993) 307-322
5. Conceptual Knowledge Markup Language, CKML. (Robert Kent, Washington State University, Christian Neuss, Technishe Hochschule Darmstadt) http://wave.eecs.wsu.edu/WAVE/Ontologies/CKML/RDF-to-CKML.html
6. Daniel Jr., R., Ianella R., Miller E.: Expressing the Dublin Core in the Resource Description Framework: Suggestions based on an early examination of the problem. Los Alamos National Laboratory. (7 A4 pages). (1997) http://www.acl.lanl.gov/~rdaniel/RDF/DC/ExpDC_2.html.
7. Freeman, J.A., Skapura, D., M.: Neural Networks. Algorithms, Applications, and Programming Techniques. Addison-Wesley Publishing Company. Reading Massachusetts. (1991) 17-18
8. Lagerstedt, R., Christiansson, P., Engborg U.: User Models in Search and Navigation Systems on the Internet". Proceedings of the Third Congress held in conjunction with A/E/C Systems'96. ASCE Technical Councils on Computer Practices. (1996) 21-27 (http://delphi.kstr.lth.se/reports/aec96.html)
9. Honkela, T., Kaski, S., Lagus, K., Kohonen, T.: Self-Organizing Maps of Document Collections. Neural Networks Research Centre, Helsinki University of Technology. (5 A4 pages) (1997). http://www.diemme.it/~luigi/websom.html
10. Mace, S., Flohr, U., Dobson, R., Graham, T.: Weving a Better Web. BYTE, March (1998) 58-68.
11. Metadata Tools and Services. Distributed Systems Technology Center. University of Queensland Australia. http://metadata.net/dstc/.
12. Modin, J.: KBS-Class: A neural network tool for automatic content recognition of building texts. Construction Management and Economics. Special issue on Information Technology in Construction. (1995) 411-416
13. Resource Description Framework (RDF) Model and Syntax W3C Working Draft 16 Feb 1998. http://www.w3.org/TR/WD-rdf-syntax/

Heating System Design Support

Piotr Cichocki[1], Maciej Gil[1], Jerzy Pokojski[1]

[1] Warsaw University of Technology, Institute of Machine Design Fundamentals
ul. Narbutta 84, 02-524 Warsaw, Poland
{pcicho, mgil, jpo}@simr.pw.edu.pl

Abstract. Computer support for certain engineering tasks such as analysis, calculation, drafting, configuration and simulation tends to be integrated into one system where all modules are coherent. Because of that a software layer should also be capable to provide communication between modules and to keep data and knowledge visible and available for every module playing a particular role in the system. This paper presents results of the development of such sample integrated environment. Though our system supports heating installation design, methods and techniques can be applicable in other domains.

Introduction

The design process is an activity whose result is the vision of a product. Decisions made during this process have strong influence on the quality of a product [4,5,10,13,16].

It is very important to see during the design process all decisive functions of a product. The designer should be enabled to build and analyze the best models of the considered phenomena, according to his merit knowledge. He should have the possibility to break the limitations of the existing methods, algorithms, computer systems, and the way they are integrated.

The maze model [4] of the design process fulfils the above claims. With the maze model we can initiate a design process from different nodes. We can conduct this process in different ways and finish in different nodes.

The paper shows an attempt to create an architecture, which goes in the direction of a maze model in heating system design.

As a first stage of this work some commercial software packages were analyzed. Nearly everywhere linear and limited (to the products of a particular producer or to particular functions, for example: heat components placing) models of design process were used. The basic differences between the packages were observed in the formalisms used for problem analyzing. After examining the existing commercial software, some modules - nodes from the maze model, which will be included in the newly created system, were specified.

We have developed software modules which correspond to the following maze model nodes: selection of heating system type, configuration of heating system components, estimation of energy transfer through the building, simulation and evaluation of newly created design. We also propose tools which facilitate the navigation through the design maze: the taskbar – which helps to activate particular node problems, the wizard – which supports routine design activities, and the navigator – which helps with the exploration through the design objects and variables.

A specially developed control module first has to identify the preferences of the future user of the heating system. According to the identified preferences this module activates other modules for a more detailed and specific dialogue. After that a control module creates a sequence of simulation problems. The results can be interesting for a particular user with a particular profile of using the heating system, in a particular house, in a particular place with the particular correctness of calculations. Finally a process of simulation is conducted. This process is done with little support of the expert system technology on the routine level [13]. The results of the simulation are stored. Later another module supports the process of estimation of the simulation results.

The module for the simulation of flow and heat phenomena in the heating installation has as an input data about the geometric description of the heating system and the physical aspect of a particular simulation. As a result we can get the temperature of every room and the energy flow in the time domain. Obviously it is not possible to simulate the work of the heating system for all the year. The results of the simulations are the basis for the estimation of global energy consumption in a season.

In the actual version the modeling of the geometry of the heating system is done with the help of a 2.5 D graphic editor with a 3D visualization module made by the authors of the system. This newly created part co-operates with the earlier developed system Pressure Drops [3,9,14] for fluid flow simulation with knowledge based modeling support [13].

Most of the modules are developed in the expert system technology and use CLIPS [6,15] environment together with Microsoft Visual Basic front-end interface. Each of the nodes can be considered as a separate, small rule-based subsystem. When the designer focuses on a particular node he has to load the appropriate knowledge base to the subsystem. While examining our system we noticed the lack of a meta-structure which could help the system developer and in future the designer to compose a suitable set of rules and facts fitting to the context of the given design problem. Becouse of that we developed special software for supporting the process of knowledge structuring. In the following chapters the basic concepts and their implementation are presented.

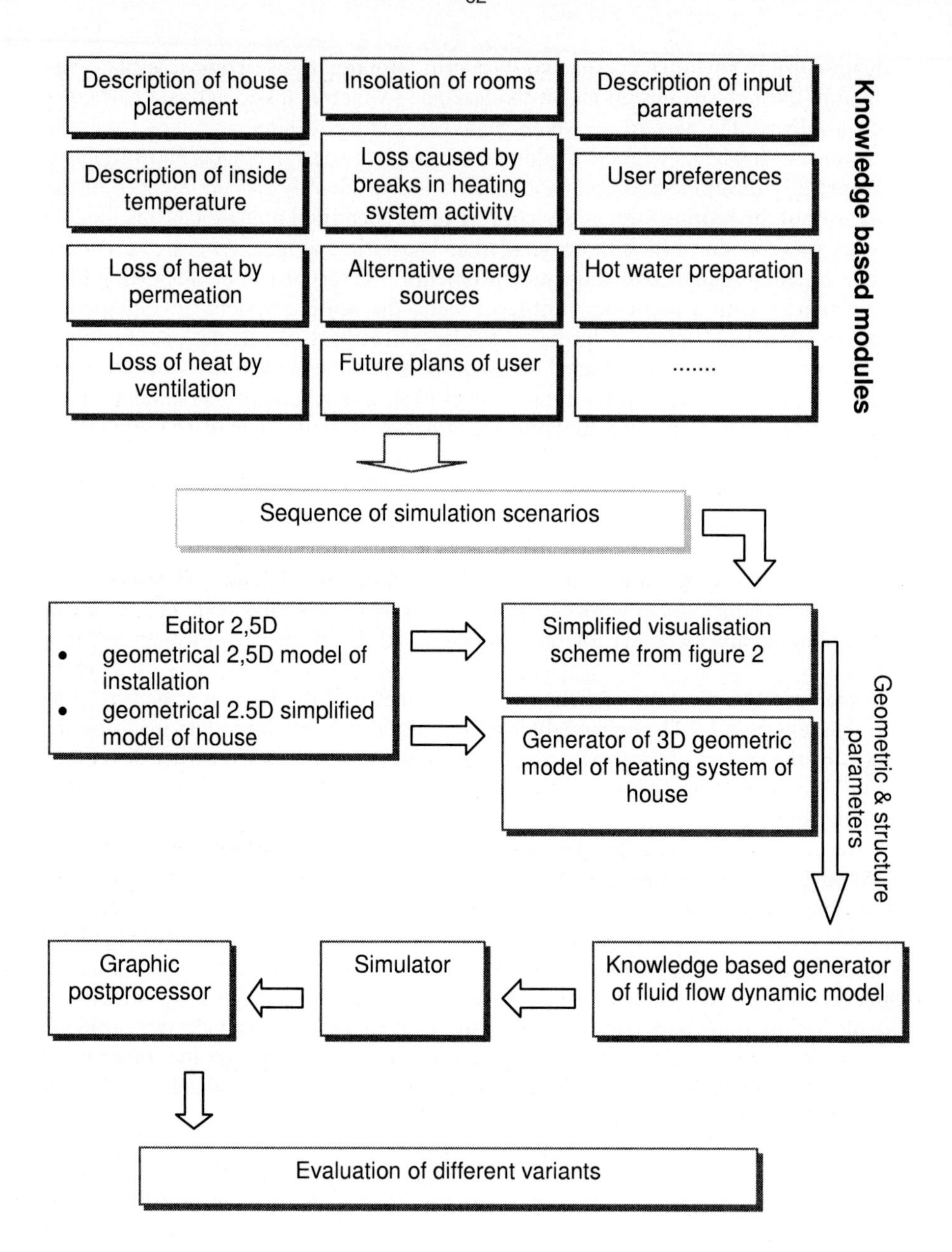

Fig. 1. Heating system design nodes. Each of them consists of a set of rules dealing with a particular topic

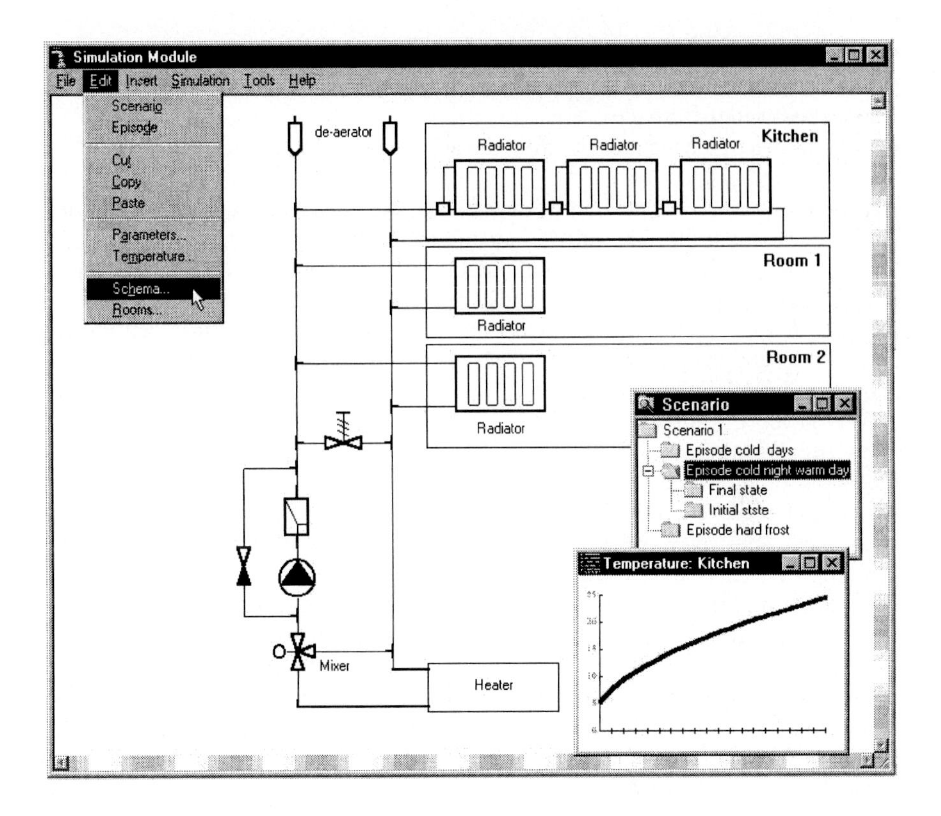

Fig. 2. Project of graphical interface for simulation module. The actual version is based on a 2.5D editor and a 3D-model generator

The structure of the system

In the stage of development of our system we observed a rapid growth of knowledge base structures (being under construction and planed for development), especially when we concentrated on realistic cases. But up to now we have considered only knowledge sources from literature and our own experiences. We haven't achieved yet the stage of regular contacts with professionals. When starting our work we didn't have any prefixed concepts of what the knowledge structure in our domain should look like. We selected some number of knowledge sources and came to the conclusion that the designer, in our case, should be supported by the following knowledge bases:

- with knowledge from professional literature, journals, papers prepared by producers, special knowledge from literature supporting modern heating system design,

- with knowledge from associated domains and relationships between domains,
- with knowledge from experts (designers, workers, service people, users),
- with knowledge from past cases,
- with knowledge which results from the application of our system.

As a first step we created an open software structure which can help us to initiate knowledge structuring and will be not limited to the possibilities of some existing tools which originally were not mentioned for design problems. As a future step, after experiments with real experts and real design problems, we want to consider the problem of appropriate software selection again. We suspect that the final result will be based on the integration of a rule based system, blackboard architecture [7,8,12] and a case base reasoning system [1,2,10].

We have a knowledge structure, which stores general domain knowledge, knowledge resulting from cases and knowledge created by experts. This central repository of rules and facts is organized as a relational database. The designer can search and retrieve professional knowledge, earlier examples, can refine them, or create new ones. New rules are automatically retained for future reuse. A special indexing system is developed which supports the process of finding associations. With the help of this tool we try to identify the knowledge structure of our domain.

The newly developed tool can successively interact with all subsystems of our system. When the user finishes his/her work with one module, it switches to another module and reloads the knowledge base. This technique is similar to blackboard architecture (now made by hand), where we have a central global database for the communication of independent knowledge sources focusing on related aspects of a particular design problem.

The software concept

In the process of building the knowledge base many knowledge chunks appear which have the form of rules and facts. Not every knowledge chunk should be activated in every problem. So rules are assigned to indexed groups. The user (designer or system developer) can activate the required sets of rules for particular sessions. He should exploit his/ her associations with the groups and design problems.

Our tool is the integration of an expert system environment with a relational database. The information stored in the database has CLIPS rule format. Each rule is stored in one record. The rules are identified by their names, and are divided into left and right hand sides. A special table stores CLIPS functions, which are used for the processing of rules.

The use of database for rule storage has several advantages. We can find rules easily, and we can use all tools available in the data base management system. The data transfer from a database to a CLIPS text file is based on querying. This way data management is much easier than pure text files processing in the CLIPS environment. Each rule can be assigned to any number of groups. The groups are corresponding with aspects in which we propose to consider a problem. This organization allows making sets of rules describing indexed cases.

While writing rules the user can use the developed dictionary of terms. The dictionary contains names of all used objects and defined functions. For a problem description we can use:

- object – attribute – value
- object – state
- object – relation – object

Objects and attributes can be used alternatively. When the description is more detailed the attributes become objects. Additionally multimedia files can be connected to the groups. Multimedia files can be presented in HTML format with any Internet browser.

After choosing the proper set of rules, and exporting them to the text file, they are loaded and processed in CLIPS environment. Using database for the storage of rules doesn't limit the tools of the CLIPS environment. We can still watch the agenda, the rule firings and the current state of reasoning. When we find a wrong rule, or when we want to add some rule, we can use an especially prepared form (in data base structure) where the rule can be edited. The edited rule can be loaded to CLIPS without interrupting its actual session. Any time the rule processing can be stopped, and the current state of the reasoning can be stored on a disk.

A special application controls the process of reasoning with some number of created knowledge bases. The process of reasoning with a particular knowledge base can be interrupted, as it consists of a number of sessions. Some reasoning processes can interfere. The edition of knowledge bases during reasoning sessions is allowed.

The authors actually work on a new module, which can parallel to the execution of the few reasoning processes monitor blackboard dependencies among reasoning processes. The monitored dependencies, together with the information about initiated, interrupted reasoning processes are shown in a special window.

Exemplary environment

The system is tested on the knowledge connected with heating systems design. Actually sets of rules dealing with the following topics are coded:

- identification of user preferences, investment conditions, generation of set of design alternatives
- determination of heat losses in building
- determination of profile of heating
- determination of alternative energy sources
- determination of type of hot water heater
- determination of future plans and house modifications
- supporting process of installation elements specification
- preparing data for module which simulates work of newly designed installation
- evaluation of installation

After preparing the whole data, simulations are launched. It allows the estimation of the effectiveness of an installation in the whole heating season.

The new application is made in the following environments:
1. relational data base in Microsoft Access environment,
2. rule based system in CLIPS environment,
3. integration modules in Microsoft Visual Basic.

In figure 4 different pieces of newly created software are presented.

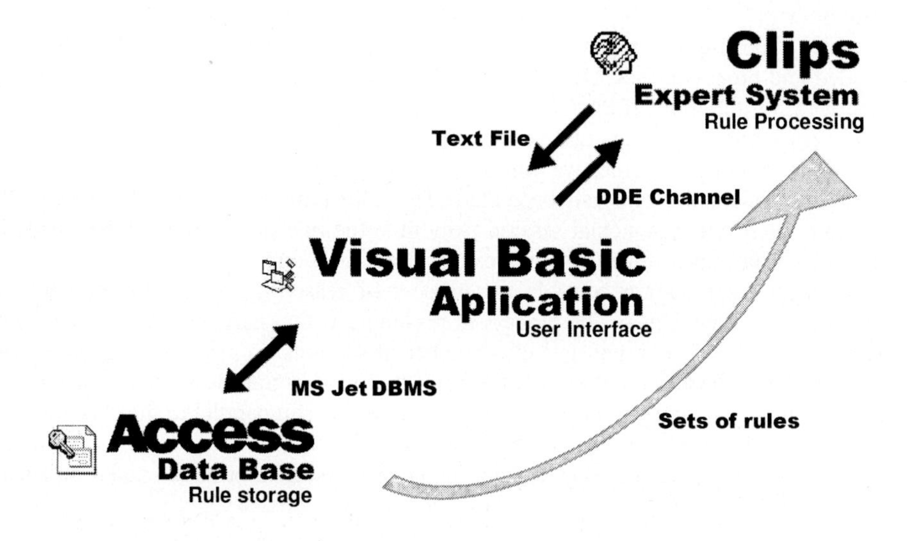

Fig. 3. Schema of application for dynamic knowledge modifcation. (MS Jet DBMS: questions handling and KB specification; Text file: results of inferencing; DDE Chanel- Clips processing commands)

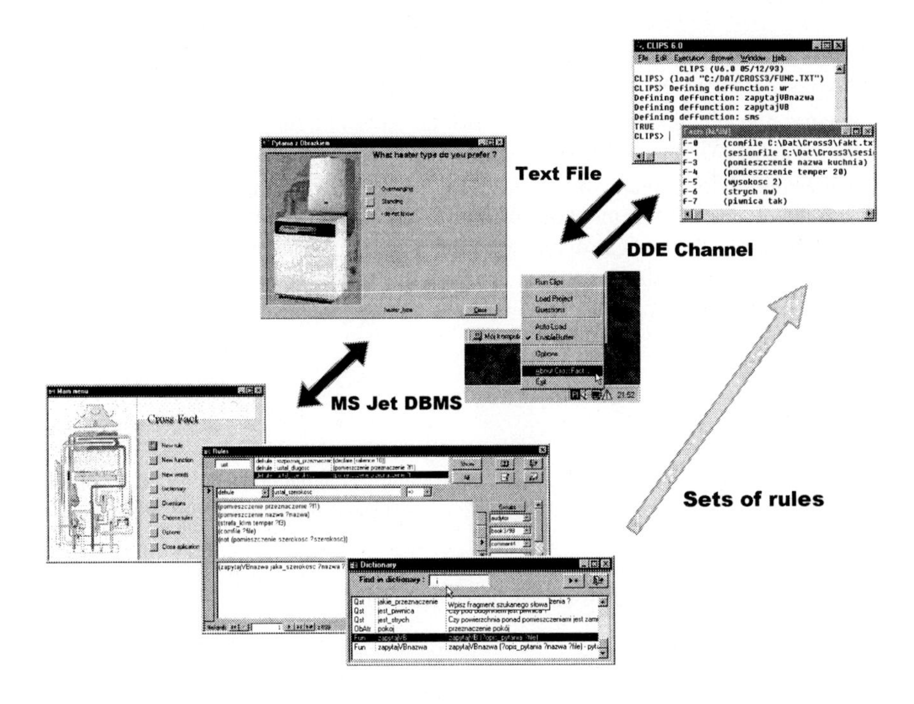

Fig. 4. Pieces of newly created software

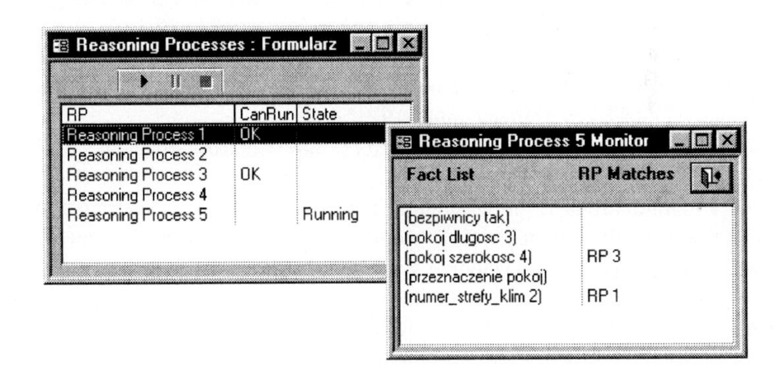

Fig. 5. The reasoning process monitor (under construction)

Conclusion

The paper presents the actual state of work on environment supporting heating system design. The environment is now on the stage of testing and knowledge base building. The considered domain are small houses and their installations.

The developed environment has universal form. Applications in two other domains (harmonic drives design, car dynamic analysis) are under construction.

The integration of the developed environment together with the module supporting mult-criteria, hierarchic, multi-level optimization is prepared.

References

1 Aamodt A., Plaza E.: Case –Based Reasoning. Foundational Issues, Methodological Variations, and System Approaches. AICom – Artificial Intelligence Communications, IOS Press, Vol. 7: 1, pp. 39-59.

2 Aamodt A. : A Knowledge Representation System for Integration of General and Case – Specific Knowledge. Proc. of IEEE Conference on Tools for Artificial Intelligence, Nov. 1-5,1994, pp. 1-5.

3 Babala D.: A Brief Description of The Compter Program INES. AB ASEA-ATOM, Vasteras, Sweden ,1989.

4 Boulanger S., Gelle E., Smith I.: Taking Advantage of Design Process Model, IABSE COLLOQUIUM, Bergamo 1995 pp. 87-96.

5 Cichocki P., Gil M., Pokojski J.: Integrated System for Heating System Design EG-SER-AI Workshop, Lahti 1997 pp. 27-32.

6 CLIPS, manual, version 6.0, COSMIC, Athens, 1993.

7 Fathi-Torbaghan M., Hoffmann A. : Fuzzy Logik und Blackboard Modelle in der technischen Anwendung. Oldenbourg Verlag GmbH, 1994.

8 Hayes-Roth B.: A Blackboard Architecture for Control. Artificial Intelligence 26, (1985), pp.251-321.

9 Maetz J.: Programm ISOM. Ein Programm zur Erstellung von Isometrien und Stucklisten. Manual Kerntechnik –Entwicklung -Dynamik, Rodenbach, Western Germany,1990.

10 Maher M. L., Gomez A.: Developing Case-Based Reasoning for Structural Design. IEEE Expert Intelligent Systems & their Applications, June 1996, pp. 42-53.

11 Maybury M.T. : Intelligent Multimedia Information Retrieval. The MIT Press, 1997.

12 Nii H.P.: Blackboard Systems at the Architecture Level. Expert Systems with Applications. Vol.7, pp.43-54, 1994.

13 Pokojski J.: An Integrated Intelligent Design Environment on the Basis of System for Flow Dynamics Analysis. International Conference on Engineering Design, 1995, Prague, pp. 1333-1338.

14 Pokojski J. : Manual for system "Pressure Drops". Kerntechnik-Entwicklung-Dynamik, Rodenbach, Western Germany, 1990.

15 The Haley Enterprise, Eclipse, manual, 1997.

16 Tong Ch., Sriram D.: (Ed.): Artificial Intelligence in Engineering Design. Vol. 1,2,3. Academic Press, 1992.

Collaborative Desktop Engineering

Edward L. Divita, John C. Kunz, Martin A. Fischer

Center for Integrated Facility Engineering, Stanford University, Stanford, CA 94305
{divita@leland.stanford.edu, kunz@ce.stanford.edu, fischer@ce.stanford.edu}

Abstract. This paper discusses conclusions from development, testing and comparison of three AI based applications that support multidisciplinary analyses of complex engineering problems. Our approach to automating multidisciplinary analyses involves integrating independent software tools (services) into unified systems that reason about shared computer interpretable models. Early desktop engineering systems integrate a fixed set of services and are limited in their flexibility to modify the types of analysis included in the integrated system. Next generation systems face the challenge of providing users flexibility to choose services freely based upon their analysis needs while providing a structured framework for controlling interactions and data sharing among the chosen services. This paper starts with a case example illustrating pre-project planning (PPP) challenges that today's practitioners confront. We then describe our approach to integration and automation as exemplified by the Facility Alternative Creation Tool (FACT), a system that utilizes structure, behavior, and function representation to implement integration and automation of PPP. We discuss how *service reasoning* exploits the function, behavior, and structure representation to process project information from multiple perspectives in a general way. We describe how *circle reasoning* enables dynamic addition and deletion of services and automated sequencing of changing sets of services. These reasoning approaches together with a PPP ontology enable greater flexibility for integrating and automating PPP. We compare FACT to two other applications, focusing on five issues: purpose, representation, reasoning, interface, and testing. From experimenting with these systems, we conclude that, like desktop publishing, desktop engineering enables significant change in engineering practice.

1 Case Example: Pre-Project Planning of a Fast-Food Restaurant

Pre-Project Planning (PPP) is the process of creating, analyzing, and evaluating project alternatives during the early planning phase to support a decision whether or not to proceed with the project and to maximize the likelihood of project success. PPP for capital facilities requires analysis from multiple perspectives: marketing; architecture; engineering; planning; cost estimating; organization; and finance.

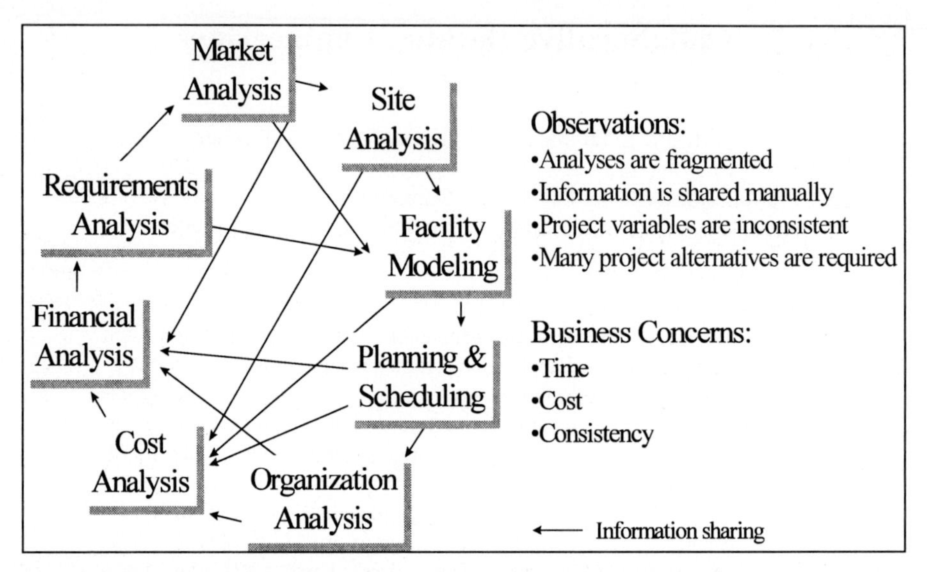

Fig. 1. PPP requires multiple analysis perspectives (e.g., Market Analysis, Site Analysis, Facility Modeling). *Analyses are fragmented,* normally performed by independent specialists using specialty software. The results of each analysis are *shared manually* among the participants via reports or drawings (e.g., Facility Modeling relies upon the Market Analysis and Site Analysis outputs to size the building appropriately; Financial Analysis relies upon Planning and Scheduling to time the cash flows, and Organization and Cost Analysis to support cost projections). We also observed that project nomenclature and units of measure for project *variables are inconsistent* from project to project and participant to participant. Further, planners must consider *many project alternatives* to identify an optimum project configuration. Business managers voice concerns that the PPP process is overly time consuming, costly, and inconsistent.

PPP decisions (e.g., choice of market, selection of site, and determination of building size) have a significant influence on a project's outcome. PPP has two outputs: first, the decision whether or not to proceed with the project under consideration ("go/no go"); second, if the decision to proceed is positive, a plan of action guiding later phases.

We investigated the PPP process of an international developer of fast food restaurants (Divita et al. 1997-B). In our study we used interviews and structured data collection to identify PPP procedures, analyses, participants, and realistic case data. We learned that business managers are concerned that the PPP process is time consuming, costly, and often inconsistent from project to project and from practitioner to practitioner. We observed that fragmentation of PPP analyses, inconsistency of project variables, manual information sharing, and the need to consider many project alternatives appear to contribute to PPP time, cost, and inconsistency. Figure 1 illustrates these observations.

2. Background

2.1 Technical and Social Collaboration

Project teams comprise many specialists, and each specialist is responsible for a portion of the project information processing involved in PPP. Fragmentation complicates PPP because the specialists who describe, analyze, or evaluate the project or its constituents are commonly distributed geographically, along with their software tools. During PPP, project participants require information generated by others to perform their work. For example, when recommending the appropriate size and capacity for a new fast food restaurant, the architect requires market information (e.g., population demographics and demand indicators) prepared by a market specialist. When estimating construction cost, the construction professional requires a description of the project scope prepared by the architect. When performing cash flow analysis and predicting expected return, the financial analyst requires cost and schedule information prepared by several of the other participants. When information changes (e.g., a different site is considered or a different building configuration is proposed), the participants must disseminate the new information throughout the team. Upon interpretation, analysis, and evaluation, the participants provide feedback about the impact of the change to the other team members.

Participants normally share information manually through *social collaboration*, reports and drawings. These human participants either know from experience what other participant(s) to go to for information or they ask questions to figure it out. These participants also share a common vocabulary that includes PPP variables and terminology. Using this vocabulary facilitates communication among the participants. A system of software tools interacting to perform a task such as PPP in an integrated and automated manner is *technical collaboration*, analogous to social collaboration (Fischer & Kunz 1995). However, the software tools in use by PPP practitioners today do not know how to interact with each other the way humans do. Prior researchers have demonstrated the benefits of using a common computer interpretable vocabulary, or ontology, to facilitate knowledge and data sharing among heterogeneous software applications (Gruber 1993). When a team of specialists work together to perform PPP, they plan their sequence of interactions based upon the information requirements associated with the project. To reduce the time and cost, and improve the consistency, of PPP through integration of software services and automation we face the engineering problem: **How can an assemblage of software tools be made to interoperate and automatically coordinate and sequence their analyses?**

2.2 Small Number of Users

During our study of fast food restaurant PPP, we observed that there is a central project manager or decision-maker who gets advice from various specialists and consultants (e.g., engineers, architects, planners, controllers, brokers). This key manager directs the PPP process and controls the assumptions and decisions required to specify a project alternative. However, the key manager lacks the time and may also lack the technical expertise and skills to perform the necessary analyses. For this reason, managers spend time and money acquiring feedback from specialists. The decision-maker simply needs accurate, reliable advice and sharable data. To continue the analogy between social and technical collaboration, just like a team of specialists, an integrated system of software services could advise a single decision-making user. To explore this type of desktop engineering, we face the engineering problem: **How can integrated systems of software services function like a room full of experts supporting a single decision-maker performing a function such as PPP?**

2.3 Related Studies

Other researchers have made progress towards answering these questions. The Combine 2 (1995) project tested prototypes that integrate a building product model with design tools and a supervisory module for HVAC and Architectural CAD design cases using on-line actors performing CAD work. The Combine supervisory module independently controls interactions among users (e.g., network communications) but does not dynamically sequence nor automate the design process. Combine 2 supports users in collaboration, but does not emphasize a framework for reducing the number of users involved in the design process. The Intelligent Boiler Design System, IBDS, (Riitahuhta 1988) integrates a CAD system with an engineering analysis system and a relational database system to perform automated design of boilers for power plants. IBDS reduces the number of users involved in Design by replacing specialists with a rule base that propagates constraints from the attributes of one component to the relevant attributes of others. The IBDS approach is functional in practice for design of boilers, but as systems become more complex, heuristic design rules become more difficult to manage. The IBDS knowledge base is specialized, and it would be laborious to modify for other types of design. Continuing international standardization efforts such as ISO 10303-1 (1994) are focused on standardizing the semantics of product representations such as shape and structure so that various design tools can communicate by using a common representation and information exchange format. This standardization is useful in providing a consistent vocabulary for describing project components, but does not provide a basis for automating or controlling interaction among software tools nor a way to relate software services to a SPM. To date, these standards have not addressed the domain of PPP.

We are interested in developing a more flexible principled approach to automating PPP, where knowledge bases are modularized within separate specialty software services, services are integrated into a unified control system, and the unified services reason about a shared project model in a consistent and structured manner. The next section describes an integrated prototype system, Facility Alternative Creation Tool (FACT). FACT derives and controls the interactions of services during the PPP process from the function, behavior, and structure attributes of a project model and the information requirements of the chosen set of services.

3 Desktop Engineering

3.1 FACT System Design

The FACT prototype performs integrated and automated PPP analysis by joining otherwise independent software tools (PPP analysis services) into a system of interoperable coordinated services that reason in a consistent way about a shared project model. We also pursue rapid feedback on modifications and simple yet effective management of project alternatives. To address these objectives, FACT implements circle architecture (Fischer & Kunz 1995) for PPP of fast food restaurants. Our implementation of the circle architecture involves a simple set of linked autonomous software services, in which case the circle output is the union of the outputs of the individual applications. Figure 2 shows a concept diagram for the PPP circle architecture.

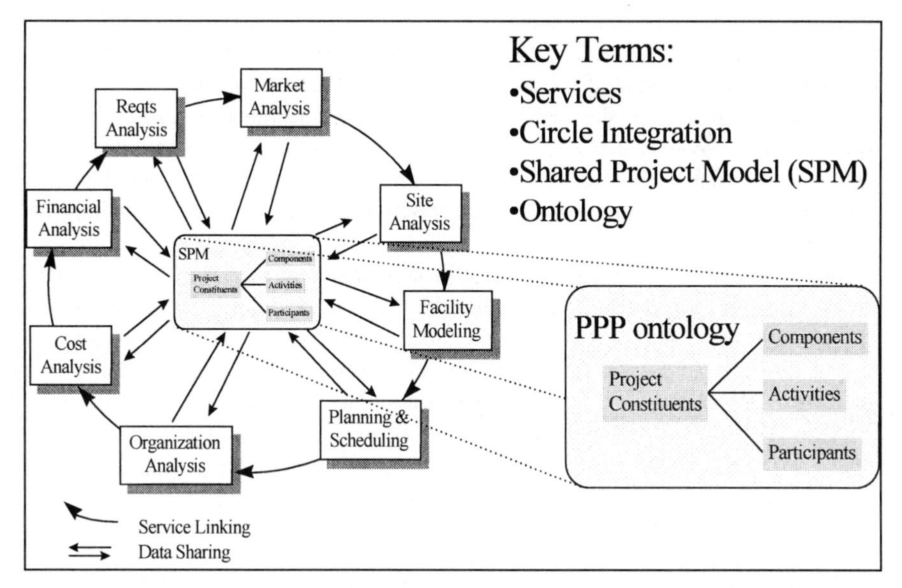

Fig. 2. Concept diagram of FACT prototype circle architecture. At the center of the system is the shared project model (SPM). For each project alternative, the SPM is derived from a PPP ontology. The nodes around the circle represent software services that work cooperatively to analyze the SPM. Services are linked together in sequence and data is shared among the services through the SPM.

3.2 The Shared Project Model (SPM)

A limitation of current PPP practice is its use of manual data sharing. Manual data sharing is slow, error prone, and inefficient. Prior research efforts demonstrate the benefits of integrated computer based shared project models (Combine 2 1995, Riitahuhta 1988). Construction Knowledge Expert (COKE) (Fischer 1991) automates constructibility feedback for the preliminary design of a reinforced concrete structure using a shared model. The Object Model Based Project Information System (OPIS) (Froese 1992) combines project planning applications around a shared object-oriented project database. The Intelligent Real Time Maintenance Management (IRTMM) system discussed later in this paper (Kunz et al. 1995) utilizes a symbolic process plant model shared among three analysis modules. FACT has a SPM as a persistent information exchange model for related services. However, in our search of literature we were unable to find structured computer interpretable content for a pre-project planning SPM. To overcome this limitation, we created a new PPP ontology.

3.3 PPP Ontology

An ontology describes the concepts and relationships that can exist for an agent or a community of agents in a domain to enable knowledge sharing and reuse (Gruber 1993). A conceptualization is an abstract view of what we wish to represent for some purpose. An ontology specifies that conceptualization explicitly. In this sense the ontology acts as a template for the creation of shared project models.

In the FACT system, the SPM is a project model instance derived from the PPP ontology. The ontology includes the elements and variables of a project that practitioners consider during PPP for fast food restaurants. We created the ontology by observing the PPP practices of a fast food restaurant developer and abstracting the essential PPP concepts and variables.

During automated PPP analysis, a SPM is instantiated from the PPP ontology and the services analyze the SPM to predict the project outcomes. By going around the circle (figure 2), a decision-maker develops a project scenario (alternative) in a consistent fashion and gets rapid and consistent feedback from the perspective of each specialty. We document a detailed case example using FACT in (Divita et al. 1997-A).

3.4 Representation

The key elements of the representation scheme of FACT are the SPM and the services. This section describes the representation scheme using an abstract simplified case facility project. The case project involves a site, a building, and three services: Site Analysis; Facility Modeling; and Cost Analysis. We implement the representation using object-oriented programming terminology. The representation scheme involving the SPM, service identifier objects, and relationships establishes the virtual project world wherein PPP reasoning can be performed in a consistent and structured manner.

The example SPM illustrated in figure 3 contains three main components: Site; Building; and Project. Each of these components is an object with attributes: Behavior, Function, and Structure. We use the behavior/function/structure terminology in a manner consistent with Gero (1990) to describe elements of the SPM. Figure 3 illustrates the behavior, function, and structure attributes for the Site, Building, and Project components of the SPM. Structures are components that describe the physical characteristics of an entity. The Site object's structure attributes include: *length, width, FAR* (floor area ratio), and *coverage*. Functions are requirements or purposes of the object. The Site object's function is *provision of land*. Behaviors are descriptions of how something behaves. Behaviors are derived from structures by means of reasoning, or analysis. The Site object's behaviors are *max building area, site area,* and *site cost*.

The SPM is a facility model that services analyze. The Behavior/Function/Structure framework provides a consistent representation that will support consistent reasoning methods that we discuss later.

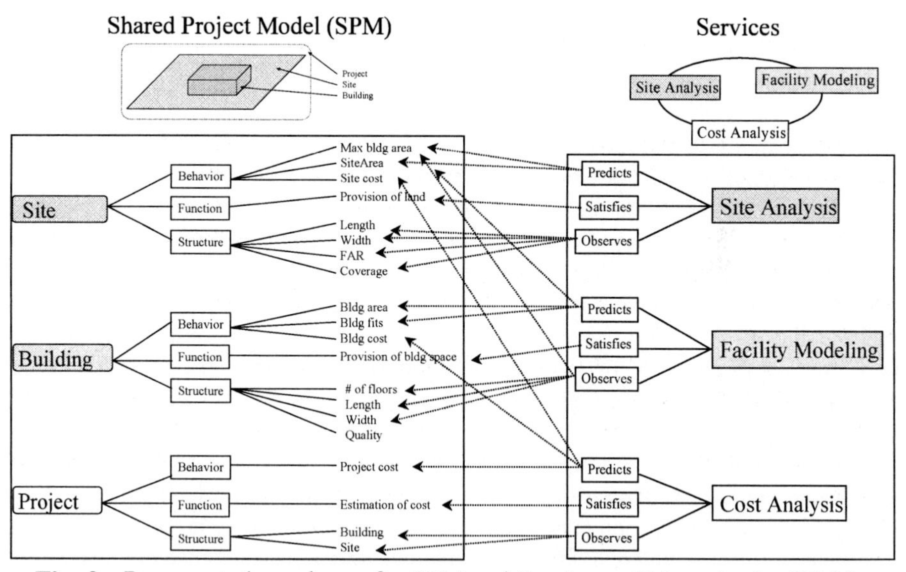

Fig. 3. Representation scheme for SPM and Services. Objects in the SPM have behaviors, functions, and structures. Services have "predicts," "satisfies," and "observes" relationships with entities in the SPM.

The figure 3 example has three services that relate to components described in the model: Site Analysis; Facility Modeling; and Cost Analysis. Each service has an identifier object that specifies how the service processes information in the SPM. The service object establishes relationships between the service and elements of the SPM. Figure 3 also illustrates the relationships among the services and the SPM. A service *predicts* behaviors through analysis. The Site Analysis can predict *max building area* and *site area.* A service *observes* any structure or behavior that is an input to the service. The Site Analysis service observes length, width, FAR, and coverage. A service *satisfies* functions by performing the type of reasoning that determines if requirements are fulfilled. For example, the Site Analysis service satisfies *provision of land.*

A key objective of our representation is to enable straightforward reasoning. The project component function, behavior, structure representation enables consistent reasoning methods (*service reasoning*) for formulation, analysis, and evaluation that will be discussed in the next section. These reasoning methods can be utilized by any service on the circle, regardless of specialty. This feature supports our goal of flexible service selection because all services can reason about the SPM utilizing a consistent reasoning framework. Also, the "predicts," "satisfies," and "observes" relationships among the services and the SPM enable simple reasoning (does not require a complex rule set or knowledge base) methods for sequencing and controlling circle integration that will be discussed in the next section. This new construct also supports the flexibility goal by providing a precise and consistent way of describing the relation between services and the SPM.

3.5 Reasoning

During automated PPP analysis, a SPM is instantiated from the PPP ontology and the services analyze the SPM to predict the project outcomes. By going around the circle (figure 2), a decision-maker develops a project scenario (alternative) in a consistent fashion and gets rapid and consistent feedback from the perspective of each specialty. We document a detailed case example using FACT in (Divita et al. 1997-A). Our engineering problem is to build an integrated system of software services that function like a set of analysts supporting a single decision-maker. While each service performs a different type of analysis, we propose that each of the services reason about the SPM in a consistent and structured way. We refer to this approach as service reasoning. A second engineering problem is to build a flexible integration architecture that allows services to be added or deleted as appropriate and that adjusts the application control structure to use the changing set of services effectively. We call this control circle integration reasoning. For service reasoning we build on the model of design as a process (Gero 1990). For circle integration reasoning we build on circle integration (Fischer & Kunz 1995).

3.5.1 Service Reasoning

This section describes the service reasoning that is implemented in FACT. In the following examples, our simplified case is reasoned about from three different perspectives: site analysis; facility modeling; and cost analysis. The example illustrates how functions transform into expected behaviors (formulation), behaviors transform to structures (synthesis), structures transform to predicted behaviors (analysis), and expected and predicted behaviors are compared (evaluation).

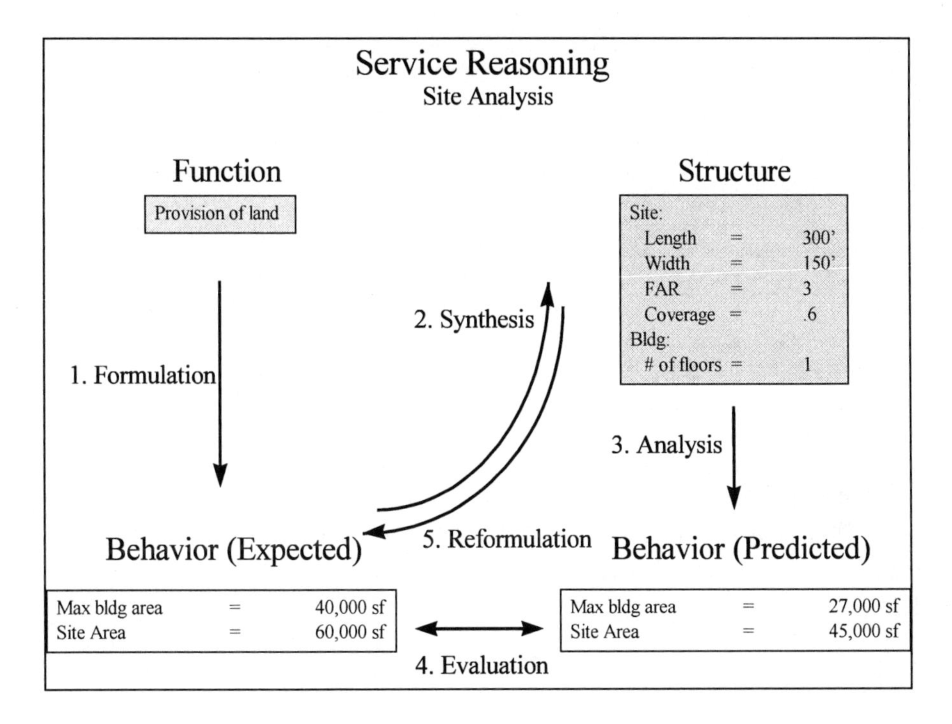

Fig. 4. Service reasoning for Site Analysis (based upon Gero 1990)

Figure 4 illustrates the first kind of service reasoning for our test case, Site Analysis. **Step 1 Formulation:** The service transforms function, *provision of land,* into expected behaviors, *max building area* and *site area.* During PPP, when searching for a parcel of land for the facility, participants must specify the building area and site area that they desire (expect). In our example, the user expects that a 40,000 square foot (sf) building on a 60,000 sf site will fulfill his requirements. **Step 2 Synthesis:** The system transforms behaviors into structures by listing the variables that must be analyzed to predict behaviors. The user proposes his project alternative by specifying the values of the structure variables. During PPP, participants consider potential sites and building parameters. In our example, the user proposes a site that is 300' by 150' with floor area ratio of 3 and allowable site coverage of .6. The proposed building is 1 story tall. **Step 3 Analysis:** The service analyzes the proposed structure to predict behaviors. The service calculates site area to be 45,000 sf. The service calculates max building area for the proposed site to be 27,000 sf (.6 x 45,000 sf x 1 floor = 27,000). **Step 4 Evaluation:** The user compares predicted behaviors with expected behaviors. The largest building that the proposed site will allow is 27,000 sf. For this case, the predicted area is significantly less than expected. However, there are alternatives. **Step 5 Re-formulation/Re-synthesis:** The user could reformulate and reduce the max building size to work with the proposed site. Alternatively, he could re-synthesize and propose another site or choose to put a building with two or more stories on the site. With feedback from the site analysis service the user can perform site analysis independently.

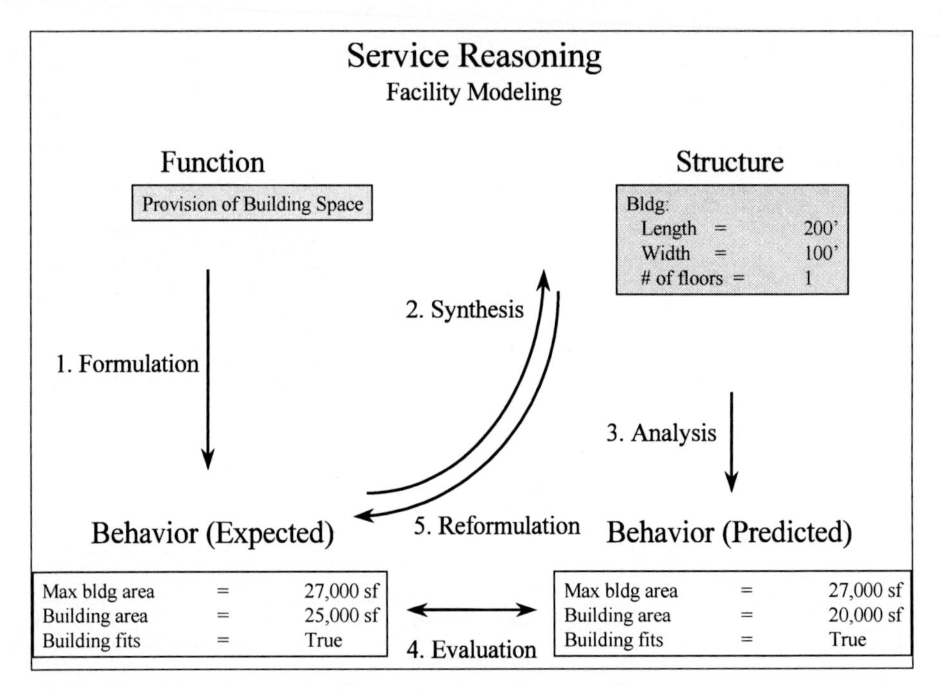

Fig. 5. Service reasoning for Facility Modeling (based upon Gero 1990)

Figure 5 illustrates the second kind of service reasoning, Facility Modeling. **Step 1 Formulation:** The service transforms function, *provision of building space,* into expected behaviors, *max building area, building area,* and *building fits*. During PPP, participants configure buildings by specifying the expected area. In our example, the user expects that a 25,000 sf building will fulfill requirements. **Step 2 Synthesis:** The user proposes a project alternative by specifying its structure. During PPP, participants consider alternative building configurations. In our example, the proposed building is 200' by 100' and 1 story tall. **Step 3 Analysis:** The service analyzes structure to predict behaviors. The service calculates building area as 20,000 sf. The service calculates max building area for the proposed site as 27,000 sf. **Step 4 Evaluation:** The user compares predicted behaviors to expected behaviors. The predicted building area is less than the expected building area, therefore the building fits on the site. **Step 5 Re-formulation/Re-synthesis:** The user could reformulate and reduce the expected building area. He could also re-synthesize and propose a larger building.

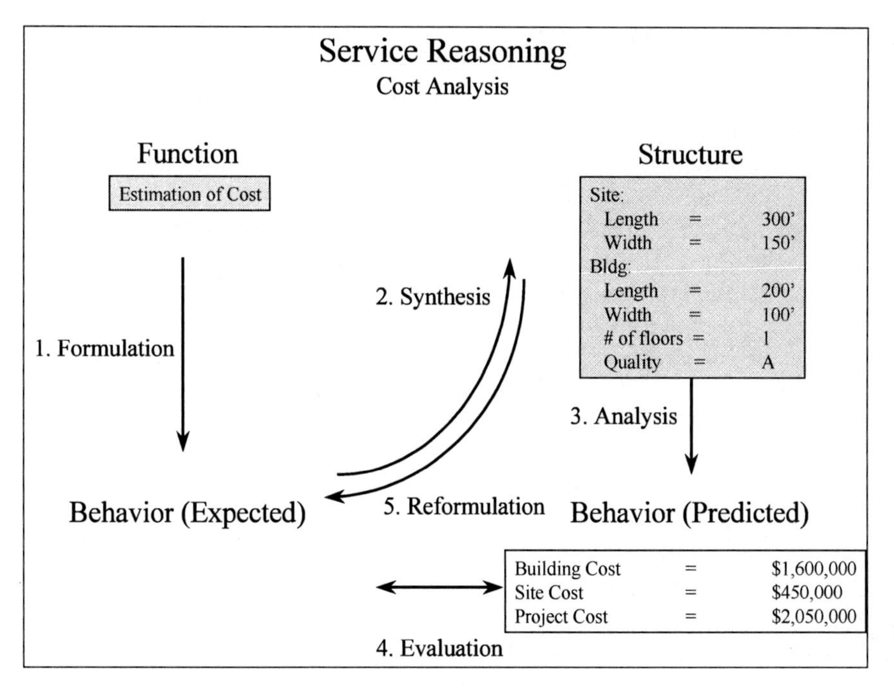

Fig. 6. Service reasoning for Cost Analysis (based upon Gero 1990)

Figure 6 illustrates the third kind of service reasoning for our test case, Cost Analysis. **Step 1 Formulation:** The service transforms function, *estimation of cost,* into expected behaviors, *building cost, site cost,* and *project cost*. During PPP, participants anticipate (expect) budgets. In our example, the user expects that a building cost budget of $1,400,000 and a site cost budget of $500,000 will fulfill requirements. **Step 2 Synthesis:** The user proposes an alternative by specifying structure. During PPP, participants propose building and site dimensions and quality levels. **Step 3 Analysis:** The system analyzes structure to predict behaviors. During cost analysis, the service references historic data to estimate project costs based upon the proposed project structure. The service estimates building costs to be $1,600,000 and the site costs to be $450,000. The service estimates the total project cost to be $2,050,000. **Step 4 Evaluation:** The user compares predicted behaviors to expected behaviors. The predicted project costs are greater than the expected project costs. The project as proposed is over budget. **Step 5 Re-formulation/Re-synthesis:** The user could re-formulate and increase budgeted building costs, or he could re-synthesize and propose different structure variables (e.g., dimensions or quality). With feedback from the cost analysis service the user can perform cost analysis independently.

This example illustrates a key benefit of the explicit function, behavior, structure representation in the SPM. Each service on the circle processes information in a structured and consistent manner. Services determine what behaviors are to be considered based upon function. Values for predicted behaviors are derived from the proposed structure.

3.5.2 Circle Integration Reasoning

Circle integration provides a concept for systematically joining software services into a unified system. The FACT system operationalizes circle integration mechanisms that implement the concept. This section briefly describes our methods for planning and controlling integrated and automated PPP. The representation and reasoning of FACT mutually support our flexibility objective.

The FACT circle integration mechanisms extend circle integration. The *Circle Broker* mechanism allows the user to choose services to add to or delete from the system. When a user proposes a function to be satisfied, the system searches the network for services that satisfy that function. When the user adds the service to the circle, the service establishes *predicts, observes, and satisfies* relationships to the shared project model. These relationships support reasoning about the sequence of services for automated PPP.

The *Circle Navigator* plans the sequence of services on the circle. The navigator uses a model-based reasoning approach to planning. The navigator evaluates the variables that each service observes and predicts with consideration for the "predicts" and "observes" relationships that exist among the variables in the SPM and other services on the circle. With this derived knowledge, the navigator sequences the services using constraint satisfaction. During automated PPP, when a structure variable is revised (by the user or another service), the navigator invokes all other services that observe the revised structure variable. The navigator also invokes any service around the circle that observes a behavior that is predicted by one of the services that observed the revised structure. Using circle reasoning, the navigator propagates the effect of a revised variable throughout the system of unified services.

These circle integration mechanisms rely upon and benefit from the function, behavior, structure representation of the PPP ontology. The "satisfies" relationship between functions and services provides a purposeful way to link services to the SPM. The "predicts" and "observes" relationships between services and behaviors/structures provides the basis for sequencing and navigating services by deriving paths based upon information requirements. Using function, behavior, and structure, these relationships are determined dynamically, "on-the-fly", as opposed to being "hard-coded" in rule-bases or methods. This enables much greater flexibility and freedom in choosing services to use in an integrated system. The FACT service and circle reasoning also simplify interface and integration requirements.

3.6 User Interface

Figure 7 shows some FACT user interface windows. The Circle Navigator interface controls circle integration. Using this interface, the user controls planning and navigation around the circle.

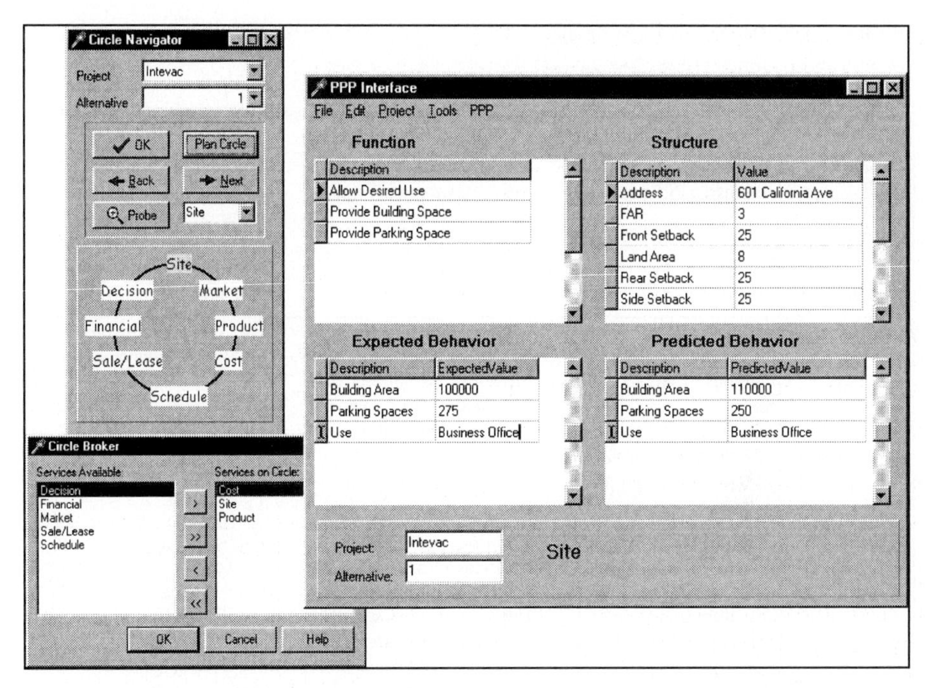

Fig. 7. FACT user interface windows: Circle Navigator, Circle Broker, and PPP Interface.

The user can also override the planned sequence and probe into any service on the circle. The Circle Broker interface allows the user to choose services from the network and add them to the circle. The PPP Interface has a layout that is similar to the model of design as a process diagrams. In this interface, the user can view the values of functions, behaviors, and structures, change them, review effects of change, and accept, modify, or undo changes. A useful extension would be to add interactive CAD and CPM displays.

3.7 Comparison of Desktop Engineering Systems

The table below compares three systems developed at the Center for Integrated Facility Engineering. Each is an example of a Desktop Engineering system whose purpose is to link multiple analyses and support different engineering perspectives. The first generation of Desktop Engineering applications - represented by IRTMM and ViRP - has a fixed set of applications services that are invoked automatically and whose sequence of invocation is set by the system developer. The FACT reasoning methods enable dynamic addition and deletion of services and automated sequencing of changing sets of application services.

Table 1. Comparison of desktop engineering systems across 5 issues

5 Issues	FACT	IRTMM	ViRP
Purpose	To perform integrated and automated PPP analysis.	To help process plant owners perform value-based plant maintenance.	To provide appropriate decision support during concurrent engineering, considering product design, construction plan, and the organization.
Representation	A shared project model (SPM) derived from an ontology of project entities classified as Function, Structure, and Behavior. The SPM is linked to each analysis service for input and output	A shared plant model consisting of plant components, activities, parameters, and resources. Three software modules reason about the plant model.	4D-CAD: Building objects in a 3D CAD model of designed facility are linked to time-based discrete schedule activities. VDT: Organization participant objects, "actors," linked to activities of a CPM.
Reasoning	Service Reasoning: Functions transform to expected behaviors (formulation), structures transform to predicted behaviors (analysis), expected and predicted behaviors are compared (evaluation). Circle Integration Reasoning: broker, plan, and navigate circle.	Model-based diagnosis, heuristic classification, case-based reasoning, hierarchical activity generation and constraint satisfaction sequencing, and decision tree expected benefit	4D-CAD: Purely graphical time-based visual simulation. VDT: discrete event simulation of an organization functioning as an information processing and communication system.
Interfaces	Graphical user interface with Circle Navigator as primary control of automation, Circle Broker as primary control of integration, and a separate interface for each service	Graphic user interfaces including schematic diagrams (P&ID) and tabular explanations of results.	4D-CAD: Graphic 3D and 4D CAD model and a graphic presentation of CPM diagram. VDT: Organization chart and CPM diagram. Gantt chart and actor in-tray output.
Testing	Bench testing of software with realistic test cases. System testing by developer, fast food restaurant managers, AEC industry practitioners, and students.	Multiple realistic industrial test cases.	Realistic test case with components, activities, actors, and relationships based on an actual retrofit of a component of semiconductor manufacturing facility.

FACT, IRTMM and ViRP (Kunz et al. 1997) share the common theme of joining software modules together to form unified systems that enhance the integration and automation of engineering processes. Our industry partners envision benefits from integrating existing software tools into systems that can give a single user more perspective on engineering problems. The research challenge exists in representing engineering problems in a structured way so that systems can reason in a principled way to solve the problem while providing flexibility to the user.

FACT and IRTMM both represent the engineering problem using software services linked to a shared model. ViRP links components of a CAD model to components of a process model (Critical Path Method precedence model) and components of an organization model. IRTMM and ViRP use a fixed set of components and software tools. For these two systems, the analysis processes and software components are fixed, which is appropriate since each has a small number of reasoning services and represents a specific engineering problem. FACT utilizes a changeable set of software services and sequences the interaction among services dependent upon what services are currently on the circle and the information requirements of each. The open system and dynamic application sequencing of FACT provide more flexibility but require that software services developers comply with the PPP ontology.

Each of the systems has a graphic user interface. The ViRP interface includes a graphic CAD model, CPM model, and organization hierarchy. These idioms are natural to CAD operators and project managers. The IRTMM system utilizes a process and instrumentation diagram (P&ID) that is a natural form of communication among plant engineers and operators. The FACT interface utilizes the "circle integration" and "design as a process" models wherein project information is generally described to the user as graphics, text and numbers. Testers find that the P&ID and CAD graphical interfaces of IRTMM and ViRP communicate naturally. Users find the textual interface elements of FACT to be less intuitive.

FACT utilizes realistic test cases developed through observation of PPP in practice. While continuing development and testing are underway, preliminary results of testing by industry and student users reveal that significantly more project alternatives can be processed in a given duration using FACT when compared with manual methods. IRTMM utilized multiple realistic industry test cases for specific plant maintenance scenarios to validate that the results generated by the system were consistent with those commonly found in practice. ViRP utilizes a realistic test case based upon an actual retrofit of a component of a semiconductor manufacturing facility to illustrate the proposed integration concepts. We plan future testing of FACT using the charrette testing method (Clayton et al. 1997) to validate the benefits of reduced time and increased project alternatives.

4 Discussion

The FACT, IRTMM, and ViRP systems are examples of Desktop Engineering systems. In our work, we relate Desktop Engineering to Desktop Publishing. We make an important distinction between *social* and *technical collaboration*. Integrated desktop systems enable collaboration among applications that support different disciplines. There is frequent and rich data exchange among interoperable applications that share access to semantically consistent models. The desktop engineering examples each include linked symbolic models of an engineering product and one or more engineering processes.

Desktop systems shift the integration issue from social integration to technical integration of multiple computer applications. For the FACT, IRTMM, and ViRP systems, we emphasize technical integration to reduce the need for social integration. The computer provides significant value as a tool for representation, manipulation and analysis of given data. It integrates multiple applications. While the traditional process has been social with multiple people doing their work over a period of time, the Desktop process generally involves one or a small number of participants who have broad responsibility for generating and presenting content.

There is distinct but modest social collaboration among different human specialists who might use different parts of an integrated system. A general theme is that social collaboration among project participants becomes quite different using one of these systems than it is in routine practice without the integrated system. Desktop publishing has made it common for a single user with limited experience to create a publication without collaborating with others. Prior to desktop publishing tools, publication required collaboration among several specialists (e.g., writer, editor, typesetter, and graphic artist). Desktop engineering changes engineering practice in a similar way. With desktop engineering systems, authority and decision-making responsibility become highly centralized, and the number of participants may become significantly smaller than in the current routine process.

We have found that each of the services of an integrated system benefits from a specialized user interface (e.g., CAD, CPM network, P&ID) that allows users to view their products in specialized, natural ways. However, some of the data represented within each interface is shared with other services. Our approach to integrating specialized services respects the value of a natural idiom interface for each application, and it adds the benefit of providing the same values for shared data since each UI takes values of shared data from a shared project model.

Like Desktop Publishing, the Desktop Engineering process has an opportunity to be qualitatively faster and qualitatively higher quality than the prior manual method. Many of the current participants in the process will find that they are adding less value and have a less-necessary role than in the past. Client users will get much of the value of the new technology, and the content provider will have a significantly enhanced responsibility for content generation.

5 References

Clayton, M. J., Fischer, M. A., Teicholz, P., and Kunz, J. C., (1997), The Charrette Testing Method for CAD Research, *Applied Research in Architecture and Planning,* vol. 2, ed. R. Hershberger and M. Kihl, Tucson, AZ: Herberger Center for Design Excellence.

Combine 2 (1995). *Computer Models for the Building Industry in Europe.* Final Report, Godfried Augenbroe Ed., TU Delft.

Divita, Edward, Jr., Fischer, Martin A., and Kunz, John (1997-A). "Integration and Automation of Pre-Project Planning Through Circle Integration with a Shared Project Model." Technical Report, Nr. 115, Center for Integrated Facility Engineering, Stanford.

Divita, Edward, Fischer, Martin, Kunz, John, and Brown, Reggie (1997-B). "Integrating and Automating Pre-Project Planning Using FACT (Facility Alternative Creation Tool)." ASCE Construction Congress V, Stuart D. Anderson (Ed.), Minneapolis, MN, Oct. 4-8, 964-971.

Fischer, Martin A. (1991). "Constructibility Input to Preliminary Design of Reinforced Concrete Structures." Technical Report, Nr. 64, Center for Integrated Facility Engineering, Stanford.

Fischer, M., & Kunz, J. (1995), "The Circle: Architecture for Integrating Software," *J. Comp in Civ. Engrg.*, ASCE, 9(2), 122-133.

Froese, T., (1992), *Integrating project management software through object-oriented project models*, PhD thesis, Dept. of Civ. Engrg., Stanford Univ., Stanford, CA

Gero, J.S., (1990), "Design Prototypes: A Knowledge Representation Schema for Design," *AI Magazine,* 11(4), (pp 26-36) Winter 1990

Gruber, T. R., (1993), "A translation approach to portable ontologies," Knowledge Acquisition, 5(2):199-220.

ISO 10303-1 (1994), ISO-TC184/SC4 ISO, Industrial automation systems and integration -- Product data representation and exchange -- Part 1: Overview and fundamental principles.

Kunz, J.C., Jin, Y., Levitt, R.E., Lin, S., Teicholz, P.A., (1995), "The Intelligent Real-Time Maintenance Management (IRTMM) System: Support for Integrated Value-Based Maintenance Planning," *CIFE Technical Report 100*, Stanford University, January 1995.

Kunz, J.C., Fischer, M.A., Kim, J.S., Nasrallah, W., Levitt, R.E., (1997), "Concurrent Engineering of Facility, Schedule and Project Organizations," *International Journal of Computer-Integrated Design and Construction (CIDAC)*, in press.

Riitahuhta, A. (1988), "Systematic Engineering Design and Use of Expert System in Boiler Plant Design," *Proceedings of the 10th International Conference on Engineering Design,* Budapest, Hungary.

Towards Personalized Structural Engineering Tools

Steven J. Fenves

University Professor
Department of Civil and Environmental Engineering
Carnegie Mellon University
Pittsburgh, Pennsylvania USA

Abstract. Structural engineering software has moved from in-house development to third - party vendors. This trend has been beneficial for the profession as a whole and permitted rapid migration to new environments, particularly in exploiting IT capabilities in communication and information sharing. However, the trend has also significantly reduced individual organizations' abilities to express their design philosophies and style in their software. The paper reviews this trend and juxtaposes it to the author's efforts towards providing customizable tools. The paper concludes with some questions that need to be addressed before moving towards more personalized computational support.

1 Introduction

One predominant theme of my work in computer-aided structural engineering over the past forty years has been the search for means whereby the collective expertise of the structural engineering profession could be pooled by means of shared design and analysis tools, while still allowing personalization or customization of these tools. Ideally, this means that common, generally accepted methods would be programmed only once, and that each organization, even each individual practitioner, could augment, extend and otherwise customize the tools to suit particular, even idiosyncratic, situations and styles of practice.

The purpose of this paper is to review the developments of the past forty years, juxtaposing the prevalent professional practices with my own efforts towards the ideal sketched above. The paper concludes with a set of questions concerning the future directions of the quest that will be outlined.

In this rather introspective paper, the references are largely limited to my own publications, which illustrate the evolution of my thinking on the theme. Space does not allow me to provide a full set of references to the many significant contributions to the emergence of today's computer-aided structural engineering environment.

2 The early days

From the earliest days of computer applications in structural engineering until roughly the mid - 1980's, the only generally available approach was custom programming. Each organization that desired some particular computational capability had to program the entire application; there was no way to differentiate between common and custom components. User groups and program exchange groups were formed quite early to provide some mechanism for sharing results. These groups formed strong and very useful social networks among the early program developers, but the actual sharing of reusable software was quite limited. The title of a 1960 paper, "Computer Program Exchange: Myth and Reality", succinctly summarizes the dilemma of the time [1].

In 1962, my colleagues Bob Logcher, Sam Mauch, Ken Reinschmidt and I developed STRESS in order to provide a common, shareable tool for the analysis of framed structures [2]. It is a reflection on the prevailing attitude of that time that several colleagues in large consulting firms considered STRESS to be an unwarranted academic intrusion into practice, as the wide availability of the tool eliminated the competitive advantage that accrued to their firms from their in-house analysis programs. The original version of STRESS contained a METHOD command. It was our intention that the STIFFNESS method initially implemented would be followed by many other methods, such as FLEXIBILITY and MOMENT DISTRIBUTION, to accommodate analysts' preferences. We quickly became convinced that this level of customization was not needed, and indeed essentially all structural analysis programs today provide only the stiffness method option.

As design applications incorporating provisions of design specifications and standards proliferated, each organization had to interpret the relevant provisions before these could be coded. In 1966, I proposed that a formal representation of the design standard provisions would not only increase the efficiency of program development, but could also reduce, if not totally eliminate, errors in interpretation and even assist standards developers in producing clear, complete and unambiguous standards [3]. Standards processing has continued to be an active research area, but in practice software developers still tend to rely on direct coding of individually interpreted standards provisions.

Returning to the original theme of pooling resources for shared program development, in 1971 I proposed that the structural engineering profession sponsor a national organization that would collect, verify, and collate program segments and distribute customized programs extracted from its collection [4]. With my colleagues Bob Schiffman and Mel Baron, I worked on the technical issues to be addressed [5], and there was sufficient professional interest that a National Institute for Computers in Engineering (NICE) was actually incorporated. The venture never became operational, as it was completely overtaken by the developments described in the next section.

3 The recent past

Starting in the mid - 1980's, two interrelated developments have drastically altered the scene. One development was the emergence of the personal computer, which has moved access to, and more importantly control of, computing resources from remote mainframes to the desk of the engineer. The vastly increased market thus created led to the second development, the maturing of a lively, competitive application software industry. A third development, less significant than the first two, has been the availability of productivity tools such as spreadsheets and equation solvers, providing declarative programming tools well suited for ad-hoc, exploratory problem solving.

These developments radically changed the environment for the production and use of structural engineering application software. Custom programming by individual engineering firms essentially disappeared, and virtually all production work is performed with acquired, third - party software. Overall, this transformation has been beneficial. The pooling of resources that I advocated in the 1970's has in fact taken place, albeit not in one central facility but at least in a small number of specialized, highly professional and intensely competitive organizations. Imagine where the profession would be today if all the quality control and software certification procedures, platform migrations, human- computer interface adaptations, provisions for interoperability linkages and, most recently, Web-enabling interfaces would have had to be performed individually on each custom program of every firm.

Software vendors respond to user demands and, by introducing new capabilities contribute to their rapid dissemination. However, they also have to respond to market forces, and can introduce only those capabilities for which there is sufficient demand. Customization capabilities in structural analysis and design programs do not appear to be in high demand, and few provisions for user customization are provided. Such provisions are, however, available in related tools, such as procedural attachment capabilities in CAD systems and macros in estimating programs.

The net effect is that today's practicing structural engineer is largely constrained to operate within the capabilities and options provided by the tools acquired by his or her organization. If a needed capability is available in another, less familiar tool, a steep learning curve is required to master that tool. As stated above, general productivity tools are available for ad-hoc exploratory programming and even for developing reusable templates, but their incorporation into office procedures and the enforcement of required quality control processes is highly problematic.

In the last decade, my students and I have explored the issue of customizability within the knowledge-based expert system (KBES) and general Artificial Intelligence (AI) paradigms. Enrique Ibarra-Anaya and I developed a prototype customizable expert system for evaluating the seismic resistance of existing buildings [6]. The system is largely procedural, but at various points windows are provided for the user to enter sets of rules for inferring the value of a parameter. For example, using an approximate rigid-body model of behavior, the system computes the number of alternate load paths in each direction, but the user has to provide a set of rules to determine whether that number is adequate, depending on the given context. George Turkiyyah and I worked on a prototype finite element modeling assistant [7]. The system enables an analyst to create a symbolic model by invoking assumptions. The assumptions "know" how to "operationalize" themselves to produce a numerical model that serves as input to a standard finite element analysis tool, as well as to evaluate the numerical results produced by the tool in order to verify that the assumptions are valid. Neither of these explorations progressed beyond the prototype stage. The first approach could, in principle, be generalized by decomposing a conventional program and inserting windows for user-defined rules at appropriate decision points. The second approach would require a major collaborative effort in building up a comprehensive collection of assumptions, as well as a very good interface for defining user-specified, custom assumptions.

4 The current scene

The major emphasis in the current research literature, as well as in many activities in practice, is on integration by means of shared product models and foundation classes. There appears to be considerably less emphasis on processes, and even less on customizable tools implementing such processes. One currently active research area, however, has the potential of providing direct support for generating and using custom process knowledge. This area is Case-Based Reasoning (CBR), which aims to provide to the user past solutions, or fragments of past solutions, that may be relevant to the user's current problem and may serve as a starting point for generating solutions to that problem (see, e.g., [8]). CBR, in itself, does not differentiate between cases on the basis of their source, that is, whether a case in the case-base was generated within the user's own organization or was acquired from an external source. However, it would seem natural that in a field such as structural engineering, where many assumptions, approximations and design styles may be equally valid, an organization would want to restrict the case-base to past solutions generated by that organization only. Viewed in this way, CBR can serve as a powerful means of storing and reusing a structural engineering organization's custom knowledge, whether that knowledge deals with design, modeling and analysis, or detailing.

My students, Hugues Rivard and Nestor Gomez, and I are developing a prototype of a tool for conceptual structural design that combines two approaches toward customization [9]. It is a module of the SEED software environment intended to support the early phases of building design for recurrent building types [10]. The SEED conceptual structural design module is based on a strict separation of the information model that contains the emerging design description from the knowledge that is used to generate that description. The generation knowledge is contained in a hierarchical graph of technology nodes, where each node represents the knowledge needed to advance the design description by one step. The technology nodes may be viewed as coarse-grain productions or rules, where the condition part specifies when a technology is applicable and the action part describes the design extension produced by the technology. However, there is no global inference or search engine, because each node dynamically identifies its potential successor nodes. If multiple technologies are applicable, the alternatives are presented to the user for selection. It is our intention to make the technologies dynamically definable to the maximum degree possible. Thus, an organization's specialized, even proprietary, knowledge could be encoded in technology nodes and made available to designers within that organization. A second component of the prototype is a case-based reasoning capability. In accordance with the general CBR philosophy of SEED [11], cases are generated as an automatic byproduct of the design process and retrieved cases are adapted using exactly the same process as in the initial design. Furthermore, we achieve essentially zero overhead for indexing cases, as most of the classifiers used for indexing are assigned by the generating technologies. For example, if at a particular design step the technology graph shows that steel, concrete and timber are applicable materials for a building element and the user chooses steel, the classifier "material = steel" appended to the element's description is automatically made an index to that element. We don't intend to place any source control on the cases saved, but since cases can only be accumulated by the design process, they will naturally represent only the user organization's practice and style.

5 Summary and some remaining questions

In this paper I have presented a brief historical survey of the evolution of structural engineering software, juxtaposed with my long quest for software that is responsive to the practitioners' need for personalized capabilities. The term AI was used only once, yet it should be apparent that a variety of AI techniques have been brought to bear on the issue, and many more need to be explored and adapted in order to find the right mix of techniques and approaches.

In lieu of definitive conclusions, I end the paper with a series of questions. First, is the degree of customization that I have been seeking really needed, or can it be expected that in due time software vendors will respond to their users' needs by incorporating the various alternative procedures or styles as options in their general-purpose software. Second, is it too late to be concerned with the issue, or have educators and practitioners learned to adapt to the available software capabilities in fashioning the way they teach and practice the art of structural engineering. More fundamentally, has the available software so standardized or "homogenized" structural engineering thinking that there simply is no demand for ways of expressing individual styles. Or is it that that engineers' differing styles deal with issues more abstract than those addressed by current software, and that customization will have to be addressed only when attempts are made to provide tools that deal with these higher-level issues. Stated succinctly, is there a problem?

If there is a sense that there is a problem, in that the current software delivery system fails to satisfy the structural engineering community's needs for personalized or personalizable tools, then a second set of questions arises. What are the approaches that have been used or could be explored in order to provide the desired capabilities? More specifically, in line with the title of this meeting, what can AI contribute to the solution? It is my expectation that this meeting can provide some answers.

References

1. Chang, J., "Computer Program Exchange: Myth and Reality", Proceedings Second Conference on Electronic Computation, American Society of Civil Engineers, New York, NY, pp. 27-34, 1960.
2. Fenves, S. J., Logcher, R. D., Mauch, S. P. and Reinschmidt, K. F., STRESS : A User's Manual, MIT Press, Cambridge, MA, 1964.
3. Fenves, S. J., "Tabular Decision Logic for Structural Design", Journal of the Structural Division, American Society of Civil Engineers, New York, NY, Volume 93, No. ST3, pp. 401-417, 1966.
4. Fenves, S. J., "Scenario for a Third Computer Revolution in Structural Engineering", Journal of the Structural Division, American Society of Civil Engineers, New York, NY, Volume 97, No. ST1, pp. 3-11, 1971.
5. Fenves, S. J., Schiffman, R. L. and Baron, M. L., "A Position Paper on Computer-Based Technology Transfer in Civil Engineering", Proceedings Colloque International sur les Systemes Integres en Genie Civil, University of Liege, Liege, Belgium, 1972.
6. Ibarra-Anaya, E. and Fenves, S. J., "A Framework for Customizable Expert Systems", Proceedings of Conference on Computing in Civil Engineering, American Society of Civil Engineers, New York, NY, pp. 298-307, 1991.
7. Turkiyyah, G. M. and Fenves, S. J., "Knowledge-Based Assistance for Finite-element Modeling", IEEE Expert, New York, NY, Volume 11, No. 3, pp. 23-32, 1996.
8. Issues and Applications of Case-Based Reasoning in Design, Maher, M. L. and Pu, P. (editors), Lawrence Erlbaum Associates, Mahwah, NJ, 1997.
9. Fenves, S. J., Rivard, H., Gomez, N. and S. C. Chiou, "Conceptual Structural Design in SEED", Journal of Architectural Engineering, American Society of Civil Engineers, New York, NY, Volume 1, No. 4, pp. 179-186, 1995.
10. Flemming, U. and Woodbury, R. "A Software Environment to Support the Early Phases in Building Design: Overview", Journal of Architectural Engineering, American Society of Civil Engineers, New York, NY, Volume 1, No. 4, pp. 147-152, 1995.
11. Flemming, U., Aygen, Z., Coyne, R. and Snyder, J. "Case-Based Design in a Software Environment that Supports the Early Phases in Building Design", in Issues and Applications of Case-based Reasoning in Design, Maher, M. L. and Pu, P. (editors), Lawrence Erlbaum Associates, Mahwah, NJ, pp. 61-85, 1997.

Complex Systems: Why Do They Need to Evolve and How Can Evolution Be Supported

Gerhard Fischer

University of Colorado, Center for LifeLong Learning & Design (L^3D)
Department of Computer Science, Campus Box 430
Boulder, CO 80309-0430, USA
email: gerhard@cs.colorado.edu

Abstract. We live in a world characterized by evolution—that is, by ongoing processes of development, formation, and growth in both natural and human-created systems. Biology tells us that complex, natural systems are not created all at once but must instead evolve over time. We are becoming increasingly aware that evolutionary processes are ubiquitous and critical for social, educational, and technological innovations as well.

The driving forces behind the evolution of these systems is their use by communities of practice in solving real-world problems as well as the changing nature of the world, specifically as it relates to technology. The seeding, evolutionary growth, and reseeding model is a process description of how this happens. By integrating working and learning in communities of practice, we have created organizational memories that include mechanisms to capture and represent task specifications, work artifacts, and group communications. These memories facilitate organizational learning by supporting the evolution, reorganization, and sustainability of information repositories and by providing mechanisms for access to and delivery of knowledge relevant to current tasks.

Our research focuses specifically on the following claims about design environments embedded within dynamic human organizations: (1) they must evolve because they cannot be completely designed prior to use; (2) they must evolve to some extent at the hands of the users; (3) they must be designed for evolution; and (4) to support this approach with World-Wide Web technology, the Web has to be more than a broadcast medium; it has to support collaborative design.

Keywords: design; evolution; domain-oriented design environments; seeding, evolutionary growth, and reseeding model (SER model); open versus closed systems

Acknowledgments. The author would like to thank members of the Center for LifeLong Learning & Design at the University of Colorado who have made major contributions to the conceptual framework and systems described in this paper. Jim Ambach and Ernesto Arias contributed to the conceptual framework and scenario development. The research was supported by (1) the National Science Foundation, Grants REC-9631396 and IRI-9711951; (2) NYNEX Science and Technology Center, White Plains; (3) Software Research Associates, Tokyo, Japan; (4) PFU, Tokyo, Japan; and (5) Daimler-Benz Research, Ulm, Germany.

1 Introduction

1.1 Necessity for Evolution

The basic assumption that complete and correct requirements can be obtained at some point of time is theoretically and empirically wrong. Many research efforts do not take into account the growing evidence that system requirements are not so much analytically specified as they are collaboratively evolved through an iterative process of consultation between end users and software developers [23]. A consequence of the "thin spread of application knowledge" [10] is that specification errors often occur when designers do not have sufficient application domain knowledge to interpret the customer's intentions from the requirement statements.

Design methodologists (e.g., [43,44]) demonstrate with their work that the design of complex systems requires the integration of problem framing and problem solving and they argue convincingly that (1) one cannot gather information meaningfully unless one has understood the problem, but one cannot understand the problem without information about it; and (2) professional practice has at least as much to do with defining a problem as with solving a problem. New requirements emerge during development because they cannot be identified until portions of the system have been designed and implemented. The conceptual structures underlying complex software systems are too complicated to be specified accurately in advance and too complex to be built faultlessly [7]. Specification and implementation have to co-evolve [47] requiring the owners of the problems [15] to be present in the development. While evolution is no panacea and creates its own problems, there are strong reasons to increase the efforts and the costs to include mechanisms for evolution (such as end-user modifiability, tailorability, adaptability, design rationale, making software "soft") into the original design of complex systems. Experience has shown [8,25] that the costs saved in the initial development of system by ignoring evolution will be spent several times over during the use of a system.

1.2 Theory about Evolution of Complex Systems

Evolution of complex systems is a ubiquitous phenomenon. Many researchers have addressed the evolutionary character of successful complex systems and of scientific endeavor. Laszlo [28] stresses that a new paradigm is emerging in many fields, leading to a replacement of earlier ideas that were based on mechanistic determinism toward new models of change, indeterminance and evolution. Popper [36] reminds us that knowledge should be open to critical examination and that the advance of knowledge consists in the modification of earlier knowledge. Dawkins [11] demonstrates that big-step reductionism cannot work as an explanation of mechanism; we can't explain a complex thing as originating in a single step, but complex things *evolve* (implying that models from biology may be more relevant to future software systems than models from mathematics).

1.3 The Evolution of Complex Software Systems

Evolution is especially essential in software systems. The assumption of complete requirements at any point in time is detrimental to the development of successful (i.e., useful and usable) software systems. Brooks [7] argues that *successful* software gets changed because it offers the possibility to evolve. Lee [29] describes many convincing examples (including the failure of the Aegis system in the Persian Gulf) that design approaches based on the assumption of complete and correct requirements do not correspond to the realities of this world. Curtis and colleagues [10] identify in a large-scale empirical investigation that fluctuating and conflicting requirements are critical bottlenecks in software production and quality. The Computer Science and Technology Board [8] provides empirical data that 40-60 percent of the lifecycle costs of a complex system is absorbed by maintenance and 75 percent of the total maintenance efforts are enhancements. Much of this cost is due to the fact that a considerable amount of essential information (such as design rationale [18,31]) is lost during development and must be reconstructed by the designers who maintain and evolve the system. In light of these data, development and maintenance have to merge into cycles of an evolutionary process making capturing of design rationale a necessity rather than a luxury.

1.4 Claims about Evolution

Software design needs to be understood as an evolutionary process where system requirements and functionality are determined through an iterative process of collaboration among multiple stakeholders (including developers and users). Requirements cannot be completely specified before system development occurs. Our previous study of design processes support the following claims:

- *Software systems must evolve; they cannot be completely designed prior to use.* Design is a process that intertwines problem solving and problem framing [43]. Software users and designers will not fully determine a system's desired functionality until that system is put to use. Process models that describe the different phases of the software life cycle need to take advantage of this fact.
- *Software systems must evolve at the hands of the users.* End users experience a system's deficiencies; subsequently, they have to play an important role in driving its evolution. Software systems need to contain mechanisms that allow end-user modification of system functionality.
- *Software systems must be designed for evolution.* Through our previous research in software design, we have discovered that systems need to be designed *a priori* for evolution. Software architectures need to be developed for software that is designed to evolve.

2 Domain-Oriented Design Environments (DODEs)

Domain-oriented design environments (DODEs) [14] are software systems that support design activities within a particular domain such as the design of kitchens, voice dialog systems, and computer networks. DODEs are a particularly good example of complex software systems that need to evolve. Design within a particular domain typically involves several stakeholders whose knowledge can be elicited only within the context of a particular design problem. Different stakeholders include the developers of a DODE (environment developers), the end users of a DODE (domain designers), and the people for whom the design is being created (clients). To effectively support design activities, DODEs need to increase communication between the different stakeholders and anticipate and encourage evolution at the hands of domain designers.

Software systems (such as DODEs) model parts of our world (e.g., the physical computer networks consisting of computers, networks, etc.). Our world evolves in numerous dimensions as new artifacts appear, new knowledge is discovered, and new ways of doing business are developed. There are fundamental reasons why systems cannot be done "right" at the beginning. Successful software systems need to evolve.

2.1 Design Environments: Limited Scope, Better Support

Not *all* problems in the development of *any* complex software can be solved by one (and always the same) approach. The more specifically we address a certain kind of complex software, the more likely will we be able to find effective support for its evolutionary development. Henderson and Kyng [25] demonstrate that enhancements extending through the lifetime of a complex system are critical. Norman [34] shows that design and evolution have many things in common. Simon [45] provides convincing evidence that complex systems evolve faster if they can build on stable subsystems.

In our work on DODEs, we build on object-oriented techniques, but transcend pure, basic object-orientation by integrating and embedding object abstractions in the specific context of *design in an evolving domain*. Like design patterns [1,21] and application frameworks [30], DODEs provide a meaningful context for evolution [13]. DODEs are even more powerful than the other above-mentioned contextualizations, as they take the domain into account.

2.2 Domain Orientation: Situated Breakdowns and Design Rationale

Domain-oriented systems are rooted in the context of use in a domain. While the DODE approach itself is generic, each of its applications is a particular domain-oriented system. Our emphasis on *domain-oriented* design environments acknowledges the importance of situated and contextualized communication and design rationale as the basis for *effective* evolutionary design. Polanyi [35] analyzes the observation that human knowledge is tacit (i.e., we know more than we can say) and that some of it will be activated only in actual problem situations. In early knowledge-based system building efforts, there was a distinct knowledge acquisition phase that was assumed to lead to complete requirements—contrary to our assumption of the seeding, evolutionary growth, reseeding (SER) model (presented later in the paper). The notion of a "seed" in the SER model emphasizes our interpretation of the initial system as a catalyst for evolution—evolution that is in turn supported by the environment itself.

2.3 End-User Modification and Programming for Communities: Evolution at the Hands of Users

Because end users experience breakdowns and insufficiencies of a design environment in their work, *they* should be able to report, react to, and resolve those problems. Mechanisms for end-user modification and programming are, therefore, a cornerstone of evolvable systems. At the core of our approach to evolutionary design lies the ability of end users (in our case, domain designers) to make significant changes to system functionality, and to share those modifications within a community of designers. It is their perception that should determine what is considered urgent to change, not the risks determined by developers. The types of changes that must occur during the evolutionary growth of a system go beyond the setting of predefined parameters or preferences and include the ability to alter system behavior in non-trivial ways. Winograd [49] argues why design environments are needed to make end-user programming feasible. We don't assume that all designers will be willing or interested in making system changes, but drawing upon the work of Nardi [33] we do know that within local communities of software use there often exist local developers and power users who are interested in and capable of performing these tasks.

3 Evolutionary Design at Work: A Scenario

The following scenario illustrates how a design environment can affect the exemplary domain of computer network design. The scenario emphasizes the importance of *end-user driven evolution.* The system described, *NetDE* (see Fig. 1 and Fig. 2), is a DODE for the domain of computer network design. NetDE incorporates several principles including:

- Domain-oriented components that provide domain designers (in this case, computer network designers) the capability to easily create design artifacts; these domain designers are the end-users of NetDE.
- Features that allow the specification of design constraints and goals so that the system understands more about particular design situations and gives guidance and suggestions for designers relevant to those situations.

- Mechanisms that support the capture of design rationale and argumentation embedded within design artifacts so that they can best serve the design task.
- Mechanisms that support end-user modifiability so that the network designers experiencing deficiencies of NetDE can drive the evolution of the system.
- Features that increase communication between the system stakeholders (i.e., designers of NetDE and the network designers using NetDE).
- The integration of communities of practice within the evolution of NetDE.

This scenario involves two network designers (D_1, D_2) at the University of Colorado who have been asked to design a new network for clients within the Publications Group in the dean's office at the College of Engineering.

3.1 Evolution of Design Artifacts: Designing a New Network

D_1's clients are interested in networking ten newly purchased Macintosh Power PCs and a laser printer. Through a combination of email discussions and meetings, D_1 learns that the clients want to be able to share the printer, swap files easily, and send each other email. D_1 raises the issue of connecting to the Internet, and was told that the clients would be interested at some point, but not for the time being. It was also made clear that the clients had spent most of their budget on the computer hardware, and did not have much left over for sophisticated network services and tools.

From our previous work in network design, we know that design specification and rationale comes from a number of stakeholders, including network designers and clients, and is captured in different media including email and notes. To be most effective, this rationale needs to be stored in a way that allows access to it from the relevant places within a design.

D_1 invokes the NetDE system. A World-Wide Web (WWW) Browser appears on the desktop presenting a drawing of the College of Engineering. Every network and subnet in the College of Engineering can be accessed by navigating through different parts of the drawing. By selecting the "New Design" option, D_1 is presented an empty NetDE page that he names "Publications OT 8-6" after the office where the clients are located. The new page becomes a repository for all of the background information and rationale that D_1 has regarding the new network. This is achieved by sending all email and text files that D_1 has to the (automatically created) email address "PublicationsOT8-6." NetDE insures that the WWW page immediately updates itself to show links to the received mails and files (Fig. 1, (1)).

Selecting the "Launch Construction Component" option opens a palette of network objects (Fig. 1, (2)) as well as a worksheet (3). D_1 starts by specifying certain design constraints to the system (4). Immediately the Catalog (5) displays a selection of existing designs that have constraints similar to those specified by D_1. Selecting one of the designs represented in the Catalog moves that design into the worksheet where D_1 can modify it. D_1 changes the design to reflect the specific needs of the Publications Group. NetDE is accessible through the World-Wide Web, so that other network designers ($D_2 \ldots D_n$) can use it, also. The existing designs may be contributions from other designers.

WWW access is crucial for maintaining a distributed community of practice. The Behavior Exchange [39] is addressing these needs. The capture of design rationale and argumentation can occur through the use of a group memory. The GIMMe system [19] explores the creation and maintenance of a group memory accessible through the WWW.

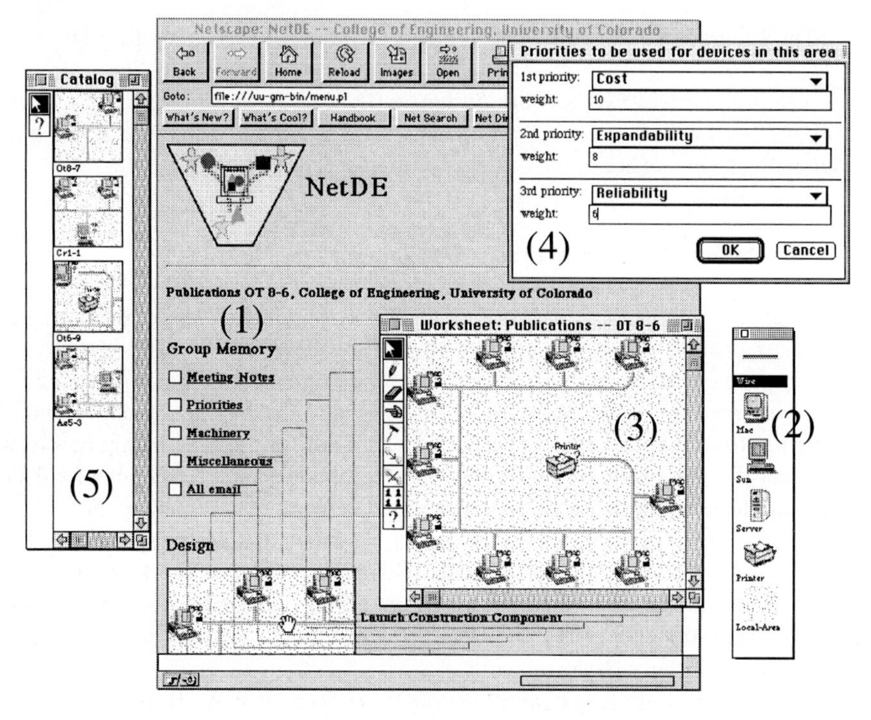

Fig. 1. NetDE in Use

Finally, NetDE provides a domain specific construction mechanism (the palette and the worksheet), and allows the specification of design constraints and goals. Using additional specification mechanisms, D_1 describes how the network will be used, and what kinds of networking services are desired. This is the first time D_1 has networked Macs, so he takes advantage of the NetDE critiquing feature, which will evaluate his design and compare it to the established design constraints. During evaluation, NetDE suggests the use of the EtherTalk network protocol, and the PowerTalk email capabilities that come standard with the Macs. D_1 agrees with this assessment because they limit the cost of the network. He finishes creating his design.

Integration of specification, construction, catalog, and argumentation components is the characteristic strength of a DODE such as NetDE. These components and their interaction are critical to the "evolvability" of the system. The process D_1 and D_2 follow (below) is an instantiation of our seeding-evolutionary growth-reseeding process model.

3.2 Evolution of a Design Environment: NetDE

Several months pass, and Publications is interested in changing its network. D_1 is not available, so D_2 is to design the new changes. D_2 receives email from Publications indicating that their network needs have changed. They want to start publishing WWW pages and will need Internet access. They will also be using a Silicon Graphics Indy computer. They have received a substantial budget increase for their network.

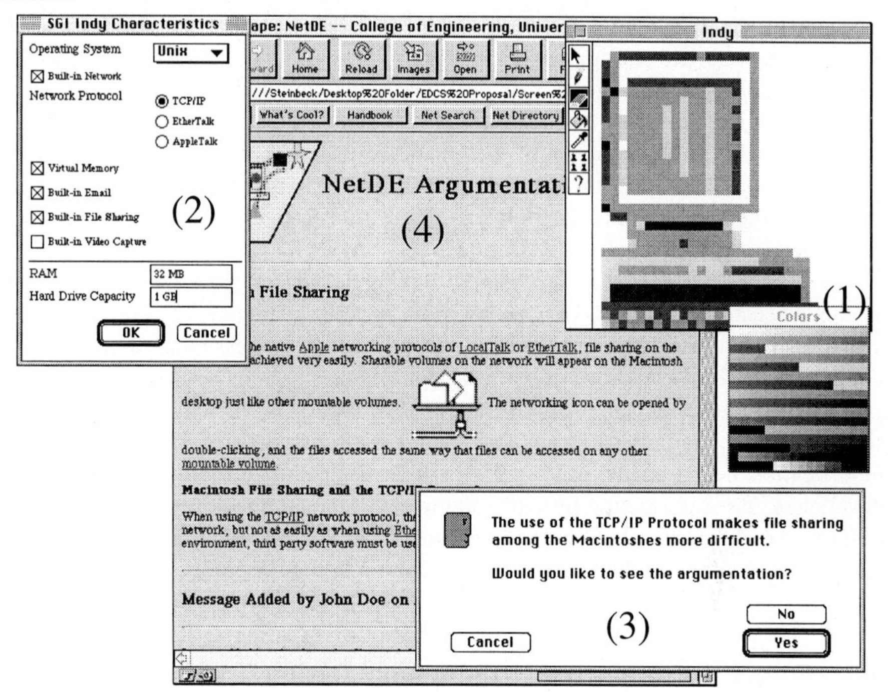

Fig. 2. The Network Evolves

First, D_2 accesses the NetDE page that describes the Publications network. She quickly reviews the current design and rationale to learn what has already occurred. She updates the design specification to reflect the fact that cost is no longer as important, and that speed has become more important. Then, she searches the NetDE palette to see if it has an icon representing the Indy. She does not find one, and realizes that it must be added. After reviewing the specs for the Indy from the Silicon Graphics Web Page, D_2 creates a new palette element for the Indy (Fig. 2, (1)), and then defines its characteristics using some of NetDE's end user modifiability features (Fig. 2, (2)). According to the company's specs, the Indy has built-in networking capabilities, and understands the TCP/IP network protocol. D_2 enters this information, and the new icon appears in the palette.

Since breakdowns are experienced by end users, they need to be able to evolve the system's functionality. This calls for the development of specialized mechanisms that allow end users to alter system functionality without having to be computer programmers. In order for NetDE to take full advantage of new objects added by the network designer, it must provide facilities that define not just the look of the new object, but also its behavior.

D_2 adds the Indy to the design, and NetDE indicates (by displaying different colored wires) that the two types of machines (Macs and the Indy) are using different network protocols. D_2 knows that Macs can understand TCP/IP protocol, so she changes the network's protocol to TCP/IP. After invoking NetDE's critiquing mechanism, D_2 receives a critiquing message indicating that the use of TCP/IP violates the easy file-sharing design constraint (Fig. 2, (3)). After reading through some of the argumentation (Fig. 2, (4)), D_2 learns that although file sharing is possible in TCP/IP with the Macs, it is not as easy as when they are using EtherTalk. D_2 decides that this is not a constraint she would like to break, and decides to ask some other network designers if there is a way to get the Indy to understand EtherTalk. D_2 learns that there is software the Indy can run to translate protocols, and she adds an annotation to the Indy object to reflect this.

A critiquing component is important in linking design rationale and argumentation to the designed artifact as well as for pointing out potential breakdowns to the designer. Very drastic changes (like the introduction of wireless communication) will not be covered by the end-user modification mechanisms. In those cases, the ability to describe system changes to environment developers is critical for maintaining communication among different stakeholders of the system. When unexpected modification needs occur, users must be able to articulate their needs and notify developers. The information provided to the environment developer will be useful to describe how the system is being used and what sort of issues the system is not addressing, leading to subsequent radical evolution of the system.

4 Computer-Supported Evolutionary Design

The above scenario illustrates both (1) the DODE itself evolves and (2) how artifacts created with the DODE evolve at the hands of end-users (such as network designers). The ability of a DODE to co-evolve with the artifacts created within it makes the DODE architecture the ideal candidate for creating an evolvable application family. In the following, we describe the SER process model as a systematic way to structure evolution [20].

The domain orientation of a design environment enriches (1) the amount of support that a knowledge-based system can provide, and (2) the shared understanding among stakeholders. Design knowledge includes domain concepts, argumentation, case-based catalogs, and critiquing rules. The appeal of the DODE approach lies in its compatibility with an emerging methodology for design [9,12], views of the future as articulated by practicing software engineering experts [8], reflections about the myth of automatic programming [42], findings of empirical studies [10], and the *integration* of many recent efforts to tackle specific issues in software design (e.g., recording design rationale [18], supporting case-based reasoning [38], creating artifact memories [48], and so forth).

4.1 Seeding, Evolutionary Growth, Reseeding—The SER Process Model for DODEs

Because design in real world situations deals with complex, unique, uncertain, conflicted, and unstable situations of practice, design knowledge as embedded in DODEs will never be complete because design knowledge is tacit (i.e., competent practitioners know more than they can say) [35], and additional knowledge is triggered and activated by actual use situations leading to breakdowns [16]. Because these breakdowns are experienced by the users and not by the developers, computational mechanisms that supporting end-user modifiability are required as an intrinsic part of a DODE.

Three intertwined levels can be distinguished whose interactions form the essence of the SER model:

- On the *conceptual framework level*, the multifaceted, domain-independent architecture constitutes a framework for building evolvable complex software systems.
- When this architecture is instantiated in a domain (e.g., network design), a domain-oriented design environment (representing an application family) is created on the *domain level*. An instantiation in the network domain is NetDE.
- Individual artifacts in the domain are developed by exploiting the information contained in the generic DODE (in the scenario this is represented by the network developed for the Publications Group).

Fig. 3 illustrates the interplay of those three layers. Darker gray indicates knowledge domains close to the computer, whereas white emphasizes closeness to the design work in a domain. The figure illustrates the role of different professional groups in the evolutionary design: the *environment developer* (close to the computer) provides the domain-independent framework, and instantiates it into a DODE in collaboration with the help of the domain designers (knowledgeable domain workers who use the environment to design artifacts; in the scenario, D_1 is a domain designer). Domain designers are the "end users" of a design environment. The artifact is eventually delivered to the client (e.g., the Publications Group in the scenario).

Breakdowns occur when domain designers cannot carry out the design work with the existing DODE. Extensions and criticism drive the evolution on all three levels: Domain designers directly modify the artifacts when they build them (artifact evolution), they feed their modifications back into the environment (domain evolution), and—during a reseeding phase—even the architecture may be revised (conceptual framework evolution). In Fig. 3, the little building blocks represent knowledge and domain elements in any of the components of the multifaceted architecture (i.e., the Indy, critics about network protocols, etc.).

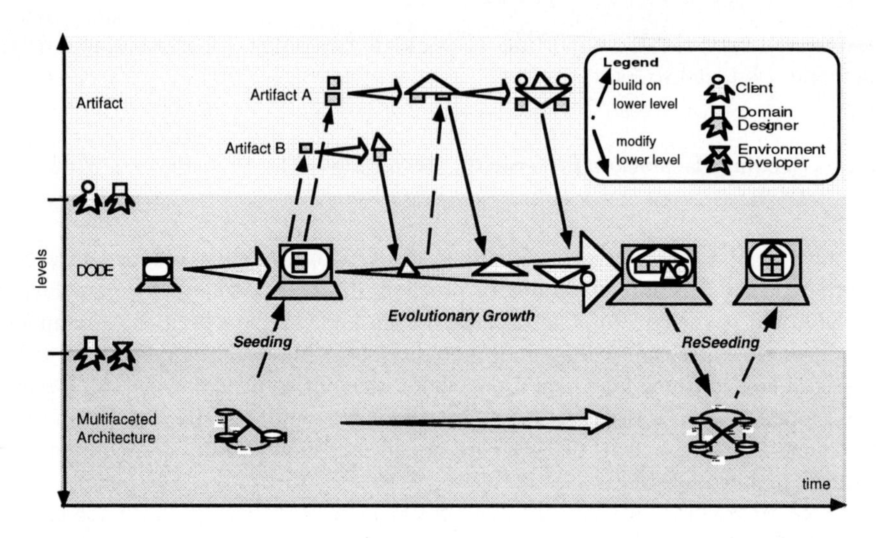

Fig. 3. The SER Model: A process model for the development and evolution of DODEs

The evolution of complex systems in the context of this process model (more detail can be found in [20]) can be characterized as follows:

Seeding. A seed will be created through a participatory design process between environment developers and domain designers. It will evolve in response to its use in new network design projects because requirements fluctuate, change is ubiquitous, and design knowledge is tacit. Postulating the objective of a seed (rather then a complete domain model or a complete knowledge base) sets our approach apart from other approaches in software engineering and artificial intelligence and emphasizes evolution as the central design concept.

Evolutionary growth. Network experts use the seeded environment to undertake specific projects for clients (such as the Publication Group in the scenario). During these design efforts, new requirements may surface (e.g., the desire to access the Internet), new components may come into existence (e.g., the Indy) and additional design knowledge not contained in the seed may be articulated (e.g., that EtherTalk supports both AppleTalk and TCP/IP protocols). During the evolutionary growth phase, the environment developers are not present, thus making end-user modification a necessity rather than a luxury (at least, as argued before, for small-scale evolutionary changes). Visual AgenTalk (VAT) addresses this problem, building on our experience with end-user modifiability and end-user programming [13,17].

Reseeding. In a deliberate effort to revise and coordinate information and functionality, the environment developers are brought back in to collaborate with domain designers to organize, formalize, and generalize knowledge added during the evolutionary growth phases. Organizational concerns [24] play a crucial role in this phase. For example, decisions have to be made as to which of the extensions created in the context of specific design projects should be incorporated in future versions of the generic design environment. Drastic and large-scale evolutionary changes occur during the reseeding phase.

5 Systems-Building Efforts

The preceding sections have identified important issues that need to be addressed to support evolutionary system development: design rationale, design-in-use, end-user modifiability, and collaboration support for a community of practice. Following prototypical systems are briefly introduced that address those issues.

5.1 Group Interactive Memory Manager (GIMMe)

The research objective behind GIMMe [19] is to support the communication between different stakeholders involved in the development process and to capture and structure this communication in order to make it immediately available as design rationale [32].

Description. GIMMe is a web-based group memory system. It helps communities of practice (e.g., project teams, interest groups) to capture, store, organize, share, and retrieve electronic mail conversations. Mail sent to a specific group alias is automatically added to an information space and categorized according to its subject line. Group members can access the information space via the Internet. It supports three retrieval mechanisms to this information space: (1) browsing in reverse chronological order, (2) browsing according to project-specific categories, and (3) retrieval by free-form text queries (using the Latent Semantic Indexing algorithm [27]. GIMMe supports users in creating, rearranging, or deleting categories and the mail belonging to them (a more detailed description of GIMMe can be found at http://www.cs.colorado.edu/~stefanie). It will allow groups to create and negotiate domain conventions and concepts over time, as well as to evolve categories that reflect the structure and vocabulary of the application domain.

How does GIMMe support evolution? In order for the users to evolve a system they have to be able to understand the design decisions that lead to the current system. GIMMe addresses the problem of how to capture this important information, structure it for later reuse, and make it available to the users. GIMMe is an effort demonstrating that design rationale emerges as a by-product of normal work. This is critical because we know from empirical evidence that most design rationale systems have not failed because of the inadequacy of a computational substrate, but because they did not pay enough attention to the question "who is the beneficiary and who has to do the work?" [24]. Our experiences with the use of GIMMe at NYNEX Science & Technology and the University of Colorado have encouraged us to pursue this idea of a growing group memory and have shown that such a system can be employed for real world problems and projects. An evolvable DODE and evolvable artifacts within a DODE will require design rationale and, therefore, GIMMe has to be an integral component.

5.2 Expectation Agents

Expectation Agents [22] were designed to support communication between end users and developers of an interactive system during actual use situations [25]. Expectation Agents observe and analyze the reactions of end users to the system by not relying on only a small subset of end users but by reaching the whole (or a large subset) of the community of practice.

Description. Expectation Agents are an active part of the system in which the end user designs artifacts. They observe the actions of the end user and compare them to descriptions of the intended use of the system. These descriptions are provided by the system developers and represent their expectations about how the system should be used (e.g., in which order certain tasks are performed). If an Expectation Agent identifies a discrepancy between how an end user uses the system and the description provided by a system developer it then notifies the developer and prompts the end user for an explanation. This explanation is then emailed to the developer as well, establishing a communication between the system developer and a specific end user.

How do Expectation Agents support evolution? Expectation Agents are used to support evolutionary growth. When developers and users create a seed (see Fig. 3) they hold a number of explicit and implicit assumptions about how the system will be used and how it supports the work practices. Expectation Agents are one way to compare those assumptions to actual use patterns.

5.3 Visual AgenTalk (VAT)

The objective behind VAT is to support end-user modifications and programming as an essential component for an evolvable DODE [13].

Description. VAT [40] (a detailed description of VAT can be found at http://www.cs.colorado.edu/~ralex) has been created on top of the Agentsheets programming substrate. Agentsheets is used as a substrate for the construction component (e.g., specifically to create interactive graphical simulations of complex dynamic systems). These simulations consist of active agents that interact with each other and exhibit a specific behavior. Combined with Agentsheets in a layered environment, Visual AgenTalk is used to define end-user programming languages that are tailorable to a particular domain, promote program comprehensibility, and provide end users with control over powerful, multimodal interaction capabilities. VAT provides mechanisms for the creation of commands so that domain-specific languages can be developed. Conditions, actions, and rules are all graphical objects, and end users can try out their programs by dragging and dropping them onto agents in a worksheet. The ability to test programs within the context of a particular agent increases the end user's ability to comprehend Visual AgenTalk programs.

How does VAT support evolution? VAT enables end users to modify and program the behavior of active agents inside a simulation environment. New agent types can be created, modified, and shared. Thus VAT promotes evolutionary growth on the hands of end users. VAT extends the construction and simulation component in order to accommodate end-user programming. VAT is used to implement expectation agents, thereby giving end users the possibility to modify the behavior of expectation agents. Agentsheets provides mechanisms that can be used by both environment developers and domain designers to define the look of individual construction objects, and VAT can be used by environment developers to define domain-specific languages to be used by domain designers to evolve the behavior of the construction objects.

5.4 New Conceptualization of the World Wide Web

The Web in its current form does not support evolutionary design. Fig. 4 presents three models that illustrates different types of Web usage [3].

Traditional Web-based use engages the Web as a Broadcast Medium (Fig. 4, Model M_1). In this model, instructional content is predetermined and placed on static Web pages. Most popular general-purpose Web tools provide support for the easy generation of this static content. In M_1, the Web serves as a distribution channel and provides few opportunities for learners to interact with the information because the content was not originally designed to be interactive. Responding to the need for feedback from consumers, many Web sites are evolving into forms that augment

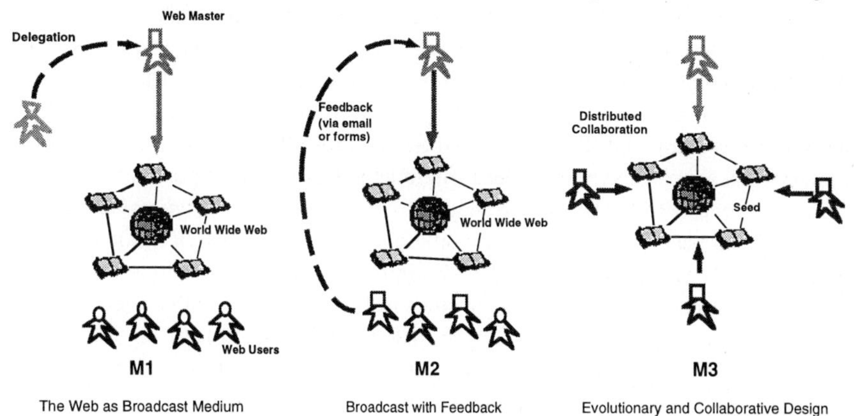

Fig. 4. Making the World Wide Web a Medium for Collaborative, Evolutionary Design

content with some communication channels. This mechanism of *broadcast with feedback* (M_2) expands the original model by providing some link from consumer to producer such as allowing learners to provide feedback and ask questions by filling out forms. Although users can react to information provided by the author, this presentation model provides little support for evolution.

The M_3 model demonstrates an essential requirement for collaboration and evolution for the Web. In M_3, users can use the Web to collaborate on projects by *actively* contributing and by learning from all contributors. The evolution of content and ideas is now the responsibility of the participating community of practice, focusing on the distributed generation of content and the reflection upon it. An M_3-type model is needed to support the SER model. When a wide variety of individuals collaborate in a cooperative forum, the unique skills of the members all become valuable resources in making the Web resources useful in the current context. The M_3 model poses a number of technical challenges, including the ability (1) to add to an information space without going through an intermediary, (2) to modify the structure of the information space, and (3) to modify at least some of the existing information.

6 Assessment

6.1 Understanding Pitfalls Associated with Evolutionary Design

To make evolutionary design a more ubiquitous activity, the forces that prohibit or hinder evolution must be understood. Examples of such forces are: (i) the resistance to change because it requires learning efforts and may create unknown difficulties and pressures, (ii) the problem of premature standards, (iii) the difficulties created by installed bases and legacy systems, and (iv) the issues of who are the beneficiaries and who has to do the work in order for evolution to occur.

The Oregon Experiment [2] (a housing experiment at the University of Oregon instantiating the concept of end user-driven evolution) serves as an interesting case study that end user-driven evolution is no guarantee for success. The analysis of its unsustainability indicated the following major reasons: (1) there was a lack of continuity over time, and (2) professional developers and users did not collaborate, so that there was a lack of synergy. These findings led us in part to postulate the need for a reseeding phase (making evolutionary development more *predictable*), in which developers and users engage in intense collaborations. With design rationale captured, communication enhanced, and end user support available, developers have a rich source of information to evolve the system in the way users really need it. Another interesting source of information for the SER model is Kuhn [26], in which general conceptual frameworks can be found to decide when the time has come to engage in a reseeding process rather than continue with evolutionary growth.

6.2 Assessment of the SER Model

The SER model is motivated by how large software systems, such as Emacs, Microsoft-Word, Unix, and the X Window System, have evolved over time. In such systems, users develop new techniques and extend the functionality of the system to solve problems that were not anticipated by the system's authors (following the observation that any artifact should be useful in the expected way, but a truly great artifact lends itself to uses the original designers never expected). New releases of the system often incorporate ideas and code produced by users. In the same way that these software systems are extensible by programmers who use them,

Open-Source Software Systems. The development of the Linux operation system [37] provides an interesting existence proof that reliable, useful, and usable complex systems can be built in a decentralized "Bazaar style" by many [41] rather than in a centralized, "Cathedral style" by a few. The Linux development model treats users as co-developers and is currently tested in a number of new areas, such as: (1) *Netscape Communicator* (for more information see http://www.mozilla.org/); (2) Gamelan (http://www.gamelan.com; the first community repositories of Java-related information allowing Java developers looking for information about what other people are doing with Java; the large number of developers who contribute to the Gamelan repository and the number of people who search for information in Gamelan provide evidence that the Java community has taken a great deal of interest in using community repositories to locate information); (3) *Educational Object Economy* (EOE; http://trp.research.apple.com/; the EOE is realized as a collection of Java objects (mostly completed applets) designed specifically for education; the target users of the EOE are teachers wishing to use new interactive technology and developers interested in producing educational software).

Domain-Oriented Design Environments. DODEs poses a major additional challenge to make the SER model feasible and workable: Whereas the people in the above mentioned development environments are computationally sophisticated and experienced users, DODEs need to be extended by domain designers who are neither interested in nor trained in the (low-level) details of computational environments. The SER model explores interesting new ground between the two extremes of "put all the knowledge in at the beginning" and "just provide an empty framework." Designers are more interested in their design task at hand than in maintaining a knowledge base. At the same time, important knowledge is produced during daily design activities that should be captured. Rather than expect designers to spend extra time and effort to maintain the knowledge base as they design, we provide tools to help designers record information quickly and without regard for how the information should be integrated with the seed. Knowledge-base maintenance is periodically performed during the reseeding phases by environment developers and domain designers in a collaborative activity.

7 Evolutionary Design—Beyond the Boundaries of Disciplines

7.1 Avoid Reinventing the Wheel

In this article, I have mostly discussed examples from the domain of software design. Software design is a new design discipline relative to other more established disciplines. Software designers can learn a lot by studying other design disciplines such as architectural design, graphic design, information design, urban design, engineering design, organizational design, musical composition, and writing. For example, the limitations and failures of design approaches that rely on directionality, causality, and a strict separation between analysis and synthesis have been recognized in architecture for a long time [12]. A careful analysis of these failures could have saved software engineering the effort expended in finding out that waterfall-type models can at best be an impoverished and oversimplified model of real design activities. Assessing the successes and failures of other design disciplines does not mean that they have to be taken literally (because software artifacts are different from other artifacts), but that they can be used as an initial framework for software design.

7.2 Evolutionary Design in Architecture

Evolutionary design is a concept of equal important in architecture [6]. For many arguments and considerations articulated in this article, the words "software systems" and "buildings", and "software designer/programmer" and architect are interchangeable. Software design being the much younger discipline could learn a lot from design methodologies developed in architectural design. Designing complex artifacts from scratch, while often considered to be highly desirable to avoid the constraints of dealing with existing structures and legacy systems, leads to artifacts which are often missing a "quality" that exists in evolving artifacts—as illustrated by cities such as Brasilia and Abuja versus cities such as London and Paris. Artifacts are embedded in time, and over time many of the determining factors influencing a design will change; and because buildings and cities are going to be modified many times, they should be designed with unanticipated future changes in mind.

The problems facing both professional groups have initiated at least some interest in each other work. Alexander's work [1] on patterns has found many followers in the software design community, specifically in object-oriented design [21]. In our own research we have created an "Envisionment and Discovery Collaboratory" to explore new computational environments that enhance communication between different stakeholders, facilitate shared understanding, and assist in the creation of better artifacts by integrating physical and computational media for design [4]. By doing so, we attempt to integrate the best of both worlds: the dynamic nature of computational media and the strength of physical media in allowing people to operate and think with tangible objects.

7.3 DODEs in Architecture

The concept of DODEs can be applied to many areas, and in our work (in close collaborations with professionals from the respective design disciplines), we have created DODES for kitchen design [14], lunar habitat design [46], and urban design [4]. Design activities embedded in computational environments are the best domains for DODEs, because the activities take place *within* the computational environment and the power of DODEs (providing critics, linking action and reflection spaces, supporting simulations, etc.) can be most successfully and most easily exploited.

8 Conclusions

DODEs are software systems that support design activities within a particular domain and are built specifically to evolve. DODEs have provided the foundations in our research to develop a theoretical and conceptual framework for the evolutionary design of complex systems illustrating (i) the importance of end user modifiability and end user programming, (ii) the capture and retrieval of design rationale, and (iii) the necessity of improved communications among different design stakeholders including system developers and end users.

Evolution of complex systems is a ubiquitous phenomenon. This is true in the physical domain, where, for example, artificial cities such as Brasilia are missing essential ingredients from natural cities such as London or Paris. "Natural" cities gain essential ingredients through their evolution—designers of "artificial" cities are unable to anticipate and create these ingredients. It is equally true for software systems for the reasons argued in this paper. A challenge for the future is to make (software) designers aware of essential concepts that originated and were explored in evolution, such as ontogeny, phylogeny, and punctuated equilibrium. Even though we are convinced that models from biology may be more relevant to future software systems than models from mathematics, we also have to be cautious: to follow an evolutionary approach in software design *successfully* does not imply that concepts from biological evolution should be mimicked literally, but rather they need to be reinterpreted in the domain of software design [5].

References

1 Alexander, C.; Ishikawa, S.; Silverstein, M.; Jacobson, M.; Fiksdahl-King, I.; Angel, S., *A Pattern Language: Towns, Buildings, Construction*; Oxford University Press: New York, 1977.

2 Alexander, C.; Silverstein, M.; Angel, S.; Ishikawa, S.; Abrams, D., *The Oregon Experiment*; Oxford University Press: New York, NY, 1975.

3 Ambach, J.; Fischer, G.; Ostwald, J.; Repenning, A., *Making the World Wide Web A Medium for Collaborative, Evolutionary Design*, At http://www.cs.colorado.edu-/~ostwald/papers/WWW97/PAPER200.html, 1997.

4 Arias, E. G.; Fischer, G.; Eden, H., *Enhancing Communication, Facilitating Shared Understanding, and Creating Better Artifacts by Integrating Physical and Computational Media for Design*, In *Proceedings of Designing Interactive Systems (DIS '97)*; Amsterdam, The Netherlands, 1997; pp. 1-12.

5 Basalla, G., *The Evolution of Technology*; Cambridge University Press: New York, 1988.

6 Brand, S., *How Buildings Learn—What Happens After They're Built*; Penguin Books: New York, 1995.

7 Brooks, F. P., Jr., *No Silver Bullet: Essence and Accidents of Software Engineering*, In *IEEE Computer* 1987, *20*, pp. 10-19.

8 Computer Science and Technology Board, *Scaling Up: A Research Agenda for Software Engineering*, In *Communications of the ACM* 1990, *33*, pp. 281-293.

9 Cross, N., *Developments in Design Methodology*; John Wiley & Sons: New York, 1984.

10 Curtis, B.; Krasner, H.; Iscoe, N., *A Field Study of the Software Design Process for Large Systems*, In *Communications of the ACM* 1988, *31*, pp. 1268-1287.

11 Dawkins, R., *The Blind Watchmaker*; W.W. Norton and Company: New York - London, 1987.

12 Ehn, P., *Work-Oriented Design of Computer Artifacts*; Almquist & Wiksell International: Stockholm, Sweden, 1988.

13 Eisenberg, M.; Fischer, G., *Programmable Design Environments: Integrating End-User Programming with Domain-Oriented Assistance*, In *Human Factors in Computing Systems, CHI'94*; Boston, MA, 1994; pp. 431-437.

14 Fischer, G., *Domain-Oriented Design Environments*, In *Automated Software Engineering* 1994, *1*, pp. 177-203.

15 Fischer, G., *Putting the Owners of Problems in Charge with Domain-Oriented Design Environments* In *User-Centred Requirements for Software Engineering Environments*; R. W. D. Gilmore, F. Detienne, Ed.; Springer Verlag: Heidelberg, 1994; pp. 297-306.

16 Fischer, G., *Turning Breakdowns into Opportunities for Creativity*, In *Knowledge-Based Systems, Special Issue on Creativity and Cognition* 1995,

17 Fischer, G.; Girgensohn, A., *End-User Modifiability in Design Environments*, In *Human Factors in Computing Systems, (CHI'90)*; Seattle, WA, 1990; pp. 183-191.

18 Fischer, G.; Lemke, A. C.; McCall, R.; Morch, A.,*Making Argumentation Serve Design* In *Design Rationale: Concepts, Techniques, and Use*; T. Moran and J. Carrol, Ed.; Lawrence Erlbaum and Associates: Mahwah, NJ, 1996; pp. 267-293.

19 Fischer, G.; Lindstaedt, S.; Ostwald, J.; Schneider, K.; Smith, J., *Informing System Design Through Organizational Learning*, In *International Conference on Learning Sciences (ICLS'96)*; Chicago, IL, 1996; pp. 52-59.

20 Fischer, G.; McCall, R.; Ostwald, J.; Reeves, B.; Shipman, F., *Seeding, Evolutionary Growth and Reseeding: Supporting Incremental Development of Design Environments*, In *Human Factors in Computing Systems (CHI'94)*; Boston, MA, 1994; pp. 292-298.

21 Gamma, E.; Johnson, R.; Helm, R.; Vlissides, J., *Design Patterns - Elements of Reusable Object-Oriented Systems*; Addison-Wesley: Reading, MA, 1994.

22 Girgensohn, A.; Redmiles, D.; Shipman, F., *Agent-Based Support for Communication between Developers and Users in Software Design* In *Proceedings of the 9th Annual Knowledge-Based Software Engineering (KBSE-94) Conference (Monterey, CA)*; IEEE Computer Society Press: Los Alamitos, CA, 1994; pp. 22-29.

23 Greenbaum, J.; Kyng, M., *Design at Work: Cooperative Design of Computer Systems*; Lawrence Erlbaum Associates, Inc.: Hillsdale, NJ, 1991.

24 Grudin, J., "Seven plus one Challenges for Groupware Developers," 1991.

25 Henderson, A.; Kyng, M., *There's No Place Like Home: Continuing Design in Use* In *Design at Work: Cooperative Design of Computer Systems*; J. Greenbaum and M. Kyng, Ed.; Lawrence Erlbaum Associates, Inc.: Hillsdale, NJ, 1991; pp. 219-240.

26 Kuhn, T. S., *The Structure of Scientific Revolutions*; The University of Chicago Press: Chicago, 1970.

27 Landauer, T. K.; Dumais, S. T., *A Solution to Plato's Problem: The Latent Semantic Analysis Theory of Acquisition, Induction, and Representation of Knowledge*, In *Psychological Review* 1997, *104*, pp. 211-240.

28 Laszlo, E., *Evolution: The Grand Synthesis*; Shambhala Publications, Inc.: 1987.

29 Lee, L., *The Day The Phones Stopped*; Donald I. Fine, Inc.: New York, 1992.

30 Lewis, T., *Object-Oriented Application Frameworks*; Prentice Hall: Englewood Cliffs, New Jersey, 1995, 344 pages.

31 MacLean, A.; Carter, K.; Lovstrand, L.; Moran, T., *User-Tailorable Systems: Pressing the Issues with Buttons* In *Human Factors in Computing Systems, CHI'90 Conference Proceedings (Seattle, WA)*New York, 1990; pp. 175-182.

32 Moran, T. P.; Carroll, J. M., *Design Rationale: Concepts, Techniques, and Use*; Lawrence Erlbaum Associates, Inc.: Hillsdale, NJ, 1996.

33 Nardi, B.; Zarmer, C., *Beyond Models and Metaphors: Visual Formalisms in User Interface Design*, In *Journal of Visual Languages and Computing* 1993, pp. 5-33.

34 Norman, D. A., *Turn Signals are the Facial Expressions of Automobiles*; Addison-Wesley Publishing Company: Reading, MA, 1993.

35 Polanyi, M., *The Tacit Dimension*; Doubleday: Garden City, NY, 1966.

36 Popper, K. R., *Conjectures and Refutations*; Harper & Row: New York, Hagerstown, San Francisco, London, 1965.

37 Raymond, E. S., *The Cathedral and the Bazaar*, At http://earthspace.net/~esr/writings-/cathedral-bazaar/cathedral-bazaar.html, 1998.

38 Redmiles, D. F., "From Programming Tasks to Solutions—Bridging the Gap Through the Explanation of Examples," 1992.

39 Repenning, A.; Ambach, J., *The Agentsheets Behavior Exchange: Supporting Social Behavior Processing*, In *Computer-Human Interaction (CHI '97)*; Atlanta, GA, 1997; pp. 26-27 (Extended Abstracts).

40 Repenning, A.; Ioannidou, A., *Behavior Processors: Layers between End-Users and Java Virtual Machines*, In *Visual Languages*; Capri, Italy, 1997; pp. 402-409.

41 Resnick, M., *Turtles, Termites, and Traffic Jams*; The MIT Press: Cambridge, MA, 1994.

42 Rich, C. H.; Waters, R. C., *Automatic Programming: Myths and Prospects* In *Computer*, The Computer Society: Los Alamitos, CA, 1988; Vol. 21; pp. 40-51.

43 Rittel, H., *Second-Generation Design Methods* In *Developments in Design Methodology*; N. Cross, Ed.; John Wiley & Sons: New York, 1984; pp. 317-327.

44 Schön, D. A., *The Reflective Practitioner: How Professionals Think in Action*; Basic Books: New York, 1983.

45 Simon, H. A., *The Sciences of the Artificial*; The MIT Press: Cambridge, MA, 1996.

46 Stahl, G., *Interpretation in Design: The Problem of Tacit and Explicit Understanding in Computer Support of Cooperative Design,* Ph.D. dissertation. UMI#9423544, Department of Computer Science. University of Colorado at Boulder. Technical Report CU-CS-688-93, 1993.

47 Swartout, W. R.; Balzer, R., *On the Inevitable Intertwining of Specification and Implementation* In *Communications of the ACM*, 1982; Vol. 25; pp. 438-439.

48 Terveen, L. G.; Selfridge, P. G.; Long, D. M., *Living Design Memory: Framework, Implementation, Lessons Learned*, In *Human-Computer Interaction* 1995, *10*, pp. 1-37.

49 Winograd, T., *From Programming Environments to Environments for Designing*, In *Communications of the ACM* 1995, *38*, pp. 65-74.

Formalizing Product Model Transformations: Case Examples and Applications

Martin Fischer[1], Florian Aalami[2], and Ragip Akbas[3]

[1] Assistant Professor, Center for Integrated Facility Engineering (CIFE)
Department of Civil and Environmental Engineering
Stanford, CA 94305-4020
fischer@cive.stanford.edu
[2] PhD Candidate
aalami@ce.stanford.edu
[3] Graduate Research Assistant
rakbas@leland.stanford.edu

Abstract. Today, product models can be built manually or by instantiating objects from a standards library. Database evolution schemas have been defined to maintain data integrity as the product model changes throughout design and construction. Based on a project case example, this paper defines mechanisms that transform a design-centric decomposition of a product model into a production-centric decomposition. The three mechanisms are (1) the introduction of temporary structures, (2) the refinement of components, and (3) the aggregation of components. These mechanisms complement the manual and standards approaches to product modeling. They are defined at the user level and use abstracted knowledge about components and activities to transform a product model as required by a particular set of engineering tasks. The challenge in defining and operationalizing the transformation mechanisms lies in abstracting the knowledge that determines when to use what mechanism and in formalizing the knowledge that creates and inserts new product model objects at the appropriate place in the product model hierarchy.

1 Introduction

After more than a decade of research into product modeling (Gielingh 1988, Björk 1989, Eastman et al. 1991, Scherer and Katranuschkov 1993), it is now imaginable that product models will become common in engineering practice. Several research efforts have demonstrated the applicability and usefulness of product models (e.g., CIMSTEEL (Crowley and Watson 1997), COMBINE (Augenbroe 1995), ICON (Ford et al. 1994)), and the International Alliance for Interoperability has proposed product modeling standards for the building industry in the form of Industry Foundation Classes, version 1.5 (IAI 1998). Most of these product modeling efforts have focused on providing a design-centric, component-based view of a project (Harfman and Chen

1993). Many researchers (van Nederveen and Tolman 1992, Howard et al. 1992, Eastman and Siabiris 1995) have pointed out the need for product model evolution and elaboration mechanisms if product models are to become a living electronic model of a project. Such an electronic model would support the exchange of structural and other design information with construction, provide accurate information for detailed structural design, enable 2D, 3D, and 4D visualizations at various levels of detail, automate quantity takeoffs for conceptual estimating and for the duration calculations of detailed construction activities, and support progress monitoring during construction. It would also provide a basis for the addition of detail as necessary during design and construction, enable rapid generation and evaluation of detailed design alternatives, and support the rapid design of temporary structures. These are just a few examples of the engineering tasks a product model should be able to support. In summary, a user should be able to generate or summarize (aggregate) detail in a product model as necessary and transform the product model quickly to suit the particular representation needs of a given task. In essence, the goal is to provide a model as accurate as possible for a particular set of tasks at as low a cost as possible. If defined and implemented at the operational or user level, product model transformation mechanisms can be an effective way of maintaining a consistent product model throughout design and construction. As a result, more tasks can be augmented with better computer support, and the number of guesses and assumptions necessary to analyze and evaluate a particular scenario rapidly can be reduced.

This paper shows how predefined construction method models support the addition of detailed components to a product model as necessary. We also present product model transformation mechanisms that use the knowledge in the method models to transform a structure- or design-focused product model into a production- or construction-focused model. As others, we argue that such mechanisms should become part of the engineering and product modeling tool-set to improve the ability of engineers and builders to develop integrated product and process models from the beginning of a project and to enable detailed constructibility simulations and evaluations for a proposed design. We will use two case examples to illustrate these points: (1) the detailed structural design and construction planning for a portion of the skin of Frank O. Gehry & Associates' Experience Music Project in Seattle and (2) the evaluation of two construction method alternatives for the construction of a small ridge lodge.

2 Case Example 1: Motivation of Problem

The skin on the Experience Music Project is supported by curved steel ribs. These ribs are in turn braced with secondary steel and support reinforcing bars. The actual skin has variable curvatures in all three dimensions and consists of shotcrete, waterproofing, insulation, and a final outside layer of thin steel plates (Fig. 1). The product model shown in Figure 2 corresponds to the information contained in the 3D model created by the architect. The product model contains objects for each of the steel ribs (e.g., Rib_A_1 and Rib_A_2) and the skin. The skin is broken down into a skin-assembly and individual steel panels. It does not break up the skin-assembly into its constituting parts such as reinforcement, shotcrete, and insulation. The structural engineer used this information to design the rib sizes and skin thickness. In this design-centric product model, the designers have specified the following information for each object or component: what type of component it is, what material it consists of, where it is, what dimensions it has, and what supports it.

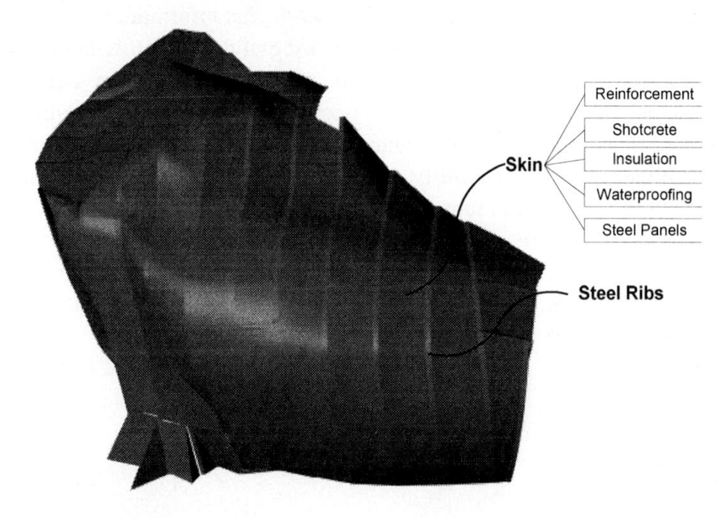

Fig. 1. Base 3D CAD model for Element 2

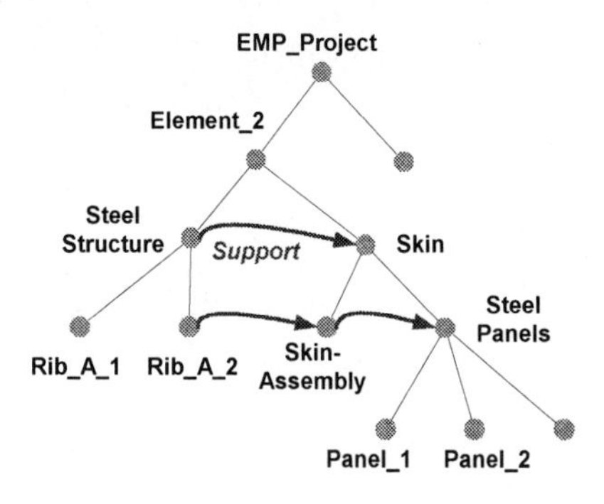

Fig. 2. Design-centric product model decomposition

The objective for the detailed structural design is to ensure structural stability when only part of the skin has been completed and the structure is loaded asymmetrically with wet shotcrete. The objective for construction is to allow the subcontractors to employ construction methods that lead to productive construction operations and a safe and efficient use of resources. These two objectives are interdependent. A particular method, sequence, and speed of construction might affects the structural reliability of the half-finished shell, and constraints arising from the detailed structural analysis might preclude the application of a particular construction method. A design-centric product model alone cannot support the concurrent design and analysis necessary to develop an efficient and safe approach to build (produce) the shell. The next few paragraphs introduce the views (or information) the structural engineer and the superintendent need to have to support their engineering tasks. We also show how a corresponding product model could provide this information.

The structural engineer has determined that the braced steel structure will be able to support the incomplete skin if it is constructed in three meter increments starting at the bottom of the structure (Fig. 3). Figure 4 shows the change to the product model required to support the view the structural engineer needs to confirm the detailed analysis of the structural reliability during construction. It requires the addition of detail for the skin component to represent the 3 meter wide strips of skin.

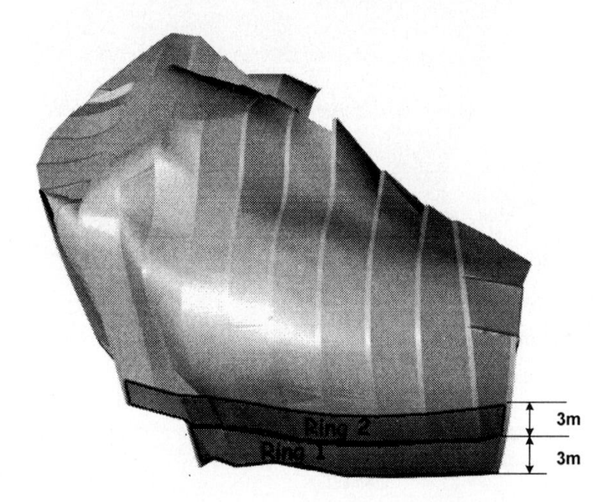

Fig. 3. CAD model for structural analysis of shotcrete application

The construction superintendent has developed four different construction methods (Fig. 5) to build the shell. Their applicability depends on the vertical grade of the skin. In perimeter areas where the skin is near vertical, scaffolding will be erected to allow workers to place reinforcement, shotcrete the skin and install the other skin components. In areas where the skin is quite steep adjacent to vertical or near vertical perimeter sections of the skin, the scaffolding will be cantilevered out to give workers access to the skin. Where the skin is relatively flat, crews will be able to work directly on the skin. In areas which are steep but away from the perimeter of the building, workers will be tied off (like for the construction of a roof). Figure 6 shows the areas of applicability for these four methods. Figure 7 shows the product model needed by the construction superintendent to plan the work methods. It shows the addition of scaffolding, the introduction of detail to the skin-assembly object, and the aggregation of steel plates into batches (zones) to synchronize the fabrication and delivery of the steel plates with the on-site production.

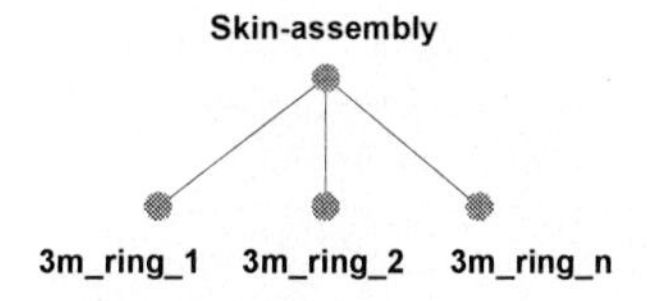

Fig. 4. Decomposition of skin-assembly object for structural analysis of shotcrete application

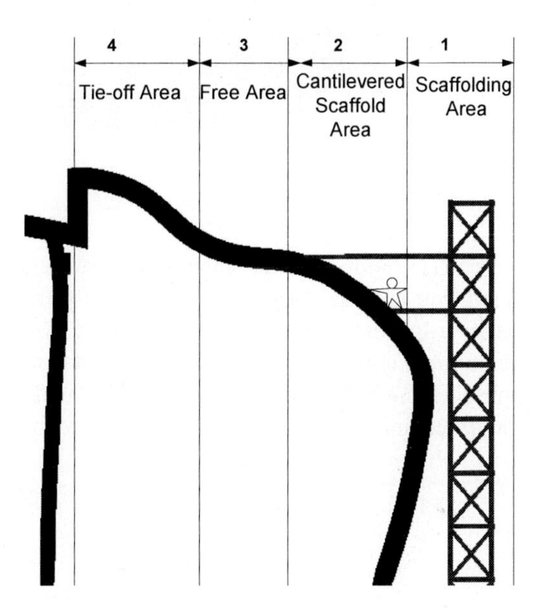

Fig. 5. Applicable Construction Methods for the construction of the shell structure

This example illustrates several of the support functions of a product model for the engineering tasks discussed above:

- The structural engineer needs accurate quantity takeoffs for various stages of construction to determine the loads on the structure and the stiffness of the structure to assess its structural reliability. Ideally, s/he would be working with the construction superintendent to explore several phasing or staging scenarios and trade off or optimize, as far as possible, the structural safety and construction efficiency.
- The construction superintendent also needs accurate partial quantity takeoffs to determine the duration of activities, design the work flow, order the materials for a day, etc. S/he needs to break up and regroup the skin into sets of pieces that are different from the structural engineer.
- The superintendent needs to add construction-specific components and details, such as the scaffolding and reinforcing details, to the electronic design model.
- The superintendent might also like to visualize the work planned for one day for the daily subcontractor meeting.

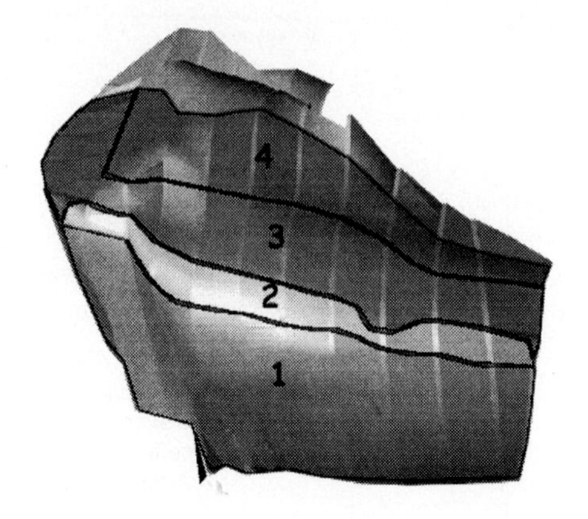

Fig. 6. The skin is broken down according to the application areas for construction methods

The static, design-centric product model cannot support these engineering tasks, since it would almost never provide a representation of the product (building) at the level of detail necessary. It would be very time-consuming, and therefore not economical, to generate multiple integrated structural design and construction alternatives by hand. The complex geometry of the example exacerbates this point, but similar problems exist for more common structures (as we will show for the ridge lodge example).

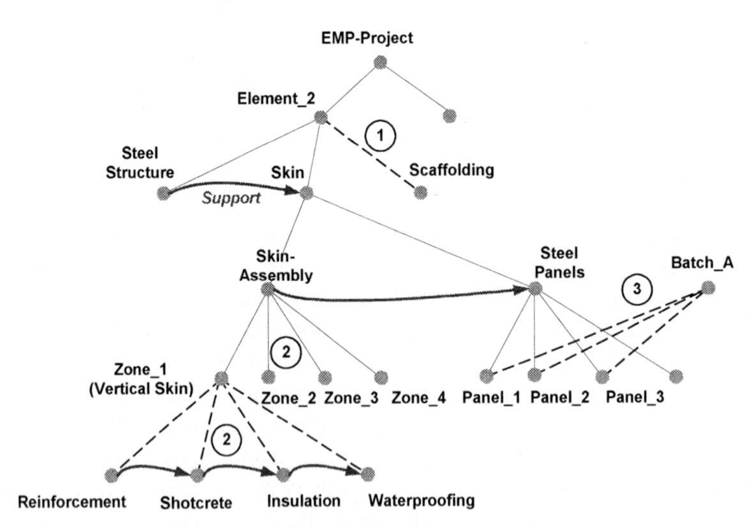

Fig. 7. Production-centric product model after the three product model transformation mechanisms have been applied

This example demonstrates the need for three product model transformation mechanisms (shown with the corresponding number in Fig. 7): (1) the introduction of temporary structures, e.g., scaffolding or shoring, (2) the refinement of components (addition of detail), e.g., elaborating a skin element into reinforcement, shotcrete, etc., and (3) the aggregation of components, e.g., zones. While there are undoubtedly other mechanisms necessary to support architectural, engineering, and construction tasks, we will focus on these three mechanisms in the remainder of this paper.

Next, we will first give a brief overview of our Construction Method Modeler (CMM) system and then describe the transformation mechanisms in more detail and discuss related research. We finish the paper with a detailed discussion of the implementation of the transformation mechanisms as part of the CMM prototype.

3 CMM - Construction Method Modeler

We have operationalized the three transformation mechanisms by embedding them in a computer aided process planning system (CPPS), CMM. CPPS systems determine the process activities needed during the manufacturing or construction process of a product (Zhou et al. 1995, Cay 1997). CMM generates process activities by employing a hierarchical construction planning process. The hierarchical construction planning process uses computer-interpretable construction method models that capture general construction planning knowledge (Fischer and Aalami 1996). The method models use an abstract construction planning vocabulary consisting of <**C**> components, <**A**> actions, <**R**> resources, and <**S**> sequence constraints to define the types of activities that make up each method. Each activity type is defined by a <**CARS**> tuple. To create a construction schedule or process model, users select the methods that apply on their project from CMM's method library. CMM then applies the selected methods to the design-centric product model and generates the activities specified in the method models, sequences the activities, and links them to their corresponding components in the product model. In summary, with these method models and using the design-centric product model as a starting point, construction engineers and superintendents can now rapidly generate a hierarchical model of the construction process with as much or as little detail for specific project phases or areas as necessary. This model now gives construction superintendents a tool to analyze a proposed design from their perspective, since there now is an explicit model of activities, components, and their relationships (which we call "4D production model"). This allows them to iterate rapidly with structural and other engineers on design alternatives.

CMM is implemented in PowerModel™ (an object-oriented programming environment developed by Intellicorp in Mountain View, CA) on SUN Sparcstations running Solaris 3.4. A runtime version of CMM is available for the Windows 95 and NT 4.0 operating systems on PCs. So far we have modeled approximately 50 different construction methods and scheduled commercial and industrial projects with up to 3,000 components in the product model. CMM is able to read a product model in the STEP Part 21 format. We have implemented product transformation mechanisms that, if triggered by a method model, can break monolithic walls into courses, add construction-specific detail to concrete elements (formwork, reinforcement, and concrete), generate soil volumes for excavation, and introduce various types of shoring.

Its output, the 4D production model, is also created as a file in Part 21 format. We are currently implementing a Java-based method modeler to allow participants in a construction project to capture construction method knowledge in construction method model templates interactively on the web. We are also planning to apply CMM on an on-going construction project to test its scalability.

Figure 8 shows how the construction planning knowledge of the superintendent related to method 1 was captured in the template defined for the CMM system. The method's domain attribute specifies the applicability of the method, in this case the *installation* of *vertical skins*. The template allows a user to define the constituting activity types for the method. Each constituting activity is defined by a **<CARS>** tuple and associated productivity and cost data. For example, shotcrete (the activity's component **<C>**) is installed **<A>** by a particular type of crew (C-1) **<R>** at a rate of about 10 m^3 per day and needs support **<S>** (which, in this case, is provided by the reinforcement). Note that the method model is different from a skeletal plan because it only predefines each general activity type and not each activity instance. When selected, it uses project-specific knowledge from the product model to instance each specific activity and to determine the number of particular activities needed.

As can be seen in the method model in Figure 8, a particular method or set of construction activities is likely to require more detail or a different aggregation of product model objects than is available in the design-centric product model. The next three subsections describe the three transformation mechanisms we have operationalized to transform a design-centric product model into a production-centric model at the required level of detail.

Domain-> Action type: Install Component type: (Vertical) Skin						
Component	Scaffolding	Reinforcement	Waterproofing	Insulation	Shotcrete	Steel Panel
Action	Raise	Install	Install	Install	Apply	Batch
Resource	L-2 Type Crew	I-1 Type Crew	R-2 Type Crew	R-1 Type Crew	C-1 Type Crew	S-2 Type Crew
Seq. Constraint	-none-	Support	Support	Support	Support	Support
Productivity	20 m/d	1 t/d	30 m2/d	30 m2/d	10 m3/d	1 Batch/d
....	....	....	....	....	....	....

Fig. 8. Definition of construction method model in CMM for vertical skin

3.1 Introduction of Temporary Structures

Temporary structures play an important part in the construction of a building. They require resources for their installation and take up space for the duration of their use. Therefore, a construction superintendent needs to take the construction and dismantling of temporary structures into account when creating and evaluating a construction schedule. Typically, temporary structures are not included in design documents or design-centric product models.

Planning with CMM, the construction superintendent selected the method shown in Figure 8 for the construction of the vertical sections of the skin. The method model contains the activity type *raise scaffold*, and CMM tries to generate this activity. To complete the 4D production model it needs to link the action *raise* to the corresponding component *scaffold*. CMM searches through the existing product model to find an appropriate scaffold component. Since it cannot find such a component, it invokes the product model transformation mechanism for the insertion of new objects into the model. From a library of component classes, it retrieves a generic scaffolding object. (If the component type specified in the method model does not exist in the library, the user would have to create it manually and add the necessary semantics. This process would be analogous to Clayton et al.'s (1996) interpretation step.) Using parametric design methods (Dharwadkar and Cleveland 1996) specified for this component type, project-context knowledge from the product model, and user input, CMM customizes the generic scaffolding object to project-specific instances. Using knowledge about the component type, CMM inserts the scaffolding object into the product model at the appropriate level (see arc nr. 1 in Fig. 7). For example, objects of type scaffolding know that they insert themselves at a level parallel to the domain component of the method. Each component type also knows how to visualize itself in the 3D CAD and 4D production models.

3.2 Refinement of Components

The design-centric product model represents the skin as one object, since this level of detail is sufficient for structural design. The superintendent needs to plan the installation of the skin in more detail. The activities in his method model act on the detailed skin components *reinforcement*, *waterproofing*, *insulation*, *shotcrete*. He has already refined the level of detail in the product model to represent the zones where the available construction methods apply. For the vertical skin sections, CMM again generates the activity instances specified by the activity types in the method model. It searches for the detailed components in the product model. Since it cannot find them, it invokes the product model refinement mechanism based on the definitions for the types of components required by the method. For example, components of type reinforcement know that they are typically part of the domain component in the method. CMM then generates the detailed components for the skin to complete the 4D production model (see arc nr. 2 in Fig. 7).

3.3 Aggregation

Each steel plate forming the outer skin on the Experience Music project is individually designed and represented in the product model. To create a manageable schedule, and to give the structural engineer input on the construction phases, the superintendent wants to aggregate several *steel plates* into erection *batches*. He reflected this in the method model by specifying that the installation of the *steel plates* needs to be done in *batches*. To generate the complete activity for the installation of the steel plates, CMM needs to know the size of the batches for the steel plates. Based on user input or on zoning parameters, CMM aggregates the individual steel plates into the appropriate batches for installation (see arc nr. 3 in Fig. 7). For example, the superintendent might choose the number of plates that can be installed in one day as the batch size.

4 Approaches to Product Model Transformations

As the Experience Music project demonstrates, the research challenge is to formalize the transformation of a product model generated during design into production-centric decompositions. One could create the product model decomposition required for a particular set of engineering tasks by hand. First, engineers and constructors could adjust the graphical model (as shown in Fig. 3 and 6) and then assign semantics to each new component and area through the process of interpretation as proposed by Clayton et al. (1996). Or, users could draw on a large library with pre-defined types of components and areas and adjust the product model first and then visualize it graphically as proposed in product model standards (IAI 1998). We find both of these approaches useful, but also limiting. The standards approach will give a user a good starting set of product model objects quickly, but it is unlikely to lead to the right model for a particular situation on a specific project. The interpretation approach will allow a user to customize a product model for a given scenario, but this might be too time-consuming to be useful. We imagine the following usage scenario for the creation and transformation of product models to support the many specific engineering tasks on a project. Standards give users a first version of a product model rapidly, transformation mechanisms then allow the customization of this model to the context of a set of tasks on a project (at least to some degree), and manual interpretation allows users to adjust the model for unique situations.

Eastman and Siabiris (1995) and Howard et al. (1992) have defined mechanisms at the database language level to add detail and new components to a product model and to group existing components into aggregations or composites. We build on the basic database transformation mechanisms defined by Eastman and Siabiris (1995) to operationalize the three types of product model transformations (Fig. 9): (1) the introduction of temporary structures, e.g., scaffolding or shoring, (2) the refinement of components (addition of detail), e.g., elaborating a skin element into reinforcement, shotcrete, etc., and (3) the aggregation of components, e.g., zones. These product model transformation mechanisms are implemented at the level of the component type that would be required by an activity in the method model, e.g., shoring. These mechanisms are defined generally, i.e., they are independent of a specific project context. They make use of knowledge from product models and construction method models (Fischer and Aalami 1996) to support the customization of a product model. Prior approaches to product model transformation have largely been object-based, i.e., they focus on maintaining the integrity of data in the evolving data models. Howard et al. (1992), e.g., proposed the primitive-composite (P-C) approach for data modeling. The P-C approach could support the product modeling required for the three product model transformation in the following way. Users would have to predefine (or take from a library) all the primitives required for the tasks at hand. This would allow them to introduce temporary structure objects and add detail where needed (e.g., break up the skin into its constituting parts). They could also aggregate primitive objects into composites (e.g., zones) as necessary. The P-C approach gives users the structure to create a context-specific product model. The primitives are like building blocks, and the composites are placeholders for groups of primitives. However, the P-C approach does not define and operationalize transformation mechanisms, and, therefore, product model transformations are still largely manual.

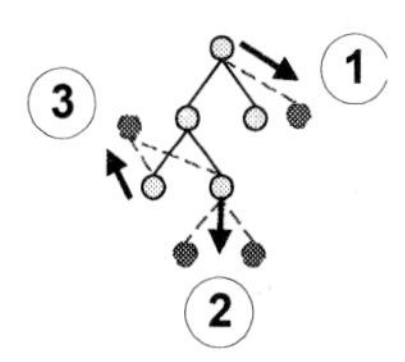

Fig. 9. Product model transformation types

Eastman and Siabiris (1995) define the EDM, a data model and database language. EDM supports product evolution through the definition of structures (e.g., *compositions*) and functions. It addresses fundamental database issues, such as the maintenance of data integrity; the defining of derivations and views; and the definition of a procedural language supporting model addition, deletion and modification. The functions *create* and *insert* support modifications of the product model through the addition of detail and the aggregation of objects into *composites*. These capabilities at the data level are needed and form the foundation for our product model transformation mechanisms.

Using the second case example, the next section discusses and illustrates the mechanisms, their implementation, and application in more detail.

5 Application of Product Model Transformation Mechanisms

The Experience Music project exemplifies the need for a production-centric product model that supports realistic constructibility analysis and 4D visualization. Product transformation is not only needed on complex projects. Even less complex projects can benefit from product transformation mechanisms that convert a design-centric representation of the project into a production-specific view. We use the Ridge Lodge example, a case study developed for one of the AEC integration courses taught at Stanford (Fruchter 1997), to illustrate the implementation of the transformation mechanisms presented in this paper. Students use the Ridge Lodge example to study the impact different construction method and design decisions have on a project. The CMM system is used for the exercise to model various construction methods and to rapidly generate 4D production models. The constructibility of the project under different scenarios is assessed by visualizing the 4D production model.

The structural engineer developed the design-centric product model (Fig. 10). To support analysis, he decomposed the project into a system-based *part-of* hierarchy. For example, he has grouped some columns and beams as frames (Frame 1) and other beams and slabs as floor systems (Roof). The granularity of the components and their grouping support his analysis needs.

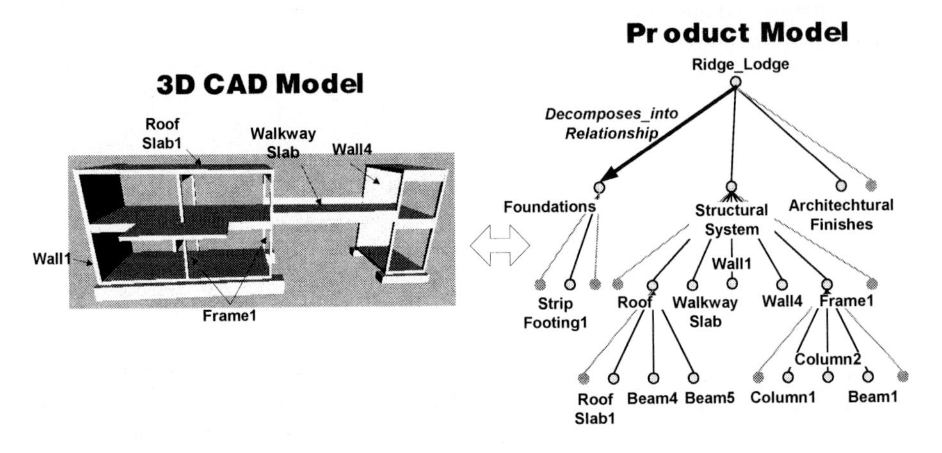

Fig. 10. Partial view of design-centric product and 3D CAD models representing Ridge Lodge

CMM uses general planning knowledge stored as construction method models and project-specific data captured in product models to assist the user in the construction planning process. CMM engages the user in a hierarchical planning process that generates plans elaborated to the level of detail needed by the planner. Construction method models are the mechanism with which the user jointly elaborates product and process models to add detail to the 4D production model.

CMM initiates the planning process by generating a seed activity. The seed activity, "Build Ridge_Lodge" is explicitly linked to the highest-level component in the Ridge Lodge product model and represents the *intent* (i.e., to build the project) of the plan. The hierarchical planning process fleshes out the seed activity. To plan the activity "Build Ridge_Lodge" in more detail, the planners apply the *build in components* construction method to the seed activity. The construction method knowledge embedded in the *build in components* method model instructs CMM to generate an activity for each sub-component the *domain* component (in this case Ridge_Lodge) decomposes into. This method of planning does not require any product model transformation; it relies on the existing definition and decomposition of components in the design-centric model. The *build in components* method can be used to further elaborate activities generated in the first round of elaboration, e.g., the activity "Build Structural System". Figure 11 shows the 4D production model after the build in components method has been applied to the activities "Build Ridge_Lodge", "Build Structural System", and "Build Roof".

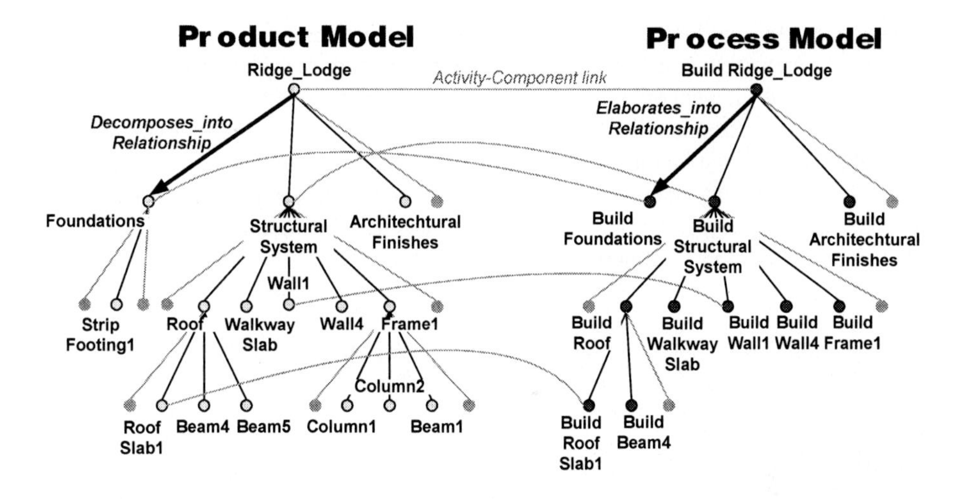

Fig. 11. Partial view of 4D production model after the activities "Build Ridge_Lodge", "Build Structural System", and "Build Roof" have been elaborated using the *build in components* construction method

Without product model transformation mechanisms, CMM would only be able to generate activities that act on the design-centric model. The *action* constituent of an activity (activities are represented as <**C**omponent, **A**ction, **R**esource, **S**equence constraint> tuples) could be varied, e.g., from *build* to *erect* or *form*, but the component constituent would always be limited to one already defined in the design-centric model. Not providing an explicit component on which an activity acts can suit some planning purposes, but does not support an explicit link between product and process model. An explicitly linked product and process model (4D production model) enables the generation of process-specific quantity takeoffs (as described in the Gehry project) and the creation of more realistic 4D visualizations that support constructibility analysis, coordination, and communication.

The planners of the Ridge Lodge project decide to assess the tradeoffs between using cast-in-place versus precast construction methods for the elevated slabs. The Walkway Slab is of most interest to them because the only access to the rear of the project is below the walkway. To plan the activity "Build Walkway Slab" in more detail, the planners select and apply the *cast-in-place slab construction using shoring system A (cip-method)* construction method model to it (Fig. 12). This construction method prescribes the setting up of shoring system A, the construction of the slab, and the dismantling of the shoring after the slab's concrete has cured. Each activity type in the method is represented as a **<CARS>** tuple in the method model.

CMM's hierarchical planning process is shown in figure 13. The process is triggered whenever a construction method model is applied to an activity. In the simplest case, e.g., when using the *build in components* method, no product model transformation is required, since all necessary components exist in the product model. When applying the *cip-method* to the Ridge Lodge project, though, transformation mechanisms are triggered because a component of type shoring system A does not exist in the design-centric product model. The first step of the transformation process is the creation of a new object and its addition to the product model. The prototype for the new object resides in general object libraries. A new object is created regardless of the transformation mechanism needed. The new object is added to the product model but its exact position in the hierarchy is not yet determined, this happens in the configuration and dimensioning stage. Using the base (design-centric) product model as a guideline, the new object configures and dimensions itself using the configuration and dimensioning algorithms it inherited form its prototype.

Domain-> Action type: Build Component type: Elevated Slab					
Component	Shoring Sys. A	Formwork	...	Concrete	Shoring Sys. A
Action	Set-up	Install	...	Cure	Dismantle
Resource	L-2 Type Crew	I-1 Crew	...	-none-	L-2 Type Crew
Seq. Constraint	Support	Support	...	TC1	TC2
productivity	30 m2/d	9.0 MHRS/M3	...	28 d	50 m2/d
....	....	....	....	....	....

Fig. 12. Partial construction method model template for *cip slab construction using shoring system A* method

Configuration involves the generation of the appropriate conceptual relationships to other components in the product model, e.g., *part-of* and *supported-by*. During dimensioning, a component calculates its own dimensions. Parameters driving the dimensioning algorithm are either obtained directly from the product model (e.g., the *domain* component that needs shoring), are stored as default values in the algorithm, or are entered by a planner. After transformation, the new component is passed on to the hierarchical planning process that then completes the joint product/process elaboration process. The hierarchical planning process is concluded with the generation of detailed activities that are linked to the new component. Figure 14 shows the 4D production model after a component of type shoring system A (Shoring System1) has been added to the product model. CMM generates the activities "Setup Shoring System" and "Dismantle Shoring System" and links them to the new Shoring System1 component.

CMM has inserted Shoring System1 under the highest-level component, Ridge_Lodge. This mode of insertion represents transformation mechanism (1), the introduction of temporary structures. The structure of the 4D production model in figure 14 reveals that the shoring system is composed of several sub-components, such as a platform and several legs. The shoring system and its sub-components are related to the rest of the project through the appropriate *supported-by* relationships (shown as arcs with arrow head endings in figure 14). This newly generated, production-centric information, namely, the enumeration of the sub-components and the topological relationships support detailed quantity takeoffs, automated activity sequencing (CMM uses the *supported-by* relationships to automate some of the sequencing), and realistic 4D visualization.

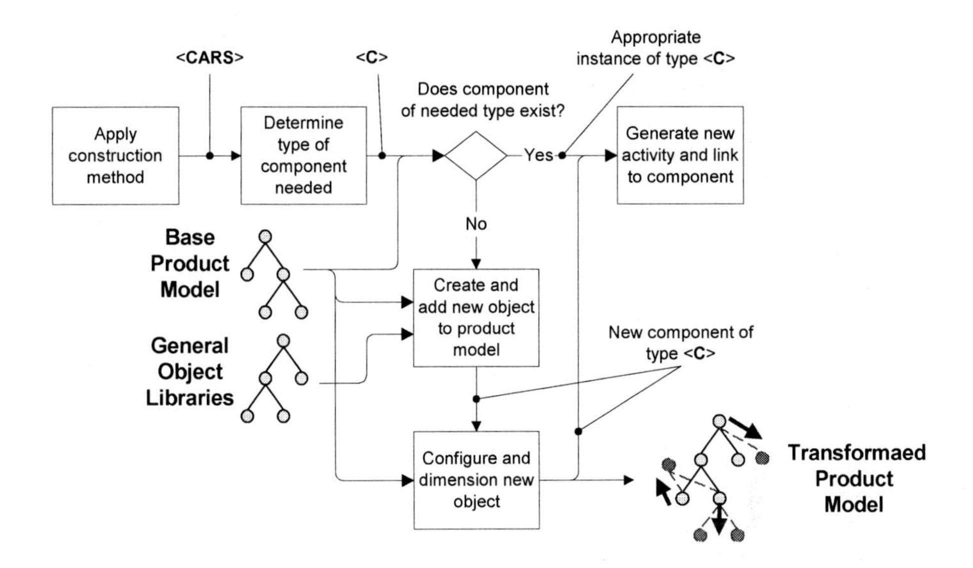

Fig. 13. Process diagram showing flow of information and steps during CMM's hierarchical planning process. CMM's planning process transforms a base product model, typically a design-centric model, into a production-centric model, if necessary

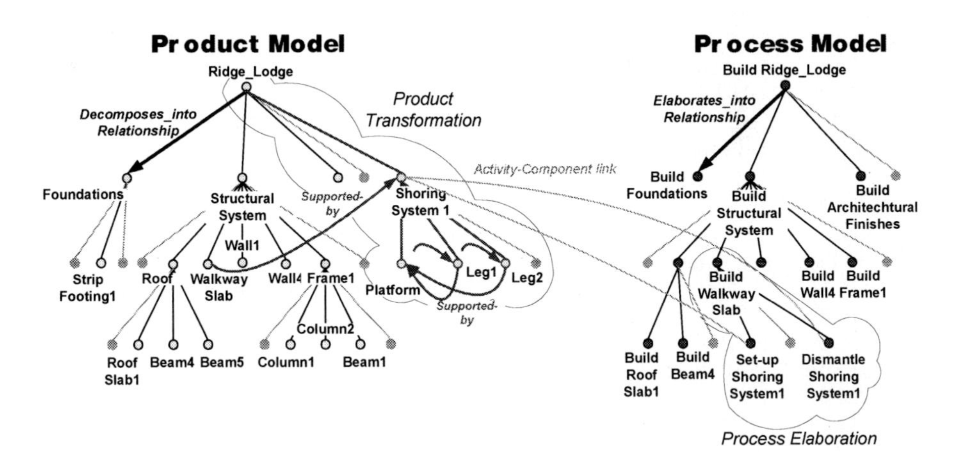

Fig. 14. 4D production model after joint product/process elaboration of "Build Walkway Slab" using *cip-method*. The component Shoring System1 and its associated sub-components are added to the product model and linked to the new detailed activities

Below is a schematic of the algorithm used by Shoring System1 to configure and design itself (stated in English). The original code is implemented in CMM using PowerModel's ProTalk language.

***Design_Yourself* method for objects of type shoring system A**

/* The purpose of the shoring system is to provide temporary support for elements. The shoring system is composed of a platform and legs. Each leg must be supported by another component in the project. Only components of type slab, beam, or foundation may support a leg. */
/* Establish project context */
```
(1) ?component_that_needs_support == component that is
acted on by domain activity;
```
/* Create new component */
```
(2) ?new_shoring_system == Make instance of object
class Shoring System A;
```
/* Configure new component */
/* Position new component in product model hierarchy-- Shoring-type components are added as a separate branch under the highest level component */
```
(3) ?new_shoring_system.Is_Part_Of == highest level
component in the project;
```
/* Define supported-by relationship for new component */
```
(4) ?component_that_needs_support.Is_Supported_By ==
?new_shoring_system;
```
/* Design new shoring component--Create and design platform */
```
(5) ?new_platform == Make instance of object class
Platform;
(6) ?new_platform. Is_Part_Of == ?new_shoring_system;
(7) ?new_platform.Dimensions == dimension_function
(?component_that_needs_support);
(8) ? new_shoring_system.Is_Supported_By ==
?new_platform;
```
/* Create and design legs—design function determines how many legs are needed based on leg type */
```
(9) while (design_function (?new_platform, Leg type) ==
True);
(10) do {
```
/* Design each leg */
```
(11)     ?new_leg == Make instance of object class Leg;
(12)     ?new_leg.Is_Part_Of == ?new_shoring_system;
?new_leg.Is_Supported_By == search_function
(?new_leg.Position, allowed component types, product
model);
(13)     ?new_platform.Dimensions == dimension_function
(?new_platform, new_leg.Position,
new_leg.Is_Supported_By);
(14)     ?new_platform.Is_Supported_By == ?new_leg;
(15) }
```

The *Design_Yourself* method shown above, though particular to components of type Shoring System A, embodies generalizable principles that apply more broadly to the three product model transformation mechanisms. To operationalize the application of transformation mechanisms during the construction planning process, we have abstracted the knowledge of when to use the *Design_Yourself* method. In CMM, user-defined and applied construction method models (Figure 13) that capture the planner's construction planning intent trigger the appropriate transformation mechanisms and also the *Design_Yourself method*, if required. User-defined method models become the mechanism with which a planner controls the database-level transformation mechanisms such as those proposed by Eastman and Siabiris (1995). Line 2 of the *Design_Yourself* method calls database functions such as *create* and *insert*. We have formalized the knowledge needed to create and configure the new product model objects. The component that is acted on by the *domain* activity (line 1 of the code) establishes the appropriate project context. The project context for a particular transformation mechanism determines where the new component is positioned in the product model (line 3), is used to establish the appropriate topological relationships (line 4), and is used to design the new component (lines 5-15). The basic structure of the algorithm is the same for the three transformation mechanisms. Where they differ is in the placement of the new component into the product model (line 3) and in the particulars of the configuration functions.

After elaborating the "Build Walkway Slab" activity, the planners choose to also apply the *cip-method* to the activity "Build Roof Slab1". Furthermore, the planners apply a *masonry block wall* construction method to the activities "Build Wall1" and "Build Wall4". The *masonry block wall* construction method is defined with respect to activities that act on the individual courses (a course is one layer of blocks) of a masonry block wall. CMM elaborates "Build Roof Slab1" in the same manner in which it elaborated "Build Walkway Slab" and in the process generates a new instance of Shoring System A for Roof Slab1. CMM can constrain the number of shoring systems of type A available for a project. CMM would calculate a longer project duration if the activities acting on the shoring were on the critical path and the shoring system is a limited resource. The elaboration of the two activities acting on walls triggers transformation mechanism (2), the addition of detail, because masonry block courses are not represented in the base model (analogous to the addition of 3m wide strips to the skin of the Gehry project). The *Design_Yourself* method for Masonry Wall Course components is similar to the one for Shoring System A.

The 4D visualization of the final production model after all of the above elaboration and transformation steps are completed is shown in figure 15. In comparison to the base 3D model (Figure 10), the final 4D visualization provides a more realistic representation of the production process both in terms of the level of detail and temporary structures. Both the appropriate level of detail and the temporary structures were automatically added to the 4D production model by CMM without the need for user input. This case example demonstrates the operationalization of transformation mechanisms in CMM. Construction method models and the state of the base product model, i.e., whether or not a component specified in the method exists, control the application of transformation mechanisms during CMM's hierarchical planning (joint product and process elaboration) process.

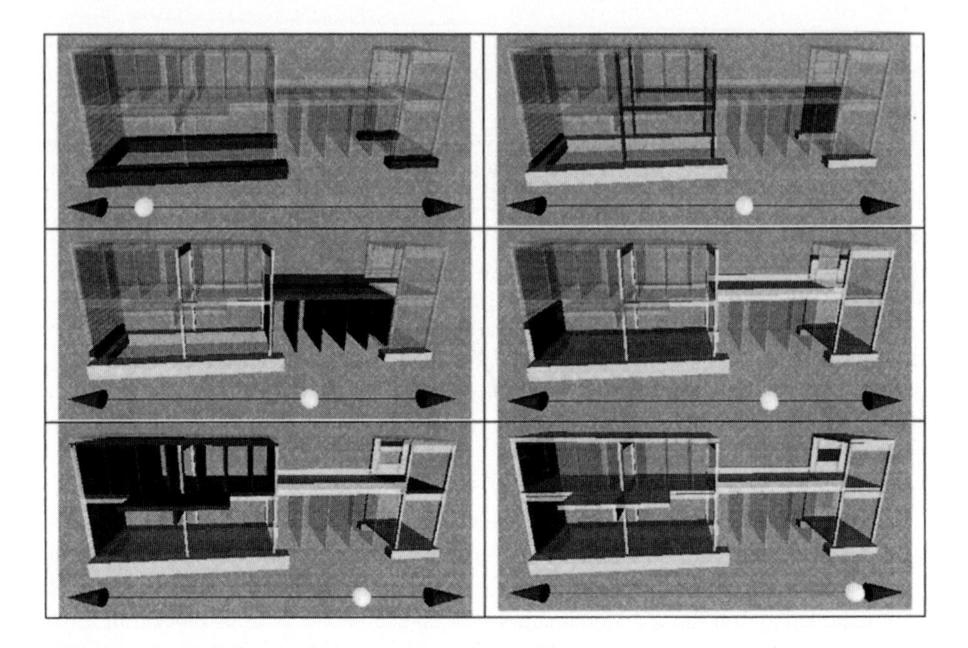

Fig. 15. 4D visualization of production model. The product model in this visualization has been transformed from the design-centric model with the addition of temporary structures (shoring) and the addition of detail (masonry block courses in Walls 1 & 4). The timeline of the 4D visualization is at the bottom of each frame

6 Conclusions

This discussion and the examples have shown how formalized product model and construction method knowledge can be used to operationalize database transformation mechanisms as defined by Eastman and Siabiris (1995). Structural engineers and superintendents can now rapidly modify the original design-centric product model to suit their particular analysis needs. This supports concurrent engineering of a structure and its construction process on the basis of accurate models at appropriate levels of detail. The transformation mechanisms we have implemented make the reasons for each product model transformation explicit. Therefore, new product models suitable for a particular set of tasks can be generated rapidly, and the resulting specific product models can be maintained intelligently as designs and construction methods change. In our experience, it is very cumbersome to maintain an appropriate product model throughout the design and construction process without such explicitly defined and implemented product model transformation mechanisms. These mechanisms require a sound implementation at the database level and a clear specification at the user or operations level. As mentioned, we anticipate that several other transformation mechanisms are needed to make product modeling an engineering reality. In our future work, we plan to define such additional mechanisms. Eventually, we should be able to build a tool that allows users to define transformation mechanisms to support the refinement and aggregation of component, activity, resource, and other project model objects as needed by a particular set of engineering tasks.

7 Acknowledgments

We thank the National Science Foundation and the Center for Integrated Facility Engineering for their financial support of this work. We are also grateful to Jim Glymph, George Metzger, Dennis Shelden, and Kristin Woehl from Frank O. Gehry & Associates in Santa Monica, CA, and to Anne Kirske, Chris Raftery, Terry Griffin, and Lisa Wickwire from Hoffman Construction's Seattle office for giving freely of their time and for providing access to the Experience Music Project as a case study.

References

1. Augenbroe, Godfried (1995). "The Combine Project: A Global Assessment." CIB Proceedings 180, W78 Workshop on Modeling Buildings Through Their Lifecyle, Stanford, CA, August 21-23, 1995, 163-171.
2. Björk, B-C. (1989). "Basic structure of a proposed building product model." Computer Aided Design, 21(2), 71-78.
3. Cay, Faruk (1997) "IT view on perspectives of computer aided process planning research." Computers in Industry, 34(3), 307-337.
4. Clayton, M.J., Kunz, J.C., and Fischer, M.A. (1996). "Rapid Conceptual Design Evaluation using a Virtual Product Model." Engineering Applications of Artificial Intelligence, 9(4), 439-451.
5. Crowley, Andrew J. and Watson, Alastair S. (1997). "Representing engineering information for constructional steelwork." Microcomputers in Civil Engineering, 12(1), 69-81.
6. Dharwadkar, Parmanand V., and Cleveland, Jr., Alton B. (1996). "Knowledge-Based Parametric Design Using Jspace." Proceedings of the Third Congress on Computing in Civil Engineering, Jorge Vanegas and Paul Chinowsky (Eds.), ASCE, Anaheim, CA, June 17-19, 1996, 70-76.
7. Eastman, C. M., A. Bond and S. Chase (1991). "A Formal Approach to Product Model Information." Research in Engineering Design, 2(2), 65-80.
8. Eastman, C.M. and Siabiris, A. (1995). "A generic building product model incorporating building type information.," Automation in Construction, Vol. 3, 283-304.
9. Fischer, Martin and Aalami, Florian (1996). "Scheduling with Computer-Interpretable Construction Method Models." Journal of Construction Engineering and Management, ASCE, 122(4), 337-347.
10. Ford, S., Aouad, G.F., Kirkham, J.A., Cooper, G.S., Brandon, P.S., and Child, T. (1994). "Object-oriented approach to integrating design information." Microcomputers in Civil Engineering, 9(6), 413-423.

11. Fruchter, Renate (1997). "A/E/C virtual atelier: Experience and future directions." Proceedings of the 1997 4th Congress on Computing in Civil Engineering Philadelphia, PA, USA, ASCE, New York, NY, USA. p. 395-402.
12. Gielingh, W. (1988). "General AEC Reference Model (GARM)." ISO TC184/SC4 Document 3.2.2.1 (Draft), TNO-IBBC.
13. Harfman, A.C. and Chen, S.S. (1993)."Component-based building representation for design and construction." Automation in Construction, Vol. 1, 339-350.
14. Howard, H.C., Abdalla, J.A., and Phan, D.H.D. (1992). "Primitive-Composite Approach for Structural Data Modeling." Journal of Computing in Civil Engineering, ASCE, 6(1), 19-40.
15. IAI (International Alliance for Interoperability) (1998). Industry Foundation Classes, Version 1.5.
16. Scherer, Raimar J., and Katranuschkov, Peter, (1993). "Architecture of an object-oriented product model prototype for integrated building design."Proceedings of the 5th International Conference on Computing in Civil and Building Engineering - V-ICCCBE Anaheim, CA, USA, Publ by ASCE, New York, NY, USA. p 393-400.
17. Van Nederveen, G.A., and Tolman, F.P. (1992). "Modelling multiple views on buildings." Automation in Construction, Vol. 1, 215-224.
18. Zhou X., Yan J., Jin Y., Ma D. and Ling Zh. (1995). "Representation of manufacturing environment and structured knowledge base in a CAPP system." Proceedings of the 1995 Database Symposium Boston, MA, USA, ASME, New York, NY, USA. P. 647-654.

Internet-Based Web-Mediated Collaborative Design and Learning Environment

Dr. Renate Fruchter
Director, Project Based Learning (PBL) Laboratory
Civil and Environmental Engineering Department, CIFE
Stanford University, CA 94305-4020

Abstract. This paper presents an overview of an on-going effort that focuses on combined research and curriculum development. It presents a tool kit of collaboration technologies developed by our group, and an education testbed for multidisciplinary and geographically distributed teams that exercise the Internet-based Web-mediated collaborative technologies tool kit. The tool kit of collaboration technologies is aimed to assist team members, project managers and owners to: (1) capture, share, publish, and link knowledge and information related to a specific project, (2) navigate through the archived knowledge and information, (3) evaluate and explain the product's performance, and (4) interact in a timely fashion. The Architecture/Engineering/Construction (AEC) course offered at Stanford University acts as a testbed for cutting edge information technologies and a forum that trains a new generation of professionals to team up with practitioners from other disciplines and take advantage of information technology to produce a better, faster, cheaper product. The paper concludes with a number of questions regarding the impact of information technologies on team performance and behavior.

1 Introduction

How can professionals in a multidisciplinary, geographically distributed team create and use information technologies to better communicate, collaborate, and coordinate?

Our observations of traditional teamwork indicate that:

- Team members develop their solutions independently as well as collaboratively.
- Each team member develops multiple alternatives. Evolution of discipline solutions and interactions among professionals are hard to document and track.
- Unsatisfactory changes prompt team members to backtrack to earlier solutions, which many times have to be recreated.
- Different discipline solutions interact with each other. The process of identifying shared interests is ad-hoc and based on participants' imperfect memories. This error-prone and time consuming process rapidly leads to inconsistencies and conflicts.
- Meetings are usually the forum in which inconsistencies are detected and resolved before the project can progress.

In addition, the current conventional project information development and management process is based on:

- Individual notebooks, recording background information and results of reasoning and calculations. Notebooks are private documents and are not shared among team members.
- Memos, generated by computers but handled as paper documents, distributed to selective team members, and filed. Paper memos can not be easily updated and are hard to retrieve.
- Graphics and other data, indexed by drawing number and date are generally hard to recover and in their paper form laborious to annotate and update.
- Documentation, in the form of successive approved versions under configuration control is filed as signed off paper documents.

The thesis of this paper is that *collaborative design and learning require computer support for*

(1) different modes and types of interaction in time and space, and
(2) various needs of content sharing, exchange, and information management.

Information technologies (IT) provide a vast array of opportunities for professionals and educators charged with providing information. The IT infrastructure in the

- Design-build project permits diverse interactions among team members, project manager and team, design-build team and owner;
- A/E/C course permits to augment communication for diverse interaction scenarios, e.g., instructor-students, instructor-student, peer-to-peer, i.e., student-student and instructor-instructor, students-instructors-practitioners.

This paper provides an overview of a collaboration tool kit composed of cutting edge IT research prototypes and commercial tools that support collaborative teamwork. This IT tool kit is exercised in an innovative multidisciplinary Architecture/Engineering/Construction (AEC) learning environment.

2 Modes of Interactions and Content Sharing

Throughout the design process all participants need to be able to express, capture and share knowledge, experiences, design intents, critiques and decisions by using (i) a shared workspace in collocated or distributed synchronous lectures and face-to-face meetings, and (ii) feedback and change notifications in collocated or distributed asynchronous work. This study characterizes collaborative design and learning as a function of *time*, *space* and shared *content*.

- *Time.* Throughout the teaching, learning, and building project process participants transition between synchronous and asynchronous types of interaction:
  (a) Synchronous collaboration occurs in face-to-face meetings. At that time, faculty and practitioners offer lectures and present case studies, and team

members define the overall design of the future building and determine the various discipline models. They communicate discipline concepts and assumptions that may have cross-disciplinary impacts.

(b) Asynchronous collaboration, in which (1) faculty and practitioners might provide feedback to students, (2) students might go over course material delivered over the Internet or via the World Wide Web, or (3) team members might work independently at concurrent or different times on detailing discipline subsystems of their project.

- *Space*. Faculty, practitioners and students get together for lectures, round table discussions, or project team meetings to review design proposals and decisions. Such face-to-face meetings can take place in a collocated setting, where all members travel to the meeting place, or in a distributed setting, where team members remain in their offices and use network applications (e.g., groupware, video conferencing) to share and exchange information and discuss their design decisions.

- *Shared content.* Project team members work on their discipline design solutions. As the design progresses, team members, faculty and industry mentors need to:
  (a) use a shared project workspace to publish shared 3D graphic building models to identify shared interests, multi-criteria semantics of graphic features and share symbolic, multicriteria critiques, explanations from all disciplines, and expert feedback as they work in a synchronous mode; and
  (b) use local discipline models and exchange design information and change notifications related to building features in which they expressed a shared interest, as they work in an asynchronous mode.

3 A/E/C Design and Learning Information Technology Tool Kit

The IT tool kit developed and tested in this study is aimed to take the *distance* out of distance learning and support collaborative building design. Internet mediated design communication, integration and organization frameworks, groupware technology and multimedia are used in the A/E/C course to support collaboration and teamwork:

- ***World Wide Web.*** The *World Wide Web* is used for team building and as a medium to disseminate and share conceptual design solutions of the design teams.
  1. *Team building on the WWW.* The "team building on the Web" exercise, is based on generic skill definitions of the A/E/C students and hypothetical project calls for bids posted on the Web. Students have to identify the specific project they can work on among the different calls for bids and publish on the Web their skills, project preference, and request for collaborators from the other disciplines. This exercise exposes students to the Web and one of its future potential commerce and business applications.

  2. *Cyberarchive for building project case studies used for role modeling.* The "Joan & Irving Harris Concert Hall" in Aspen Colorado was used as a case

study presented by Dr. G. Luth, from KLA Inc. and the teaching team. The project team of the music hall consisted of architect Harry Teague, structural engineer G. Luth, acustitian E. Cohen, detailer D. Rutledge, contractor Shaw Construction. The WWW and MediaWeaver [14], a graphical database, were used to create a Web-based information archive that describes the case study project and can be shared and accessed by both faculty and students "any time, anywhere." MediaWeaver provided a computational infrastructure to capture, index and search graphical information consisting of pictures and AutoCAD files which can be shared over the Internet. Students could learn more about the discipline issues of the case study by searching the project database on a particular discipline of interest (e.g., architecture, structure, construction) and at different levels of detail (e.g., music hall interior view, structural conceptual layout, retaining walls, excavation).

3. *Shared WWW Project Workspace.* A shared WWW workspace was created for each A/E/C project team to archive, share, access, and retrieve project information that ranged from sketches, VRML product models, Word documents, Excel spreadsheets, AutoCAD drawings, email notes, and CAD related change notifications (Figure 1). The WWW workspace includes private workspaces for individual team member, shared workspace, and hypermail archive that contains a log of all the electronic interactions between team members, students and faculty members, and students and industry mentors. Each A/E/C student team managed its shared WWW project workspaces and provided access permissions to faculty and industry mentors on an as-needed bases. This emulates the industry environment with Intranet and Extranet permissions to specific project information.

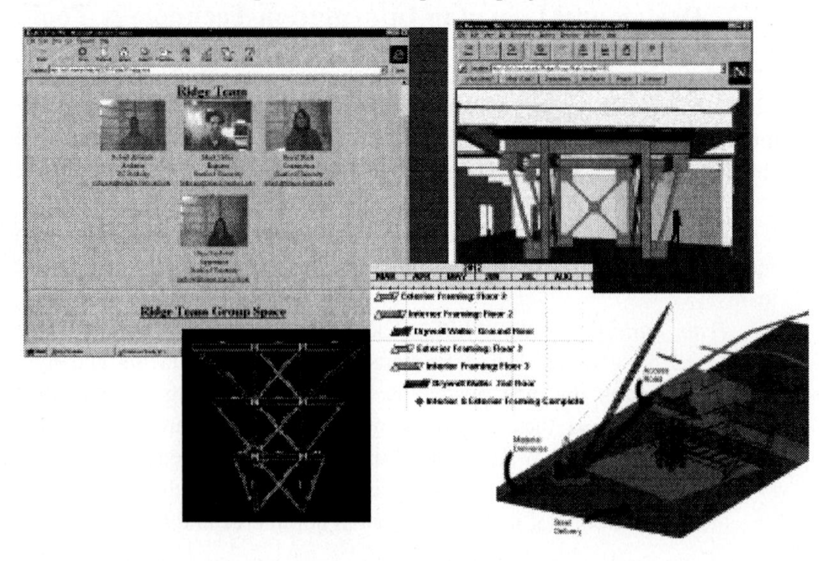

Fig. 1. Illustration of Shared WWW Project Workspace

4. *Digital Lectures and Meetings.* The course lectures and meetings are digitized and stored in a digital lectures and meetings WWW space. This enables the students to review lectures and meetings. They can go through the material at their own pace and revisit ideas, concepts, or debates related to critical issues. The digital modules are created using an experimental technology called Vxtreme.

- **Internet-mediated synchronous and asynchronous collaboration.** *Interdisciplinary Communication Medium, ICM* [8], is used as an integration environment to support the development of a shared building model that uses a graphic representation in AutoCAD as the central interface among designers (human-to-human) and as the gateway to tools/services (human-to-machine) in support of interdisciplinary design. ICM accommodates and integrates many perspectives within a design-build enterprise and allows team members to:

    - *augment shared 3D CAD product model* with the (1) team members' design intents, interests, and responsibilities, and (2) formal and informal design rationale, knowledge and information
    - *gather networked information* by using the discipline models to customize their search for additional discipline information,
    - *analyze and evaluate* the discipline models to derive behavior and compare it to function,
    - *explain* the results to other members of the team,
    - *infer shared interests and route change notifications* for proposed changes with regard to a modified feature or perspective.

Gathering network information using the *WWWCoach* tool [9], analyzing and evaluating designs using networked services and explaining evaluation results using knowledge based tools like *Comfort* for pasive energy conservation critique and *Egress* for floor plan egress evaluation [8], *QLRS* [10] for qualitative structural analysis, CMM [11] for constructibility evaluation, are beyond the scope of this discussion. These tasks and the tools supporting them have been presented in a previous paper [8].

The information and knowledge related to the shared product model is organized by ICM as the follows:

Graphics Objects contain Drawing Interchange File (DXF) representations of the graphic model entities.

Interpretation Objects encapsulate features for a particular perspective. An *Interpretation Object* has two primary attributes: a list of *Feature Classes* and a list of *Feature Objects*. Feature Classes provide an ontology to describe the semantic meaning of the graphics within a context. This ontology can be defined or augmented by the user at run-time. The list of *Feature Objects* is edited by the user to contain the instances from a particular graphic model which are relevant to an interpretation.

Feature Objects capture the link between graphic entities and symbolic entities. We define a feature to be a constituent element of a design that has meaning to a designer within a particular context. The basic components of a *Feature Object* are a *Feature Class*, an identifier or *Feature Name*, and a list of *Graphics Objects*. Other information objects can be linked to *Feature Objects* such as *Note Objects*, *HyperLink Objects*, and *Notification Objects*. *Feature Objects* allow graphic entities to have multiple meanings within different interpretations.

Person Objects serve as a record of the project participants and their declared roles and interests. A *Person Object* consists of the designer's name, a user-name, a user-password, an email address, a list of responsibilities, and a list of interests. *Person Objects* can be added, updated, and deleted by the users. The lists of interests and responsibilities are used by ICM to infer which team members should be sent email notifications about changes to a portion of the design

Note Objects contain text written by the project members. *Note Objects* are used to capture the design rationale or other design related information that a designer traditionally records in notebooks, memos, etc. Notes are encapsulated in *Feature Objects* to describe design requirements or intents. ICM's *Note Browser* allows the user to browse and search *Note Objects* in order to locate specific *Feature Objects* or *Interpretation Objects*.

HyperLink Objects provide a mechanism to link a *Feature Object* to sources of information. ICM currently handles references to World Wide Web (WWW) pages and electronic images. A feature in the graphic model can be linked to component specification sheets, code pages, structural details, schedule and cost information available on the WWW or a photo of a prototype. Additional functions enable the user to launch AutoCAD sessions with specific 3D product models from within the shared WWW project workspace of a team.

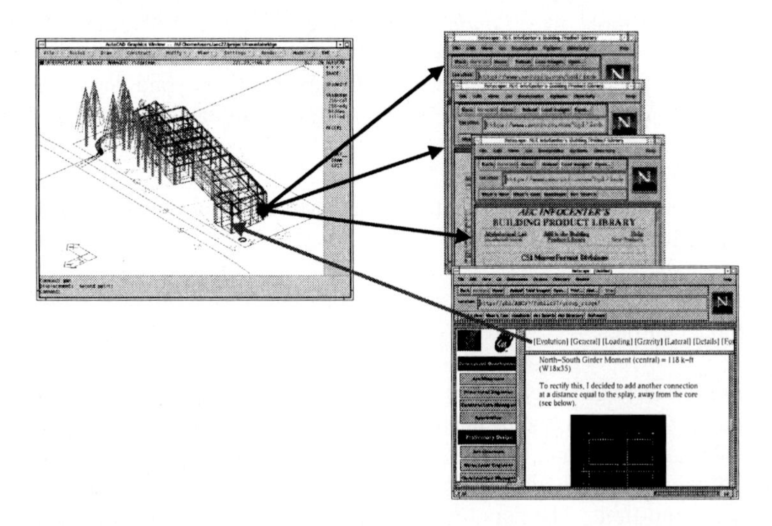

Fig. 2. Use of Shared 3D CAD Product Model and WWW Project Workspace as Navigation Vehicles with Formal Information Captured in Diverse WWW Documents

ICM facilitates the use of the shared 3D CAD product model and the project WWW workspace as navigation vehicles that enable the team members, project manager, or owner to access and retrieve project information and knowledge (Figure 2).

Notification Objects record the communications among the designers and are routed in asynchronous mode. These notifications can be used to solicit feedback, to give approval, to broadcast change notifications, or to initiate negotiations. A *Notification Object* consists of:

- *Feature Objects*, the focus of the notification
- affected *Interpretation Objects*, share an interest in the *Feature Objects*
- *Person Objects,* the mailing list, and
- a *Note Object,* describes the rationale or situation.

Notification Objects are stored as a part of *Feature Objects* in the shared product model.

ICM enables A/E/C team members to explore the different cross-disciplinary issues among architectural and structural form modeling, and constructibility.

- **Videoconferencing and Desktop Sharing.** Multi-disciplinary design teams are usually geographically distributed, which implies a large time budget allocated for traveling to meeting places. A number of software applications have been tested and used in this learning environment:

  Two software applications, *XMX* [12] and *xmove* [13], are used in the Unix environment to address some of the space barrier and the need face-to-face meetings in a geographically distributed setting. These enable sharing of both graphic and symbolic information. XMX provides a master/slave style and xmove a floor control strategy. The use of *groupware*, such as XMX and xmove in combination with telephone for audio communication enabled desktop sharing and collaborative teamwork for the geographically dispersed teams, that is, between Berkeley and Stanford students. These groupware tools enabled the team members to share and work concurrently on the same desktop, Xwindow, or application (e.g., AutoCAD). The team members can share design proposals, interpretations, critiques, and explanations as they negotiate alternative solutions.

  Video conferencing, i.e., Kodak systems, ProShare, and NetMeeting provide the medium that can take the *distance* out of distance learning. They provide capabilities that enable sharing of AutoCAD, Word, Power Point, etc. applications. This provides the necessary medium for the A/E/C teams to have virtual face-to-face meetings between A/E/C team members, A/E/C design teams and "owners," industry mentors, and faculty who were in different geographic locations, to discuss the design solution in real time. I have successfully used ProShare videoconferencing over ISDN or Internet line in four different distance learning scenarios:

  1. *Face-to-Face Meetings in Cyberspace.* A/E/C team members at Stanford and Berkeley, since the structural engineer and construction management students where at Stanford and the architecture student at Berkeley.

2. *Distant Learning Lectures.* Students and faculty from Stanford and Berkeley participated in real-time interactive lectures in a geographically distributed setting.
3. *Office hours in Cyberspace.* The videoconference with desktop sharing session enables the student team and faculty to focus on what-if scenarios, manipulate, and edit the content of documents and AutoCAD models.
4. *Final Project Presentations.* The videoconference technology enabled practitioners and faculty members to participate in the final project presentation session of the A/E/C course.

4 The Education Testbed

The Architecture/Engineering/Construction (AEC) course represents an education testbed. This effort was initiated at Stanford's Civil Engineering Department six years ago, to develop, test, implement and refine an A/E/C virtual atelier learning space that takes the "master builder's atelier" as a model for bringing the A/E/C disciplines together again. This effort is the result of a partnership between academia and industry.

Similar efforts are being pursued in other departments [1] and at other universities, e.g., CMU [2], MIT/Cornell/Hong Kong and other [3], Georgia Tech [4], Penn State [5], [6], University of Sydney [7]. Most of these efforts are discipline centric, e.g., a group of architects collaborating on a design project, and engage either graduate or undergraduate students in projects. The A/E/C course at Stanford builds a number of key bridges, i.e., between undergraduate and graduate programs, among A/E/C programs, between academia and industry.

The AEC course takes a multi-site, cross-disciplinary, project-based, team-oriented approach to teaching. It engages students, faculty members, researchers, and practitioners from the three disciplines. The A/E/C virtual atelier is based on a **PBL** pedagogical approach, where PBL stands for Problem, Project-, Product-, Process-, People-Based Learning. The PBL is a process of teaching and learning that focuses on problem based, project organized activities that produce a product for a client. PBL is based on re-engineered processes that bring people from multiple disciplines together.

One of the innovative features of PBL is represented by the role each of the participants play, i.e.:

- undergraduate and graduate students play the roles of apprentice and journeyman, respectively. They experience team dynamics as members of a multidisciplinary teams.
- faculty members play the role of "master builders." Their role is changing from the traditional teacher who delivers the course material in class to the *coach*.
- industry representatives play the role of mentors and sponsors. They become active participants in the teaching process and education of the next generation of practitioners.

The AEC student teams represent the core atom in this learning environment. Each AEC team consists of an architect, a structural engineer, a construction manager, and one or two apprentices. The students come from the different programs, departments, and universities. The course currently engages students and faculty from Stanford University, UC Berkeley, Georgia Tech, and Cal Poly as San Louis Obispo. For instance, the most complex AEC team this year had the architect at Georgia Tech, the structural engineer at Cal Poly, and the construction manager at Stanford University.

The core activity in this learning environment is a building project with a program, a budget, a site, a time for delivery, and a demanding owner. Faculty, A/E/C team members and practitioners are engaged in diverse working, teaching and learning settings, such as, lectures, face-to-face team meetings as well as in independent work.

The IT tool kit is exercised to facilitate the communication and cooperation within multidisciplinary teams. Computer tools play a key role in bridging all the participants of PBL together through shared workspaces on the World Wide Web, integrated tools that enable information sharing and exchange, and videoconferencing.

Finally, to support PBL, the classroom setting is changed to provide a flexible learning space. The space can be reconfigured by faculty and students on an as-needed bases to accommodate the different learning and teaching activities described above, such as computer lab activities, individual work, teamwork, presentations, and interaction in a geographically distributed setting (Figure 3).

(a) (b)

Fig. 3. Examples of PBL Configurations:
(a) Individual Workspace, (b) Group Workspace

5 Preliminary Observations

5.1 Engaging Learning Environment

Information Technology (IT) augmented learning can play an adjunct role to the traditional classroom instruction or be a primary delivery system for distant learning courses. IT can provide the following pedagogical benefits:

- immediacy - especially compared to the print-based correspondence courses,
- sense of group identity - the shared project workspace becomes a meeting place for students, faculty, and industry,
- improved dialogue - students interact and articulate their issues more than in traditional classroom settings,
- improved capture, sharing, access and retrieval of logged activities,
- active learning - students participation improves.

There is evidence that IT can be used to enhance the interaction and course content. The impact of IT on teaching and learning includes key transitions:

- from passive to engaged learners: the dominant model of learning has been for the student to passively absorb knowledge disseminated by the professors and textbooks. With IT students can move away from passive reception of information to the active engagement in the construction of knowledge.
- from coverage to mastery: traditional courses give students problems that they can solve using theory and knowledge taught in traditional discipline courses, i.e., *know-what* and *know-how*. The use of IT in project-based learning can guide students to discover disciplinary and interdisciplinary objectives, and thereby to develop *know-why* knowledge in an interdisciplinary context.
- from the classroom problems to real world projects: too often students walk out of the class ill-equiped to apply their new knowledge to real world problems and contexts. Conversely, too frequently the course examines concepts which are out of the context of real world projects. IT can help break down the walls between classroom and the real world.
- from text to multiple representations: linguistic expressions can be augmented by multidisciplinary and multimedia representations.
- from isolation to interconnection: IT helps students move from a view of learning as an individual act done in isolation toward learning as a collaborative activity. In addition, ideas and concepts are examined in multidisciplinary contexts.
- from products to process: we can move past a concern with the products of academic work to the processes that create knowledge. Students learn how to use IT tools that facilitate the process of scholarship.

Because of the widespread access provided by IT, the way we teach and pass information to learners around the world will change from the traditional teacher/classroom environment to a virtual classroom with no walls. It will engage educators and practitioners from interdisciplinary backgrounds and diverse cultures, institutions and companies.

5.2 Challenges

Computer support for collaborative teamwork is still in the early stages of development. Robust integrated systems are needed that link commercial applications in a seamless fashion and support interoperability and information exchange and communication in heterogeneous network computer hardware and software.

WWW environments and networking programs allow today information sharing among team members. However, linking applications on an as needed bases still needs major developments in the industry sector and standards for interoperability. Key industry efforts that address these issues are under way, such as the Industry Foundation Classes (IFC) effort initiated by Autodesk, the Industry Alliance for Interoperability (IAI), and STEP.

The A/E/C virtual atelier course is an innovation; like most innovations, its implementation is hardly straightforward. Numerous barriers have to be resolved. Despite public support to the contrary, professors recognize that certain behaviors, i.e., research, publication, fund raising, are far more rewarding than teaching. In order to encourage such endeavors in other institutions, the institution has to put in place the necessary support and reward system. Development, field testing, and revision of PBL courses is time, resource and budget intensive. Faculty members have to be ready to invest the effort to develop and manage a PBL course, establish the IT infrastructure, train support staff. Having departmental and institutional support is crucial. Finally, engaging practitioners in this type of teaching and learning endeavor will prove beneficial to both the education and industry environment. Practitioners play an active role in educating the new generation or professionals. They offer their expertise and real world project case studies, and are exposed to cutting edge IT tools that can provide a competitive advantage to their companies.

6 Concluding Remarks

Assessing and responding to the rapidly changing information and collaboration technologies and their impact on the global business environment are continuous challenges that industry and education are faced with. The development, testing, implementation and assessment of collaboration technologies in this on-going study indicates that *collaborative design and learning require computer support for:*
(1) different modes and types of interaction in time and space, and
(2) various needs of content sharing, exchange, and information management.

As we establish new learning and working environments that exercise collaboration technologies a number of questions arise and have to be addressed in future research. These include:

- How should IT be treated in any deployment plan? I conjecture that information technology deployment in both academia and industry should not be treated as a “fix” to a problem, a project, or an organization, but rather as a strategic plan that takes into consideration four key factors, i.e., technology, culture, economics, and politics.

- How does each IT change the behaviors of the individual, team and organizations? I conjecture that any information technology will change the communication patterns in the organization and will require new communication protocols not only at the technology level but at the social and business model level.

The AEC education program offers an excellent testbed to conduct empirical studies of the impact of specific IT on the interaction among people and the formalization of new communication patterns and protocols.

Acknowledgments. The author expresses her appreciation to H. Krawinkler, M. Fischer, Bob Tatum, from Stanford University; M. Martin, from the Architecture Department at UC Berkeley, P. Chinowsky from Georgia Tech, V. Vance from Cal Poly San Louis Obispo, the industry mentors G. Luth from KL&A, T. Neidecker from Barnes Construction, D. Fischer from Fluor Daniel, A.M Duffy from Bechtel, K. Woehl from Frank O. Gehry's Office, R. Alvarado from C.S. Slater Assoc. Inc. for their participation in this year's course. The author wishes to thank the dedicated TAs, and the five generations of pioneering students who worked with us. The author gratefully acknowledges the industry support group of the *"Computer Integrated A/E/C"* course: Autodesk, Sun Microsystems, Intel, and IntelliCorp for providing the necessary software and hardware. This research is funded by the NSF Synthesis Coalition, Stanford President's Fund, Stanford's Commission on Technology in Teaching and Lerning, and the Center for Integrated Facility Engineering (CIFE), at Stanford University.

URL: http://www-leland.stanford.edu/group/CIFE/ce222/
http://pbl.stanford.edu

References

1. Toye, T., Cutkosky, M.R., Leifer, L., Tennenbaum, M., Glicksman, M.J., "SHARE: A Methodology and Environment for Collaborative Product Develoment, " *2nd IEEE Workshop on Enabaling Technologies Infrastructure for Collaborative Enterprisess,* 1993.

2. Fenves, S., "An Interdisciplinary Course in Engineering Synthesis," Proc. *Second ASCE Congress of Computing in Civil Engineering*, ed. J.P. Mohsen, ASCE, New York, 433-440, 1995.

3. Chen, N., Kvan, T., Wojtowicz, J., Bakergem, D., Casaus, T., Davidson, J., Fargas, J., Hubbell, K., Mitchell, W., Nagakura, T., Papazian, P., "Place, Time, and The Virtual Studio, Proc. of *ACADIA 94 Conference*, ed. A. Harfmann and M. Fraser, 115-131, 1994.

4. Vanegas, J. and Guzdial, M., "Engineering Education in Sustainable Development and Technology," Proc. *Second ASCE Congress of Computing in Civil Engineering*, ed. J.P. Mohsen, ASCE, New York, 425-432, 1995.

5. El-Bibani, h., Vande, M., Gowda, P., Branch, B., Groth, J., Schuld, K., "Information Technology and Education: Towards the Virtual Integrated Architecture/Engineering/Construction Environment," Proc. *Second ASCE Congress of Computing in Civil Engineering*, ed. J.P. Mohsen, ASCE, New York, 420-424, 1995.

6. Riley, D., "Educating the Modern Master Builder," Proc. *Second ASCE Congress of Computing in Civil Engineering*, ed. J.P. Mohsen, ASCE, New York, 449-452, 1995.

7. Khedro, T., Genesereth, M., Teicholz, P. "FCDA: A Framework for collaborative distributed multidisciplinary design", *AI in Collaborative Design Workshop*. Menlo Park, CA, AAAI, 1993.

8. Fruchter, R., "Conceptual Collaborative Building Design Through Shared Graphics," IEEE Expert special issue on "AI in Civil Engineering," Vol 11 Nr. 3, 33-41, June 1996.

9. Fruchter, R. and Reiner, K.,*"Model Centered World Wide Web Coach,"* ASCE 3rd Congress of Computing in Civil Engineering, ed. J. Vanegas and P. Chinowsky, Anaheim, June 1996, pp1-7.

10. Fruchter, R., Krawinkler, H. and Law, K.H., *"Qualitative Modeling and Analysis of Lateral Load Resistance in Frames,"* AI EDAM Journal, AI EDAM 7(4), 1993, pp239-256.

11. Fischer, M., and Aalami F. *"Scheduling with Computer-Interpretable Construction Method Models."* Journal of Construction Engineering and Management, ASCE, . 1996, 122(4), 337-347

12. Bazik, J., "XMX: A X Protocol Multiplexor," Department of Computer Science, Brown University, Providence, RI 02912.

13. Slomita, E., "xmove: A Pseudoserver for Client Mobility," X Consortium, 1994.

14. S.X. Wei "An Open Testbed for Distributed Multimedia Databases," Proc. AAAI Workshop on Indexing and Reuse in Multimedia systems, 1994.

Wearable Computers for Field Inspectors: Delivering Data and Knowledge-Based Support in the Field

James H. Garrett, Jr.

Department of Civil and Environmental Engineering
Institute for Complex Engineered Systems
Carnegie Mellon University
Pittsburgh, PA, USA
garrett@cmu.edu

Asim Smailagic

Department of Computer Science
Institute for Complex Engineering Systems
Carnegie Mellon University
Pittsburgh, PA , USA
asim@cs.cmu.edu

Abstract. Field inspectors, such as bridge inspectors, oftentimes have to perform their tasks in harsh environments and need to have their hands free. Thus, they cannot carry reference materials. The inability to access material during the inspection effort influences the efficiency with which field data is collected and its quality.

Field inspectors, such as bridge inspectors, oftentimes have to perform their tasks in harsh environments and need to have their hands free. Thus, they cannot carry reference materials. The inability to access material during the inspection effort influences the efficiency with which field data is collected and its quality.

A prototype wearable computer for supporting bridge inspectors has been developed, which provides users unlimited access to data and knowledge in the field. The application has been designed to recognize the field context and provide intelligent support for field data collection and decision making.

This describes the requirements of, and the prototype for, a wearable computer to support bridge inspectors while in the field. As this wearable computer for bridge inspectors is still a work in progress and very much an early prototype, opportunities for using AI to provide intelligent assistance and support to bridge inspectors is discussed.

1. Introduction

Field inspectors oftentimes have to perform their monitoring and inspection tasks in harsh environments and under conditions that do not lend themselves current forms of computing support. Due to the nature of the inspection, the inspectors need to have their hands free to support their climbing and inspecting activities. They cannot carry reference manuals, previous inspection reports, blueprints, or coding manuals (i.e., manuals used to translate what the inspector sees into a specific coded level of damage) or other information with them onto the bridge. They may only consult their manuals and the previous reports after leaving the site and returning to their inspection vehicles or to the office. In addition, field inspectors typically prepare sketches and note templates to guide their inspection and to facilitate the recording of what they see in the field. Hence, during most current inspection processes, inspectors record what they see, then later compare that information with blueprints, coding manuals, previous reports, etc. The inability of an inspector to access this supporting information and knowledge during the inspection effort influences the efficiency with which field data is collected and its quality.

Bridges are one major type of infrastructure requiring a significant amount of inspection. In 1971, the National Bridge Inspection Standards (NBIS) [*1*] were promulgated in response to several catastrophic bridge collapses that could be traced to undiagnosed problems with these bridges. State and local governments began to systematically inspect their bridge population and collect vital condition information. The inspection and monitoring of bridges has continually evolved to the point where states and local governments are now implementing Bridge Management Systems (BMSs) to store and retrieve data on their bridge inventory and to support maintenance, repair, redesign and replacement decision making. However, effective bridge management is heavily dependent on field inspectors being able to collect detailed accurate condition information on all of the individual elements of a bridge and enter this data into a bridge management system database. The Bridge Inspector's Training Manual/90 [*2*] states, "A very careful inspection is worth no more than the records collected during that inspection."

To collect bridge condition data, the bridge inspectors must walk and climb on the bridge to assess its condition. While advanced condition assessment technologies, such as video imaging, infrared thermography and ground-penetrating radar are being used for certain infrastructure applications, such as pavement inspection and bridge deck evaluation described in [*3*], visual inspection is still the primary means of data collection and will likely be for the foreseeable future [*4*].

From this discussion, it is obvious that field inspectors, such as bridge inspectors, can benefit from some form of computer-based support while in the field. There are nearly 590,000 bridges in the National Bridge Inventory over 6.1 m (20 ft.) in length that must be inspected at least every two years [*5*]. There are many other bridges with spans less than 6.1 m (20 ft.) that agencies must still maintain at a cost to state taxpayers. While advanced bridge management systems are being created, deployed and used to collect this data for inventories of bridges and support the systematic identification and prioritization of needs, the bridge inspectors in the field are still primarily using paper-based notes to support, and record the results of, their inspection processes. Inspectors are currently limited in terms of the material they can reference while in the field and the tools they can use to capture the nature and location of defects found on bridge elements. The computing needs cannot simply be addressed by giving them a laptop with desktop-oriented software to use in the field. This approach has been tried and simply does not work (see Section 3).

To properly support an inspector in the field, the hardware and software must be designed to recognize the field context and provide intelligent support for field data collection and decision making. For example, for bridge inspectors, the necessary information about a bridge, at the right level of detail, must be collected ahead of time, as it is now, and entered into, or made accessible from, the computer. Some of this information will indicate where damage has been previously located, while other information may be recommendations about where to look on the bridge for damage, such as fatigue-prone details. Intelligent support is needed to assist the inspector in finding the needed information while in the field. It is not enough just to put a desktop application in the hands of field personnel using a laptop or palmtop computer. Field inspectors cannot be standing there blindly searching through pages of information displayed on their computer, it is both unsafe and impractical. Knowledge based systems for locating possible damage sites and assisting inspectors in assessing their states of damage are also needed. If a defect or error is found, the location and nature of damage must be accurately captured , recorded and eventually entered into a bridge management system's database. To provide this level of computing support in the field requires: (1) a hardware system that is unobtrusive and designed based on the field inspector's tasks and mode of operation; and (2) that the software system be designed to support their specific tasks, provide knowledgeable advice about these tasks and provide intelligent, efficient forms of human-computer interaction that takes account of the field context of inspection.

Researchers at Carnegie Mellon have developed several generations of wearable computers specifically for inspection-oriented applications. The wearable computer provides users computing support in the field as they conduct other operations that require full use of their hands. In all of these applications, the emphasis has been on mobility, mostly hands free operation and either recording or displaying technical information. In one specific application, wearable computers were employed to assist U.S. Marines in conducting a limited technical inspection (LTI) of amphibious vehicles. This application of wearable computers lead to cost savings in terms of personnel needed (50% reduction), inspection time needed (40% reduction), and post-processing time needed for generating reports from the inspection data (30% reduction), for a total of 70% reduction in time from inspection through logistics data entry. In a second application, wearable computers were used to assist in the inspection of KC-135 aircraft, yielding a 50% reduction in data entry time. A third system supported maintenance of airport people movers providing remote access to over 120,000 engineering drawings, searching of previous maintenance reports, and telephonic and visual collaboration with remote personnel via a microphone and 3 oz. hand-held camera. In these applications, the reference material and inspection processes were the same for all instances of amphibious vehicles, KC-135 aircraft, and people movers, respectively.

Field inspection operations in civil and environmental engineering, such as bridge inspection, have similarities to, and differences with, the previously described applications. The bridge inspection process is similar in that the inspector does need to be supported in the field while doing an inspection. However, bridge inspection is different from these previous applications in that each bridge, and the corresponding steps in its inspection process, may be unique. The advice that is provided about places to look for potential damage will have to be determined using a knowledge-based reasoning process, such as that performed by the Bridge Fatigue Investigator [6]. The information needed about each bridge will have to be collected ahead of time, as it is now, and entered into, or made accessible from, the wearable computer. A second significant difference concerns the site. The previous inspection applications were performed in a maintenance facility. If a defect or error was found, it was tagged in some way and the defect remained tagged until a maintenance activity could be performed. Such tagging in the field is not feasible for bridges (with the exception of defects likely to lead to catastrophic failure, when an inspector can close or post a bridge immediately); the location and nature of damage must be accurately captured, recorded, and eventually entered into a bridge management system's database. Maintenance decisions are then made by looking at the condition of the larger inventory of bridges; resource limitations may force some maintenance to be put off until a future time.

This paper discusses the design and implementation of a prototype wearable computer to support bridge inspectors in the field as they inspect the individual elements of a bridge and record their findings. The early prototype has been designed from scratch to take account of the nature of the bridge inspection task and the context in which it occurs. The prototype supports the inspectors tasks of reviewing previous inspection reports and completing the current report in the field by recording comments, making sketches and taking and annotating photographs. The inspector is able to naturally interact with the software and hardware via three different modes: speech, handwriting, or a virtual keyboard. The thesis of this paper is that while this tool begins to provide intelligent support for field inspectors, there exists significant opportunities to provide additional intelligent support to bridge inspectors in the field. This intelligent support should yield a decrease in cost and an improvement in quality of bridge inspections.

The remainder of this paper is organized as follows. The second section provides a brief overview of the bridge inspection process. The third section discusses several previous attempts to provide computing support for bridge inspectors. The fourth section provides a brief description of two previous applications of a wearable computer to military inspection processes. The fifth section presents the requirements for, and the design and implementation of, a prototype wearable computer for bridge inspectors. The sixth section discusses several ways in which AI could be used to improve the support provided to field inspectors. The last section is a summary.

2. Overview of the Bridge Inspection Process

The Bridge Inspectors Training Manual/90 [*2*] provides an in-depth description of the bridge inspection process; this section briefly summarizes this description so as to provide a context for the application of wearable computing to bridge inspection. The five basic duties of the bridge inspector are: 1) planning the inspection; 2) preparing for the inspection; 3) performing the inspection; 4) preparing the report; and 5) identifying items for repair and maintenance.

Planning an inspection involves developing a sequence and schedule for the inspection tasks, organizing field notes that will need to be taken and anticipating the traffic control needs and other needs during the inspection. In preparing for an inspection, the inspector first reviews the sources of information available about the bridge, such as "as-built" bridge plans, previous inspection reports, maintenance and repair records, geotechnical data, etc. Previous inspection reports are an important source of information, because they document the condition of the bridge in previous years and can indicate where special attention might be needed. Maintenance and repair records are also valuable in that they identify which of the defects found during the previous inspections were addressed and what additional maintenance was conducted.

So as to be able to recall this previously reviewed information while in the field, the inspector prepares a set of notes, forms and sketches before going onto the bridge. Any sketches used during previous inspection procedures are photocopied and carried with the inspector for use and reference during the inspection process; extra copies are usually carried in case one gets damaged in the field. If previous sketches are not available, then generic sketches of deck sections, diaphragms, bracing members, etc. are usually created and carried onto the bridge for use in recording the location of defects seen.

Once the inspector has planned the inspection, reviewed the documentation and prepared the field notes he or she will carry onto the bridge, he or she then performs the inspection. The inspector attempts to record the location and nature of every defect seen on the bridge. Accurate records of these observations are obviously extremely important to the success of the entire inspection process. When inspecting individual elements of a bridge, the bridge inspector is looking for a wide variety of defects, such as corrosion, cracking, splitting, connection slippage, deformations due to overload, and collision damage. Every one of these types of damage needs to be accurately located on the bridge and recorded in the field notes; the inspector needs to record both quantitative and qualitative information about these defects and the elements on which they are found.

After completing the inspection of the bridge, the inspector must submit a report, in a standardized form using specific codes for the condition level of every element of the bridge. The median number of data items collected and reported per bridge (over all states in the U.S.) is 270, with some states collecting as many as 700 pieces of data on their bridges and some as few as 115 [*7*]. This report is generated after the inspector leaves the bridge and is based on the field notes taken. This report contains fields for recording the physical condition of all elements making up the bridge and for recording the information about the inspection process conducted. The coding for the condition level of an element is performed by reviewing the field notes taken on that element and then referring to a coding manual, such as the AASHTO Commonly Recognized (CoRe) Element Guide [*8*]. For each element, the quantity of material in each condition state must be reported (and thus measured while in the field) [*9*]. After completing this data collection effort, the data is entered into a Bridge Management System where it becomes data upon which preservation and improvement needs are prioritized and scheduled.

3. Current Approaches to Computer Support for Bridge Inspectors

While most inspectors are still using pencil and paper on the site, some states and inspection agencies are experimenting with using computers on-site during the inspection process. All of the computing now provided on-site appears to be based on laptop, notebook and palmtop computers using a pen-based interface. For example, the Massachusetts Highway Department is using a system called IBIIS to store and manage all of their bridge documents [*10*]. As part of this system, IBIIS provides inspectors with a video camcorder and a notebook computer. "The camcorder is used to take video and still photographs and the notebook computer is used to enter the NBIS rating data for each bridge into a database application and commentary into a word processor" [*10*]. Another application is a system developed by the University of Central Florida for the Florida Department of Transportation (FDOT) [*11*]. The system consists of both a field and office set up with a pen-based notebook computer used to collect all field inspection data.

The use of a notebook computer in the field is inconsistent with the inspection process described in Section 2. Turner and Richardson [*7*] report that some states have experimented with having its inspectors enter inspection data directly into personal computers, but found that it is awkward to carry portable computers while walking around on a bridge. They also indicate that some experiments with small computers strapped to the inspectors arm or attached to a clipboard have been conducted by several states, but the 'keyboard and viewing screen are so small that data entry can be very difficult" [*7*]. Inspectors need to be able to use their hands for safe and effective inspection of a bridge and need to be able to view the data they are entering in less than optimal viewing conditions.

In the next section, we discuss the wearable computer paradigm that provides nearly hands-free operation of a computer for field data collection and provides a much more usable output interface.

4. Wearable Computer Technology and Applications

Since 1980, technology has been devoted to shrinking the size and weight of personal computers without substantially changing the way users interact with their computing environment. Conventional input/output devices place an ultimate limit on the size and weight of personal computers. The size of today's personal computers is limited by the conventional typewriter-like keyboard whose dimensions have not changed substantially for over one hundred years. Both size and weight are limited by displays the size of notebook paper intended to be viewed from several feet away. Since the size of the display places a lower bound on the personal computer's energy consumption, weight is dictated primarily by the weight of the energy storage devices such as batteries

Carnegie Mellon University invented the term "wearable computers" by building VuMan 1 as the first wearable computer in 1991 *[12]*. The Wearable Computer Laboratory at the Engineering Design Research Center has built seventeen generations of wearable computers [*13*, *14*, *15*]. These generations have been initially designed and implemented in the *Wearable Computer Project Course* taught by one of the authors. *In these applications, the emphasis has been on mobility, mostly hands free operation and either recording or displaying technical information.* The Wearable Computer Laboratory has been defining, and will continue to define, the future for not only computing technologies but also for the use of computers in daily activities. The application domains range from inspection, maintenance, manufacture, navigation to on-the-move collaboration, position sensing, global communication, real time speech recognition and language translation, and medical. Carnegie Mellon has composed an integrated interdisciplinary team that has conducted field trials of over a dozen systems over the past five years. The Carnegie Mellon team's approach to wearable computing is notably different and more applied than that taken by MIT. The wearable computing being developed at MIT can be classified as affective wearable computing, wearable sensory systems, computer-mediated reality, and wearable fashion clothing *[16, 17, 18]*.

The wearable computer paradigm is one where useful and effective computing support is provided to a user while in the field and conducting other operations that require nearly hands-free operation of that computer. The user interfaces with the wearable computer via a number of input and output devices. For output, head-worn and hand-held video devices have been used. For input, we have used three different devices [*19*].

- A custom dial input. This device was used in the check list application. It is also suitable for use for any hypertext application. The hot links in the hypertext are known to the application and maintained as a circular list of the information on the screen. A rotation of the dial results in moving from one link to the next. A selection results in either recording the current piece of information or entering the link to retrieve another screen.
- A joystick. This device provides general two-dimensional input that was useful for positioning on the graphic based input where location is important.
- Speech. This modality allows the user to speak words and commands, which are then translated into text and commands for the computer.

The development process has been a rapid prototyping process, utilizing a user-centered design process [*20*]. The user-centered design process for these wearable computers involves viewing the context of use of the computer, and then designing the system. The computer hardware, input devices and supporting software are all developed with close interaction with the end-user.

One of wearable computing applications, VuMan3, was used to assist a Marine in filling out a checklist as he is inspecting different components of an amphibious vehicle [*21*]. Each component is marked to be suitable or to need repair in the checklist. The inspection requires the Marine to crawl under the vehicle, to get into very tight compartments and to use both hands to manipulate components being inspected (see Fig. 1).

The main characteristics of this application were identified as:

- technical information is contained in a large set of manuals, with cross-references, schematics, diagrams, text;
- it is necessary to access these manuals while repairing or inspecting a vehicle; and
- it is inconvenient to handle/consult the manuals during an inspection/repair activity.

An example of a maintenance document is the Limited Technical Inspection (LTI). An LTI is a 50-page document including :

- a check-list of about 600 items, one for each part of the vehicle (e.g. front left tire, rear axle, windshield wipers);
- selecting one of four possible options about the status of the item: Serviceable, Unserviceable, Missing or On Equipment Repair Order (ERO); and
- entering further explanatory comments about the item as needed, for example the part is unserviceable due to four missing bolts.

VuMan3, a wearable computer designed for performing maintenance on complex military vehicles, employed a novel input device (a rotary dial) and a ring of selection buttons. Each small rotation of the dial moved the user onto selectable interface regions on a head-worn display. Because the dial allowed discrete, accurate selection, technicians could use the computer on the move. (This product would have been very difficult to operate with a free-moving mouse and cursor.) Options to be selected on a screen were logically arranged in a circular list. Rotation of the dial one position clockwise changes the highlighted option to the next one clockwise in the circular list. The same applies for counterclockwise rotation of the dial. Depression of one of the buttons performs the function specified by the highlighted option, enters the highlighted information into a database, provides auxiliary information, selects a hypertext link, or selects an option on a World Wide Web page. Application software was modified to reflect the dial design decisions and the final result was a new user interface concept: circular input, circular visualization.

Fig. 1. VuMan3 Wearable Computer

The LTI checklist available on VuMan3 consists of a number of sections, with about one hundred items in each section. The users would have to manually go through each of these items by using the dial to select "next item", or "next field." The Smart Cursor feature represents built-in intelligence in the user interface, and was designed to help automate some of this navigation. This approach is accomplished with the use of the following:

- An input pattern recognizer, which keeps track of what fields the user selects on a given screen, and that is called a "working set." If the working set remains the same over two or more screens, the navigation system starts moving the cursor automatically to the fields in the working set. In essence, this is a macro recorder that runs continuously during the user's work session, and uses a heuristic about when to repeat recorded keystrokes.
- A domain-specific heuristic, developed through studies of how users usually navigate through LTI hypertext documents (e.g. their behavior in the presence of multiple options).

VuMan3 was used in a field trial for Marine heavy vehicle Limited Technical Inspection and compared with the current practice trial results. Half the personnel were needed; 40% less time was needed for inspection; and 30% less post processing time was needed. In total, a 70% reduction in time from inspection through logistics data entry was seen.

5. A Wearable Computer for Bridge Inspection: Requirements and Design

A prototype lightweight portable computer system has been developed to support bridge inspection workers anytime and anyplace. The objective in developing this system was to provide bridge inspectors with wearable computers, like VuMan3, that are hardened to function where they may be dropped or hit and that have the capabilities and an appropriate user interface to support faster and better bridge inspection.

5.1 Requirements

In Section 2, the major stages of bridge inspection process were described as planning, preparing for, conducting, and reporting the results of, an inspection. In preparing for an inspection, bridge inspectors retrieve and review previous inspection and maintenance reports, make sketch templates of elements on the bridge, assemble the new inspection report forms, and identify and review the elements being monitored. When on site and conducting the inspection, inspectors need to be able to review the previous inspection reports, add comments and sketches of element damage to the current report, take and annotate photos of this damage, and relate them to the relevant parts of the inspection report. While doing all of these tasks, the inspector is usually climbing on or under the bridge and usually needs his hands free. Hence, one major requirements of a wearable computer for this application, is that it must allow the inspector to easily, i.e., naturally and with near hands-free operation, access previous reports and add comments, sketches and photos to the current inspection report.

As was discussed in Section 3, conventional palmtop computers are clearly not suitable because they have limited input/output interfaces. Most PDAs require two hands to operate, making them even less desirable for supporting inspectors. To maximize the effectiveness of a wearable computer for supporting bridge inspection, *the user interface design must be carefully matched with user tasks*. In the previous wearable computer applications, a user-centered design approach was taken, which means that the functionality and user interface are developed in close concert with the eventual end users of the system.

For this application, bridge inspectors were informally interviewed and observed in the field to determine the information they need and record while in the process of an inspection. Prior to attending these interviews, the Bridge Inspectors Training Manual/90 [*2*] was studied to gain a foundation upon which to build an understanding of the bridge inspection process. To orient these early interviews, a prototype based on a previous wearable computer system was developed and demonstrated to several different groups of inspectors, using a set of bridge drawings. We asked them about the tasks they perform and how well such a system would support those tasks. We asked them about their preferences for methods of display (e.g., head-mounted display or body-worn display) and the methods of interaction (e.g., mouse, rotary dial, touchpad and stylus). These inspectors provided a large amount of critical feedback on the usefulness and usability of the prototype. They recommended that the user interface be redesigned to make it more appropriate and natural for their tasks. For example, the buttons on the interface were too small in their opinion to be used while in the field. They asked for voice input and speech-to-text translation; they stated emphatically that they do not want to type anything in the field and that they found handwriting to be less desirable than speaking. These inspectors made it very clear that they must be able to quickly and easily identify the location of a defect and record a textual description of it, take and store a photo of it, and sketch the defect as they often do now (even when taking a photo because often times the photos are not taken in the best of conditions). Another aspect brought up by the inspectors was that for larger inspection projects, the inspections are conducted by multiple inspectors. Hence, a wearable computer for supporting larger bridge inspections must be designed to allow these inspectors to communicate and coordinate their actions while on the bridge. Currently, they use radios to do this. A wearable computer should allow them to not only communicate words, but text, reports, images, drawings and sketches. It was clear from this initial prototype, and the feedback given, that these wearable computers for bridge inspectors must be developed from the beginning with close interactions with bridge inspectors.

5.2 System Design

Based on the requirements derived from talking with and observing bridge inspectors, the students in the Spring 1998 Wearable Computer Project Design Course designed and built a prototype wearable computer for supporting bridge inspectors. The hardware for the system, known as MIA (Mobile Information Assistant), consists of the following components (see Fig.2): 1) a 3 lb. computer, consisting of a 133 MHz Pentium processor, a full duplex sound processing chip, a 1.2 GB rotating disk, a serial port for connection to a digital camera, a 6.5 inch black and white (diagonal) VGA display, a microphone port, and a PCMCIA wireless communication card supporting spread spectrum radio at 2 Mbps; 2) two lithium-ion batteries, with 2000 mAh capacity each; a noise-canceling microphone; 3) a digital camera; and 4) a van-based unit that provides a battery charger and a laptop to use in wirelessly backing up the data collected on the wearable unit.

The software for the system consists of the following components: 1) a graphical user interface that presents overlapping panels with tabs for viewing previous inspection reports, the current inspection form, the collection of sketch templates, and a photo album (see Fig. 3); 2) a set of speech commands that mirror the buttons on the GUI, which allows the user to invoke commands via speech and thus in a hands-free mode; 3) a database for storing the contents of a bridge file: previous inspection reports, current inspection report, monitoring systems, sketches and photos; 4) a tool for sketching (see Fig. 4); 5) a tool for viewing/editing photos (see Fig. 5); 6) a tool for creating text comments that supports speech recognition, handwriting recognition and a virtual keyboard; and 7) a component for handling the wireless file transfer requests to and from the van unit.

As with VuMan3, this system has been designed to provide an interaction that intelligently supports the user in the field. For example, when a previous report or a current report is accessed, the interface presents an expandable, hierarchical view of the report to allow faster movement between report sections (see left window in Fig 3). When the user selects one of the nodes in this view, which is associated with an element, such as a girder or a bearing, the comments, photo icons and sketch icons associated with this element are all simultaneously displayed. When the photo and sketch icons are selected, they are displayed in the appropriate software component for editing. The user need not be concerned with how or where the photos and sketches are stored and what the filenames are. The user can easily and naturally enter text comments about, and annotations on, the photos and sketches using speech, handwriting or the virtual keyboard. The interface based on the tabbed panels was chosen so as to allow the user to easily move between previous reports and the current report, maintaining the same relative location in each report, so as to be able to easily copy and paste previous comments and sketches into the current report.

The prototype supports the basic functions of an inspector in the field: recording comments about, sketching, and photographing elements of a bridge. The system is currently undergoing field evaluations by several of the PENNDOT inspectors, but no results of these tests are available at the time of submittal of this paper. However, we do know from our early interactions with these inspectors that the system can be made easier to use. While the system was designed to provide simple, easy to use, and nearly hands-free support for bridge, there are several additional forms of intelligent support that should eventually be offered within this system to make it easier to use while in the field and more useful.

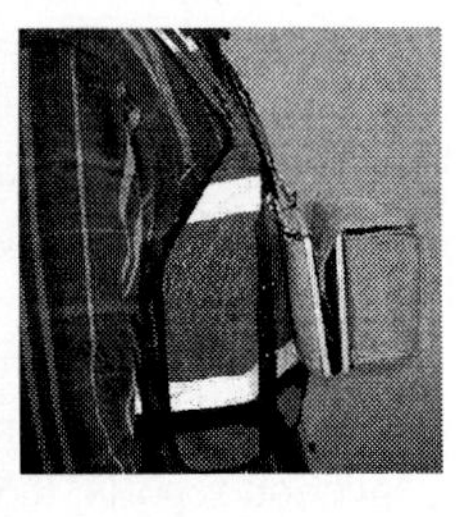
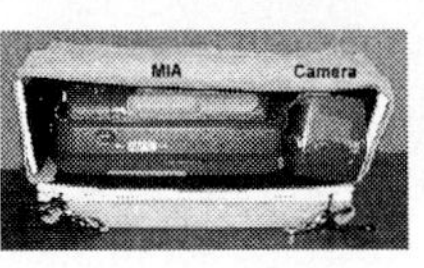

Fig. 2. Wearable Computer Hardware and Field Packaging

6. Potential AI Applications

Intelligent support is needed to assist the inspector in finding the needed information while in the field. It is not enough just to put desktop applications in the hands of field personnel, which has actually been the practice so far in the bridge inspection industry. The application has to be designed to recognize the field context and provide intelligent support for field data collection and decision making. Inspectors cannot be standing in the field while searching through pages of information displayed on their computer, it is both unsafe and impractical.

Knowledge based systems for locating possible damage sites and assessing their states of damage should also be provided on this wearable computer. For example, Wilson and Fisher have created a knowledge-based system, known as the Bridge Fatigue Investigator (BFI), to identify the locations on a bridge where the inspector should look for fatigue cracks in steel structures [6]. Such KBES are needed to help inspectors identify the locations of other types of damage in elements specific to materials such as prestressed concrete, reinforced concrete and laminated timber.

If a defect or error is found, the location, nature and extent of the damage must be accurately described. Again, the inspector could greatly benefit from an intelligent assistant that would present the inspector with a detailed, context-specific description of the procedure to follow in assessing the damage once it has been discovered. For example, a KBES could help an inspector determine exactly where and how many times the rust and scale should be scraped from the girders on a superstructure so as to measure the remaining section. Inspectors simply do not have the time to inspect every single location of rust on even a small bridge. Structural engineers may have this type of knowledge, but the inspectors may not. For this task, it may be possible to take images of the bridge and use image analysis to identify the potential locations of damage due to corrosion, which could then be provided to a KBES with a model of the bridge and the requisite expertise needed to reason about which locations should be measured in detail for section loss.

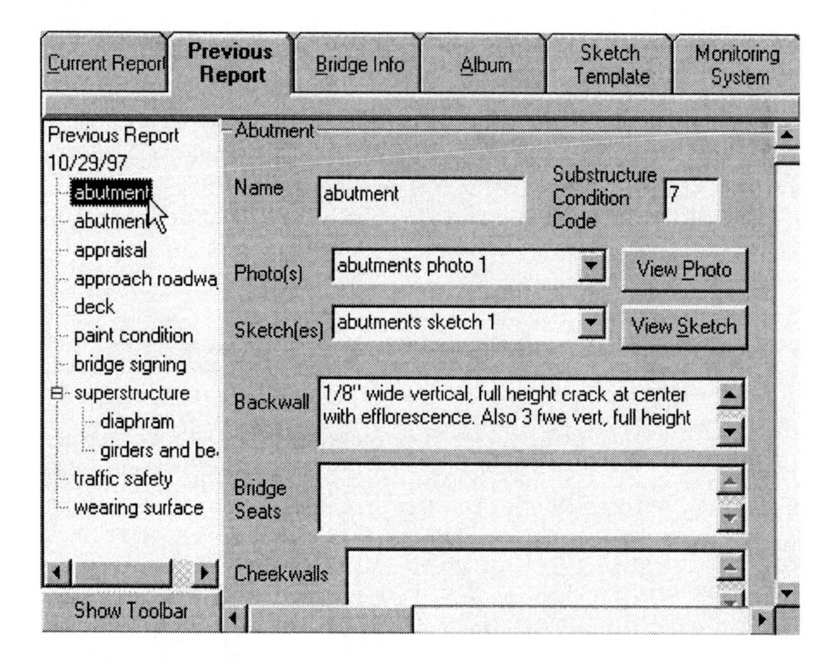

Fig. 3. Tabbed GUI showing "Abutment" Form

Case-based reasoning could be used to help inspectors identify damage locations and extent on the bridge being inspected. Based on the locations and extent of damage on other similar bridges, under similar loading conditions, exposed to similar weather conditions, a case-based reasoning system could relate these damage locations to the current bridge. This approach could be used in the absence of a mechanistic prediction of possible damage, such as the BFI described in the previous paragraph.

Oftentimes, one of the most difficult tasks in an inspection is to assign the condition values for an entire component, such as the superstructure or a deck, based on the damage detected on isolated elements, such as a girder. These aggregate condition values are the main descriptors of bridges in the NBIS. The fact that they are subjective and will vary from inspector to inspector has led to a call for much more objective assessment approaches. To increase the uniformity of evaluation, a rule-based system, combined with multimedia descriptions of the meanings of condition values, could be provided to the inspectors to help them in assigning these condition values more uniformly.

In addition to providing additional knowledge and advice to the inspectors, AI techniques could also be extremely useful in making the system much more usable while in the field. For example, inspectors would likely benefit from some form of Smart Guide (or Cursor) in the user interface, which would help put the user in the right part of the inspection report based on where they've been and their past experience with this type of bridge (similar to the smart cursor concept on the VuMan3). A domain-specific heuristic, developed through studies of how users usually navigate through the bridge inspection documents, could be employed. The knowledge about possible high-level navigation patterns could then be encoded in a navigation support tool that would automatically move the user to this next position in the form suggested by past experience embodied in this heuristic. The system could then also use a heuristic to decide when to apply the navigation assistant and when to let the user do the selecting without intervention.

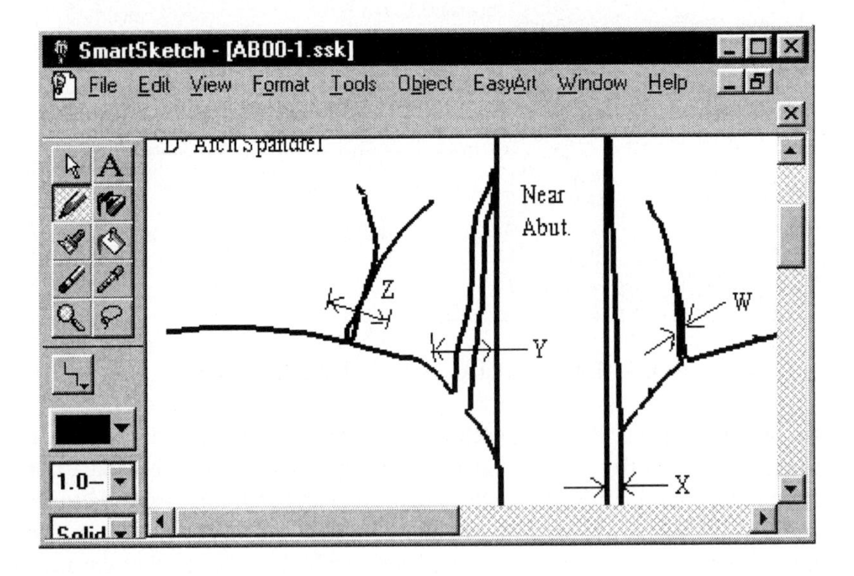

Fig. 4. Sketch Tool Interface

When filling out the specific parts of the reports, many of the inspectors comments are describing bridge elements and the type, location, and extent of damage. Hence, one could build a grammar of the various comments inspectors usually make and use this grammar, in concert with statistics about the relative frequency of usage in the current inspection context, to anticipate the comments the inspector is trying to create and finish their comments for them. The system could also track its accuracy and stop completing comments if the inspector has to continually edit the completed comment. This would be taking the smart cursor idea on the VuMan3 a bit further, but with the same intention, i.e., to save time. One point that the inspectors continually made when they were interviewed was that they wanted to spend more time inspecting bridges and less time in filling out paperwork.

Fig. 5. Photo Editing/Viewing Tool Interface

As with filling out comments, the inspectors could use some assistance in generating sketches while in the field. Currently, they use sketch templates or draw them from scratch. While in the field, sketching is not usually easily afforded. Hence, as with comments, it would be quite useful to have a grammar defined for the types of sketch components and configurations an inspector uses during a specific inspection context. With this grammar, the system could observe what the inspector is doing in the sketch and attempt to offer alternative completions of the sketch from which the inspector may select. This anticipatory sketch support could be made to keep track of its performance and learn when to make suggestions and when to remain silent and let the inspector finish his drawing. A similar approach was developed by Yang, et al., using a neural network approach, for determining when CAD users were drawing by hand a pre-drawn object that existed in the library [22]. The users were shown a set of possible pre-drawn objects that might match what they were attempting to draw and could select one of them if in fact that was what they were trying to do. A similar approach would likely be found quite useful by bridge inspectors, most of whom profess to not sketching well in the field.

7. Summary

To collect bridge condition data, the bridge inspectors must go to the bridge to assess its condition. While on the bridge, the inspectors need to have their hands free to support their climbing and inspecting activities and cannot carry reference manuals and other information with them onto the bridge. Researchers at Carnegie Mellon have developed several generations of wearable computers specifically for inspection-oriented applications. The wearable computer provides users computing support in the field as they conduct other operations that require full use of their hands. Recent applications of this technology to military inspection processes has led to a significant reduction in overall time spent on the inspection and documentation process.

The bridges in the National Bridge Inventory must be inspected at least every two years. If wearable computers can be developed and used to support bridge inspectors, significant decreases in the cost of bridge inspections are possible. The quality of the inspections is also expected to improve due to the greater access by inspectors to information and knowledgeable advice while they are in the field. A prototype wearable computer for supporting bridge inspectors is being developed. The next phase of this project will involve field testing the developed prototype with Pennsylvania state bridge inspectors. Only then will the true usefulness and impact of such a wearable computing device be determined. If the inspectors won't use it, it will not have much of an impact (as evidenced by current approaches to supporting bridge inspectors). By taking a user-centered approach to this problem, the resulting wearable computer should be found usable and useful by bridge inspectors. By starting out with a simple, yet useful and usable system, it is our hope that we can get the inspectors to gain respect for the usefulness of these wearable computers. Once this respect is earned, it will be possible to then provide additional, more intelligent support to these inspectors, such as providing knowledgeable advice about where to inspect, assistance is assessing the element conditions more uniformly, or assistance in actually using the device much more efficiently.

8. Acknowledgements

The authors would like to thank Prof. Daniel Siewiorek, the primary instructor in the Wearable Computer Project course, Mary Courtney, Brian Gollum, and the 20 students in the Spring 1998 course: Bikram Baidya, Anitha Balasubramanian, Neeraj Bansal, Lyren Brown, Rebecca Buchheit, Justin Hildebrandt, Barry Huie, Jason McDowall, Gautam Kharkar, Pace Lin, Chris Lumb, Alex Lozupone, Rob Migliore, Amy Roch, Julie Rodriguez, Eric Seshens, Robert Slater, and Jirapon Sunkpho. The authors would also like to thank Mr. Lou Ruzzi, Mr. Marty Neaman and Mr. Wally Stadtfelt from the Pennsylvania Department of Transportation District 11 office, and Mr. Pat Kane and Mr. Ray Hartle from the Michael Baker Corporation for their input into the development of this prototype. Funding for this project was provided through the Pennsylvania Infrastructure Technology Alliance and the Institute for Complex Engineered Systems at Carnegie Mellon University.

9. References

1. National Bridge Inspection Standards. Section 23, CFR, Part 650.3, Wash., D.C., 1971.
2. Hartle, R. A., W. J. Amrhein, K. E. Wilson, D. R. Baughman, and J. J. Tkacs. Bridge Inspector's Training Manual 90," Report No. FHWA-PD-91-015, FHWA, 1991.
3. Maser, K. Condition Assessment of Transportation Infrastructure Using Ground Penetrating Radar. Journal of Infrastructure Systems, Vol. 2, No. 2, 1996.
4. Czepiel, E. Bridge Management Systems Literature Review and Search. Northwestern University BIRL Industrial Research Laboratory, Technical Report No. 11, March 1995.
5. Better Roads, November 1996.
6. Wilson, J. L., S.J. Wagaman, J.W. Fisher, F.A. Harvey, G. Sadavage and T.J. Jaworski. The Hypermedia Bridge Fatigue Investigator. Proc. 11th Annual Intl. Bridge Conference, Pittsburgh, PA, June 1994, Paper IBC-94-12
7. Turner, D. S. and J. A. Richardson. Bridge Management System Data Needs and Data Collection. Transportation Research, Circular No. 423, TRB, pp. 5-15, April 1994.
8. American Association of State Highway Transportation Officials. AASHTO CoRe Element Guide, Draft version expected to be published in 1997.
9. Hearn, G., D. M. Frangopol, T. Szanyi, and S. Marshall. Data and Data Interpretation in Bridge Management Systems," Proceedings of Structures Congress, Vol. 1, American Society of Civil Engineers, 1996, pp. 245-252.
10. Leung, A. Perfecting Bridge Inspecting. Civil Engineering Magazine, 1996, pp. 59-61.
11. Kuo, S. S., D. A. Clark, and R. Kerr. Complete Package for Computer-Automation Bridge Inspection Process. Transp. Research Record, No. 1442, 1994, pp. 115-127.
12. Smailagic,A. and D. P. Siewiorek. The Design and Implementation of the VuMan Wearable Computer. EDRC Technical Report, Carnegie Mellon, 1992.
13. Siewiorek, D.P., A.Smailagic, J.Lee, and A.Tabatabai. An Interdisciplinary Concurrent Design Methodology as Applied to The Navigator Wearable Computer System. Journal of Computer and Software Engineering, Vol. 2, No. 2, 1994.
14. Smailagic, A., Siewiorek, D.P. The CMU Mobile Computers: A New Generation of Computer Systems. Proceedings IEEE COMPCON 94 International Conference, 1994.
15. Smailagic, A., D.P. Siewiorek, D. Anderson, C. Kasabach, J.Stivoric. Benchmarking an Interdisciplinary Concurrent Design Methodology for Electronic/Mechanical Systems. Proc. ACM/IEEE Design Automation Conference, pp. 514-519, June 1995.
16. Starner,T. and S. Mann. Augmented Reality through Wearable Computing. Presence, Vsence, Vol.6, No.4, 1997.
17. Mann,S. Smart Clothing: Wearable Multimedia Computing and Personal Imaging. Proc. of ACM Multimedia, pp. 163-174, 1996.
18. Paradiso, J. and N. Gershenfeld. Musical Applications of Electric Field Sensing. Computer Music Journal, Vol.21, No.2, 1997.
19. Smailagic, A., and D. P. Siewiorek. Modalities of Interaction with CMU Wearable Computers. IEEE Personnal Communications, Vol.3, No.1, February 1996.
20. Smailagic, A., and D. P. Siewiorek. The CMU Mobile Computers and their Applications. Mobile Computing (Eds. H. Korth and T. Imielinski), Kluwer Academic Pub., 1996.

21. Smailagic, A., and D. P. Siewiorek. A Case Study in Embedded-System Design: The VuMan 2 Wearable Computer. IEEE Design and Test of Computers, Vol. 10, No. 3, 1993.
22. Yang, D., J. L. Webster, L. A. Rendell, D. S. Shaw and J. H. Garrett, Jr., " Symbol recognition in a CAD environment using a neural network." International Journal of Artificial Intelligence Tools (Architectures, Languages, Algorithms), Vol. 3, No. 2, pp. 157-185, June 1994.

Conceptual Designing as a Sequence of Situated Acts

John S Gero

Key Centre of Design Computing
Department of Architectural and Design Science
University of Sydney NSW 2006 Australia
john@arch.usyd.edu.au

Abstract. This paper introduces conceptual designing within an F-B-S framework. It then goes on to describe a number of models of designing before introducing the notions of situatedness and situated acts. The remainder of the paper describes the role of situatedness and situated acts in conceptual designing. It attempts to show that a number of otherwise difficult design phenomena are modelable using situatedness and situated acts. A demonstration example concludes the paper along with some of the research issues this view of designing brings with it.

1 Introduction

Designing is taken to be a mixture of activities and tasks but the vast majority of views of designing are that it is an activity, i.e. it involves distinguishable processes which occur over time. [We will use the word 'designing' as the verb and the word 'design' as the noun in order to distinguish between these two, rather than use the word 'design' for both and utilise the context to disambiguate the meanings.] This commencing idea about designing is not to imply that notions such as 'inspiration' play no role but rather that they are not the bulk of the design activity. It is common to distinguish classes of design activity and group them. One common grouping is into conceptual and detail designing.

An important characteristic of conceptual designing that is missing in detail designing is that in conceptual designing not all that is needed to be known to complete a design is known at the outset, i.e. part of the process of designing involves finding/determining what is needed. It is common to distinguish non-routine from routine designing based on this conception. This is not to imply equivalence between conceptual and non-routine designing and between detail and routine designing but rather to suggest that they share common ideas.

The foundation of models of designing using concepts from artificial intelligence and systems theory is that any model is composed of variables and processes. One common description of an outline for designing is the Function–Behaviour–Strucure

(F–B–S) framework [1]. The variables required in a design process can be grouped into the three categories of function, behaviour and structure and the various design processes connect them and transform one into the other, Figure 1. Design processes available from this figure include:

formulation:	$F \rightarrow B_e$
synthesis:	$B_e \rightarrow S$ via B_s
analysis:	$S \rightarrow B_s$
evaluation:	$B_s \leftrightarrow B_e$
documentation:	$S \rightarrow D$
reformulation - 1:	$S \rightarrow S'$
reformulation - 2:	$S \rightarrow B_e'$
reformulation - 3:	$S \rightarrow F'$ via B_e

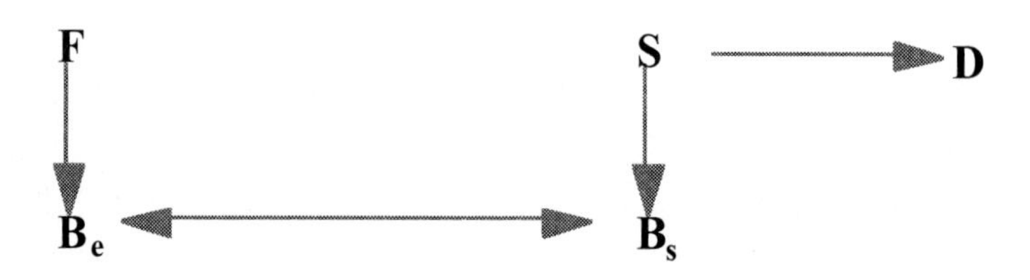

Fig. 1. The *Function–Behaviour–Structure* framework as a basis for models of designing, where B_e = expected behaviour; B_s = behaviour derived from structure; D = design description; F = function; –> = transformation; and <–> = comparison (after [1])

Another foundational concept in developing models of design is the concept of a state space. Here a state space is a representation of all the possible states that could exist if all the design processes legally operated on all the variables. It can be seen, here, as a representation of all possible solutions. These two sets of ideas: F–B–S framework and state spaces provide the opportunity to describe and develop a variety of models of designing. The remainder of this paper is concerned with very briefly describing a number of models of designing before moving to the notion of situatedness as an extension of existing design models, an extension which has the capacity to broaden our conception of the role of artificial intelligence in designing.

2 Models of Designing

In this section we introduce a number of well-known and lesser-known models of designing in order to be in a position to draw a distinction between them and designing as a sequence of situated acts.

2.1 Designing as Search

Search as a computational process underlies much of the use of artificial intelligence techniques when applied to designing [2], [3], [4]. The basic and often implicit assumption in designing as search is that the state space of possible designs is defined a priori and is bounded. The state space to be searched maps onto structure space in the F-B-S model and the criteria used to evaluate states map onto behaviours. The designing processes focus on means of traversing this state space to locate either an appropriate or the most appropriate solution (depending on how the problem is formulated). The advantages of modeling designing as search include the ability to search spaces described symbolically rather than only numerically. The assumption that the space is defined prior to searching relegates this model to detail or routine designing.

2.2 Designing as Planning

Planning here is taken from its conception in artificial intelligence as the determination of the sequence of actions required to achieve a goal state from starting state. It is a natural consequence of the existence of a well-structured search space. Planning has been used to model design [2], [5]. It also takes the same assumptions that designing as search does and therefore can only be considered as a model of routine designing.

2.3 Designing as Exploration

Designing as exploration takes the view that the state space of possible designs to be searched is not necessarily available at the outset of the design process. Here designing involves finding the behaviours, the possible structures and/or the means of achieving them, i.e. these are only poorly known at the outset of designing [6]. Exploration may be viewed in two ways. It may be viewed as a form of meta-search: the designer searches for state spaces amongst the set of possible predefined state spaces. It may viewed as a form of construction where each new state space bears some connection to the previously constructed state space(s). This form of exploration cannot be reduced to meta-search. Exploration connects with the ideas of conceptual or non-routine designing: not specifying or even being able to specify at the outset all that needs to be known to finish designing. Designing has been recognized as belonging to the class of problems called “wicked” problems [7].

2.4 Other Models of Designing

Other models of designing based on artificial intelligence or cognitive science concepts are generally either a specialization or a generalization of the models described above. Often they focus on some aspect of the model, often it is a procedural aspect. Of particular interest here are two concepts: “reflection in action” and “emergence”. The first of these refers to the notion that a designer does not simply design and move on but rather reflects on what he is doing and as a consequence has the capacity to reinterpret it. Schon [8] has called this a designer "carrying out a conversation with the materials". Implicit in these important ideas are the seeds for what will be described in Section 3. Emergence, which is a related concept to reflection, is "seeing" what was not intentionally put there [9], [10]. Reflection and emergence have evidentiary support from protocol studies of designers [11].

3 Situatedness and Constructive Memory

The lack of the models listed in Sections 2.1, 2.2 and 2.3 to adequately model our current view of designing has brought the need to develop models which include such concepts as reflection and emergence and processes which match those of exploration in Section 2.3. Work in cognitive science and related areas has developed two sets of ideas that have the capacity to augment, rather than displace, our current models to bring them closer to our needs. The two sets of ideas fall under the areas of "situatedness" and "constructive memory".

Situatedness [12] holds that "where you are when you do what you do matters". This is in contradistinction to many views of knowledge as being unrelated to either its locus or application. Much of artificial intelligence had been based on a static world whereas design has as its major concern the changing of the world within which it operates. Thus, situatedness is concerned with locating everything in a context so that the decisions that are taken are a function of both the situation and the way the situation in constructed or interpreted. The concept of situatedness can be traced back to the work of Bartlett [13] and Dewey [14] who laid the foundations but whose ideas were eclipsed for a time.

Constructive memory holds that memory is not a static imprint of a sensory experience that is available for later recall through appropriate indexing [15]. Rather the sensory experience is stored and the memory of it is constructed in response to any demand on that experience. In this manner it becomes possible to answer queries about an experience which could not have been conceived of when that experience occurred. "Sequences of acts are composed such that subsequent experiences categorize and hence give meaning to what was experienced before" John Dewey [14]. This view of memory fits well with the concept of situatedness. Thus, the memory of an experience may be a function of the situation in which the question, which provokes the construction of that memory, is asked. These two short introductions to situatedness and constructive memory suffice to allow us to now utilise these ideas in the development of our understanding of designing.

4 Situatedness, Constructive Memory and Designing

If we claim that conceptual or non-routine designing involves more than the searching, in however a well structured a fashion, of a state space of possible designs a question arises about what bases may there be to support the idea that there exist processes which do more than work within a single defined state space. We already have some processes that modify the state space of possible designs although not necessarily in a situated manner. These processes include analogy [16], case-based reasoning [17], and emergence [10] amongst others. What these processes lack is a unified framework within which they may be understood to operate. Further, they do not accord well with a view which is based on situatedness and constructive memory which we claim offers an opportunity to develop a model which accords with our current understanding of designing.

4.1 Situatedness and Designing

In conceptual designing the designer works with his experiences, his knowledge and his conception of what is in front of him – the situation – in order to determine what may be described more formally as, the variables which go to contribute to the function, behaviour and the structure of the resulting design. The particular behaviour and structure variables are not only chosen a priori but are produced in response to the various situations as they are encountered by the designer. What the designer has done previously, both prior to this design and during the current process of designing affects how the designer views the situation and what memories he constructs and brings to bear on the current situation.

Figure 2 shows graphically the notion of how a situation affects what can be "seen". In Figure 2(a) where two black human-like heads in profile are drawn, a white vase can be seen to emerge. Here, the two human-like heads provide the situation within which the emergence occurs. However, when only one black human-like head in profile is drawn no vase emerges. Here, the single human-like head in profile provides the situation. Clearly, in this example, the situation controls the emergence.

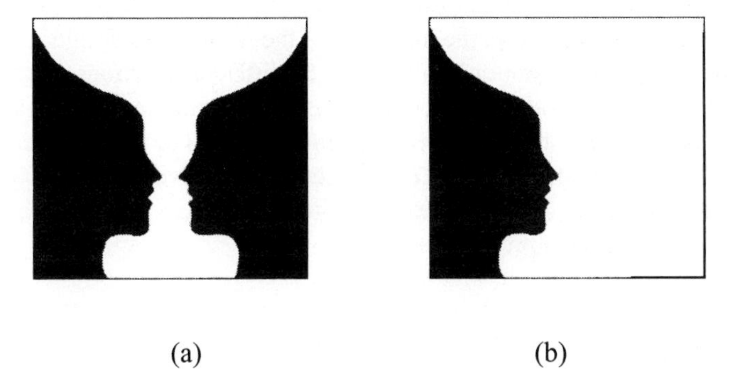

(a) (b)

Fig. 2. (a) Two black human-like heads in profile, reflections of each other create the Fig. 2.situation where a white vase can be seen to emerge; (b) a single black human-like head on the same background does not create the same situation and therefore no emergent vase can be found

The notion of situatedness is not necessarily tied to any particular representation (such as the graphical example in Figure 2). However, each representation has the potential to provide different situations and as a consequence different interpretations of what the situation is. Figure 3 provides an example of multiple representations of the floor plan of a building. Twelve alternate representations are shown but many others exist. Some of the representations favour certain interpretations over others. For example Figure 3(b) number 4 is readily situated to be interpreted as a figure plus ground which can be easily reversed along the lines of the example in Figure 2. The other representations do not lend themselves to this situation. Similarly representation Figure 3(b) number 6 can be situated to be interpreted as a grid. Grids can be moved and the entire direction of the resulting design may well be changed with this interpretation. Situations can provide a context within which a designer can interpret or reinterpret his developing design.

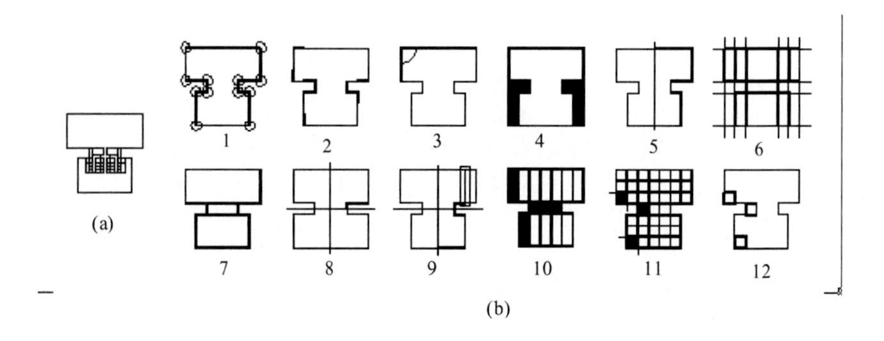

Fig. 3. Multiple representations of a building floor plan: (a) primary floor plan; (b) 1 to 12, multiple possible representations of the floor plan shown in (a) [18]

What is the role of situatedness in designing? Situatedness can be seen as a means by which the designer changes the trajectory of the developing design. Different situations provide different opportunities to move in different directions. Just as in Figure 3 what is the situation and what is being focussed on with the situation as background or context is not given but is a function the interpretation of the designer of what is in front of him. This is an important notion because it provides insight into why conceptual designing often leads in unexpected directions. Also, it may be an explanation of why designing is not a predictable act.

4.2 Constructive Memory and Designing

The interpretation of a situation is a function of what the designer knows and how the situation is represented. In the constructive memory view of the world each representation of a design, the most common one being an unstructured sketch or drawing, provides opportunities for the designer to reinterpret what is there and, therefore, to produce new ways of looking at what is there. This is akin to Schon's "conversation with the medium" [8]. Figure 4 shows, in a graphical form, the notion of constructive memory in design. A representation, which maps onto the notion of an experience, is interpreted using some structuring process (in some areas of design these are called "feature detectors"). As a consequence a new interpretation of what was there is produced, this maps onto memory. That new interpretation is added to the experience and is now available to be reinterpreted later as if it were part of the original experience.

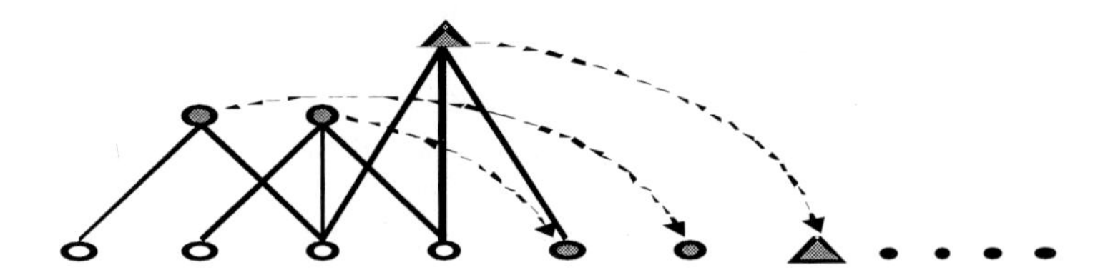

Fig. 4. The original design representation (experience), ○, are used to produce new interpretations of the design, ●, then the original and new interpretations are added as new representations and may be used later to produce further new interpretations, ▲, and so on

Consider, as an example, an initial shape that is being modified by a design system during the process of designing. The system has a representation of the shape that is the equivalent of the sensory experience. A memory of that shape can be constructed by acquiring some structure from that or an alternate representation of the shape. Let us represent the shape using a qualitative representation based on Q-codes [19]. Q-codes are a symbolic chain representational system where the symbols, with their values, represent qualitative aspects of the shape. Figure 5 shows some of the interpretations obtained by searching for structures in Q-code representations of various shapes. These structures map on to memory construction and can be added to the "experience" of those shapes. The labels are provided by humans.

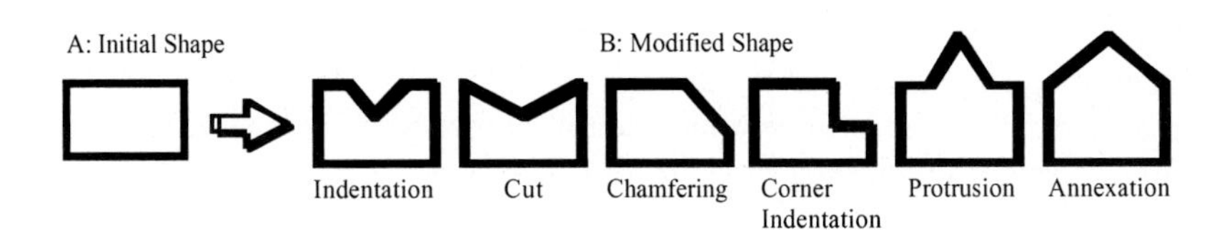

Fig. 5. Some interpretations, with their semantic labels, of modified shapes obtained by searching for structures in their Q-code representations [19]

The acquisition of structure is not the same as simply filing away responses. Structure derived from representation is "value adding" in the sense that knowledge which was not previously available has been produced and added to the system as part of the representation. This knowledge itself is situated. It carries with it aspects of the situation within which it was acquired. In this sense the process of situated learning in design is different to a simple application of machine learning.

4.2 Conceptual Designing as a Sequence of Situated Acts

The concurrence of situatedness and constructive memory provides the basis for the development of a model of conceptual designing that is closer to our current views. The model, still founded on the F–B–S framework, allows us to address the processes that were previously not well addressed: reformulation processes. Reformulation, see Section 1, is the process which in some way changes what the design is about. It has three loci: the range of possibilities of structures which can be produced is changed; the range of behaviours for which a structure is designed for is changed; or the functions for which a structure is designed for is changed.

Reformulation type 1 (S –> S') is the best understood of the three reformulation processes and is the most explored. Case-based reasoning in design and structure analogy are examples of such processes although neither is necessarily a situated process in the sense described in this paper [17], [16]. Here new structure variables are introduced into the current design from outside it. The effect of this is to change the state space of possible structures.

Reformulation type 2 (S –> B_e') occurs when new behaviour variables are introduced into the current design from outside it. The effect of this is to change the state space of possible behaviours. This may have the effect of changing the location of the selected structure within the structure state space or it may require the addition of further structure variables in order to produce a satisfactory design. Much less work has been in this area although there is currently research being undertaken which uses concepts from co-evolution which can be seen a way to approximate this process [20], and other work which uses analogy to locate and insert new behaviours.

Reformulation type 3 (S –> F') occurs when new functions are introduced into the current design from outside it. The effect of this is to change the state space of functions. This may have the effect of changing the expected behaviours, if it does then it may, but not necessarily, require changes in the structure state space.

All three types of reformulation are often likely to be situated – they all commence with an existing structure, S_e, as the driver. Access to S_e is only available after it has been produced. The process of reformulation is an act, each new structure (new in terms of new values for existing structure variables or new structure variables) potentially provides the opportunity for a different reformulation. In this sense conceptual designing can be treated as a sequence of situated acts.

5 Examples

Let us consider two examples. The first concerns the notion of re-interpretation through rerepresentation and presnts the results of an implementation. The second concerns an example of emergence in structural engineering.

Consider the example of a system being presented graphically with a single triangle. First it is rerepresented as a set of possible features which then make it possible to construct a variety of 'memories' of it depending on which of the features is used in the memory, Figure 6.

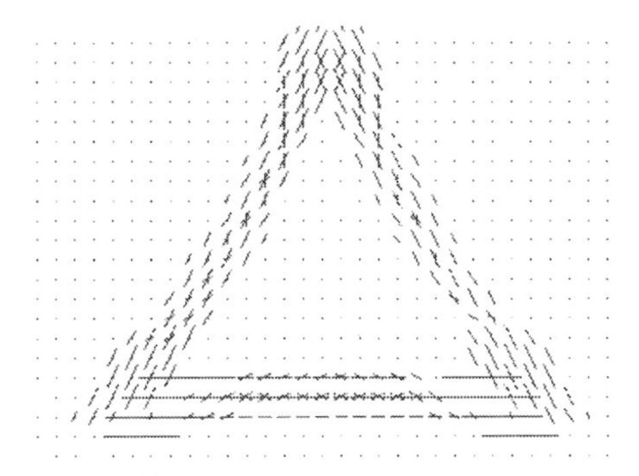

Fig. 6. The image of a single triangle being rerepresented as a set of possible features in the form of boundary contours.

That triangle and two others are located in space and, as a unique function of that situation, another triangle emerges using a model of the human vision system [21], Figure 7.

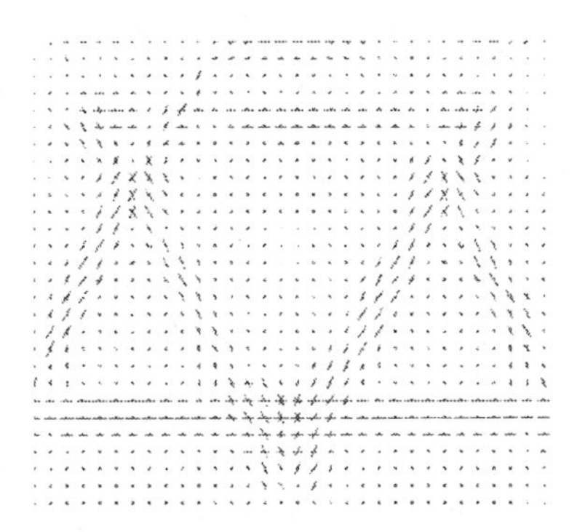

Fig. 7. An emergent triangle can be "seen" from the output of a situated vision system

The emergence of the triangle is a consequence of the results produced by the system but emergence is not built into the system. The system is a pre-attentive vision system, it is up to an "observer" outside the system to to "see" the newly emerged triangle, which can now be turned into a memory. The observer need not be a human.

Consider now a structural engineer designing the framing for a tall building. The engineer commences with a series of parallel two-dimensional frames, Figure 8. With these frames the engineer is carrying the wind load from the primary wind direction.

As a consequence of the way the engineer sees these frames he designs and analyses them as two-dimensional frames.

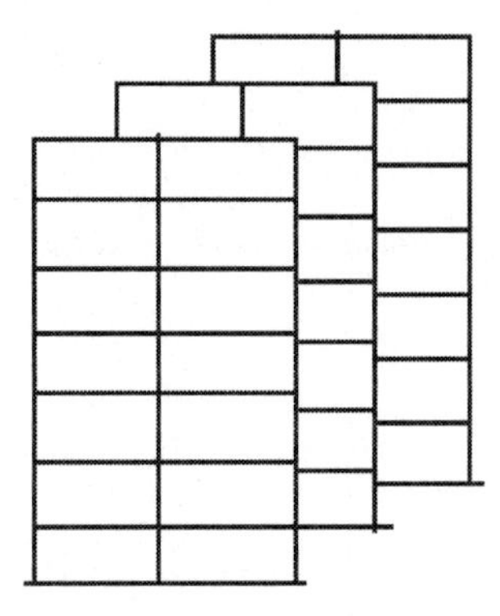

Fig. 8. The structural engineering component of a multistorey building being synthesised as a series of two-dimensional parallel frames

After the primary frames have been synthesised and the member properties determined, the engineer now attends to the lateral bracing by placing bracing beams at each floor connecting congruent joints of adjacent frames, Figure 9.

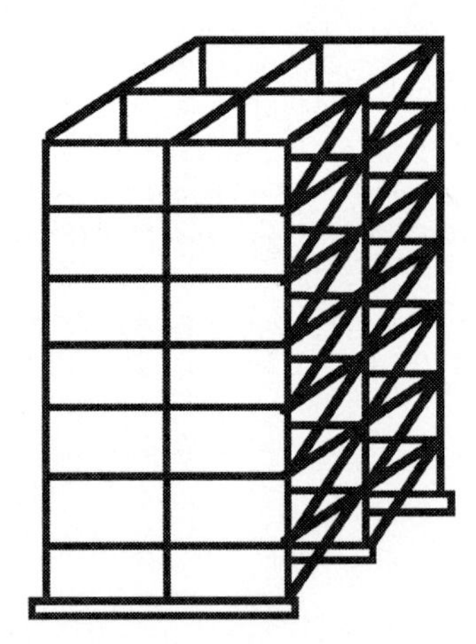

Fig. 9. Lateral bracing put in place by the engineer between the parallel two-dimensional frames

However, as the engineer inserts the bracing, he notices that the bracing produces a frame at right angles to the main frames and he decides to use the bracing as a frame. Further, having decided that there are now two sets of frames at rightangles to each other, he notices that the external frames can now be viewed as the facades of a tube building, Figure 10. As a consequence he examines the possibility of redesigning the entire lateral and vertical loadbearing system as a tube structure. This clearly has involved a re-representation of the wind bracing from bracing to lateral frames. Then, from the original frames and these lateral frames a tube structure emerges.

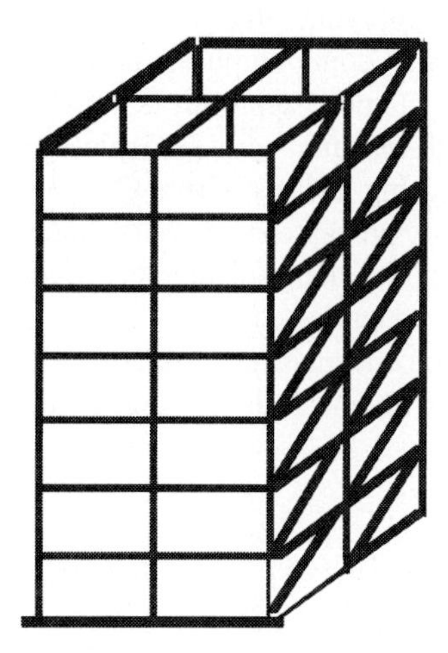

Fig. 10. The two sets of frames are now viewed as forming a tube structure

6 Discussion

Adding the notions of situatedness and constructive memory to the F–B–S framework provides the basis for a model of conceptual or non-routine designing. This model has the potential to meet our expectations of a model of conceptual designing. It is capable of dealing with those unique aspects of conceptual designing which involves working with incomplete information at the outset (the "wicked" problem syndrome) and providing the opportunity for radical changes in the trajectory of the development of the design as designing proceeds, Figure 11.

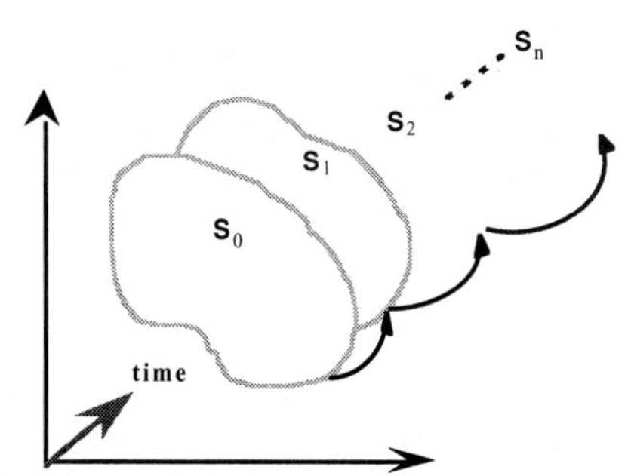

Fig. 11. The notion of conceptual designing as a sequence of situated acts modelled as a sequence of state spaces, which are interpreted as the situation; these state spaces change over time as the designer acts

What lies outside of this model is a set of unique processes capable of carrying out the three classes of reformulation within a situated and constructive memory approach. The methods of learning and applying situated knowledge are only now being developed, although there is currently some emphasis on developing approaches to handle emergence. The constructive memory approach is still in its infancy in the design research community, although we can readily see a place for it in expanding the role of case-based reasoning in designing and elsewhere [22]. What is needed is further research to develop the necessary processes to produce computationally feasible systems, which have the capacity to operate alongside human designers and to aid them in their designing in order to produce better designs.

Schon [23] summed up the concept of situatedness in designing succinctly as: "He shapes the situation ... his own methods and appreciations are also shaped by the situation".

Acknowledgment

This work has been supported by various grants from the Australian Research Council. The ideas in this paper have benefited from discussions with members of the Key Centre of Design Computing, particularly Vladimir Kazakov, Jarek Kulinski, Gourab Nath, Terry Purcell, Rabee Reffat, Rob Saunders, Tim Smithers, and Masaki Suwa.

References

1. Gero, J. S.: Design prototypes: a knowledge representation schema for design. AI Magazine. (1990) 11(4) 26–36.
2. Coyne, R. D., Rosenman, M. A., Radford, A. D., Balachandran, M. B. and Gero, J. S.: Knowledge-Based Design Systems. Addison-Wesley, Reading (1990)
3. Russell, S. and Norvig, P.: Artificial Intelligence. Prentice-Hall, Upper Saddle River, NJ (1995)
4. Dym, C.: Engineering Design. Cambridge University Press, Cambridge (1994)
5. Hauser, M. and Scherer, R.: Application of intelligent CAD paradigms to preliminary structural design. Artificial Intelligence in Engineering. (1997) 11(3) 217–229
6. Logan, B. and Smithers, T.: Creativity and design as exploration. In: J. S. Gero and M. L. Maher (eds): Modeling Creativity and Knowledge-Based Creative Design. Lawrence Erlbaum, Hillsdale, NJ (1993) 139–175
7. Rittel, H. and Webber, M.: Dilemma in a general theory of planning. Policy Sciences. 4 (1973) 155–160
8. Schon, D.: The Reflective Practictioner. Harper Collins, New York (1983)
9. Holland, J.: Emergence. Addison-Wesley, Reading Massachusetts (1998)
10. Gero, J. S.: Creativity, emergence and evolution in design: concepts and framework. Knowledge-Based Systems. 9(7) (1996) 435–448
11. Suwa, M., Purcell, T. and Gero, J. S.: Macroscopic analysis of design processes based on a scheme for coding designers' cognitive actions. Design Studies (1998) (to appear)
12. Clancey, W. J.: Situated Cognition. Cambridge University Press, Cambridge (1997)

13. Bartlett, F. C.: Remembering: A Study in Experimental and Social Psychology. Cambridge University Press, Cambridge (1932 reprinted in 1977)
14. Dewey, J.: The reflex arc concept in psychology. Psychological Review. 3 (1896 reprinted in 1981) 357–370
15. Rosenfield, I.: The Invention of Memory. Basic Books, New York (1988)
16. Qian, L. and Gero, J. S.: Function-behaviour-structure paths and their role in analogy-based design. AIEDAM. 10 (1996) 289-312
17. Maher, M. L., Balachandran, M.B. and Zhang, D. M.: Case-Based Reasoning in Design. Lawremce Erlbaum, Hillsdale, NJ (1995)
18. Gero, J. S. and Reffat, R.: Multiple representations for situated agent-based learning. In: B. Varma and X. Yao (eds). ICCIMA'97. Griffiths University, Gold Coast, Queensland, Australia (1997) 81-85
19. Gero, J. S. and Park, S-H.: Computable feature-based qualitative modeling of shape and space. In: R. Junge (ed.). CAAD Futures 1997. Kluwer, Dordrecht (1997) 821-830
20. Poon, J. and Maher, M. L.: Emergent behaviour in co-evolutionary design. In: J. S. Gero and F. Sudweeks (eds). Artificial Intelligence in Design '96. Kluwer Academic, Dordrecht (1996) 703-722
21. Carpenter, G.: The Adaptive Brain. Elsevier, Amsterdam (1987)
22. Carpenter, G. and Grossberg, S.: Pattern Recognition by Self-organizing Neural Networks. MIT Press, Cambridge, MA (1991)
23. Schon, D. Educating the Reflective Practitioner. Jossy-Bass, San Francisco (1987)

Some Personal Experience in Computer Aided Engineering Research

Kincho H. Law

Department of Civil and Environmental Engineering
Stanford University
Stanford, California 94305-4020, U.S.A.

Abstract. Symbolic computing and artificial intelligence (AI) have found many task specific applications in engineering. The recent emergence of network computing, taking advantage of communication and information management technologies, has broadened the scope of engineering software to support collaborative engineering. This workshop paper describes some of the recent and current research by the author and his colleaques in the development of preliminary analysis and conceptual design tools and in the areas of information management and software interoperability to support collaborative engineering and enterprise integration.

1 Introduction

Engineering software development has experienced significant paradigm shifts during the last couple decades. In the 1960s and 1970s, procedural programming techniques dominated the development of civil and structural engineering software. Potential applications of artificial intelligence (AI) technologies in structural engineering have been discussed since the early 1960s [2, 45]. Although there were a number of attempts, it is not until the early 1980s, with the availability of "easy to use" tools, that symbolic modeling and AI computing became popular in structural engineering research and practice. The mid-1990s brought forth the rapid advances in distributive computing, taking advantage of communication and information management technologies. The technological trend in distributive computing will no doubt have significant impact in the next generation engineering software platforms and engineering practice. This workshop paper is not intended to be a position paper about a particular theme in AI and information technology. The purpose of this workshop paper is to briefly describe some of author's research experiences and interests in developing AI tools in structural engineering and information management frameworks to support collaborative engineering and enterprise integration.

The use of AI in structural engineering has been widely reported in the literature [13]. This workshop paper first reviews some of the author's research in developing computer aided tools for structural analysis and design. Much of these research have been targeted on the development of task and domain specific tools and focused on preliminary analysis and conceptual design of structures. In this paper, a diagrammatic reasoning approach for qualitative structural analysis and a behavior-based methodology for conceptual design of moment resistant frames are presented. One common theme in these works is to combine first principle domain knowledge with appropriate heuristics and reasoning strategies to facilitate the solution process.

Collaborative decision making is part of the design process and information flow is fundamental to the success of collaboration [2]. While various AI-based tools, such as blackboard and agent architectures and other techniques, have been proposed to support collaborative engineering, relatively fewer studies have been devoted to the engineering information management and integration. It is beyond the scope of this paper to provide a detailed review of research in collaborative engineering since the literature on this subject is abundantly available elsewhere (see, for example, References [40, 43] of this proceedings). The purpose of this workshop paper is primarily intended to discuss some of the author's research related to data management and software interoperability to support collaborative engineering and enterprise integration.

2 Preliminary Analysis and Conceptual Design of Structures

In structural engineering, preliminary analysis and conceptual design are two example problems which are of interests among the applications of AI technologies. In this section, a diagrammatic approach to qualitative analysis of structures is described. Furthermore, a design support system for the conceptual design of moment resistant frames is presented. The solution strategies for these two problems are significantly different and they are selected to reflect how these problems are to be solved. It is important that, for a computer aided structural engineering tool, the solution process reveals the necessary information that would aid engineers to understand the behavior of a structure.

2.1 A Diagrammatic Approach to Qualitative Structural Analysis

"The qualitative solution to framed structures is a significantly important component of the overall understanding of structural behavior [6]." Traditional numerical structural analysis programs should be more appropriately considered as verification tools and are useful only when a structure has been "designed" (otherwise, without the information about member properties, many such programs would not run even if the structure is statically determinate!). On the other hand, it is often possible for human to qualitatively analyse a frame structure even though the exact dimension and member properties are not provided. There have been many investigations into extending the methodologies developed in qualitative physics research [33] to the problem of qualitative structural analysis [5, 16, 39, 44]. Most of these works focused on symbolic and/or mathematical modeling of structures, which are computationally expensive and sometimes intractable. Human engineers, however, are often able to qualitatively determine the deformation of a simple structure through visual diagrams.

A diagrammatic reasoning approach with rule bases containing knowledge about force, moment and deflection has been attempted by Iwasaki et.al. [30]. The goal of symbolic or diagrammatic reasoning programs is to make inferences by manipulating and inspecting the internal representations of the information. Symbolic reasoning programs make inferences through a purely descriptive representation of the knowledge of the domain and the problem itself. Diagrammatic reasoning programs, on the other hand, represent at least some of the information, especially geometric information, in a more depictive form that reflects the geometric and topological structure of what is represented. We define diagrammatic representation not only in terms of the distinction between the depictive and the descriptive but also in operational terms. An important difference between a symbolic representation and a diagrammatic one is that the information represented explicitly in a symbolic program is not necessarily what is explicit in a picture. Even though symbolic approach can be used to reason about diagrams, the reasoning process is often indirect. Diagrammatic approach employs directly the kinds of operations that humans perform visually with a picture.

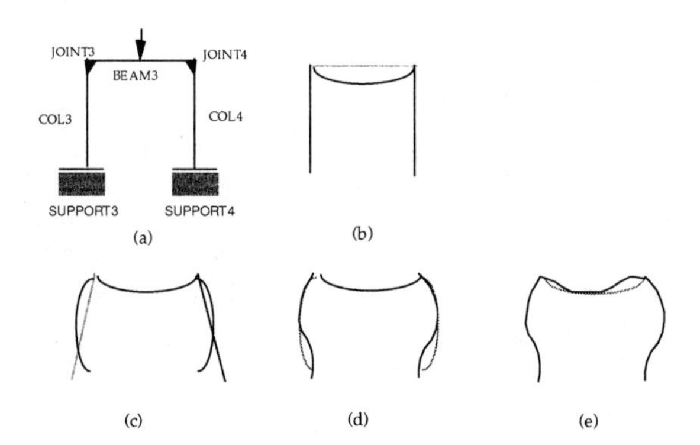

Fig. 1. A Diagrammatic Approach to Qualitative Structural Analysis

Given a load on a structure, an intuitive solution process is to modify the (deformed) shape of the structural member under the load (see Figure 1). We then inspect the modified shape to identify the places where constraints for equilibrium and geometric compatibility conditions of the structure are violated. Those constraint violations are corrected by modifying the shape of connected structural members, propagating deflection to other parts of the structure. This process is repeated until all the constraints are "qualitatively" satisfied. Diagrammatic reasoning approach follows the same solution process, the system uses its engineering knowledge to propagate constraints on the diagram of the structure and inspects and modifies the picture until a final shape is produced. Although the resulting picture is qualitatively consonant with the problem solution, it is not (nor does it intend to be) mathematically accurate or to scale.

The system architecture includes both symbolic reasoning and diagrammatic reasoning components. The symbolic reasoning component, the *Structure Layer*, contains the symbolic representation of the structural components and the qualitative structural engineering knowledge about various types of structural members, joints, supports, and the constraints they impose on the structure. It also includes a constraint-based inference mechanism to make use of the knowledge. The latter, the *Diagram Layer*, includes an internal representation of the frame structure as well as a set of operators to manipulate and inspect the shape. The diagram layer does not contain any structural engineering concepts; however, the types of manipulation and inspection operators provided for the layer do reflect the requirements of the domain. There is a translator between the structure and the diagram layers to mediate the communication between the two layers. When the structure layer posts a constraint or a command, the translator translates it into a call to an operator in the diagram layer that can directly act on the representation of the shape to manipulate or inspect it. The result is again translated back to concepts that the structure layer understands.

Figure 2 illustrates the type of communication that takes place between the structure layer, the diagram layer and the translator. Given a load placed on the middle of the beam of the frame structure as shown in Figure 1, the beam is bent in the same direction as the load, and translates into an operation showing the deflected shape of the beam. Continuing the interpretation process, the structure layer infers that since the joints are rigid the beam and columns must maintain the same angle; in this example, the members are perpendicular to each other before and after the application of the load. Querying the graphic entities reveal the actual angle between the two lines and the results are communicated to the structure layer that the constraint is not satisfied. A command is then issued to satisfy this constraint while keeping the beam fixed, which is translated into the angle between the beam and the columns are at 90 degrees. Communication between the structure and the diagram layers continues until all the constraints are satisfied.

As opposed to symbolic approach, diagrams are employed throughout the solution process to control the reasoning as well as to select proper heuristics to solve a problem or a sub-problem. The use of diagrams is computationally more efficient than other comparable symbolic systems, which often involve setting up applicable equilibrium equations and tries to solve them qualitatively. In addition, the solution process is much more instructive in helping the user to gain intuitive understanding of how the qualitative solution is being derived.

2.2 A Design Support System for the Conceptual Design of Moment Resistant Frames

Computer programs that perform routine structural design calculations are readily available and widely used by the design profession. However, there are few practical tools that assist engineers in the conceptual design of structures. In our

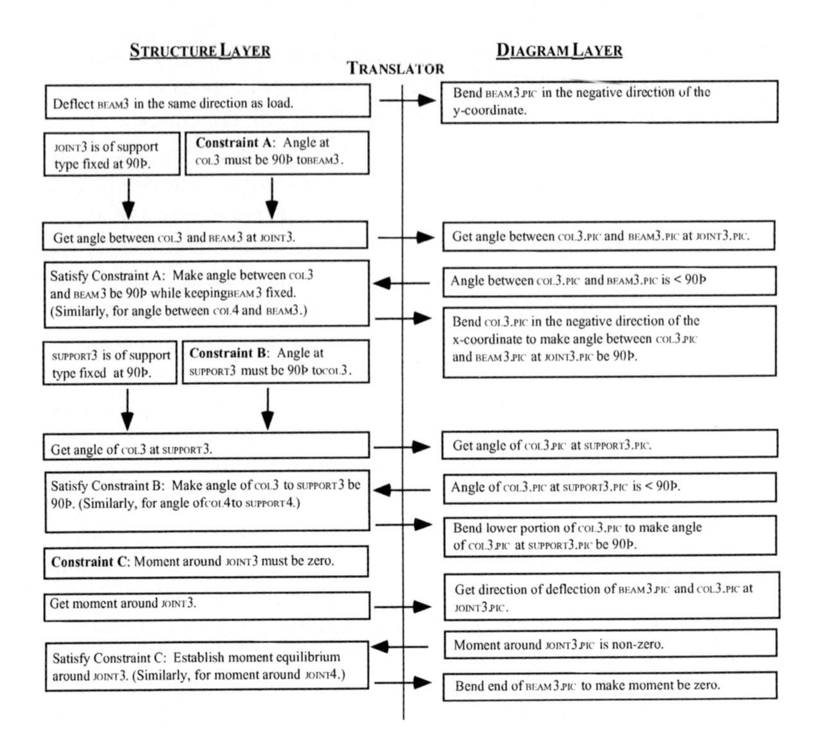

Fig. 2. A Diagrammatic Solution Process for Qualitative Structural Analysis

research on knowledge-based design support systems, the approach has been to explicitly incorporate first principle knowledge regarding the behavior of structures; heuristics or causalities are applied primarily to facilitate the solution process [19, 31].

In order to develop practical and flexible conceptual design systems that can have a wider range of applicability, it is imperative to consider explicitly the physical concepts imbedded in the function and behavior of structures [38]. The conceptual design of moment resisting steel frames involves the selection of structural members that would satisfy both strength and drift requirments given the building configuration and the gravity and lateral (earthquake) load conditions. Our research objective is to demonstrate how function (transfer of loads) and behavior (e.g., dominant modes of deformations) can be used effectively to develop a design support system for conceptual structural design [19].

For a building system that uses steel moment resisting frames (MRFs) to resist lateral (earthquake) loads, besides gravity and strength considerations, a MRF must be designed to limit lateral deflections. Both flexural and shear modes of deformation of the structural system need to be identified and controlled. The flexural drift condition is due to the bending behavior of the frame which causes axial deformations in columns. The shear drift condition is caused by horizontal shear forces in columns, which in turn produce bending moments and flexural deformation in beams and columns. In tall MRF buildings the flexural and shear modes of deformations may be of equal importance for the structural behavior under lateral loading. The complexity thus involves controlling these two modes of deformations in addition to satisfying strength requirements.

For the design support system, the basic approach adopted in the behavior based design methodology is decomposition and integration. First, the interstory drift in each story is separated into flexural and shear story drift components and the structure is decomposed into substructures. This decomposition allows us to reduce large complex systems to manageable substructures and to attain the desired performance more effectively. For flexural story drift control the required column sizes can be obtained by considering the MRFs as cantilevers subjected to lateral loading. For shear story drift control the required column and girder sizes can be obtained based on the member behavior in floor frame substructures. Finally the individual solutions at the decomposed level are integrated into the global design solution based on the following strategy:

1. Strength design for gravity and earthquake loads is performed first and a complete initial set of member sizes is determined.
2. Flexural drift limitation, specified as a fraction of total allowable story drift, is imposed to adjust the size of the columns. The larger of the member sizes from the strength design or flexural drift control are retained.
3. Shear drift limitation, which is the difference between the total allowable story drift and the computed flexural drift in each story, is considered to optimize the girders and columns per each joint and floor substructure.

The methodology has been described in details in Reference [18].

The development of a conceptual design support system would have been very difficult if it was not for the advances in software technology such as object oriented programming (OOP) methodology. One advantage of OOP for the development of a structural design system is the capability to build a symbolic model of a structure. Structural components can be defined as objects that contain their own information or a method to obtain their information. Each object also has information about other objects that it associates with and the methods to retrieve their information. Basic relationship types, such as "is-a", "part-of", "connectivity" and "reference", can easily be defined among the physical objects. In this design support system, an if-needed strategy is employed as the main mechanism for the iterative design process. One common operation is to inquire about a slot value stored in some object instance. If the slot value exists, the object instance returns the value; otherwise the object instance computes the value and returns it as if the slot had already contained the value. This feature can be implemented in an OOP environment such as CLOS with the auxiliary operators of "after" and "before" attached to a method.

The if-needed strategy has significantly facilitated the prototyping effort. In the if-needed strategy the programmer does not need to consider the existence of values in a slot since the strategy automatically finds its way to determine the slot value. That is, the programmer can focus on the programming of a specific module or an object class without paying attention to the other modules or object classes. New objects or new slots are defined as they become necessary. This top-down programming approach continues until the slot value of the desired object becomes known and available. In addition, the if-needed strategy can be used to implement the inherently iterative design process, where the sizes and internal forces of each member vary from one design level to another. In the if-needed strategy, for each iterative design level, the design process is activated by inquiring about member sizes and member weights. At the next design level, we simply delete the values of the slots so that the calculations for the new values of the slots are automatically activated.

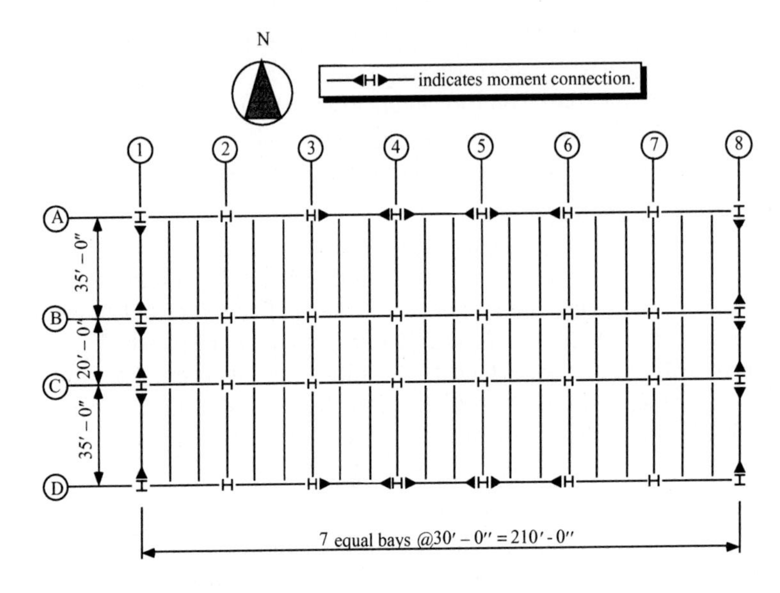

Fig. 3. Floor Plan for a 15-Story Building

Figure 3 shows a floor plan for a 15-story building. Using the design support system, the columns and girders selected for the 3-bay MRF on line A of the plan view are shown in Figure 4. The results generated from the conceptual design support system compare favorably with the design by an expert engineer [14]. Approximate analysis methods (moment distribution, cantilever and portal methods) are employed to predict member forces, member deflections, and interstory drifts. The use of these behavior based analysis methods provides approximate results pertinent to specific behavior phenomena and, thus, permits better control of the design process than could be achieved through the use of rigorous finite element methods. The explicit knowledge about the behavior of structures has also proven to be helpful in revealing the governing criteria of the design, for example, whether the members selected are controlled by shear or flexural drifts. Figure 5 shows the elevation, displacements, and interstory drifts of the perimeter MRF, where the thinner lines indicate that the member sizes are governed by shear drift control and the thicker lines indicate that the mem-

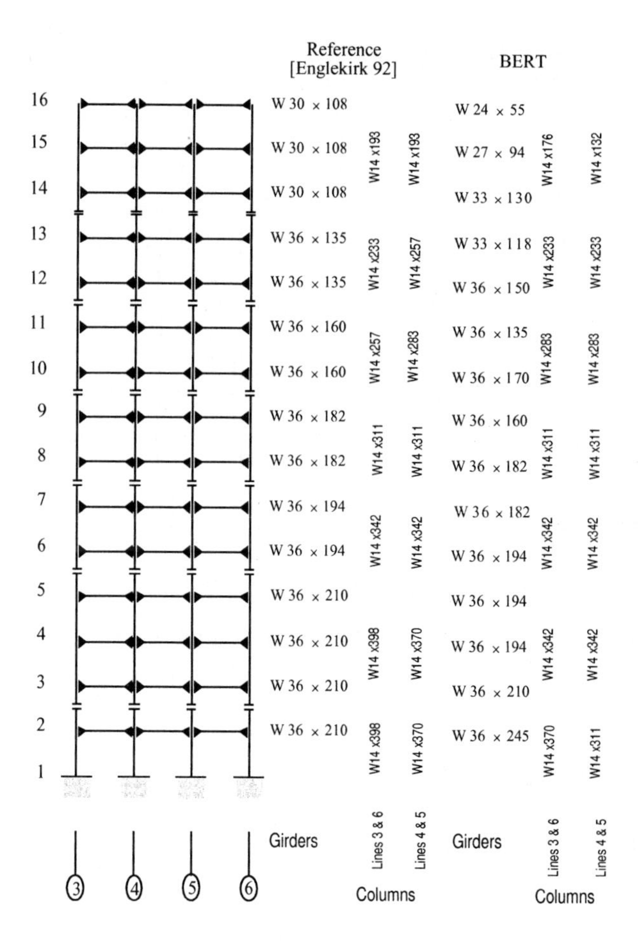

Fig. 4. Structural Members Selected by the Design Support System

ber sizes are governed by strength design. Also shown are the displacement and interstory drift due to the flexural and shear components. With this information, we can clearly observe the dominant factors in the design of the members.

3 Data Exchange and Data Management for Collaborative Engineering

A civil engineering project typically involves multiple disciplines and different organizations collaborating on the project. The enormous amount of design data generated during the design process by each participating party often complicate the coordination of such collaborative effort. Proper data exchange and data management to help ensure consistency and integrity of design information is an important issue in a computer-based collaborative engineering environment. This section describes our works on utilizing product data standards as the

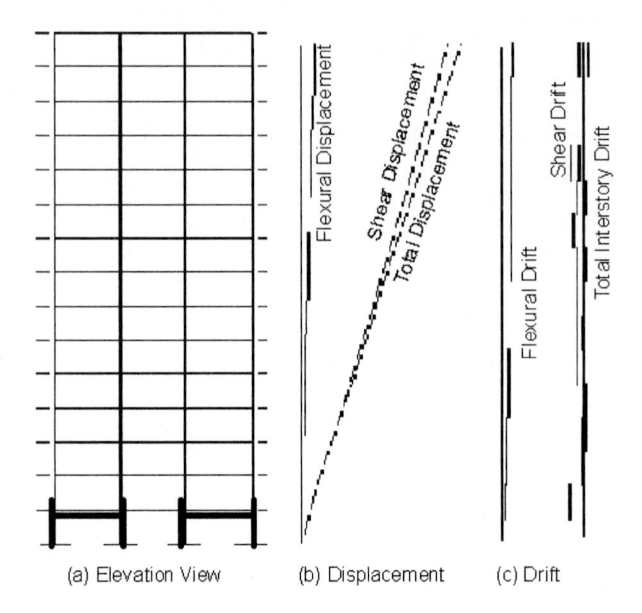

Fig. 5. Design Information for Flexural and Shear Drifts

basis to support network-based application programs and on a data management framework for design change control.

3.1 Data Exchange

The need for efficient data exchange between heterogeneous CAD databases and application programs is well known [20]. Traditionally, pre- and post-processors are developed to translate the data between two application programs using a mutually agreed exchange format. The major disadvantages of using pre- and post-processors between programs are the potentially large number of translators and the extensive software maintenance required. Using a common database does reduce data redundancy and the number of processors required. Database research for building design and engineering has been a subject of interests [8–10]. The use of semantic data modeling techniques to define design information of a facility project in a single repository has drawn many research efforts [4, 12, 35–37]. Using a common database allows easy integration of a new tool with other design tools in a "unified" but "close" system within an organization; this approach, however, does not provide the flexibility to support collaboration among multiple disciplines and organizations. Another approach is to exchange data using multiple communication formats via a common interface language [22].

Engineering companies are increasingly seeking ways to integrate their internal applications with those of potential partners and customers. One major problem confronting such integration stems from the challenges of exchanging and then possibly sharing data across the company boundaries. The International Organization for Standardization (ISO) has been actively pursuing the development of STEP, standards for computer-interpretable representation and exchange of product data [29]. The methodology of STEP has been described in Reference [11] and software tools are commercially available to integrate STEP product models with databases and other application programs [46]. Another notable effort in the development of building design data standards is by the International Alliance of Interoperability (IAI), which aims at developing a set of industry foundation classes (IFC) as a universal library of commonly defined objects throughout the lifecycle of a facility, from design to operation and maintenance [28].

As standard product models emerge, data exchanges have begun to be utilized in many engineering disciplines and have slowly migrated into civil engineering. Using standard product models can significantly facilitate communication between application programs and the development of distributive applications. We have been experimenting with the use of the IFC model to develop an on-line building code checking prototype implemented in a client/server framework (see Figure 6) [25]. In this prototype, the user (the client) can send a CAD model with the IFC definitions stored in an EXPRESS file across the network to a code checking program (residing on a server) which has direct mapping of attributes and relationships of IFC objects to its internal data structure. The client then waits for the server's notification when the code-compliance analysis is complete, and the results are posted to a web page, summarizing the results and providing a linkage to the code document governing the design. The existence of the standard IFC model has greatly facilitated the research and development of this network-based client/server application. We have also been experimenting with using a STEP model as a product repository in a process plant application [3].

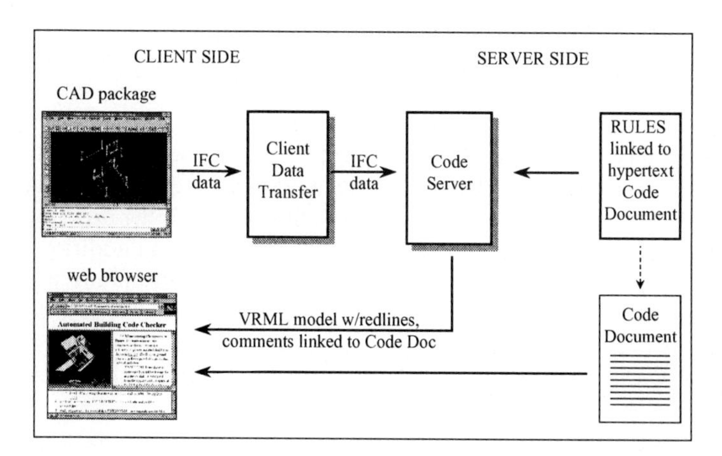

Fig. 6. Mechanics of a Prototype Client/Server Code Checking Program

While data exchange standards have enhanced information sharing among application software, data exchange protocol alone is not sufficient to support multidisciplinary collaborative effort throughout the product life cycle. For example, each STEP's application protocol (AP) represents a praticular product model for a specific engineering application. APs are not necessarily compatible with each other to support collaborative design or interoperability even in the same application domain (for examples, between a plant schematic diagram (AP Part 221) model and a plant spatial configuration (AP Part 227) model, or between a structural steel frame model (AP Part 230) and a finite element analysis (IAR Part 104) model). Current standards are focused on data exchange of a detailed design model but are not adequate nor designed to support conceptual product design process.

3.2 Data Management for Design Change Control

In practice, design changes are inevitable over the various phases of a civil engineering project, even during the time of construction and afterward. The scenario that when a designer who makes a change "immediately" informs every other designers in a team or interested parties may work for small projects but become impractical (and even "annoying") for any reasonable size projects. One major challenge is to maintain compatible design information and to help design coordination among the participants to accomodate design changes. While collaborative design and engineering has been a popular subject for over a decade, relatively few research studies have been dedicated to the issues related to management of design changes [41].

One problem that we attempted to address in our research was how to allow multiple design parties who work on a project to collaborate in a way that is consistent with industry practice [27]. Designers from different disciplines are often affiliated with different organizations. For the most part, designers from each participating discipline typically develop aspects of the project. Individual designs are then aggregrated to describe an overall project design. Besides being multidisciplinary in nature, design is an evolutionary and iterative process. Each designer typically generates, in parallel, several design alternatives, some of which are incrementally refined and/or modified until a satisfactory solution is obtained. The iterative design process requires maintaining descriptions of the entity at intermediate design stages of the design process. In addition, individual designers, despite their cooperative spirit, often desire autonomy and to retain control over the information they shared with others. In practice, designers independently evaluate several design options before sharing a more persistent description with the design team. Throughout the process, consistency and integrity of the shared data must be maintained.

In this study, we examined a three-layered closely couple data management framework of versions, assemblies (analogous to workspaces) and configurations [32, 34]. We focused on the support for the following interrelated elements:

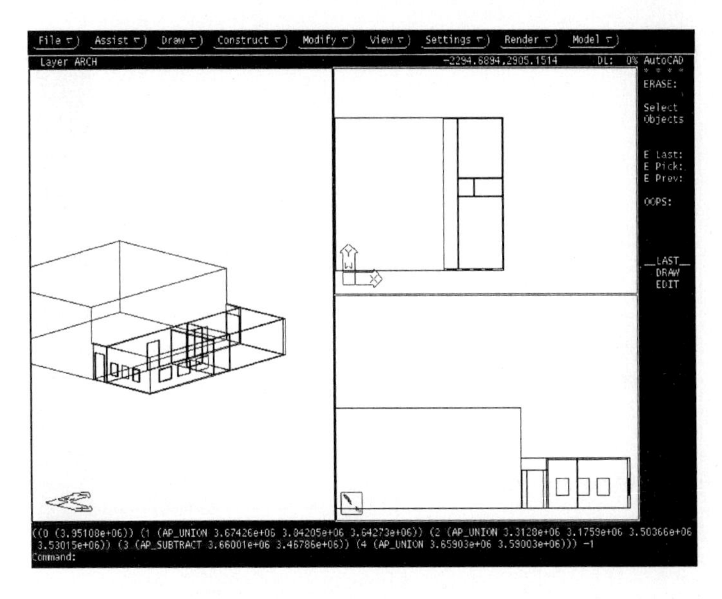

Fig. 7. An Initial Architectural Plan

1. For each discipline, a collection of *Versions* of the design entities in a database that belongs to the discipline. We referred the collection of entities as an *assembly* (or workspace).
2. *Configurations* consisting of one "assembly" from each discipline, a set of global constraints that applies to the entities contained in the assembly, and the violations of these constraints.

The framework handles multiple states of multidiscipline design components, including whether they are "frozen" (nevermore editable), and whether they are "published" (available outside their discipline).

Figure 7 shows an example of an initial architectural plan (an assembly) of a small office building. The configuration overlaying the designs from each of the particpating disciplines to describe an overall project is shown in Figure 8. In this framework, assemblies represent complex entities or designs in a discipline. Complex entities are in turn formed by aggregating instances of primitive entities in the CAD database. An evolving description of an entity is represented in a version hierarchy; each version in the hierarchy contains specific descriptions of the entities in that discipline. Operations are defined to efficiently manage changes among versions of a primitive entity. For example, when the floor plan is changed as shown in Figure 9, the changes of individual entities can be instantiated and computed as shown in Figure 10. That is, a version is a unit of granularity whose consistency can be evaluated. The close coupling of the version, assembly and configuration levels enables the computed changes in the entity version to be recursively combined to represent changes at various assembly (workspace) and configuration (project) levels. Individual design is modified

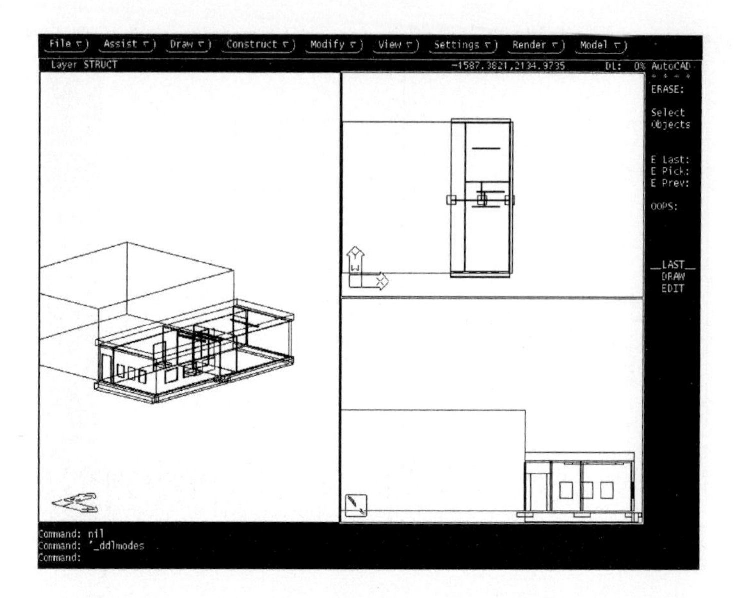

Fig. 8. Initial Configuration combining Architectural, Mechanical and Structural Plans

independently to accomodate the design changes. The final configuration for this design example is shown in Figure 11 as to ensure the designs from various disciplines are consistent.

This inherently distributed data management framework can efficiently support design project coordination through asynchronous communication of changes among designers, as well as project monitoring through systematic tracking of

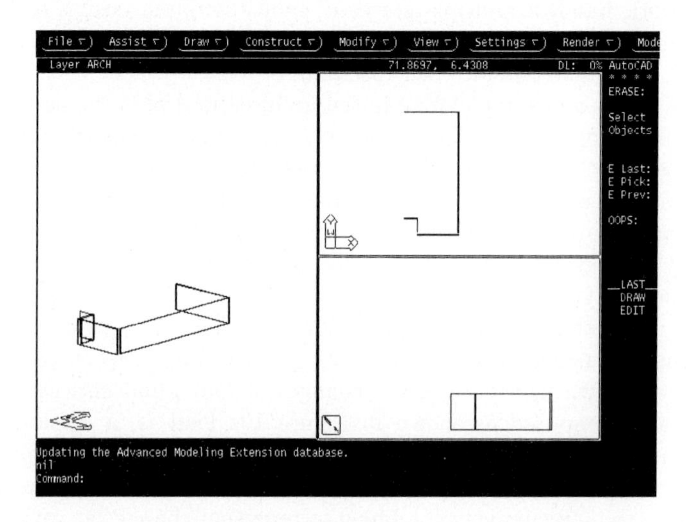

Fig. 9. Modification of the Architectural Floor Plan

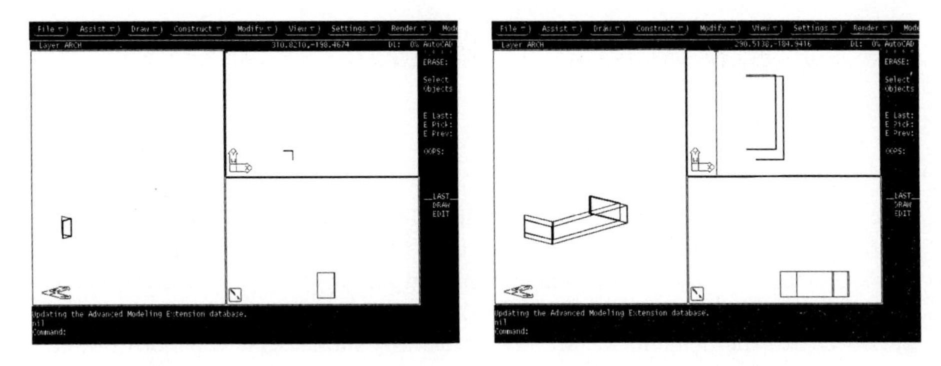

Fig. 10. Compute Changes of Entities Modified in the Floor Plan

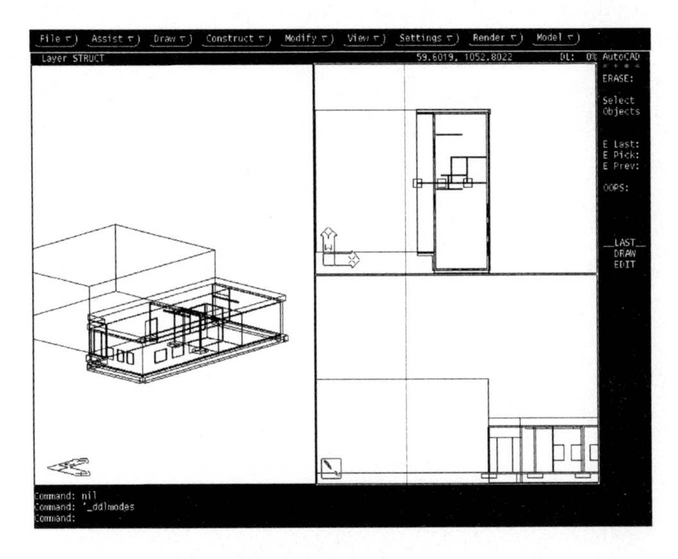

Fig. 11. Final Configuration of the Design

evolving project descriptions. Although the model does not explicitly transfer change notifications among designers, by publishing a design description a designer shares it with the rest of the team. Individual team members can then use appropriate operator to efficiently determine the net changes between two published designs. In addition, by characterizing the changes between two project designs as an aggregation of the differences among their components, the overall progress of a project can be monitored in terms of the relative progress of the project participants. Most importantly, the model affords designers the flexibility to work independently while collaborating with each other on a project. A configuration provides a framework for designers to share information by publishing their individual designs; the model gives designers control over the descriptions they published. This parallels current practice in professional design environments where designers typically insist on retaining control over the information they share.

4 Software Interoperability

One issue that becomes particularly acute when organizations engage in collaborative activities is interoperability betweeen the applications and infrastructure services deployed within an organization and among the collaborating parties. The technological heterogeneity and business characteristics of engineering companies create the demand for an interoperability model that will both preserve proprietary information and yet achieve acceptable software integration. That is, a flexible interoperability model must address more than data sharing but software interoperability, without the need for the participants involved to actually share a data warehouse and data storage protocols.

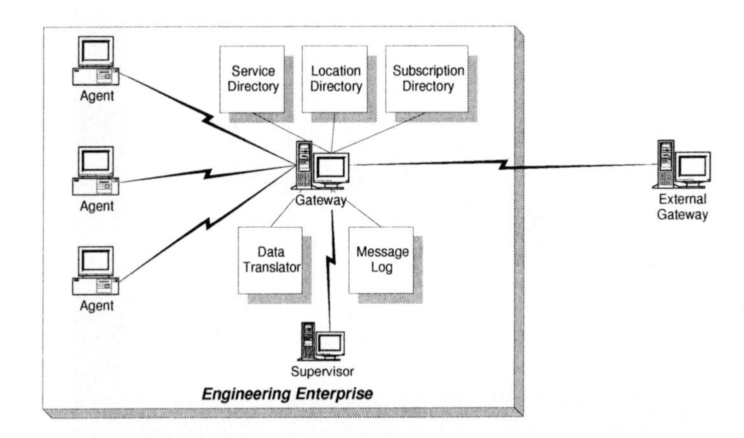

Fig. 12. An Architectural Diagram of a Software Interoperability Model

We are experimenting with a model for interoperation of computer programs among multiple organizations, where the users and applications have distinct behavior such as those commonly exist in engineering [23]. The assumption is that applications do not exchange behavioral models since these are often proprietary and generally do not have any relevance beyond the applications themselves. Instead, only "form and functional" data is exchanged, which is in keeping with traditional knowledge sharing in engineering industries. Figure 12 shows a gateway architecture designed to interconnect sophisticated applications with functional services and subscriptions. Agents (i.e., software applications) connect with other agents through gateways. Gateways are used to provide security, to manage intercommunication and to protect proprietary information about the agents when they export services outside the company to agents in other companies.

To illustrate, Figure 13 shows an example of the interoperability scenario that we would like to support using this gateway model: (1) An engineer at company 'X' using application 'A' requires a component and locates a supplier. (2) A request for the component is made. (3) The gateway acts as the proxy for the request and connects with the gateway at the vendor's site. (4) The second gateway translates the request into one compatible with the vendor's internal system. (5) The vendor's server locates the necessary component data and returns it to the gateway (6) The gateway then in turn maps the information into an open format and returns it to the first gateway. (7) The first gateway sends the incoming data to application 'A'. If application 'B' is sharing the component data with 'A', it is notified about the transaction. (8) The gateway also contacts its peer at company 'Y' since application 'C' may also share the component data with applications 'A' and 'B'.

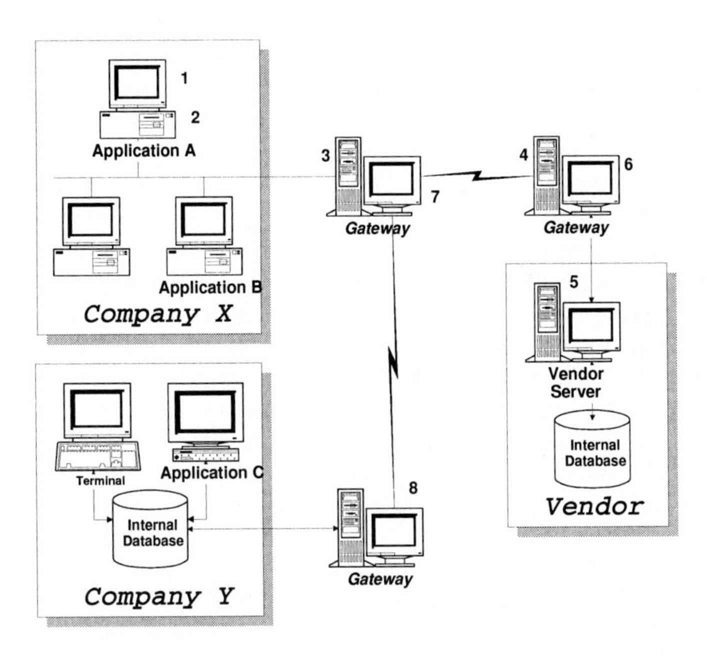

Fig. 13. An Example of a Software Interoperability Scenario

To date we have completed a prototype of the interoperability model and are currently preparing for testing, which may well lead to some architectural modifications. The objective is to design a secure system that can protect proprietary information as well as to provide a flexible system that can be configured to support customized communication, data exchange and interoperation of software applications on a project basis.

5 Summary and Discussion

In this paper, the author briefly summarizes some of his research interests in the application of AI and information technologies in structural and facility engineering. While most commercial analysis and design programs are primarily developed as a "black box" environment, the author believes that, at least for preliminary analysis and conceptual design, the tools should be designed to provide better understanding about the true behavior of a structure. Qualitative analysis may potentially provide the needed insight as well as reveal the intent of the load transfer mechanisms. Conceptual design tools should include first principle behavior knowledge that could help a structural designer to understand the governing behavior modes of the structure. With increasing attention in performance based structural engineering, much needed research remain in order to fully realize such computer assisted design support tools.

Developments in data exchange, distributed data management and software interoperability are part of the vision of a "practical" collaborative environment that can support a virtual enterprise (which is analogous to the notion of "Virtual Master Builder" [15, 42].) Facility engineering and construction projects often involve multiple organizations which are often assembled on a project basis. With the advances in internet/intranet, wireless and mobile computing, supporting technologies are now readily available to rapidly assemble and deploy teams from different organizations and to manage project activities via a virtual environment. The increasing growth in global business development and the reality of continuing company outsourcing and fast-track design build projects, virtual engineering enterprise may possibly emerge as a new model in facility engineering and construction projects. Other developments in collaborative design environment [17], Internet-based CAD [26] and web-based trading service environment [1] will further expedite this phenomenon.

One key practical issue in a virtual enterprise environment and collaborative engineering is security and control of proprietary information, knowledge and resources. Companies and IT vendors are actively investigating the technical requirements that must be met in order to support collaborative environment that can span multiple organizations [7]. Overly protected rigid environment, such as "firewalls", often make collaboration difficult [21]. A framework that not only provides the proper security mechanisms (including authentication and authorization) but also supports and encourages collaboration across organizational boundaries is needed for the success of a virtual engineering enterprise and collaborative engineering.

Acknowledgement

The author would like to express his sincere thanks to his colleaques and his current and former students who have contributed to the works described in this paper. Research supports for these studies have been partially sponsored by the Center for Integrated Facility Engineering (CIFE) at Stanford University. The opinions expressed in this paper, however, are solely those of the author and do not necessarily reflect the opinions of his colleaques, his students or CIFE.

References

1. J.A. Arnold, P. Teicholz, J. Kunz, "An Approach for the Interoperation of Web-distributed Applications with a Design Model," (sub. for publ.) *Aut. in Const.*.
2. T. Au, "Heuristic Games for Structural Design," *J. of Stru. Div.*, ASCE 92(ST6):499-509, 1966.
3. T. Binford, T. Chen, J. Kunz, K.H. Law, *Computer Interpretation of Process and Instrumentation Diagrams,* Tech. Rep. 112, CIFE, Stanford Univ., 1997.
4. B.C. Bjork, "Basic Structure of a Proposed Building Product Model," *CAD*, 21(2):71-78, 1989.
5. L.M. Bozzo, G.L. Fenves, "Reducing Ambiguity in Qualitative Reasoning," *Proc. Fifth Int. Conf. on Comp. in Civil and Bridge Engr.*, ASCE, pp. 1259-1266, 1993.
6. D. Brohn, *Understanding Structural Analysis*, Oxford Professional Books, 1984.
7. A Report on *InfoTEST International Enhanced Product Realization Testbed*, D.H. Brwon Associates, Inc., Port Chester, NY, 1997.
8. *The 1984 Workshop on Advanced Technology for Building Design and Engineering*, Building Research Board, NRC, 1985.
9. *The 1985 Workshop on Advanced Technology for Building Design and Engineering*, Building Research Board, NRC, 1986.
10. *The 1986 Workshop on Integrated Data Base Development for the Building Industry*, Building Research Board, NRC. 1987.
11. B. Danner, *Developing APs Using the Architecture and Methods of STEP*, Tech. Rep. NISTIR 96-1439, NIST, 1996.
12. A.M. Dubois, F. Parand, "COMBINE Integrated Data Model," *CIBSE Nat. Conf.*, pp. 96-108, 1993.
13. C.L. Dym, R.E. Levitt, *Knowledge-Based Systems in Engineering*, McGraw Hill, 1991.
14. R.E. Englekirk, *Steel Structures, Controlling Behavior Through Design*, Wiley, 1994.
15. S.J. Fenves, "The Virtual Master Builder: Is it feasible? Is it desirable? Is it inevitable?" Special Sem., Stanford Univ., Feb. 26, 1998.
16. R. Fruchter, Y. Iwasaki, K.H. Law, "An Approach for Qualitative Structural Analysis," *AIEDM*, 7(3):189-207, 1993.
17. R. Fruchter, "Conceptual, Collaborative Building Design Through Shared Graphics," *IEEE Expert*, pp. 33-41, June, 1996.
18. H. Fuyama, H. Krawinkler, K.H. Law, *Computer Assisted Conceptual Structural Design of Steel Buildings,* Rep. No. 107, J.A. Blume EQ Engr. Center, Stanford Univ., 1993.
19. H. Fuyama, H. Krawinkler, K.H. Law, "A Computer-Based Design Support System for Steel Frame Structures," *The Structural Design of Tall Buildings*, 3:183-200, 1994.
20. *Computer Aided Building Design*, Rep. No. LCD-78-300, GAO, 1978.
21. L. Gong, "Enclaves: Enabling Secure Collaboration over the Internet," *Proc. of The Sixth USENIX Security Symp.*, pp. 149-159, 1996.
22. J.H. Grubbs, L.M. Leach, K.H. Law, "Data Exchange for Collaborating Structural Design Programs," *J. Comp. in Civil Engr.*, ASCE, 2(2):144-159, 1988.

23. C.T. Howie, J.C. Kunz, K.H. Law, "A Model for Software Interoperation in Engineering Enterprise Integration," *Proc. ASME Design Engr. Tech. Conf.*, 1997.
24. C.S. Han, J. Kunz, K.H. Law, "Making Automated Building Code Checking a Reality," *Facility Management J.*, pp..22-28, Sept./Oct., 1997.
25. C.S. Han, J.C. Kunz, K.H. Law, "A Client/Server Framework for On-line Building Code Checking," (accepted for publ.) *J. Comp. in Civil Engr.*, ASCE.
26. C.S. Han, J.C. Kunz, J. Wang, K.H. Law, "Internet CAD: Leveraging Product Model, Distributed Object and World Wide Web Standards," (under prep.)
27. H.C. Howard, et.al., *Versions, Configurations and Constraints in CEDB*, Working Paper 31, CIFE, Stanford Univ., 1994.
28. IAI, *Industry Foundation Classes,* Release 1.0, Vol. 1-4, 1997.
29. ISO, *Industrial Automation Systems-Product Data Representation and Exchange,* ISO/DP 10303, 1989.
30. Y. Iwasaki, S. Tessler, K.H. Law, "Qualitative Structural Analysis through Mixed Diagrammatic and Symbolic Reasoning," in *Diagrammatic Reasoning: Computational and Cognitive Perspectives on Problem Solving with Diagrams*, AAAI Press, pp. 711-730, 1995.
31. D. Jain, H. Krawinkler, K.H. Law, G.P. Luth, "A Formal Approach to Automating Conceptual Structural Design – Part II: Application to Floor Framing Generation," *Engineering with Computers,* 7:91-107, 1991.
32. K. Krishnamurthy, K.H. Law, "A Data Management Model for Design Change Control," *Concurrent Engineering: Research and Applications*, 3(4):329-343, 1995.
33. Y. Iwasaki, "Qualitative Physics," in *The Handbook of Artificial Intelligence*, Addison-Wesley, Vol. 4, pp.323-414, 1989.
34. K. Krishanmurthy, K.H. Law, "A Data Management Model for Collaborative Design in a CAD Environment," *Engr. with Comp.*, 13(2):65-88, 1997.
35. K.H. Law, M.K. Jouaneh, "Data Modeling for Building Design," *Proc. Fourth Conf. on Comp. in Civil Engr.*, ASCE, pp. 21-36, 1986.
36. K.H. Law, D.L. Spooner, M.K. Jouaneh, "Abstraction Database Concepts for Engineering Modeling," *Engr. with Comp.*, 2(2):79-94, 1987.
37. K.H. Law, T. Barsalou, G. Wiederhold, "Management of Complex Structural Engineering Objects in a Relational Framework," *Engr. with Comp.*, 6:81-92, 1990.
38. G.P. Luth, H. Krawinkler, K.H. Law, *Representation and Reasoning for Integrated Structural Design,* Rep. No. 55, CIFE, Stanford Univ., 1991.
39. J.L. Martin, W.M.K. Roddis, "Integrating Qualitative, Quantitative Reasoning in Structural Engineering," *Proc. Fifth Int. Conf. on Comp. in Civil and Bridge Engr*, ASCE, pp. 1235-1242, 1993.
40. F. Pena-Mora, "A Collaborative Negotiation Methodology for Large Scale Civil Engineering and Architectural Projects," *this Proceedings.*
41. A. Mokhtar, C. Bedard, P. Fazio, "Information Model for Managing Design Changes in a Collaborative Environment," J. Comp. in Civil Engr., ASCE, 12(2):82-92, 1998.
42. V.E. Sanvido, S.J. Fenves, J.L. Wilson, "Aspects of Virtual Master Builder," *J. of Prof. Issues in Engr. Education and Practice*, ASCE, 118(3):261-278, 1992.
43. R.J. Scherer, "AI Methods in Concurrent Engineering," *this Proceedings.*
44. D.J. Schwartz, S.S. Chen, "Order of Magnitude Reasoning for Qualitative Matrix Structural Analysis," *ASCE Fifth Int. Conf. on Comp. in Civil and Bridge Engr.*, pp. 1267-1274, 1993.
45. W.R. Spillers, "Artificial Intelligence and Structural Design," *J. of Structural Division*, ASCE 92(ST6):491-497, 1966.
46. STEP Tools Inc., Rensselaer Technology Park, Troy, NY, 1997.

Knowledge Discovery from Multimedia Case Libraries

Mary Lou Maher and Simeon J. Simoff

Key Centre of Design Computing
Department of Architectural and Design Science
University of Sydney NSW 2006 Australia
mary,simeon@arch.usyd.edu.au

Abstract. Case-based reasoning and knowledge discovery are two independent fields in AI, which together can provide a design support environment for structural enigineers during the synthesis of new designs. Case-based reasoning relies on the representation of previous design cases for reminding designers of relevant past experience. Knowledge discovery is a way of finding patterns in data that can be considered new or generalised knowledge. By combining the two AI techniques, a case library can be the source of past episodic information as well as a source for discovering new patterns. We discuss the development of a multimedia library of sturctural design cases and the use of knowledge discovery techniques on multmimedia data to provide an environment for assisting in the development of new structural designs. We demonstrate the text analysis part of knowledge discovery from the SAM multimedia case library.

1 Knowledge-based systems - a shift from expertise to experience

Since the early seventies considerable research in artificial intelligence (AI) have been focused on the implementation of various knowledge-base computing models. The creators of expert systems (ES) succeeded in the development of deductive computer reasoning schemes. They brought the rule-oriented description of human *expertise* and rule-based inference techniques to a level of industrial standard. These systems require an explicit model of the subject domain. They achieve their high performance using extensive knowledge bases rather than mathematical models and numerical algorithms, thus dealing with various sorts of plausible reasoning. However, not only does the efficacy of the whole concept decrease with the increase of knowledge base size, but the inclusion of new domain data and validation of existing rules are difficult. Simultaneous incorporation of tedious general knowledge and expertís heuristics cause additional strain on the already complicated knowledge engineering process. The ìfamousî ES trio: PROSPECTOR, MYCIN, DENDRAL, operated in domains where there were good underlying models, based either on physical and chemical models or statistical inference. There has been limited success with the expert system approach in structural engineering design, possibly due to the lack of well accepted domain models of design synthesis. In the field of structural engineering, expert systems that have been developed have typically been done in Universities, sometimes in consultation with experts in practice, but rarely have they been used in practice.

The idea of using analogy, recalling previous *experience*, is a welcome alternative to the knowledge-base paradigm. Case-based reasoning (CBR) is a technique which implements this idea. CBR makes use of data that represents previous problem solving episodes when solving a new problem (1). The collection of data forms the *case base* or *case library*. Sometimes this data is already available in existing databases, thus reducing substantially the knowledge engineering effort. Considering computer implementations of the CBR paradigm, the representation of cases requires an abstraction of the experience into a form that can be manipulated by the *reasoner*, where the reasoner comprises procedural or heuristic modules for retrieving and selecting relevant cases and for adapting a selected case for a new problem. Sometimes the reasoner is assumed to be the user, rather than a computational process, where the computational support provides information about the cases but does not adapt or combine the cases.

The application of case-based reasoning to structural design (2, 3), has shown that in the early systems, attribute-value pairs and object-oriented representations were the dominating approaches to case representation. An alternative to the strict format of object-oriented and attribute-value representations is case models based on hypermedia representations. The hypermedia representations comprise weakly structured data such as text in free or table format and other multimedia data types such as images, video, sound, etc. Another characteristic of hypermedia is the use of links, where the links can connect information within a case, between different cases, or links to data that lies outside the case library.

Case-based reasoning, as a design process, is illustrated in Figure 1. A case library provides several examples of designs and the basis, in the form of an indexing scheme, for finding relevant designs to a new design problem. A new design problem provides some information that serves as the basis for recalling one or more design cases. A selected design case can then be adapted to be a new design. The resulting new design can be added to the case library, allowing the library to grow with use.

We have developed a multimedia case library of buildings that focus on structural design (4). The library is referred to as SAM, for its use in teaching Structures And Materials to undergraduate architecture students. In developing SAM, we consider the issues raised by the need to organise the material within a multimedia case library of structural designs, while presenting the material using multimedia. Specifically, we consider:

- the need to represent and manage *complex design* cases,
- the need to formalise a typically *informal body of knowledge* or experiences.

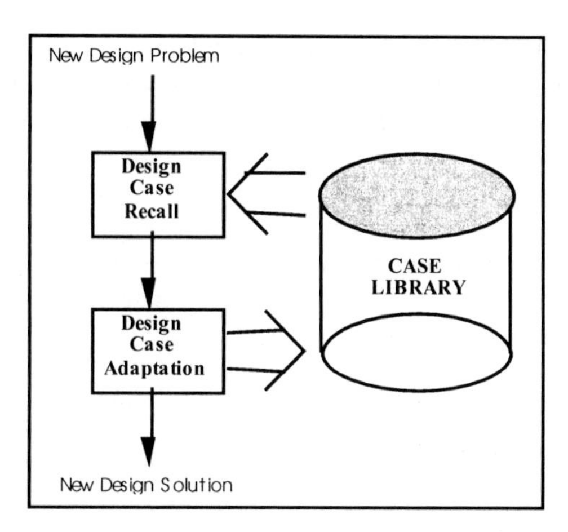

Fig. 1. Design process using case-based reasoning

Design in any domain usually involves the development and understanding of complex systems. The complex representations needed to adequately capture a design case have introduced challenges to CBR systems. The CBR paradigm assumes that there is a concept of "a case", but in most design domains this concept is not simply "a case" but a complex set of experiences and decisions resulting in a complex system. Three approaches to addressing complexity are:

- a case is a *hierarchy of concepts*, or subcases
- a case is represented by *different views*
- a case is presented as *multimedia*

A general approach to addressing domain complexity is the representation and re-use of parts of cases, typically organised as hierarchies of ìsubcases.î This supports case-based reasoning because subdividing designs in this way allows reasoning to focus only on the relevant parts of a design. By processing only some of the knowledge associated with a case, reasoning can become more efficient. The development of a case-base that has a hierarchical structure usually requires defining a typical decomposition of a design experience.

The use of different views of a design case recognises that a design can be understood from different perspectives. In this approach, a single, complex design project is represented as multiple cases. The use of multimedia can make it easier for the user to understand complex design cases - icons, images, sketches, etc. can highlight and illustrate corresponding text or tabular information.

The lack of formal knowledge in design affects both the ability to define a formal and consistent representation of design cases and the role of adaptation as a human-centred activity or an automated process. The development of CBR for design domains in which there is little formal or theoretical knowledge has been pursued by either formalising knowledge that previously was not formalised, eg by using an object-oriented representation, or by identifying a representation of the design cases to support human reasoning rather than automated reasoning, eg by creating a multimedia presentation.

Resolving the issue of ìwhat is in a design case?î is done in many different ways. Contrary to the initial observation that case acquisition should be straightforward, most design stories told by designers or found in design documents are not easily formed into cases that can be indexed and classified for reuse. A systematic approach is needed to identify a uniform representation and to parse design stories into these formats.

The representation of design cases in SAM follows the structural design principals that are taught in the Structures and Materials course. The overall organisation of design information falls into three categories:

1.project information,
2.functional decomposition of the structural design, and
3.structural system types.

These three categories are reflected in the navigation aids provided on each page as links to other parts of the case description. Figure 2 shows a typical presentation of the case description as the front page of the Grosvenor Place building. The front page is described further below. The functional decomposition of the structure is reflected in the links to the vertical load system, the lateral load system, and the footings. The structural systems types are shown a links along the third row, in Grosvenor Place the primary structural systems used are slab, columns, and core.

Fig. 2. Front page of Grosvenor Place

The project information is presented on the "front page" of each case, providing a tabular overview of the project. The information includes:

Project data - this describes the general information about the building including the people involved and the overall geometry.

Design requirements - this describes the context in which the building was designed considering the city planning issues, the architectís decisions about the overall shape and function of the building, the specifications for load and foundation conditions, etc. This information represents the considerations in the transition from the design brief to the structural design requirements.

Design alternatives and solution - this describes the different structural solutions considered by the engineers and the reasons for selecting the one implemented.

Justification - this describes the reasons for many of decisions that resulted in the final structural system.

References - acknowledges the sources of information.

The functional decomposition of the structural design is reflected in the set of "pages" that refer to the vertical, lateral, and footings systems. Each "page" further decomposes the information related to each functional system, for example, the vertical load systems can be further decomposed into gravitational and uplift load systems. Structural efficiency issues are dealt with at a number of levels - load transfer strategy, load paths, and structural actions, as illustrated for the vertical load resisting system in the Sydney International Aquatic Centre in Figure 3.

Strategy The building is rectangular in plan with the shorter dimension being 67.2 m. The strategy implemented by the primary structural system is to transfer the vertical loads in the direction of the shorter dimension, and hence over the shorter span for the building. The loads that reach the western edge are then collected and transferred vertically down to the foundation at five locations. The loads that reach the eastern edge are collected and transferred horizontally over a span of 138.5 m and then transferred down to the foundation at two locations.

The loads applied to the secondary structural system of the roof are transferred to the arch members of the diagrid as a series of concentrated loads approximately 600 mm apart, the spacing of the purlins which form the part of the secondary system that is in direct contact with the diagrid arches. The load paths and load flow pattern, through the structure, due to a typical concentrated load depends on the stiffness distribution that exists in the structural arrangement. The 1 m deep stiffening trussed arches with direct connection to the buttresses being the stiffer elements would tend to attract more of the load than v-shaped trusses spanning 25 m between the buttresses. A typical load path would thus be first along the arch, on which the load is applied, the load path would then branch out along the arches that intersect this arch and the loads would find their way to the stiffening trussed arch and the v-shaped truss, from there the loads will flow towards the buttresses on the western edge and towards the light supporting trusses on the eastern edge. The buttresses provide vertical downward load paths to the foundation. The light trusses, on the western edge provide upward load path to the transfer arch, from where the load path follows the arch to the foundation.

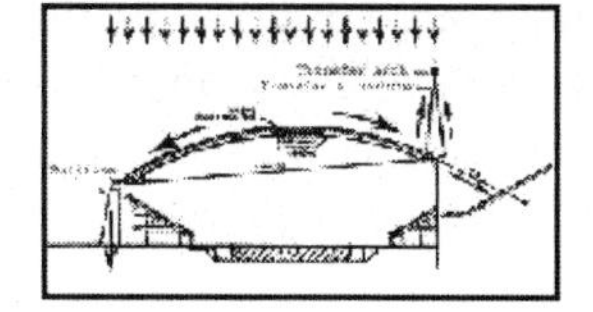

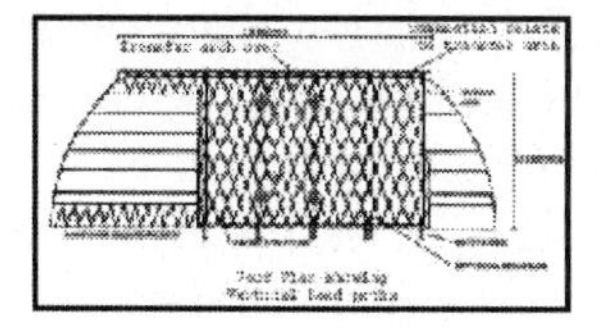

Fig. 3. Lateral load resisting system in the Sydney International Aquatic Centre

The structural systems are also described as separate ìpagesî. Each structural system is introduced generically as a type of structure, and then described in the context of the geometry and load conditions of the building case study. Each structural system is presented in terms of how it contributes to the functional decomposition. For example, the core system in Grosvenor Place is described as providing both lateral and vertical load resistance.

The indexing scheme for SAM is based on a declaring of the structural types and materials used in the design on a form in the case library. Adding a new case to SAM requires manually entering data about the case into a separate database used for indexing. This assumes that the interesting aspects of all structural design cases can be identified ahead of time. Alternatively, the designer can do a search for a particular string, or word, used in the case description. In SAM, adapting a design case is left to the designer. There is little support other than an editing environment for generating a new design description.

2. Knowledge engineering - a shift from interviewing to mining and discovery

Case-based knowledge engineering inherited the methods used in the development of expert systems. The indexing scheme is generated either through the knowledge acquisition technique of *interviewing* the expert to identify the critical features (Jackson, 1990), or through machine learning techniques to identify the most discriminating features by induction. The current approach to developing the adaptation knowledge is similar: the expert provides the rules and/or models of relevance. As a result, the shift in knowledge-base paradigm from expertise to experience identified new problematic issues related to the knowledge intensive indexing, retrieval, adaptation and maintenance of cases. However, the specialist knowledge is not easily captured, the process is time-consuming, painstaking and complicated, requiring carefully developed questioning strategies, observational procedures and analysis methods. A logical consequence is the increasing attention of the case-base reasoning community towards machine learning algorithms [5].

A promising way to override these difficulties is to employ data *mining* and knowledge *discovery* (KD) techniques, recently developed for identifying useful implicit information coded in databases [6]. Viewing knowledge engineering as a discovery process means examining a data source for implicit information that one is unaware of prior to the discovery and recording this information in explicit form. This spans the entire spectrum from discovering information of which one has no knowledge to where one merely confirms a well known fact. Current KD methods are developed mainly for structure-valued data [7, 8].

Discovering implicit knowledge in case bases is substantially different to data mining [9]. The data organisation units in database mining are the *data columns*. Inside the case base the organisational unit is the *case*, which comprises a variety of data types and formats. Knowledge discovery then, in our use of the term, involves finding patterns in primarily unstructured, multimedia data.

There are various discovery methodologies and discovery algorithms, based on procedures drawn from inductive logic, statistics, cluster and discriminate analysis, word combinatorics, and machine learning. We iteratively apply two approaches to knowledge discovery:

1. **data-driven exploration** where we do not specify what we are looking for before starting to examine the case data. For example, we parse the text in the case library looking for a vocabulary that can become the basis for the formal semantic models for indexing and adaptation.
2. **expectation-driven exploration** where we initially specify what we are looking for, i.e. we formulate an hypothesis, which is either refined during the exploration, partially or completely reformulated or finally rejected. For example, we specify an ontology of the design domain as a hypothesis for generating a classifier tree and allow the ontology to be modified based on the content of the case library.

3. A model of knowledge discovery from multimedia case libraries

We consider *knowledge discovery in case-based reasoning as machine learning where the training set is replaced by a case library*. This idea is illustrated in Figure 4. Most machine learning techniques operate on carefully constructed training sets, where both the content and the format are determined specifically for the machine learning technique. In knowledge discovery, the source of information for the discovery algorithms is typically developed for purposes other than machine learning; in our application, the case library was developed as appropriate content and representations for case-based reasoning.

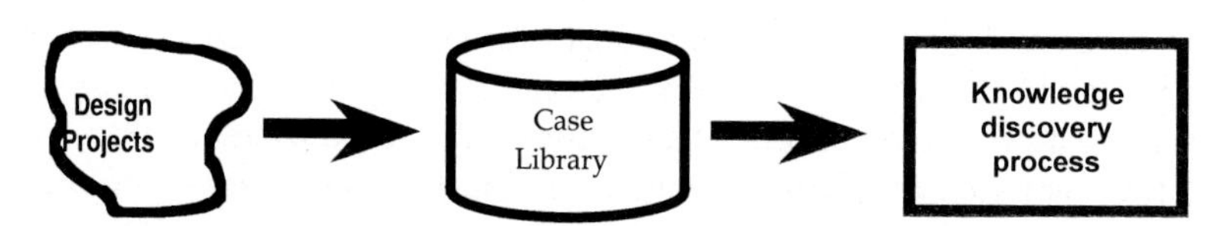

Fig. 4. Source of data for knowledge discovery in design case libraries

Our model for knowledge discovery in design case libraries, as illustrated in Figure 5, takes a hypermedia library of cases, uses a number of knowledge discovery techniques in two phases to generate domain knowldge. We are using SAM as an existing multimedia representation of structural design cases and apply knowledge discovery techniques that can find patterns in the cases. Specifically, we are interested in finding patterns in the cases that can assist with *indexing* and *adapting* cases as a way of supporting the designer in being reminded of a previous experience and being informed when an adaptation lies outside the experience base.

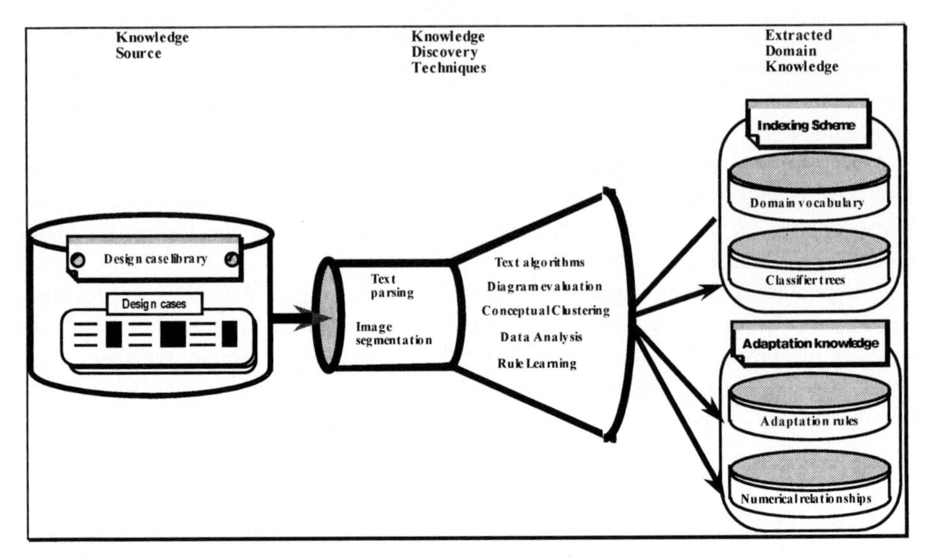

Fig. 5. Overall process for knowledge discovery in design case libraries

The two phases of knowledge discovery roughly correspond to the data-driven and expectation-driven approaches described earlier. In the first phase, the data-driven phase, a parser is applied to the case library to extract a set of relevant features. In the second phase, the features abstracted from the cases are used as input to various machine learning techniques to contribute to the indexing scheme and/or adaptation knowledge. Parsing cases to find relevant features produces a vocabulary and frequency of occurrence. Given the multimedia nature of the cases, we have identified two types of parsers: text parsers and image segmentation and content parsers.

Text parsing: Virtually all techniques for knowledge discovery in text data are based on the assumption that relevant concepts co-occur in the same text. This co-occurrence of word types in text is justified by estimating different text characteristics [10] and similarity statistics. Similarity coefficients are often obtained between pairs of distinct terms based on coincidences in term assignments to examined texts. When pairwise similarities are obtained between all term pairs the terms with sufficiently large pairwise similarities are grouped into common classes. Techniques for text parsing include: waisindex [11], the morphological parser of the Wordnet lexical database [12], some text algorithms [10] and combinatorial pattern matching methods.

Image segmentation and content parsing: These techniques include derivation and computation of image attributes to establish relations between the objects on the image. These relations can be used for evaluating diagrams of the images [13] and building a visual thesaurus, which consists of selected image primitives. The analysis takes account of colour histograms, texture, structural and relational semantic information, the distances between the primitives and the distances between the relations. Techniques for image parsing include: feature detection in raw and transformed image data [14], probabilistic and fuzzy ranking, and converting diagrams into meaningful symbolic descriptions [13]

Once a set of features has been abstracted from the case description, there are several relevant techniques for transforming the features into knowledge patterns. These techniques look more like machine learning techniques. The input, or training set, can be carefully constructed from the features found during the parsing phase. Although the content is based on the case library, the format can be determined by the needs of the particular learning technique. The idea of expectation-driven discovery means that we start with an hypothesis of the knowledge to be discovered, such as an indexing scheme, and the learning algorithm modifies the knowledge. We have identified classes of learning techniques that can be applied for knowledge discovery.

Conceptual clustering: These techniques are used in the machine learning community to synthesize generalised clusters from a training set of examples. In our project we will use the vocabulary from the parsers above as the training data for conceptual clustering. The result of the application of these techniques will be a classifier tree that can partition the case library according to meaningful terms. Techniques for conceptual clustering are: CLUSTER [15], or more recent approaches specifically for clustering design concepts [16].

Data analysis techniques: These techniques will operate on the numerical valued parameters to find approximated linear equations that reflect the data in the cases. Candidate techniques for learning about numerical relationship are: EFD: Empirical Formula Discovery [16], Interval Analysis [17] and Function Discovery [18].

Rule learning techniques: These techniques will be applied to the feature-valued pairs (numeric, boolean, symbolic). The idea is to compute the feature-differences that exist between cases and examine how these differences relate to differences in case solutions. Candidate techniques for rule learning are described in [19, 5].

4. Discovering an indexing scheme for SAM

Typically, the indexing scheme in multimedia case libraries is based on an underlying attribute-value representation. Usually this representation is a flat table, where all attributes participate in the indexing and retrieving procedures. A more sophisticated hierarchical schemata, where the attributes are grouped by some criteria, provides more flexible access, either only to a group of relevant attributes or to all attributes, but through a different path, again with respect to their relevance to the query. Usually, the structured attribute-value layer is crafted once forever, independently from the multimedia description. This approach does not utilise the advantages of the richness and vagueness of multimedia descriptions, restricting the expressive power of the queries to traditional matching of a finite set of attributes. A current alternative is the use of queries, expressed as a set of keywords against the free text included in the multimedia representation. The efficiency of this approach depends on the familiarity with the specific domain vocabulary. Rather than relying on it we prefer to build indexing schemata which allow preserving the advantages of the multimedia representation.

We have started with text analysis and focus on extending and modifying an indexing scheme.[1] A common approach for discovering terminological patterns from documents is to parse the document text, identify the content-carrying terms, cluster the inter-term dependencies, and build a conceptual hierarchy. We build on previous work in automatically learning design concepts using machine learning techniques [16] and follow the approach used by Dong and Agogino [20], where they automatically learn a belief network for mechanical engineering design from design documents. The process involves a data-driven approach in which a vocabulary is extracted from the design cases. The relevant vocabulary is based on terms that appear together in the same document, assuming that they connote similar meaning. The vocabulary can then used to construct classifier trees, numerical formulas, and rules using heuristic initial classifier trees - constituting an expectation-driven approach.

Practically speaking, the indexing scheme organises and partitions the knowledge in the case library. The indexing scheme is made up of a vocabulary of words relevant to the design domain, in this project the domain of structural design of buildings. Examples of the words in the vocabulary are: lateral load, load path, arches, beams, reinforced concrete, etc. Theoretically, the indexing scheme can be considered an *ontology*, or a way of categorising the knowledge in the case library. Originating in lexical semantics, the notion of ontology entered the field of artificial intelligence (AI) as a formal system that aims at representing domain concepts and their related linguistic realisations by means of basic elements. AI researchers brought a considerable variety in the interpretation of the term, spanning it from a simple hierarchy of keywords to domain theories and sophisticated frameworks for knowledge representation.

In a broad sense ontological knowledge includes various forms of encyclopedic knowledge about the domain, the common-sense, rhetorical and metaphorical knowledge and expressions[21]. In case-base applications ontological knowledge has the potential for increasing the efficiency of retrieval algorithms and the explanatory power of case-base front ends in natural language.

Constructing linguistic ontological systems is in some sense similar to the object-oriented analysis. Usually the sense of a word is defined at least by one set of semantically equivalent terms, a unique combination of synonyms [12], that distinguish it from other words in the language. The whole set of values of all features automatically composes the thesaurus, called the *basis* of the ontology. In general, ontology allows handling multiple meanings of a word by associating each word to more than one combination of synonyms. In addition to the basis, linguistic ontology includes variety of hierarchical and non-hierarchical semantic relations, which can be express as *basic predicates*. These relations are actively involved in query formulation. Such an approach towards ontological data mining allows us to include the experience of the developments in linguistic databases.

[1] We have not yet implemented the image parsing techniques.

Table 1. Example of text statistics for the Vertical Load Resisting System of Sydney International Aquatic Centre

Word	Frequency	Percentage
Load	19	7.9
Loads	15	6.2
Structural	12	5.0
System	11	4.5
Arch	10	4.1
Diagrid	8	3.3
Edge	8	3.3
Buttresses	7	2.9
Foundation	7	2.9
Transfer	7	2.9
Path	6	2.5
Primary	5	2.1
Shorter	5	2.1
Transferred	5	2.1
Arches	4	1.7
Gravitational	4	1.7
Light	4	1.7
Secondary	4	1.7
Strategy	4	1.7
Vertical	4	1.7
Western	4	1.7

There are several possible ways of organising the knowledge in a design case library. Given the lack of a formal model, it is appropriate that a collection of ontologies be used, each one providing a particular view of the knowledge. In this example we focus on the development of the basis of an ontology. The story usually starts with initial quantitative analysis of the text data. Usually the frequency of use of the words that are candidates for the ontology thesaurus is greater than certain threshold. There are no particular rules for the selection of this threshold. Table 1 shows all the words that occur more than 3 times in the vertical load resisting system page of Sydney International Aquatic Centre, part of which is shown in Figure 3. The union of the terms found across the case library provides the initial material for the basis of an ontology.

Following these basic statistics we use cluster analysis for establishing some relations between the terms. At this stage it is advisable to use non-hierarchical cluster analysis, which does not distinguish between the name of a category and its members and does not place each term into one and only one category, but can place any term into as many different categories as it presumably belongs. Figure 6 shows some of the established associations for the term "building".

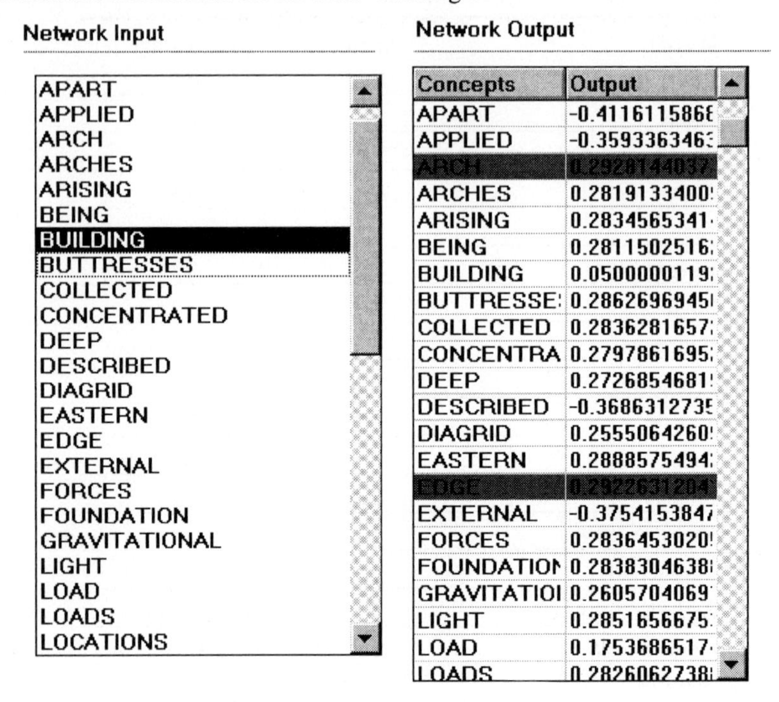

Fig. 6. An example of established association between terms

In this form the vocabulary itself and the relations are far from being applicable for indexing and retrieval. We need to build ontology in the form of computable structure. The conceptual graphs, introduced by Sowa [22] provide the backbone of this structure. Conceptual graphs allow us to incorporate, if necessary, existing lexical databases, under the condition that we restrict the labels of the relations in our graphs to be lexical items, in other words, relations, like "is-a" and "part-of", although popular in the information modeling community, are forbidden.

Thus, having the thesaurus and the relations we can define an ontology-based indexing schema as an oriented connected graph with the following labeling conventions:

- each node represents a concept, or a particular instance of a concept. For example, "building" is a concept, when "building - Sydney International Aquatic Centre" is an instance of this concept;
- each arc, representing a relation between concepts, is a concept itself. For example, an arc labeled "wide-span" can represent the relation between "building" and "arch".

Figure 7 and 8 illustrate the way we combine conceptual graphs with ontology representation. Figure 7 illustrates two ontologies: one that categorises according to the structural features that discriminate the types of structural systems (Figure 7a) and one that categorises according to the function and types of structural systems used in the buildings (Figure 7b). Note the branch "building-arch" in Figure 6, whose discovery is illustrated in Figure 7a. Thus, arcs, labeled with nouns denote the relational interpretation of such nouns.

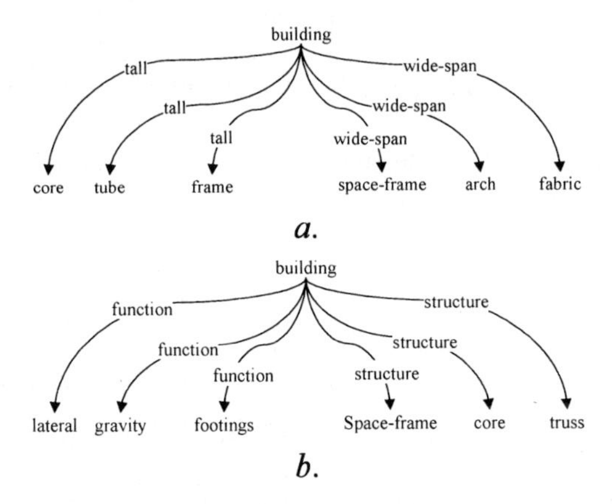

Fig. 7. Examples of simple ontologies for structural design cases.

Figure 8 shows another simple ontology. This ontology is a description of the term "tube", thus it can be attached to the corresponding node in Figure 7a. The "model" node in Figure 8 includes an instance of the concept "model".

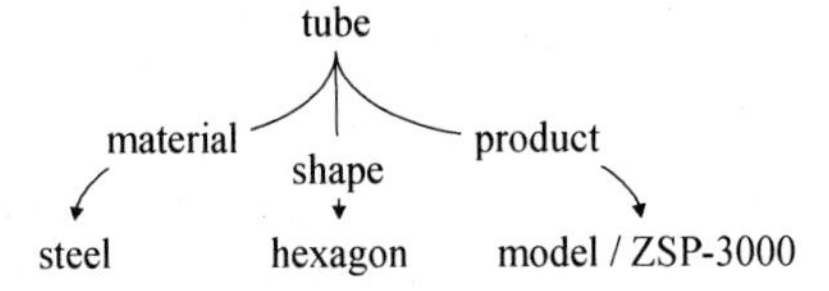

Fig. 8. Example of simple ontology, including particular instance of a concept.

Each ontology is almost a ready indexing scheme. What is missing is the link to the case. It is convenient to attach the case URL to the top of the graph, in our example in Figure 7, to the "building" node. The ontology-guided retrieval is based roughly on the same idea - the query is represented as a conceptual graph according to the taxonomic constraints of the ontology.

5. Improving the readability of SAM cases

Another outcome of the text analysis is the quality of text description of cases. Usually texts in case bases are composed by several authors with different backgrounds2. This fact puts forward the matter of readability of free text descriptions in the case base.

---

[2] For example, each case in SAM has been composed by different team of students.

Tuning readability is based on several linguistic indices, which combine two estimates:

- *word complexity*, determined either as a function of the number of syllables per word (e.g., in Flesch and Flesch-Kincaid estimators), or characters per word (e.g. in Coleman-Liau and Bormuth estimators);
- *sentence complexity*, estimated as a function of the number of words per sentences.

We illustrate this idea one of the cases in SAM case library. We apply two integrated estimators:

- Flesch Reading Ease score, which rates the text on a 100-point scale. The higher the score, the easier it is to understand the document. Typical text descriptions score usually within [65; 75].
- Flesch-Kincaid Grade Level score, which uses the U.S. grade-school level to rate texts. A score of 7.0 means that a seventh grader is able to understand such text. Typical text descriptions score usually within [7; 8].

The results, presented in Table 2, show that this case requires improvement of the text presentation. Buttresses description requires special attention.

Table 2. Two scores for the text description of Sydney International Aquatic Centre

Page	Flesch Reading Ease	Flesch-Kincaid Grade Level
Transfer Arch	62.7	5.8
Buttresses	37.1	9.5
Diagrid	60.7	6.2
Footings and Foundations	67.9	5.2
Case Front page	48.9	8.0
Lateral Load Resisting System	64.6	6.0
Vertical Load Resisting System	52.7	7.8
Average	56.4	6.9

6. The symbiosis between case-based reasoning and knowledge discovery

We have presented a case-based reasoning approach to supporting sturctural design and have investigated what can be discovered in symbolic case libraries. Discovery knowledge from case data is still in its infancy. Part of the success of case-based reasoning as an interactive knowledge-based computing model is credited to its less demanding knowledge engineering and instant learning loop embedded in the model.

Thus we propose further enhancement of the model by integrating knowledge discovery techniques in its learning loop.

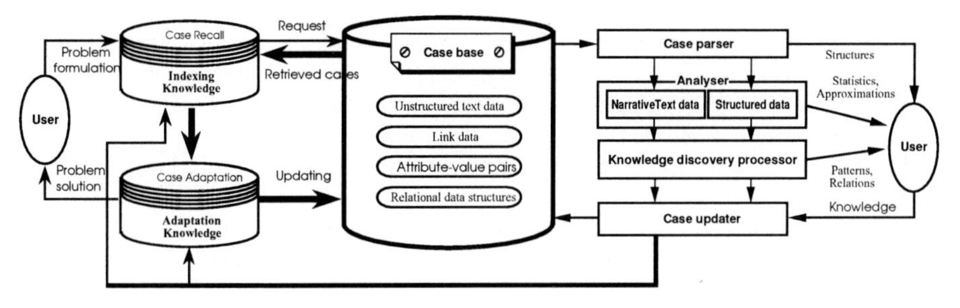

Fig. 9. The KD-enhanced case-based reasoning model

The overall enhanced model of case-based reasoning, as shown in Figure 9, illustrates the dynamics of case manipulation, analysis, mining, knowledge formulation and case update. In this model the user is involved both in the development of a new design case through problem formulation and development of a solution, as shown on the left side of the figure. In addition, the user is involved in assisting with the knowledge discovery process by monitoring the case updater. The implementation of this model requires further theoretical investigations, connected with the development of ontology for structural design and the application and appropriateness of existing learning techniques.

On the indexing side, the potential advantages of extended case-based reasoning model include:

- Possibility to generate indexing schemata which use natural language terms;
- Possibility to associate multiple indexing schemata to one case library.

On the retrieval side, the major advantage comes from the complete terminological flexibility in formulating queries due to the ontology-guided semantic transformation of the initial query and its match against the ontology-based indexing scheme.

Acknowledgment

This work has been supported by a grant from the Australian Research Council.

References

1. Aamodt, A. and Plaza, E. (1994). Case-based reasoning: Foundational issues, methodological variations and system approaches. *AI Communications*, **7**(1), pp. 39-59.
2. Maher, M.L., Balachandran, B., Zhang, D.M. (1995). *Case-Based Reasoning in Design*, Lawrence Erlbaum Associates, New Jersey.

3. Maher, M.L. and Pu, P. (1997). *Issues and Applications of Case-Based Reasoning in Design*, Lawrence Erlbaum Associates, New Jersey.
4. Maher, M.L. (1997). SAM: A multimedia case library of structural designs. In Y.T.Liu, J-Y. Tsou and J-H. Hou (eds), *CAADRIAi97*, Huís Publisher Inc., Taiwan, pp. 5-14.
5. Hanney, K. and Keane, M.T. (1996). Learning adaptation rules from a case-base. In I. Smith, B. Faltings (eds), *Advances in Case-Based Reasoning*, Springer, Heidelberg, pp.179-192.
6. Chen, M-S., Han, J. and Yu, P.S. (1996). Data mining: an overview from a database perspective. *Knowledge and Data Engineering*, **8** (6), 866-883.
7. Fayyad, U.M., Piatetsky-Shapiro, G. and Smyth, P. (1996b). From Data Mining to Knowledge Discovery in Databases, *AI Magazine*, **17** (3), 37-54.
8. Williams, G. and Huang, Z. (1996). A Case Study in Knowledge Acquisition for Insurance Risk Assessment using a KDD Methodology, Data Mining Portfolio - TR DM 96023, CSIRO.
9. Maher, M.L. and Simoff, S. (1997). Knowledge discovery in multimedia design case bases. In B. Verma and X. Yao (eds), Proceedings ICCIMA'97, Griffith University, Gold Coast, pp. 6-11.
10. Crochemore, M. and Rytter, W. (1994). *Text Algorithms*. Oxford University Press, New York.
11. Pfeifer, U., and Huynh, T. (1994). FreeWAIS-sf, ftp://ls6www.informatik.unidortmund.de/pub/wais/freeWAIS-sf.1.0tgz.
12. Miller, G. A., Beckwith, R., Fellbaum, C., Gross, D. and Miller, K. (1993). *Five Papers on Wordnet,* Cognitive Science Laboratory, Princeton University, CSL Report 43.
13. Gross, M. D. (1995). Indexing visual databases of design with diagrams. In A. Koutamanis, H. Timmermann and I. Vermeulen, *Visual Databases in Architecture*, Avebury, pp.1-14.
14. Russ, J. C. (1995). *The Image Processing Handbook*, CRC Press, Bota Raton, Florida.
15. Maher, M.L. and Li, H. (1994). Learning Design Concepts Using Machine Learning Techniques, *Artificial Intelligence for Engineering Design, Analysis, and Manufacturing*, **8**(2):95-112.
16. Voschinin, A.P., Dyvak, N.P. and Simoff, S.J. (1993). Interval methods: theory and application in design of experiments, data analysis and fitting. In E. K. Letzky, (ed.), *Design of Experiments and Data Analysis: New Trends and Results*, Antal, Moscow, 1993, pp.11-51.
17. Wu, Y.-H. and Wang, S. (1991) Discovering functional relationships from observational data. In G. Piatetsky-Shapiro and W. J. Frawley (eds), *Knowledge Discovery in Databases*, AAAI Press/ The MIT Press, Cambridge, Massachusetts, pp. 55-70.
18. Fayyad, U.M., Piatetsky-Shapiro, G., Smyth, P. and Uthurusamy, R. (eds) (1996a). *Advances in Knowledge Discovery and Data Mining*, AAAI Press.
19. Dong, A. and Agogino, A. (1996). Text analysis for constructing design representations, in J Gero and F Sudweeks (eds) *Artificial Intelligence in Design ë96*, Kluwer Academic, Holland, pp. 21-38.
20. Saint-Diszier, P. and Viegas, E. (1995). An introduction to lexical semantics from a linguistic and a psycholinguistic perspective. In P. Saint-Diszier and E. Viegas (eds) *Computational Lexical Semantics*, Cambridge University Press, pp.1-29.
21. Sowa, J. (1984). *Conceptual Structures: Information Processing in Mind and Machine*. Addison-Wesley, Ready, Massachussets.

22. Jackson, P. (1990). *Introduction to Expert Systems*, (2nd ed.), Addison Wesley, Reading, MA.
23. Michalski, R.S. and Stepp, R. (1983). Learning From Observation: Conceptual Clustering, in Michalski, R.S., Carbonell, J.G., and Mitchell, T.M. (eds) *Machine Learning: An Artificial Intelligence Approach*, Morgan Kaufmann.

Customisable Knowledge Bases for Conceptual Design

John Miles[1], Lynne Moore[2] and Ian Bradley[3]

Cardiff School of Engineering, Cardiff University, P.O. Box 686, Cardiff CF2 3TB, UK
[1] MilesJC@cf.ac.uk
[2] MooreCJ@cf.ac.uk
[3]sceib@cf.ac.uk

Abstract. Previous work by the authors has resulted in the development of a new style of knowledge representation which is specifically designed to meet the requirements of knowledge bases for practising designers. The functionality of this style of KB is being extended and enhanced and this work is described. Also an examination of the use of spreadsheets for formulating and maintaining design KBs is presented, the rationale being that spreadsheets are a familiar form of software in Engineering design offices and hence would prove easier for designers to use than specialist software.

1 Introduction

If computer support for conceptual design is to be accepted within design offices, then it is our contention that such systems will need knowledge bases (KBs) which are capable of being maintained by people who are not specialist knowledge engineers. Most design consultancies are too small and impecunious to afford knowledge engineers and yet they rely on their expertise to earn a living. If their computer systems are to reflect this expertise, they will need to contain customised knowledge bases which are developed incrementally over a long period of time. The development process will be a dynamic and continuous process. Given current technology, the only solution to this problem is to develop design KBs which are simple and easy to maintain so that selected staff within the design office can undertake the task.

Our experience and that of others [1] has also shown that designers will not accept systems which control the conceptual design process. Indeed such systems are inappropriate because it is not possible to build a KB with sufficient reasoning power for it to be able reliably to solve even routine design problems. Moore [2] shows that even a carefully designed and evaluated KB will only cope with 80 to 90% of typical domain design problems. Furthermore, Cadogan [3] has found that designers, especially at the conceptual stage, were reluctant to use software whose solution strategies were incomprehensible and/ or unverifiable. For example, the more conscientious designers will check a new FE analysis package against known solutions to determine its limitations.

Thus 3 major requirements for conceptual design decision support have been identified :-

- That the system KB should be easy and simple to maintain,
- That the system should not control the decision making,
- That the reasoning processes should be transparent.

The system known as BRIDGE2 [4] directly addresses these 3 requirements. It includes a new form of KB known as a Component Instantiation Model (CIM). Within the CIM are entirely separate items of knowledge relating to individual design solutions each being referred to as a Component Instantiation Structure. As fig.1 shows, each CIS has at its centre a design solution with on the left, negative constraints (i.e reasons for not choosing that solution) and on the right positive constraints (i.e. reasons for choosing that solution). The CIM consists of a collection of such solution based independent KBs (fig.2). In the case of the work of Rees [5], the chosen domain is beam/ slab bridge design and the CIM is split into two (deck and substructure) to reflect the domain ontology.

The simple structure of each CIS enables one to see at a glance whether the constraints are correct and consistent. The fact that each CIS is independent removes problems associated with clashes of knowledge and the order in which knowledge is presented (e.g. rule ordering). These factors facilitate maintenance by people who are not expert knowledge engineers. The system employs a critiquing style of interaction, similar to that described in [6] and this coupled to the robust form of knowledge representation avoids the brittleness problem although problems might arise if more complex constraints are present.

The use of constraints rather than, for example, rules seems to fit better with the ethos of design. Much of design consists of finding solutions which satisfy constraints, with the latter determining the boundaries of the design space. Although it is possible to represent constraints in rule based form, they are made more explicit in the CIM. Also the solution based form of the CIM removes the need for meta and search knowledge, thus further simplifying the KB.

The inferencing procedure is described in detail in [5] and [7] but briefly, the inferencing strategy used involves two basic actions, classifying and evaluating components.

Classifying involves the following actions :-

Whenever the user enters information into BRIDGE2, the inferencing procedure undertakes an exhaustive search of CIM. The first task is to determine whether any negative constraints have been satisfied. For each CIS where this occurs, that component is listed as being unsuitable. Next the CIM is searched again but only for those components who did not register a negative result in the first pass. If any positive constraints are satisfied then the component is listed as being suitable for the given problem. BRIDGE2 keeps a list of which components have received positive and negative ratings from the inferencing procedure. Classification takes place every

Negative Constraints	Solution	Positive Constraints	
		Limited construction depth	Actual clearance < required clearance
Span range between 0-16m	M beam	Aesthetic consideration	Continuous soffit required
		Skew angle > 40	

Fig. 1. An Intermediate Representation of the M-beam Component Instantiation Structure

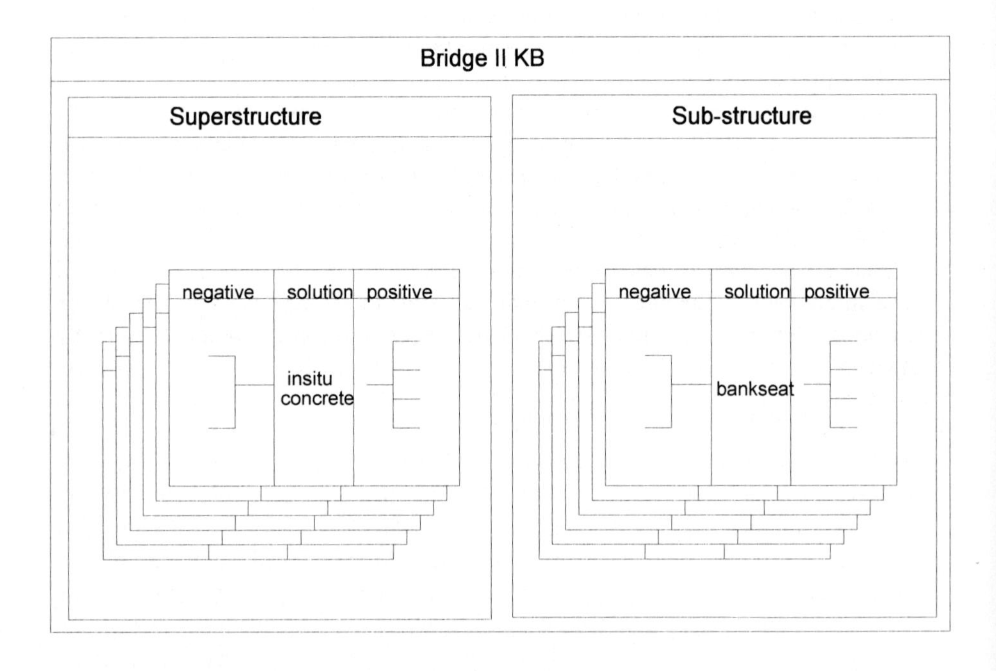

Fig. 2. The Bridge2 CIM Structure.

time that information is added and changed by the user. Thus if all the components have been chosen and the user then alters a design parameter, the list will be updated.

The evaluation procedure is as follows :-

When the design is sufficiently advanced for the designer to choose a designsolution, the user's decision is compared against the Component List obtained from the classification. If an unsuitable component choice is detected then the critiquing procedure is invoked. This warns the user of the problem and suggests all components whose positive constraints have been satisfied as alternatives. If however the user's choice does not violate a negative constraint but no positive constraints have been satisfied, the system notifies the user that insufficient design information has been added.. If the user's solution does not violate any negative constraints and satisfies the relevant positive constraint then the evaluation process is terminated.

Dividing the inferencing into two reflects the two types of user input to the system: design parameters and design solutions. The classification process uses the design parameters entered by the user to identify suitable components. The evaluation process uses the results of the classification to evaluate the designer's chosen components.

The KB maintenance facilities of BRIDGE2 allow the user either to amend the existing knowledge via a knowledge editor or to add new solutions and associated constraints via a knowledge manager. So far as is known, this ability to cope with new solutions in this manner is unique for a design system.

There are some limitations on the system's current capabilities. For example it can only cope with new solutions that fit the domain of beam/ slab bridges, it has a fixed vocabulary which limits the terms which can be included in the CIM and each CIS can only contain linear constraints with simple comparisons using operators such as "=", "<" and ">" or text. Essentially BRIDGE2 is a prototype and like all prototypes while it contains advances in important areas, it also has some further aspects which remain to be addressed.

This paper discusses an ongoing project which directly examines ways of removing some of these limitations and also explores some of the alternative strategies to those implemented in BRIDGE2. As Ullman et al [8] describe many designers tend to work on their first idea and adapt it to suit the problem to be solved, rather than considering all the available options. Hopefully we are avoiding this pitfall by looking at possible variations to the CIM of BRIDGE2.

2 BRIDGE3 - Motivation

For BRIDGE3, we are using the same domain as BRIDGE2, conceptual bridge design for beam/ slab bridges. BRIDGE2 successfully showed that it is possible to have user maintainable design KBs but the form of KB implemented in the system is relatively simple. Design frequently involves more complex relationships than those allowed by BRIDGE2. Also it is possible to base the CIM on constraints as the central feature rather than solutions.

In all our work on design systems, we have spent a considerable amount of time testing our ideas with designers and studying current design practices. One feature that is universally present in engineering design offices is the ready acceptance and frequent usage of spreadsheets. Most designers use these on an everyday basis, sometimes for quite complex calculations. They like the adaptability and the ease of programming which spreadsheets offer. If something is accepted and approved of then one should at least examine the feasibility of making use of this where ever it is logical to do so. We have tended to date to custom build all the features of our systems but if there is a ready made alternative that engineers are familiar with, possibly we should take note of this.

Thus when thinking about a successor to BRIDGE2, it was decided to examine the following :-

- More complex forms of constraint in the CIM,
- Alternative forms of the CIM,
- The possible use of spreadsheets and linking these to the CIM.

It is recognised that there are other aspects of BRIDGE2 which require further work and these are the subject of other research projects.

3 Role of Spreadsheets

BRIDGE2 is written in Visual C++ and it was intended to use this as the main language for BRIDGE3. Therefore when determining which spreadsheet to link to the system the obvious choice is EXCEL. The link can then be implemented using OLE.

A specific OLE technology known as OLE Automation can used, which essentially allows a program to expose its services and functions to other programs as well as to people. Making EXCEL programmable in this way means that its components or automation objects can be utilized or integrated with the automation controller, BRIDGE3. Instead of EXCEL being controlled by a user through the EXCEL interface, it can then be controlled by BRIDGE3 through an Automation interface [9].

Initially it was intended to call EXCEL from the CIM, as required. The spreadsheet would then provide the functionality so that rather than being limited to simple linear constraints, one could use the power of EXCEL. While we were considering this, the thought occurred, why not create and store the whole CIM in EXCEL. This then provides a ready made editing facility coupled to powerful reasoning and calculation engines which are familiar tools to engineering designers. This conceptual leap once made seems simple and obvious. It also avoids the need to write much of the code for the CIM.

With the entire CIM stored in EXCEL, the Automation protocols were written to act as a data transferring mechanism for the CIM to BRIDGE3 when required, using the EXCEL specific functions.

4 Solution View - Constraint View

The CIM in BRIDGE2 uses the solution as the central feature of each CIS. Our reasoning at the time was that designers are most concerned with solutions and presenting the CIM in this way would enhance their understanding of each CIS. However, this is a an assumption on our part and we felt we should explore other options. One obvious alternative to a solution view is to use a constraint centred view of the CIM. Fig 3 shows the constraint centred view of a section of the CIM. For contrast, fig.4 shows the solution view in a similar format.

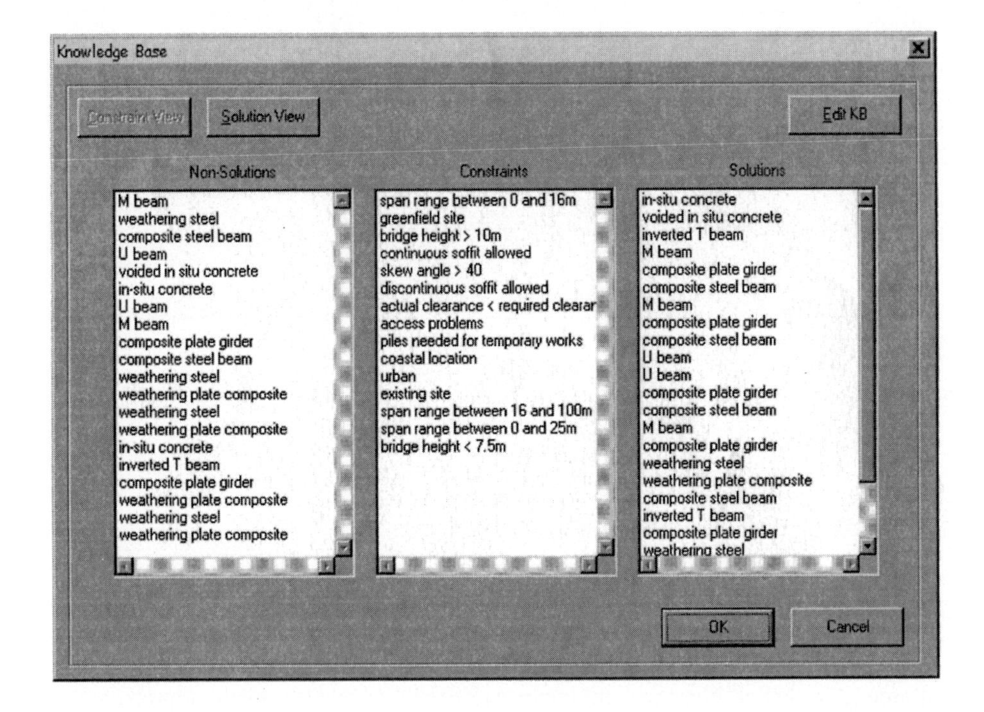

Fig. 3. Constraint View (constraints shown centrally with solutions either side)

As can be seen, although not quite as easy to understand as the BRIDGE2 form where each CIS is displayed separately, the ability to examine the CIM using 2 different viewpoints does offer some advantages. For example in the constraint view, the user is easily able to assess the environmental knowledge contained within the CIM (e.g. coastal location, urban site, existing site).

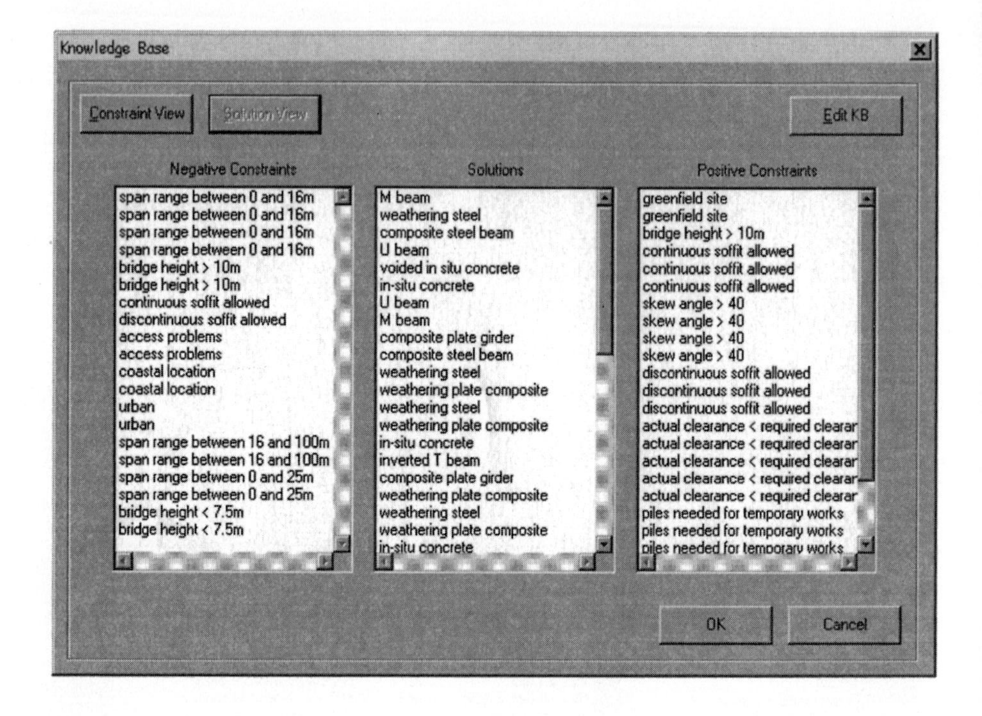

Fig. 4. Solution View (Solutions central with all constraints shown either side)

5 Further Constraints and Accuracy

One major aim of the current work is to look at the ability of the CIM style of knowledge representation to cope with more complex constraints. A number of questions remain unanswered at present. For example, is it possible to build a CIM for a more complex problem without incurring difficulties due to constraint propagation? How robust is a CIM when the constraint structure becomes more complex? How complex can a CIM become and yet be user maintainable? As yet the system is insufficiently complete for us to answer these questions.

One area which we have debated a lot is the appropriate level of accuracy for calculations. We have been advocating for some years [10] that where possible one should use algorithmic techniques even where humans would currently use heuristics. Where algorithms are currently used in the early stages of design, it is usual to use relatively simple, less accurate methods (e.g. one might use simple beam bending theory, rather than moment distribution or numerical analysis). There is a school of thought that at the early stages of design where solutions are ill defined, accurate calculations impart a false sense of confidence in the validity of the solution and that the current practice recognises this.

The counter argument is that the current practices have arisen because of time constraints and the need for human designers gradually to increase the complexity of the design. If the design was expressed in a complicated form in its initial stages, with current techniques a lot of time would be needed to create an adequate description and this is not feasible with manual techniques. Hence until the designer has narrowed the choice of possible options, it is not economic to work to a high level of accuracy. In contrast, computers can generate and handle complex representations with ease and so even at the very early stages of the design process, one could introduce advanced algorithmic techniques. This would then provide the most accurate answer possible, thus reducing the level of uncertainty in reasoning about the design to that which is present in the parameters. Hence even though the designer may not have determined the design parameters to a high level of accuracy one has not introduced further inaccuracy through the algorithm.

Our current thinking is to implement a compromise, with a slight increase in accuracy over manual practices. How effective this is will be partially evaluated when the system is tested by practising designers. It will not unfortunately be possible to assess the impact of different levels of accuracy.

6 Discussion

The work described is an ongoing project and as yet the ideas have not been tested. However some useful variations on the style of KB developed by Rees [5] are described. Also the ability to build user maintainable KBs using EXCEL is a novel idea and one which we hope to test in design offices in the not too distant future. From a research point of view, using the functionality of EXCEL saves a lot of programming effort, allowing us to concentrate on more fundamental issues. From the user's point of view, EXCEL offers familiarity and should make the software more user friendly.

References

1. Smith, I.F.C.: Interactive design - Time to bite the bullet. In Kumar, B., (ed): Information processing in Civil and Structural Engineering design. Civil Comp Press, Edinburgh UK (1996) 23-30.
2. Moore C.J.: An expert system for the conceptual design of bridges. PhD thesis Cardiff School of Engineering, Cardiff University (1991)
3. Cadogan J.V.: Constraint based decision support for conceptual design. PhD thesis Cardiff School of Engineering, Cardiff University (1998)
4. Rees, D.W.G., Miles, J.C., Moore, C.J.: A second generation KBS for the conceptual design of bridges. In: Topping B.H.V. (ed) Civil-Comp Press, (1995) 17-24.
5. Rees, D.W.G.: An non-prescriptive intelligent design system. Cardiff School of Engineering, Cardiff University (1996)
6. Fischer, G., Mastaglio, T.: Computer-Based Critics. IEEE 22nd Hawaii Conf on System Sciences, Decision Support and KBS track, IEEE Computer Soc (1989) 427-436.

7. Moore, C.J. Miles, J.C. Rees, D.W.G.: Decision Support for conceptual bridge design. AI in Engineering 11 (1997) 259-272.
8. Ullman, D. G. Stauffer, L.A. Diettrich, T.G.: Preliminary Results of an Experimental Study of the Mechanical Design Process. In Waldron, M. B. (ed) NSF Workshop on the Design Process, Ohio State University (1987) 143-188.
9. Brockschmidt K.: Inside OLE (2nd ed), Microsoft Press (1995)
10. Miles, J.C. Moore, C.J.: Some thoughts on conceptual design: heuristics and their utility in knowledge based systems. Structural Engineering Review 5(2) (1993) 87-96

Articulate Design of Free-Form Structures

William J. Mitchell[1]

[1] Dean, School of Architecture and Planning, Room 7–231,
Massachusetts Institute of Technology, Cambridge, MA 02139, USA
wjm@mit.edu

Abstract. Traditionally, structural design has been conceived as a process of adapting well-known prototypes to specific contexts under requirements for repetition and symmetry. Now, shape grammars provide a much more flexible way of specifying design languages for exploration, CAD/CAM techniques loosen requirements for repetition and symmetry, and simulated annealing supplies an effective way to direct searches for structural solutions under these conditions. This paper discusses the problem of free-form structural design, shows how shape annealing techniques (which combine shape grammars and simulated annealing) can successfully be applied to them, and concludes by speculating about future structural design systems which will combine simulated annealing with the affordances of the World Wide Web.

1 The Problem of Free-Form Design

As some recent projects like Frank Gehry's Guggenheim Museum in Bilbao (figure 1) vividly demonstrate, CAD/CAM technology now makes it feasible to construct large-scale, complex, free-form buildings at acceptable cost and on reasonable construction schedules. Powerful geometric modeling software supports the design and documentation of free-form curved elements. Modern structural analysis software (powered by abundant computational resources) allows the behavioral feasibility of these elements to be verified. Computer-controlled fabrication machinery facilitates production of complex, non-repeating components without crippling cost and construction-time penalties. And sophisticated surveying and positioning techniques simplify the process of assembling these elements on-site to produce highly asymmetrical, irregular, three-dimensional compositions.

We have, then, come a long way from the conditions that prevailed when the Sydney Opera House was designed and constructed. Utzon's original sketches (figure 2) had shown it as a breathtaking composition of free-form concrete shells of varying sizes. But this design proved impossible, at that point in history, to develop into constructed reality. First, it was too laborious and difficult to develop, detail, and document the design using pre-CAD hand-drafting techniques. Second, the structural analysis problems would have pushed the limits of available methods. And third, without flexible, computer-controlled fabrication machinery, construction would have been extremely difficult and costly. As it turned out, Ove Arup brilliantly saved the project by showing that the shell forms could effectively be approximated by triangular patches on the surfaces of spheres (figure 3); this regularized the design to a geometry of circular arcs. Even then, though, the construction process became an extraordinarily lengthy and costly one, and the "shell" forms ended up as thick, massive elements rather than the light, floating shapes that had originally been envisaged.

Fig. 1. Complex, free-form, curved-surface geometry of Frank Gehry's Guggenheim Museum, Bilbao

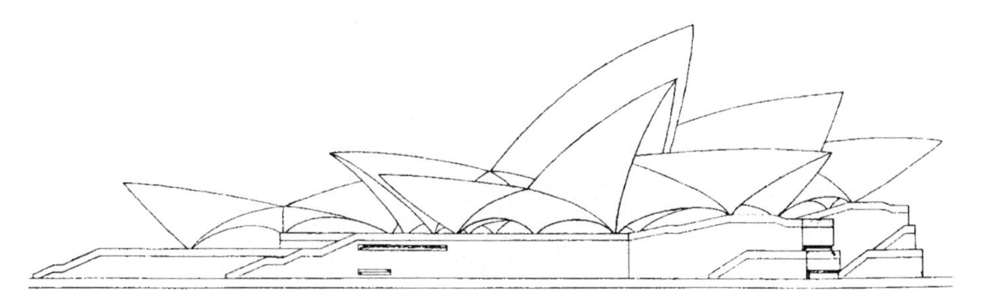

Fig. 2. Jørn Utzon's original concept for the Sydney Opera House, showing free-form curved surfaces.

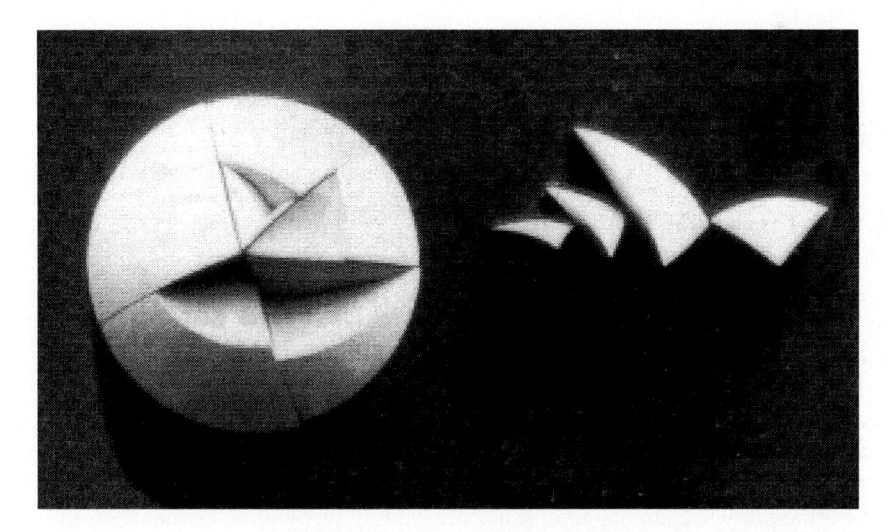

Fig. 3. Utzon's surfaces approximated to triangular patches on the surfaces of spheres.

Today then, as cutting-edge designers are beginning to realize, a whole new world of architectural form is opening up. And this raises the question of how to design large-scale, free-form structures in efficient, elegant, and articulate ways. (It should be noted, incidentally, that the steel structure of Bilbao was not treated as a design element; it was handled very pragmatically, and concealed under the skin (figure 4). Thus it does not directly play a role in creating that building's unique visual character.) This paper explores possible approaches to the task of producing structural designs (specifically in the domain of discrete structures) that are not only sound and efficient but also original and beautiful, then presents a new, computer-assisted technique known as shape annealing, and finally demonstrates the application of shape annealing to some typical problems.

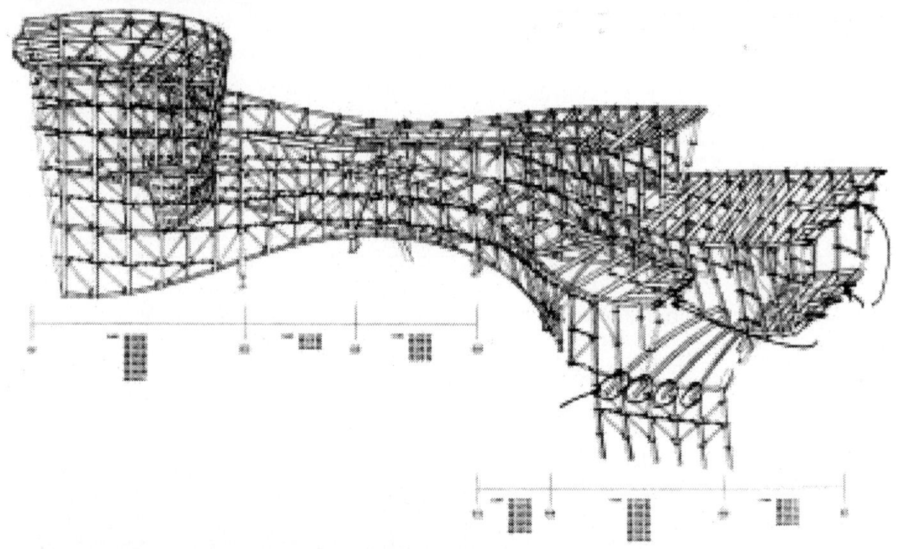

Fig. 4. Concealed structural framing of Bilbao.

2 The Limitations of Traditional Design Approaches

It is possible to some extent, of course, simply to extend traditional design approaches to free-form structural design problems. Not surprisingly, though, this strategy runs into difficulties.

For a start, traditional approaches to structural design are largely based on presumptions of regularity, symmetry, and repetition of identical parts. But these are precisely the presumptions that no longer hold in the context of free-form shapes and CAD/CAM fabrication.

Furthermore, traditional approaches are heavily typological in character. In other words, there are well-known types and subtypes of trusses, domes, space-frames, and so-on, and structural design problems can therefore be formulated in terms of choosing appropriate types for conditions at hand, then instantiating these with parameters adjusted to fit particular contexts. In free-form design, though, established structural prototypes may no longer serve as useful starting points — particularly if typologies are defined too narrowly and rigidly. A more subtle and flexible conception of type — one that generalizes more widely from known examples of successful designs — is needed.

And finally, the introduction of computers into traditional structural design processes does little to encourage innovation. On the contrary, available computer tools mostly require designers to specify initial design concepts as starting points for analysis, refinement of the concept, and possibly optimization within the framework of that beginning idea.

3 The Fred-and-Ginger Approach

An alternative possible strategy is to employ nonlinear geometric transformations to distort regular structures into free-form shapes. Curved-surface modeling software now makes this quite a straightforward operation.

Frank Gehry's "Fred-and-Ginger" office building in Prague (figure 5) illustrates the potential efficacy of this approach. The free-form glass tower at the corner can be understood as a distorted cylinder. If this element had remained as a cylinder, the framing elements of the curtain wall, and the glass panels, would have repeated regularly. In its distorted form, the framing elements and the panels became non-repeating, but this was not a problem; CAD/CAM fabrication of the steelwork and the glass (directly from the designer's CAD model, rather than through the medium of traditional shop drawings) allowed reasonably economical construction.

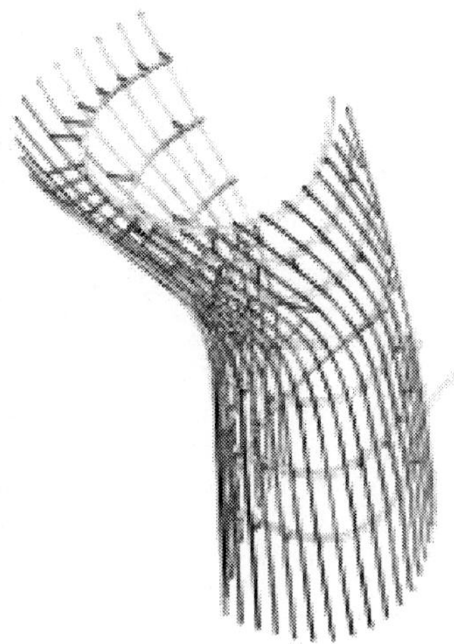

Fig. 5. A free-form glass curtain wall, at the corner of Frank Gehry's "Fred and Ginger" office building in Prague, may be regarded as a distorted cylindrical surface.

Thus we can take the well-known repertoire of planar, box-shaped, pyramidal, cylindrical, spherical, and conical structural forms, and we can generalize them simply by applying new kinds of geometric transformations to produce families of free-form instances. In other words, application of nonlinear geometric transformations to well-known structural prototypes generates equivalence classes of free-form variants. We do not have to worry too much that these free-form versions introduce greater analysis, component fabrication, and on-site erection complexities, since we now have sufficiently powerful technologies to handle those complexities.

4 The Syntactic Approach

This simple transformational approach clearly works up to a point, but it is limited in its capacity to deal effectively with design subtleties. Consider, for example, the task of covering a large, complex, free-form, curved-surface roof with small pieces of sheet metal. You could take a square or triangular grid and map it on to the surface, but this would yield panels with size and shape varying according to no particular constructional logic. It would not readily yield a good solution to the problem of satisfying panel size and curvature constraints. And the arbitrariness would probably be visually unsatisfying.

A simple alternative is to employ some sort of recursive subdivision algorithm, such that subdivision occurs until panel size and curvature constraints are satisfied. The result is a complex, beautiful surface pattern in which there are relatively large panels where curvature is slight, and there are small panels where curvature is great (figure 6). This sort of subdivision would be enormously laborious —virtually impossible, in fact — to produce by hand, but it is a straightforward computer task.

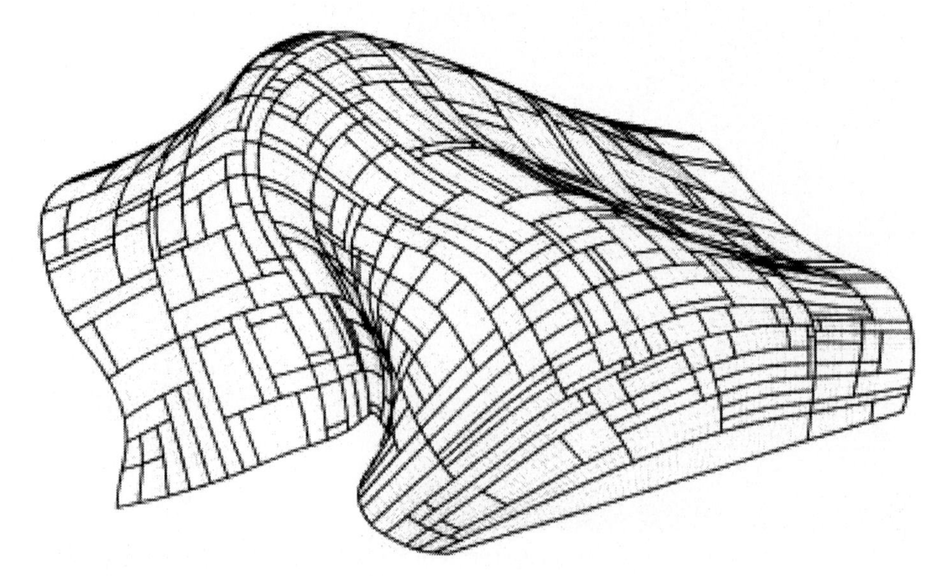

Fig. 6. Recursively subdivided curved surface (courtesy of Dennis Shelden).

More complex subdivision and framing tasks can be approached by writing shape grammars [9] that encode appropriate constructional principles. Many years ago, for instance, George Stiny published a shape grammar for producing Chinese "ice ray" lattice designs [8]. These lattices are produced by recursively subdividing polygonal (usually initially rectangular) openings with straight lengths of timber running from edge to edge of facets. Stiny's ice-ray grammar is readily programmable and executable by computer, and it produces endless variants on this theme (figure 7).

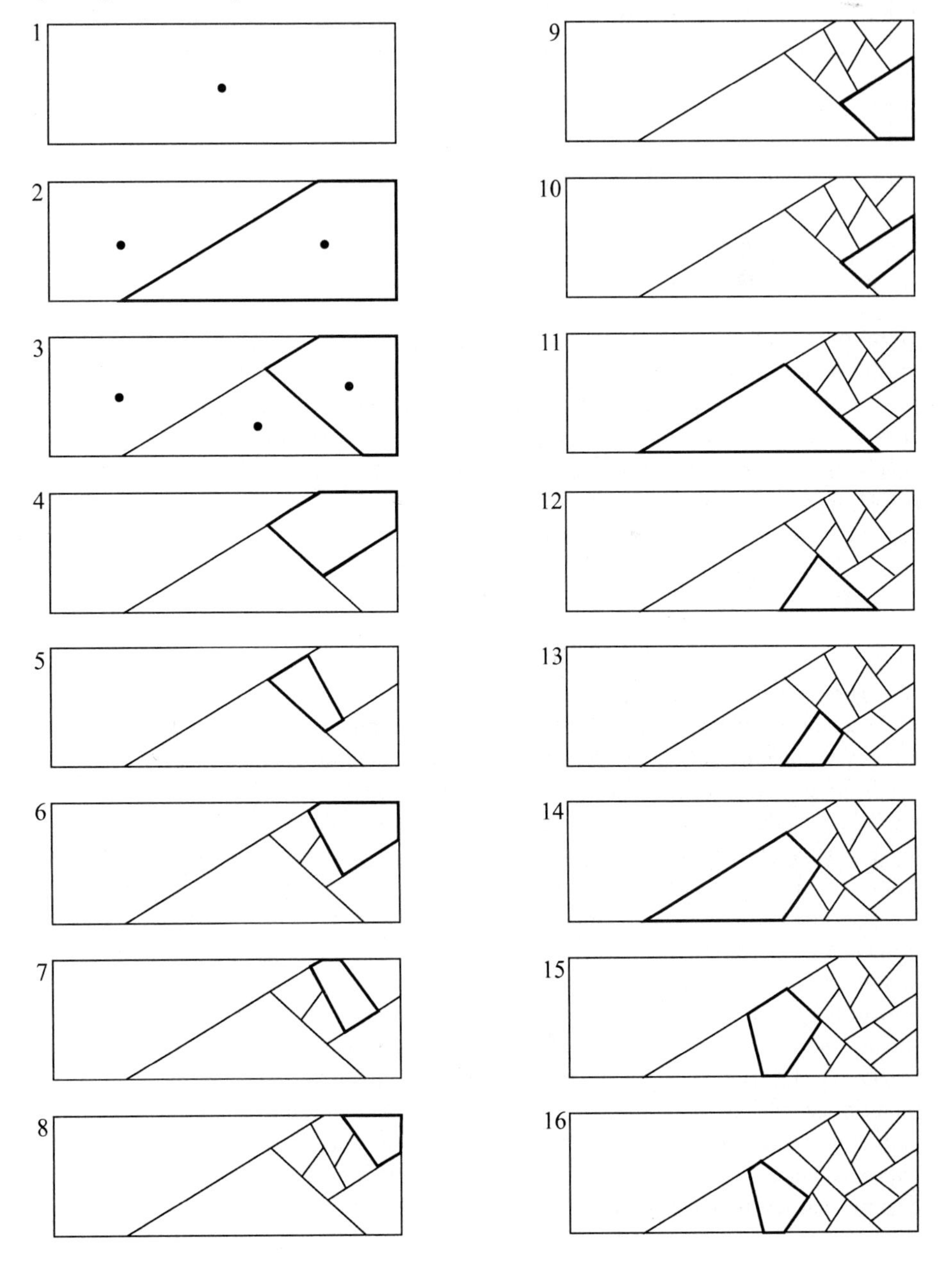

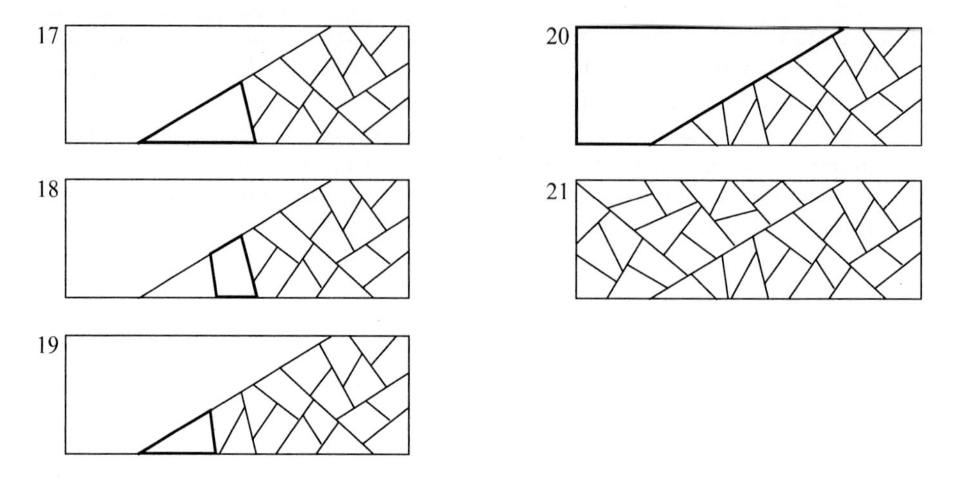

Fig. 7. Derivation of a design by Stiny's "Ice Ray" grammar.

More recently, Kristi Shea has demonstrated that the underlying principles of trusses, space frames, framed domes, and so on can also be captured by simple shape grammars [3, 4, 5, 6, 7]. These grammars can, of course, produce the standard sorts of regular, symmetrical instances (figure 8). More interestingly, they can be applied to free-form situations — for example, domed roof structures over highly irregular plans — to produce workable configurations in those contexts.

5 Automated Design of Free-Form Structures

Given a shape grammar (or some other type of algorithm) that will produce structural configurations for free-form contexts, together with software for analyzing the behaviors of those configurations, the practical design task becomes one of generate-and-test. In principle, it can be solved by means of software that automatically generates possible configurations, analyzes their behaviors, and selects member sizes accordingly. You can formulate the design problem as one of enumerating feasible solutions for consideration, or you can specify an objective function and search for optimal or good sub-optimal solutions. In the latter case, of course, an effective control strategy is required.

In effect, the intellectual labor of producing innovative designs needs to be divided up among the shape grammar rules, the analysis procedures, the objectives and constraints, and the control strategy. The shape grammar rules can effectively encode knowledge of how to put together structurally feasible and aesthetically pleasing compositions; they specify the design language to be explored. The analysis procedures model relevant behaviors of designs in the language, and the parameters of these procedures express material properties, boundary conditions, and so on. The objectives and constraints explicitly express key design goals. (Very often, the shape grammar rules implicitly express other goals, as well.) And the control strategy directs the search process towards designs that satisfy these goals.

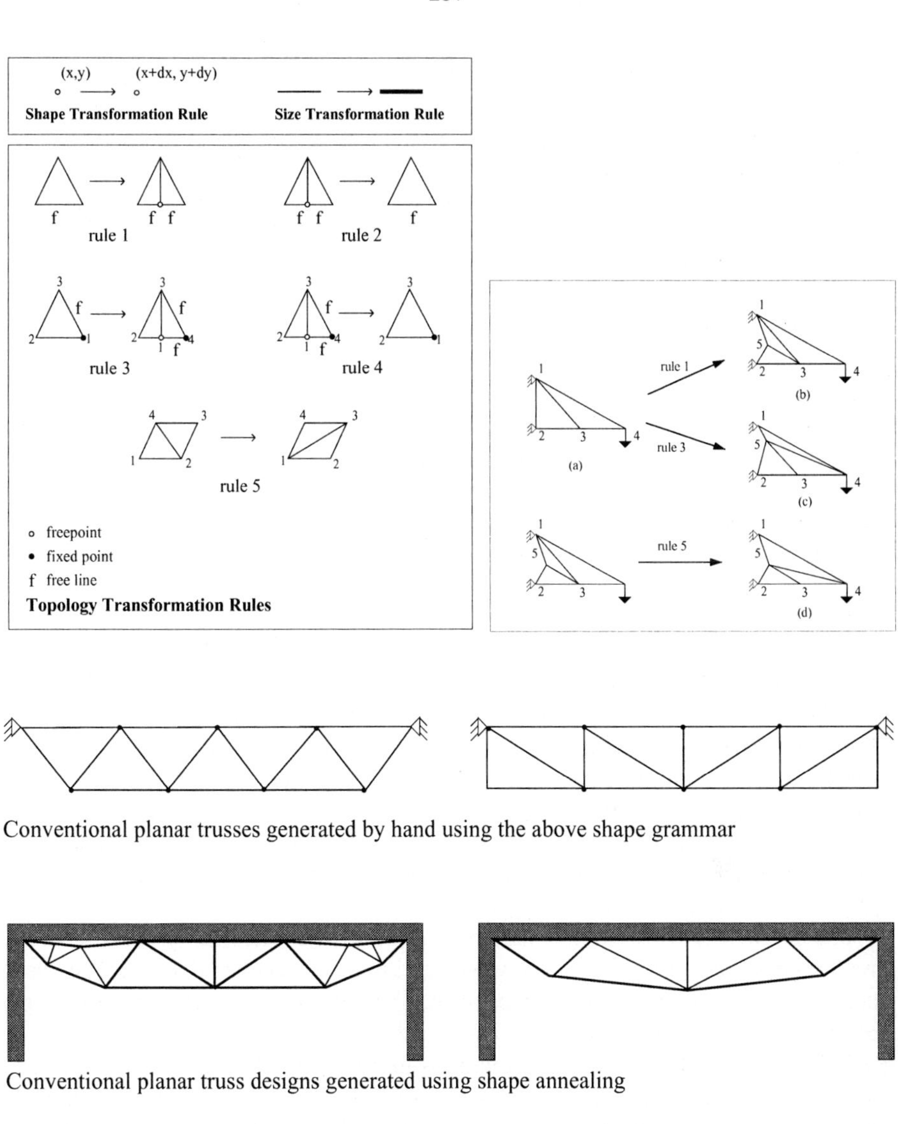

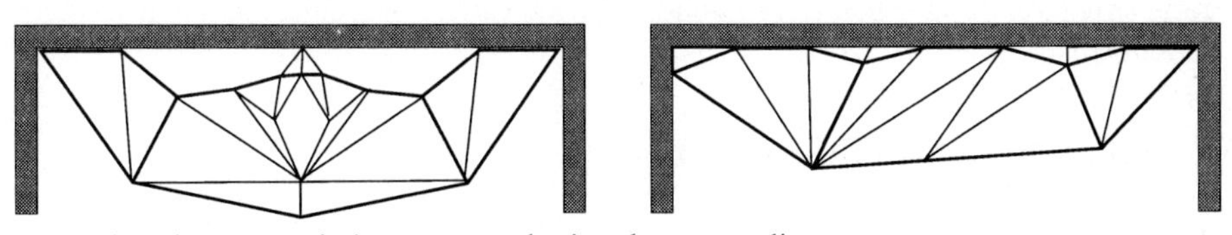

Innovative planar truss designs generated using shape annealing

Fig. 8. Kristi Shea's planar truss grammar, and examples of regular and irregular truss designs that it generates.

6 The Shape Annealing Approach

Shape annealing techniques [1, 2] divide up the task in just this way. They combine shape grammars which generate structural design possibilities with standard structural analysis software and stochastic optimization using simulated annealing. They produce optimally directed designs that follow the structural and aesthetic principles encoded in the shape grammar rules, meet the specific spatial and other requirements of particular contexts, and are efficient and economical. It has been demonstrated that they can produce both efficient versions of traditional, expected types of solutions, and also — more interestingly — large numbers of sound, efficient free-form solutions that otherwise would never have been imagined.

The examples of discrete dome designs shown in figure 8, generated by Kristi Shea, were produced by shape annealing in this way. The underlying shape grammar assured that each had the member connectivity of a geodesic dome. In the search process, member placements resulting from shape-rule applications, node locations, and member sizes were all varied. Each resulting design covers a circular floor area of 30m diameter, and rises to a height of 9.25m. And each supports self load, plus a single point load in the center.

The wide variations if the forms result from variations in the specifications of structural, architectural, and aesthetic purpose — which then guided the search process in different directions. Figure 9(a) responds purely to considerations of structural efficiency, while 9(b) maximizes the enclosed space and minimizes external surface area. Figure 9(c) minimizes the number of different member cross-sections, while 9(d) minimizes the number of different member lengths. Traditional repetition and symmetry constraints can be applied; thus 9(e) enforces 8-fold symmetry about a central axis, while 9(f) ignores this requirement. Further details of these designs are given in Shea 1997 [3] and Shea and Cagan 1997 [4].

Except where repetition and symmetry constraints are applied, this process generally produces free-form designs — even where the space to be covered is a regular shape such as a circle, as in these examples. And it applies equally well to covering both regular and free-form plan shapes.

By substituting different shape grammars for the grammar that generates pseudo-geodesic domes, the approach can be extended to other types of discrete structures. In general, then, shape annealing provides a robust method of generating efficient, free-form structural designs that can respond to a wide variety of architectural and aesthetic requirements.

7 Design Systems of the Future

The shape grammars at the foundation of the shape annealing methodology have the very useful property of modularity; they are structured as lists of discrete rules, so rules can readily be added and deleted to specify new languages of designs. Furthermore, cross-product languages can be specified by allowing the outputs of one shape grammar to serve as the inputs of another.

This meshes well with the distributed character of the World Wide Web. In the future,

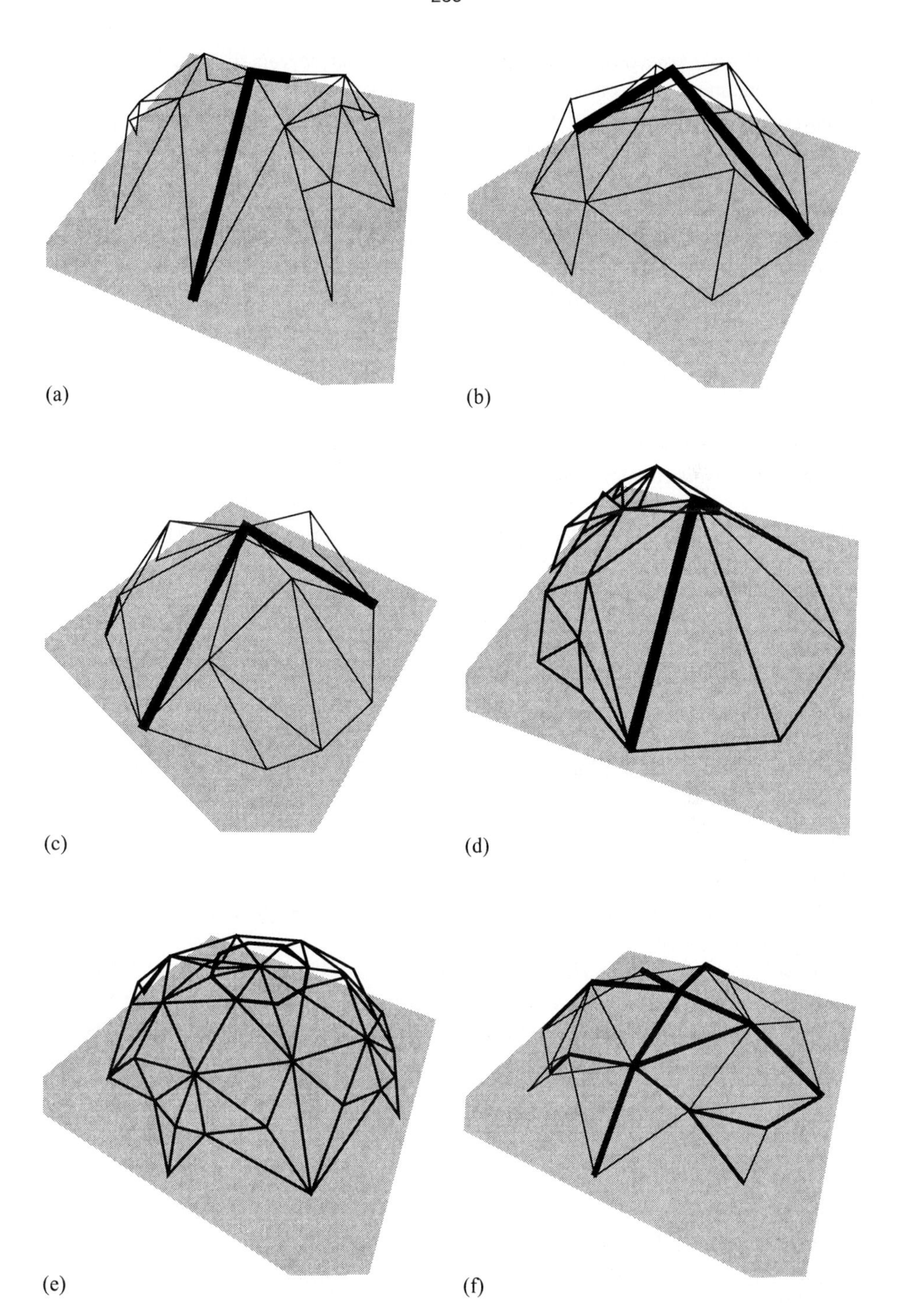

Fig. 9. Pseudo-geodesic free-form dome designs generated by shape annealing.

then, I anticipate emergence of "design webs" in which servers distributed around the world maintain growing collections of recombinable shape rules and shape grammars. These collections will together constitute a continually evolving design knowledge base. Design software operating in this distributed environment will send out software agents to seek out rules and grammars that may be applicable to problems at hand, and it will then search the dynamically specified languages that result from these rule selections to produce specific design proposals. Because of the intrinsic power of the shape annealing approach, and the evolving character of distributed collections of shape rules, the results will often surprise and delight us.

References

1. Cagan, J., and W.J. Mitchell.: Optimally Directed Shape Generation by Shape Annealing. Environment and Planning B, **20** (1993) 5-12

2. Cagan, J., and W.J. Mitchell.: , A Grammatical Approach to Network Flow Synthesis. In: Formal Design Methods for CAD, IFIP TC5/WG5.2 Workshop, Tallinn, Estonia, June, IFIP Transaction 3, B-18 (1994) 173-189

3. Shea K.: Essays of Discrete Structures: Purposeful Design of Grammatical Structures by Directed Stochastic Search. Ph.D. Dissertation, Carnegie Mellon University Pittsburgh, PA (1997)

4. Shea K., and J. Cagan.: Innovative Dome Design: Applying Geodesic Patterns with Shape Annealing. Artificial Intelligence for Engineering Design, Analysis and Manufacturing. **11** (1997) 379–394

5. Shea, K. and J. Cagan.: Generating Structural Essays from Languages of Discrete Structures. To appear in: Artificial Intelligence in Design '98. (1998)

6. Shea K., and J. Cagan.: The Design of Novel Roof Trusses with Shape Annealing: Assessing the Ability of a Computational Method in Aiding Structural Designers with Varying Design Intent. Accepted for publication in: Design Studies. (1998)

7. Shea, K., Cagan, J., and S.J. Fenves.: A Shape Annealing Approach to Optimal Truss Design with Dynamic Grouping of Members. ASME Journal of Mechanical Design. **119** 3 (1997) 388-394

8. Stiny, G.: Ice-Ray: A Note on the Generation of Chinese Lattice Designs. Environment and Planning B. **4** (1977) 89–98

9. Stiny, G.: Introduction to Shape and Shape Grammars. Environment and Planning B. **7** (1980) 343–351

Applying Quantitative Constraint Satisfaction in Preliminary Design

C.J.Moore, J.C.Miles and J.V.Cadogan

Cardiff School of Engineering, Cardiff University, PO Box 686, Cardiff CF2 3TB, Wales, UK
MooreCJ@cf.ac.uk

Abstract. This paper describes a Decision Support System for supporting bridge designers in the early stages of foundation design. The system has been developed using information acquired from practising bridge engineers, geotechnical engineers and estimators. The system is limited to the design of abutments, spread and piled foundations for small to medium span road bridges.

The design of piled foundations is conducted with the support of recognised pile design software. The spread foundation and abutment design has been developed in two stages, initially using the Strategy Detection Algorithm (SDA) and further to that the Graphically Assisted Constraint Satisfaction Method. The transition between the stages was instigated by evaluating the prototype in industry. Both methods and the evaluation are briefly described.

The system is developed in Visual C++, following an object based representation and utilising a Windows style interface.

1 Introduction

The preliminary design of foundations is recognised as a complex and demanding task. It requires the co-ordination of many engineering skills possessed by different engineering disciplines: for example, information is required from both geotechnical and bridge engineers. Decisions taken at the early stages can have significant implications later in the design. Computational support is needed at this early stage of design, yet such support is currently unavailable.

Evaluation of Decision Support Systems (DSS) in industry has shown that these systems are appropriate for supporting conceptual design processes (Moore and Miles 1995). The DSS approach adopted permits a combination of algorithmic and symbolic programming which allows the maximum benefit to be obtained from the computer/designer interaction (Rees et al 1995 and Price et al 1995). Earlier work has shown that preliminary foundation design is a domain that is ideally suited to the application of the DSS principle (Price et al 1995).

This paper describes the development of a prototype DSS for preliminary foundation design, aimed at supporting bridge designers. The system contributes to the above suite of DSS being developed at Cardiff.

2 Outline of the Research

This research aims to develop decision support for bridge engineers in the preliminary design of bridge foundations. This prototype computer system accepts information on the ground conditions, suggests suitable foundation types and provides advice on the preliminary sizing of these foundations. The system presents this information as a coherent set of parameters and decisions. These decisions are supported by providing increased levels of information, but the responsibility for the design lies with the user. The system allows the user to see the impact of the decisions on the overall design, with particular attention being paid to the affect on the design constraints. The DSS explores the implications of various design decisions and the user is given the opportunity to reassess any decisions in an attempt to overcome any difficulties which have been introduced.

The prototype system provides guidance on which parameters or decisions need to be amended if conflicts or constraints are to be avoided. Initially, a Strategy Detection Algorithm (SDA) was developed. Further to the evaluation of this prototype in industry, a simpler approach was found to be more effective, namely the Graphically Assisted Constraint Satisfaction Method.. Both methodologies and the evaluation which led to the change in direction are discussed.

The domain of the DSS for bridge foundations is limited to the conceptual design of small to medium span road bridges. The design of spread footings and piled foundations are covered. Other types of foundations are not currently considered as the majority of road bridges throughout the UK use only the above forms of foundations.

The information contained within the system has been identified by a process of Knowledge Acquisition (KA) and Knowledge Elicitation (KE) using literature, Codes of Practice and practising engineers with structural, geotechnical and estimating backgrounds. The KA work has shown that existing design processes for conceptual foundation design combine four fundamental knowledge domains:

- geotechnical knowledge
- structural knowledge
- estimating knowledge
- construction knowledge

all of which have been investigated during the course of this project. This combination of knowledge enables the system to provide the designer with improved access to design information.

Engineers have also been involved with the iterative development of the prototype, providing evaluation and constructive criticism where necessary. This involvement has enabled the scope and structure of the system to be refined to enable the DSS to support current design practices. The industrial collaboration is detailed elsewhere (Cadogan et al 1996).

3 The Decision Support System

The DSS created takes the form of two fundamental class hierarchies; the Data Input Hierarchy and the Constraint Management Hierarchy. The overall structure of the DSS is depicted in Figure 1.

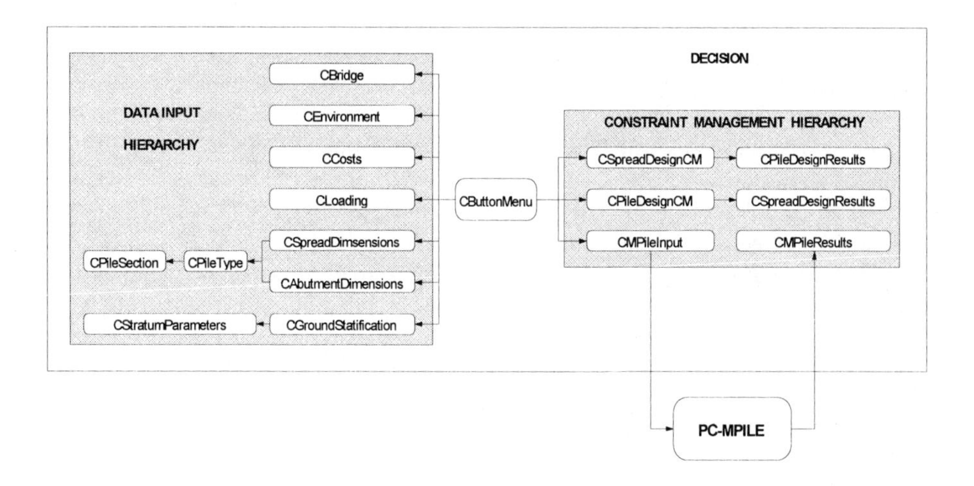

Fig. 1. DSS Class Hierarchies

CButtonMenu is the root class that provides the functionality for manoeuvring between each hierarchy. Each instance of CButtonMenu stores a complete set of basic parameters for a problem. The Data Input Hierarchy controls the way in which these parameters are acquired by the system from the user. Instances of each of the ten classes in the Data Input Hierarchy are represented as input screens (i.e. dialog boxes). The Data Input Hierarchy is described elsewhere (Moore et al 1997, Cadogan 1998).

The Constraint Management Hierarchy controls the design of spread footings, abutments and pile groups. The design of piled foundations is a particularly difficult task. Figure 1 illustrates the way in which the developed DSS interacts with PC-MPile: an existing, industrially recognised package for piled foundation design. At the outset of the project, industrial collaborators stated that this package was the recognised standard in the UK for piled foundation design. Therefore, it was preferable to incorporate this package in the DSS. However, when examining the use of PC-MPile with the industrial partners, it was realised that the input and output of MPile were too complex and unwieldy for the early stages of the design process. Therefore, in conjunction with the engineers, summarised versions of the input and output have been produced. This has reduced input time considerably whilst maintaining all of the necessary date required. The PC-MPile package now produces output that depicts a sensible pile group design in a format that can be taken directly by the DSS for use later in the design process.

3.1 Knowledge Representation

Design is essentially an iterative search process (Tong and Sriram 1992, Cadogan 1998). Constraint satisfaction is a non trivial task and tools which support this are of importance, especially in conceptual design (Serrano and Gossard 1992). A knowledge representation framework for effective constraint management is put forward here. The knowledge representation framework described applies to both the Strategy Detection Algorithm (SDA) and the Graphically Assisted Constraint Satisfaction Method.

Certain criteria will confine the designer when developing candidate solutions, thus making the conceptual design process essentially a constraint satisfaction process. Within the non pile design space, there are many sources of constraints ranging from soft to hard, depending on their level of influence and flexibility, as defined elsewhere (Waldron et al 1989, Kumar 1995). Generally speaking, constraints are numerous, complex and sometimes contradictory as well as dynamic in nature.

This is indeed the case for the domain described here: a large number of constraints exist within the foundation design domain.

3.1.1 Hard Constraints. There are numerous hard constraints applicable to the design of foundations and abutments. However, seven of these are considered by the collaborating engineers to be critical at the conceptual design stage:

1. Sliding factor of safety
2. Overturning factor of safety
3. Eccentricity
4. Bending moment at Ultimate Limit State
5. Shear at Ultimate Limit State
6. Bearing Capacity
7. Settlement

For spread footings, lateral loads may not always be applied and so in these cases constraints 1-3 are not applicable. Abutments, due to the soil retained by the stem will always have lateral loads and so the seven constraints apply at all times.

Each constraint defines a mode of failure which must be checked as part of the design process. The constraints are imposed by Codes of Practice. Therefore, satisfaction of these criteria can be viewed as essential to the design process, thus effectively forming ‘hard constraints’ (Waldron et al 1989).

3.1.2 Soft Constraints. As stated earlier, constraints which are nor considered to be critical are also present in the domain. Soft constraints can be relaxed depending on the design conditions and the preferences of the designer. For example, when determining the allowable bearing capacity, a safety factor (Fos) is used to safeguard against natural variations in he shear strength of soils, uncertainties in the accuracy of the analytical methods and excessive settlement caused by yielding. High levels of judgement are apparent when selecting this the Fos value (Tomlinson 1985). Thus the constraints on Fos are soft as the opinions of engineers will vary as will the ground conditions for different designs.

3.2 Developing a Constraint Management Strategy

Ideally, all the constraints should be represented and managed by the DSS. However, at the outset the size of the domain which was to be covered needed to be reduced to a manageable size in order to develop a prototype system which could undergo evaluation. Therefore, the above criteria were identified as important and were thus concentrated on. This limited domain has been used as the basis of this work and influence networks describing the domain have been developed on this assumption (Cadogan 1998).

The first stage of the prototype development resulted in a Strategy Detection Algorithm (SDA) which dealt with five of the seven critical constraints listed above:

1. Overturning Factor of Safety
2. Sliding Factor of Safety
3. Eccentricity hard constraint
4. Bending Moment at Ultimate Limit State
5. Bearing Pressure hard constraint

These are described in detail elsewhere (Cadogan 1998).

It was envisaged that once the SDA had been evaluated the other two hard constraints and some soft constraints would be included. However, the evaluation showed that a different approach was needed. This new approach was therefore developed also using the above five constraints. Both approaches are described in later sections.

Satisfying the five hard constraints requires a search of such a static design space. Consequently the search has formalised methods of evaluation and a defined, closed set of possible designs (Gero and Kumar 1993). In effect, the designer operates within a static constraint satisfaction design space where the differences between the designs can be characterised largely by the values selected for the design variables. The complexity of hard constraint satisfaction arises from understanding the influence of such variables on the constraints. An effective way to represent such complex interactions is to use the Network Model of Design as originally proposed by MacCallum 91982). This model has been adopted as the basis of the knowledge representation.

3.2.1 Network Model of Design. The Network Model of Design is based on the manipulation of relationships between components in a design space. There are two types of networks models of design: influence networks and dependency networks (MacCallum and Duffy 1987). Dependency networks represent the design characteristics of interest in a domain as a set of nodes joined by arcs where the direction of reliance is shown by an arrowhead. Influence Networks are the inverse of Dependency Networks, where the direction of the arrowhead signifies an influence of one node on another. Thus the influence / dependency of a node on others can be determined by tracing through all the intermediate paths in the network following the direction of the arrows connecting the nodes.

The Network Model of Design has been adopted as the basis of the knowledge representation in the DSS described here. The type of network model which is appropriate for a domain is linked to the search required in a design space. After analysing the foundation design domain, it was realised that influence networks which include Basic Parameters (BP's) and Derived Parameters (DP's) were appropriate (Cadogan 1998). An example Influence Network for this domain is shown in Figure 2.

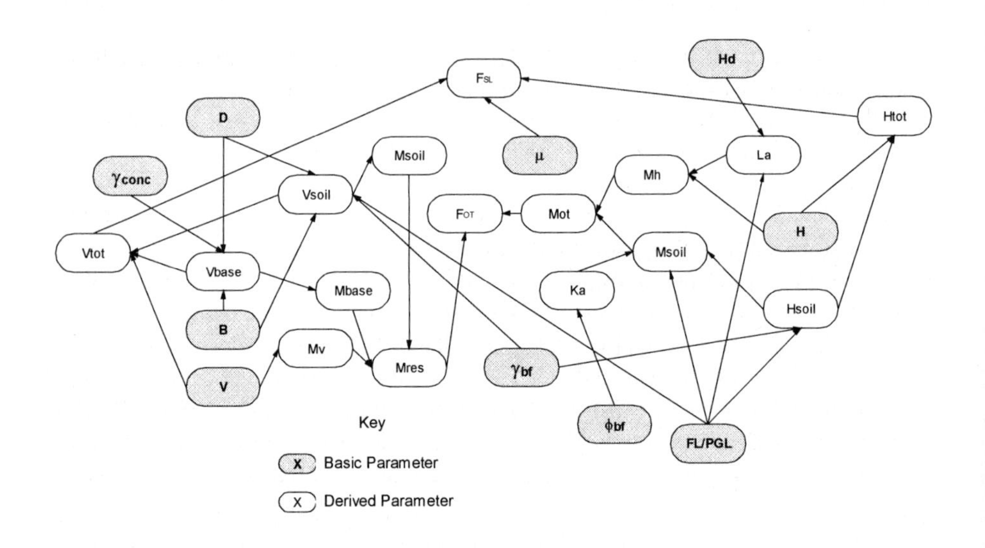

Fig. 2. Influence Network

A Basic Parameter is a parameter which cannot be traced to any other parameter in the network to derive its value, e.g. a spread footing width (B) and coefficient of friction (μ). A Derived Parameter (DP) is a parameter that can be traced to other parameters, either basic or derived (Kumar 1995, Rafiq 1988), e.g. in the above influence network overturning moment M_{ot} and weight of base V_{base}.

Following an analysis of the constituent influence networks, it was realised that a conflict strategy should always involve changing a Basic Parameter (BP) in the domain since all constraints are derived from a static set of derived parameters which are in turn derived from a static set of basic parameters. Thus an explicit search through the influence networks for each state of the design is not required at run time hence ensuring a rapid reaction time. The DSS is able to store pre-defined lists of influencing BP's for each constraint. If a constraint should fail, a constraint satisfaction strategy involves returning to the appropriate set of influencing BP's, changing a BP value and propagating it forward through the influence network in order to satisfy the failed constraint. In order to facilitate this approach to constraint satisfaction, it was necessary to investigate how much the influencing BP's needed to be changed in order to satisfy failed constraints. Therefore, for several designs, each BP was changed individually and the effect noted. Two broad types of effect were identified: detrimental and beneficial (Cadogan 1998).

The type of effect was found to be dependent on the BP's relationship with the constraint. There are six types of relationships which can exist between BP's and DP's in influence networks. These are:

- *Direct Relationships:* apply where no other nodes are involved so that a change in one node can be directly inferred in another (Kumar and Topping 1990)
- *Indirect Relationships:* apply where other nodes are involved so a direct inference is not possible unless all intermediate nodes are constant (Kumar and Topping 1990)
- *Fixed Relationships:* which are rigid in their form, either because they follow physical laws or codes of practice (MacCallum and Duffy 1987)
- *Flexible Relationships:* apply where a model changes as a design progresses and so flexible relationships may exist (MacCallum and Duffy 1987)
- *Multiple Relationships:* apply where there may be a number of valid relationships for estimating a value (MacCallum and Duffy 1987)
- *Singular Relationships:* apply where there is or only needs to be one applicable valid relationship for estimating a value

However, not all apply to this domain. Multiple and flexible relationships do not apply. Indirect relationships only apply to the Graphically Assisted Constraint Satisfaction Method. Nevertheless, the complexity of the remaining relationships means that it is not possible to explicitly list all possible solutions. Instead, solutions are inferred once the current state of the design has been assessed. This has been achieved in two different ways: the SDA and the Graphically Assisted Constraint Satisfaction method, both of which are briefly described below. More detailed descriptions are provided in Cadogan 1998.

3.3 The Strategy Detection Algorithm (SDA)

Generally the SDA satisfies constraints in the following way: once the BP's have been instantiated via the input dialog boxes, the values are propagated forward through the influence network to determine the DP's. Once all the DP's have been assessed, the constraints are checked.

Lists are stored in the DSS for each constraint, which state the influencing BP's. Thus for each failed constraint, the system returns to its list of influencing BP's. For each listed BP, the SDA is invoked in order to detect a strategy which will 'solve' the current constraint problem.

A simple algorithm is used to determine a strategy which needs to be applied to each BP in a list. A strategy has three components:

- *Type of change required in a BP*: The type of change is either an increase or a decrease in the value of a BP. The SDA tests each BP by firstly increasing and then decreasing the original value by increments and then he new values are propagated through the influence network to determine the type of change required.
- *Amount of change required in a BP:* Once the type of change has been determined, each BP is selected individually and increased and/or decreased using the appropriate increments. With every iteration, the new value associated with a BP is propagated through the influence network, and the updated DP's are then checked against the 'failed' constraint. If the required remedy has still not been achieved, another incremental cycle is carried out. This procedure is repeated until the constraint has been satisfied, or until the value of the BP is out of the allowable range. This is checked for by a consistency module which is included in the DSS. The allowable ranges have been decided by a combination of codes, literature and KE.
- *Cost of the change:* A change in some parameters can introduce a change in design cost. To estimate such values requires a knowledge of historical data as well as experience (Bullock per comm. 1994). Only direct costs have been considered in the DSS, as these are quantifiable. The rates associated with these values are included in the DSS but are changeable by the user.

Once strategies have been detected for each failed instance, results are presented. If a user changes a BP the process begins again to ensure other constraints are not in conflict with the change.

Other theoretical work in this area uses more complex mathematical analysis (Serrano and Gossard 1992). However, here the problem has been simplified to allow for practical application of the approach. The simplification allows one BP for one constraint to be considered each time the SDA is invoked. Also, the calculations for the constraints are not lengthy and so, even though there exists more complex forms of analysis, this simplification is appropriate for the accuracy of the problem being considered. The SDA is described in detail elsewhere (Cadogan 1998).

A simplified illustration of The SDA approach is illustrated in Figure 3.

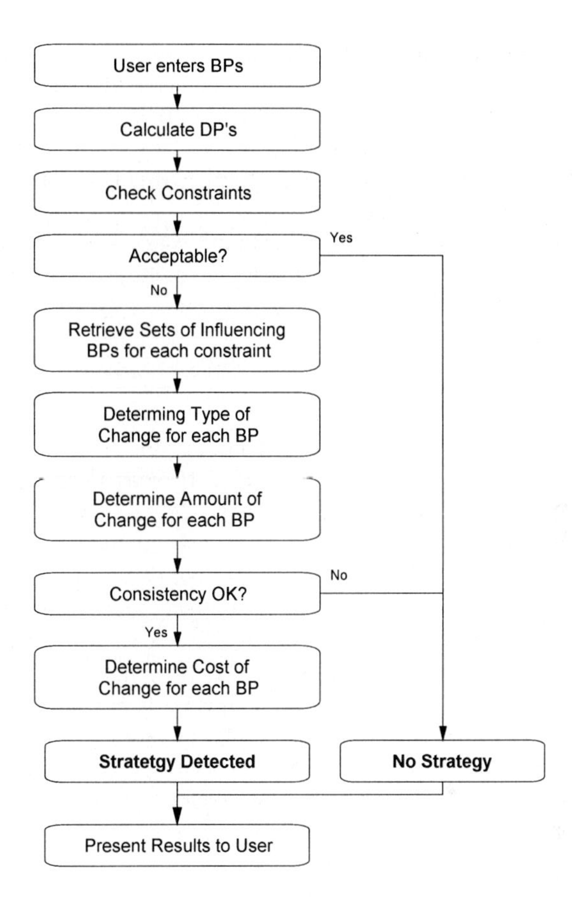

Fig. 3. Strategy Detection Algorithm

3.3.1 Output from the SDA. An example output screen from the SDA is shown in Figure 4. This figure shows that each results screen is divided into 3 sections as follows:

- *Design Results*: The left part of the screen displays the main results of the calculations. Not all DP's are displayed in order to avoid information overload.
- *Design Solutions*: The right part of the screen displays a list of all the BP's considered by the SDA. Since more than one BP may be available to satisfy the constraints, all possibilities are presented. As there is no definitive solution, no attempt is made to promote one procedure over another.
- *Warnings*: The central part of the screen displays the total number of constraints that have 'failed' and the current constraint for which design solutions are displayed. The user can move through the problem and the screen updates to display the corresponding BP change strategies for the corresponding constraint.

At any point, the user can select a BP and change its value. On returning to the main result dialog box, the new DP's are recalculated, the constraints checked, the SDA implemented (if required), and the new results presented.

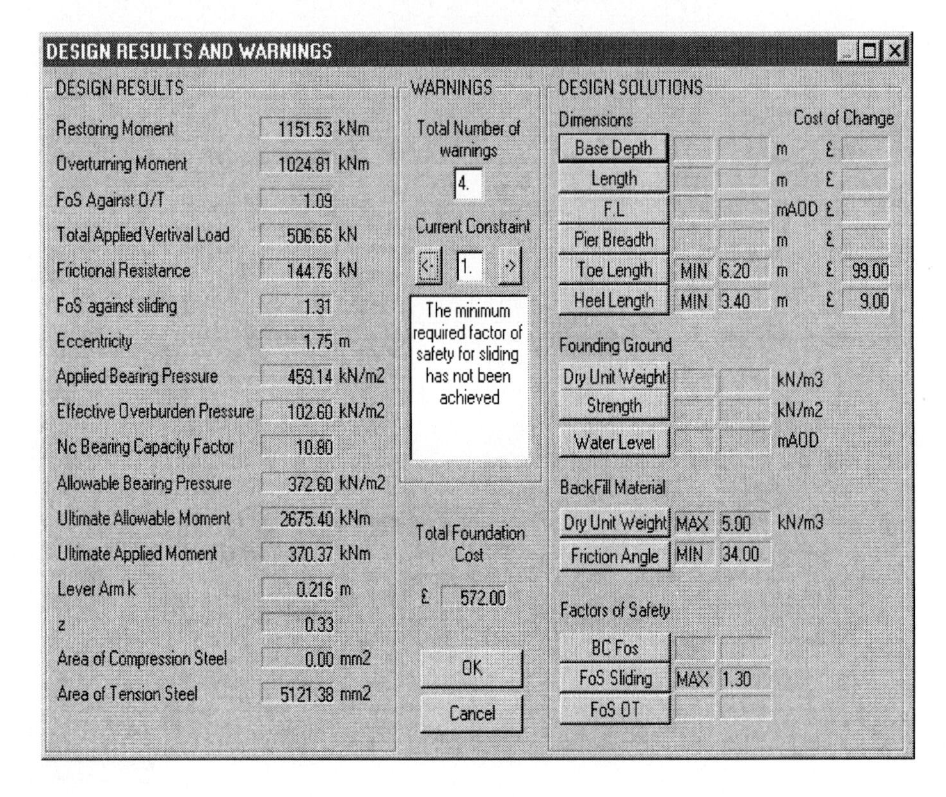

Fig. 4. Example of SDA Results Screen

4 Industrial Assessment of the Prototype Including the SDA

Prior to the industrial evaluation the DSS was tested for robustness using twelve designs provided by the collaborators. The DSS produced results in accordance with the designs and was therefore considered robust enough to undergo further evaluation. Four in house engineers tested the system and, apart from minor bugs, found the system to be acceptable and robust. These engineers had not been involved with the development of the system and commented on its potential usefulness.

Further to these initial checks, the system was released to the collaborating company for evaluation. Twelve bridge engineers were involved in the evaluation. A one hour presentation was used to introduce the system which enabled initial impressions of the prototype to be gained. These initial impressions were positive:

with the engineers commenting on the usefulness of the break down of the design problem into smaller components. The engineers also liked the iterative and incremental way in which the design parameters could be changed.

The senior bridge engineers commented that it is unlikely that safety factors, backfill parameters and ground parameters would be changed during conceptual design; the most likely parameters to be changed were the foundation dimensions. This corresponded with tests on the SDA which found that strategies for the foundation dimensions were most commonly detected (Cadogan 1998). A few minor criticisms of the standard of phraseology were made but overall the evaluation showed that the style of the DSS was welcomed and the system was deemed to be beneficial.

However, the major criticism was that although the SDA responded successfully for the majority of designs, for a few constraints, no strategies were found by the SDA. Since the constraints represented in the SDA are hard, they cannot be removed from the design space. Therefore, in order to rectify this, multiple BP analysis is required. However, it is not possible to use the SDA for constraint satisfaction for multiple BP analysis. The dimensions involved are contiguous variables and so trying every possible solution introduces the problem of combinatorial explosion (Barr and Feigenbaum 1981). One solution is to discretise the dimensions using a grid (Baykan and Fox 1992). This is unacceptable as is removes parts of the decision making process from the user. Therefore, in order to overcome this limitation, domain specific heuristics need to be incorporated to 'prune' the number of solutions considered (Maher 1987). Such heuristics focus and drastically limit the search in a large problem space hence helping to counteract combinatorial explosion.

In order to investigate the heuristics required, further interviews with practising engineers were arranged. This resulted in the development of a new approach.

5 Graphically Assisted Constraint Satisfaction - The New Approach

As discussed above, as a result of the industrial evaluation, some limitations in the SDA approach were observed which could only be rectified by the introduction of multiple parameter analysis. The approach adopted to facilitate this uses a simple but effective technique developed as a result of further KE.

5.1 The Heuristics Adopted

It was evident that the engineers interviewed frequently adopted a 'look right factor' when assessing a foundation design, i.e. whether a design had the correct relative proportions. It was necessary to assess this 'look right factor' before any heuristics could be applied. By definition, this factor is difficult to define as it involves personal judgement . However, it was found that the engineers interviewed could assess this whether a foundation 'looked right' almost immediately and would subsequently decide on appropriate heuristics in order to facilitate it. Therefore it was questionable whether the heuristics per se needed to be included in the system: instead a new style of interface has been developed which allows the bridge engineers to apply the heuristics they choose easily and efficiently, thus avoiding the need to explicitly state the heuristics within the DSS. This ensures that the designer can choose to apply his/her heuristics in an appropriate way and also reduces the possibility of changing the meaning of the heuristics by forcing them into an unsuitable representation style.

OLD APPROACH	NEW APPROACH
Weak Method	Strong method
Constraint Satisfaction Analysis - one BP for one constraint	Constraint Satisfaction Analysis - multiple BP for multiple constraints
User directed exploration of design where SYSTEM suggests strategies	User directed exploration of design where USER suggests the strategies

5.2 The New Approach

As with the SDA, the five hard constraints identified by the collaborators were used as a starting point for the development of the prototype. In the new approach, results for each of the five constraints are presented to the user, together with the individual cost of each foundation and the total cost of the foundations. The style of the Graphical User Interface (GUI) allows the user to assess the 'look right factor' of a design by showing scaled cross sections of abutments and spread footings on screen. The GUI also allows users to implement their heuristics by directly manipulating these cross sections using the mouse.

As the mouse is used to alter the dimensions in order to achieve the 'look right factor' the design and costs for the selected design are continuously re calculated and the results checked against the constraints. If any of the constraints fails, the appropriate results are highlighted. Thus, interactive graphics have been represented which the user can directly manipulate. This approach allows the user to see the numerical and spatial implications of changing a design and its dimensions. As dimensions are changed, the changes are continually incorporated into the influence network, the constraints checked and the costs determined. The new set of results is then presented to the user. All this occurs in real time and thus constraint management has moved to a more human centred approach.

Having completed the prototype which incorporated the GUI approach to constraint satisfaction, further evaluation was conducted. This is detailed elsewhere (Cadogan 1998) but in summary the evaluators liked the new, flexible approach and found the GUI style of interaction preferable and more robust than the SDA.

6. Conclusions

As part of a suite of DSS for bridge design, a DSS is being developed for the conceptual design of bridge foundations. The system is based on information gained during extensive KA using available literature and Codes of Practice as well as KE with practising designers. Ongoing evaluation throughout the project has facilitated the iterative development of a functional and applicable DSS. The DSS produced supports the bridge designer in the following ways:

- by providing non prescriptive design advice
- by providing increased access to design information
- by offering an PC-MPile/DSS interface which presents piled foundation information in a summarised format which is therefore efficient yet familiar
- by informing the designer of the impact of their decisions on the design as a whole
- by recommending ways in which the impacts of their decisions could be reduced.

Constraint management has moved away from the SDA to a more human centred approach. the system now offers a fast and flexible way of searching the design space. Interactive graphics and instant numerical feedback supports the designer's use of heuristics and allows the application of the 'look right factor'. This simplified approach to constraint management has proved to be successful both in terms of the design environment and the reaction of the engineers evaluating the system.

7. Acknowledgements

The authors would like to thank Mott MacDonald and Partners, Croydon, Sir William Halcrow and Partners, Cardiff, Ove Arup, Cardiff and MRM, Bristol for their assistance in this project. They would also like to thank Mr. Francis Bullock and Dr. Rob Francis for their support.

References

1. Brown, D. and Chandrasekaran, B.: Investigating Routine Problem Solving. In: Tong, C. and Sriram, D. (Ed.): Artificial Intelligence in Engineering Design Volume 1 - Design Representation and Models of Routine Design, Academic Press Inc, (1992) pp. 221-250
2. Bullock, F.: Bullock per comm., Discussion of Bridge Estimating Practices (1994)
3. Cadogan, J.V.: Integrated Exploratory Process of Design (1998)
4. Cadogan, J.V., Miles, J.C. and Moore, C.J.: Comp. Support for the Design and Costing of Bridge Foundations. In: Kumar, B. (Ed): Information. Proc. in Civil & Struct. Eng. Design, Civil Comp Press (1996) pp 113-120
5. Gero, J.S., Kumar. B,: Expanding Design Spaces Through New Design Variables. Design Studies, Vol. 14, No. 2, April (1993) pp. 210-221
6. Kumar, B.: Knowledge Processing for Structural Design. Computational Mechanics Publication (1995)

7. Kumar, B. and Topping, B.H.V.: Non-Monotonic Reasoning in Structural Design. In: Civil Engineering Systems. Vol. 7, No. 4, Dec (1990), pp. 209-218
8. MacCallum, K.J.: Creative Ship Design by Computer. In: Rogers, D.F., Nehrling, B.C. and Kuo, C. (Ed.): Computer Applications in the Automation of Shipyard Operation and Ship Design, North-Holland Publishing Company, Volume 9, (1982) pp. 55-62
9. MacCallum, K.J. and Duffy, A.: An Expert System for Preliminary Numerical Design Modelling. Design Studies, Vol. 8, No.4, October (1987), pp231-237
10. Miles, J.C. and Moore, C.J.: Practical Knowledge Based Systems for Conceptual Design. Springer Verlag. ISBN 3-540-19823-7. (1994) pp 243
11. Moore, C.J. and Miles, J.C.: Integrated Computer Systems for Decision Support in Bridge Design. In: Sharpe, J. (Ed): AI System Support for Conceptual Design. Springer Verlag. London, (1995) pp 227-240
12. Price, G.E., Moore, C.J. and Miles, J.C.: A Decision Support System for Conceptual Bridge Design. In: Topping, B.H.V. (Ed): Developments in CAD and Modelling for Struct. Eng. Civil Comp Press. ISBN 09487 49 33 (1995) pp 53-58
13. Rees, D.G.: A Non-Prescriptive Intelligent Design System. Ph.D. Thesis presented to the University of Wales, College of Cardiff, February (1996)
14. Rees, D.G., Miles, J.C. and Moore, C.J.: A 2nd Generation KBS for the Conceptual Design of Bridges. In: Topping, B.H.V. (Ed): Developments in CAD and Modelling for Struct. Eng. Civil Comp Press. ISBN 09487 49 33, (1995) pp 17-24
15. Serrano, D. and Gossard, D.: Tools and Techniques for Conceptual Design. In: Tong, C. and Sriram, D. (Ed.): Artificial Intelligence in Engineering Design Volume 1 - Design Representation and Models of Routine Design, Academic Press Inc, (1992) pp. 71-116
16. Tong, C. and Sriram, D.: Chapter 1 - Introduction. In: Tong C and Sriram D (Ed.): Artificial Intelligence in Engineering Design Volume 1 - Design Representation and Models of Routine Design, Academic Press Inc, (1992) pp. 1-53
17. Waldron, M.B., Waldron, K.J. and Owen, D. H.: Use of Systematic Theory to Represent the Conceptual Mechanical Design Process. In: Newsome, S.L., Spillers, W.R. and Finger, S. (Ed.): Design Theory '88, Proceedings of the 1988 NSF Grantee Workshop on Design Theory and Methodology, Springer-Verlag, (1989) pp. 36-48

Agents in Computer-Assisted Collaborative Design

Divine T. Ndumu[1] and Joseph M.H. Tah[2]

[1]Intelligent Systems Research, BT Laboratories,
MLB1/pp12, Martlesham Heath
Ipswich IP5 3RE, UK
ndumudt@info.bt.co.uk

[2]Project Systems Engineering Research Unit,
School of Construction, South Bank University,
Wandsworth Road, London SW8 2JZ, UK
tahjh@sbu.ac.uk

Abstract. For researchers working at the boundary between artificial intelligence and engineering design, the notion of "agents" working in a collaborative manner to assist the design effort is not new. What therefore is the new emerging discipline of "intelligent software agents" and how do agents in that discipline differ from our contemporary notion of computational design agents? Do the differences, if any, help advance computer-assisted collaborative design? Furthermore, what challenges does the intelligent agent approach pose the computer-assisted design research community? This paper attempts to answer these questions by briefly reviewing agents research, emphasizing its potential applications in the architecture, engineering and construction industry in general and engineering design in particular. We argue that the agent-based approach provides a useful metaphor for reasoning about design systems, as well as contributing new tools and techniques for facilitating the collaborative design process. Finally, with two examples from construction supply chain provisioning and building design, we demonstrate some of the advantages that an agent-based approach brings to computer-assisted design, and highlight the main challenges posed to the design community by the approach.

1 Introduction

The architecture, engineering and construction (AEC) industry has been generally lethargic in its uptake of new information technology (IT) solutions. The industry is very multi-disciplinary, with a high degree of inter-dependence amongst activities. The activities are typically complex creative processes, requiring huge amounts of information processing, problem-solving and decision-making. These factors make most IT solutions to the industry's problems to be themselves complex, typically focused on narrow domains, and generally are alien to the industry practitioners. Artificial intelligence (AI) and classical engineering solutions such as knowledge-based decision support systems and the finite element method have managed, to some degree, to become established in the industry, providing real utility in niche areas such as design, diagnosis and analysis. However, such solutions are generally incompatible with one another, and to a large extent, have failed to address the key need of the industry — that of support for collaboration between its multi-disciplinary practitioners. What seems to be required is a framework that facilitates interoperability between the many, diverse and heterogeneous IT systems that are prevalent in the industry today.

Agent-based computer systems are currently popular in the AI community. They are supposed to help reduce the problem of information overload [7], and more importantly, to facilitate interoperability of distributed heterogeneous systems [8]. This paper aims to assess the agent metaphor as a support mechanism for collaborative engineering design. Specifically, it strives to answer the following questions:

1. What are "intelligent software agents" and how do they differ from our contemporary notion of computational design agents?
2. Do the differences, if any, help advance computer-assisted collaborative design by providing new metaphors, paradigms, frameworks, or strategies for constructing design systems?
3. What challenges does the intelligent agent approach pose the computer-supported design research community?

We argue that the agent-based approach provides a useful metaphor for reasoning about design systems. Specifically, it emphasizes certain attributes such as autonomy, social ability, responsiveness, proactiveness and high-level communication, which are germane in a collaborative design context. These attributes introduce new concepts (for example, cooperative and/or competitive negotiation), which in turn, bring in new tools and techniques that can augment traditional multi-agent design strategies. We argue, in essence, that the agent-based approach provides a metaphor and framework for channeling results from diverse AI disciplines into practical design systems.

The breakdown of the rest of this paper is as follows. Section 2 addresses the question "what is an agent?", highlighting the similarities and differences between intelligent software agents, and computational agents in contemporary engineering design. In Section 3, we highlight the basic requirements for effective collaborative problem-solving between heterogeneous systems; and in Section 4 we briefly summarize current research on agent-based design. Section 5 aims to provide a detailed assessment of the agent-based approach to collaborative design, along with the challenges it poses the design community. This is performed through the use of two implemented examples, one in construction supply chain management and another in building design. Section 6 discusses some of the main issues raised in the paper, and Section 7 concludes the paper.

2 What Is an Agent

In spite of the ubiquitous use of the word 'agent' in the AI community, a consensus definition of what exactly an agent is has failed to emerge. In this paper, we adopt the definition by Jennings & Wooldridge [4] of an agent as '*a self-contained program capable of controlling its own decision-making and acting based on its perception of its environment, in pursuit of one or more objectives*'. Typically, an agent works for and/or on behalf of a user, and acts to support the user in achieving his/her objectives. Jennings & Wooldridge specify four main attributes that determine agenthood:

- **autonomy**: the ability to function largely independent of human intervention,
- **social ability**: the ability to interact 'intelligently' and constructively with other agents and/or humans,
- **responsiveness**: the ability to perceive the environment and respond in a timely fashion to events occurring in it, and
- **pro-activeness**: the ability to take the initiative whenever the situation demands.

Nwana & Ndumu [1,2] present a complementary view of agenthood that also includes **learning ability**, i.e. the ability to improve performance over time.

In general, an agent is any software possessing a majority of these attributes, and sufficiently complex that its behavior is best explained at the knowledge level [5]. That is, by ascribing to the agent an intention and interpreting its actions as goal-directed behavior designed to achieve that intention. Two issues are important here. Firstly, for an external system (or observer) interacting with such an agent, the internal processes of the agent are opaque to the system; thus, the system can only *infer* the goals of the agent from observing its actions on the external environment. Furthermore, the behavior of the agent can only be explained with reference to the inferred goals, the environmental context in which the agent is operating and any inferred or observed knowledge of the operating history of the agent (such historical knowledge is important when explaining agents that learn). Secondly, because of their autonomy, it is safe to assume that each agent has its own *private* knowledge, knowledge representation, reasoning and learning abilities. Thus, in order for communication between agents to be independent of their private internal languages, it should be either (i) high-level – using a common shared ontology of domain concepts; or (ii) stigmergic – where implicit non-language communication proceeds through the mutual interactions of the agents in their shared environment.

The primary factors motivating agent research are the distributed, heterogeneous and dynamic nature of scarce resources (e.g. computing, information, know-how). This demands technology that respects the natural autonomy of these resources, but allows them to inter-operate in order to solve problems beyond their individual capabilities.

2.1 Agents in Contemporary Collaborative Design

The definition of agents proffered above shares close similarities with human agents in real-world design contexts, but differs in significant ways from the computational 'agents' in most current computer-supported collaborative design systems. (In this paper, we use the term 'agent' to refer to computational (software) entities only. When used in the context of a person, we preface it with human, as in *human agent*.)

Current real-world medium- to large-scale design projects involve a number of experts collaborating to perform a design. For example, a building design project might involve a structural engineer, an architect, a building services engineer, and a quantity surveyor. Each expert brings to the team specialist knowledge of their domain, and has their own private domain-specific (or even person-specific) internal reasoning, learning and knowledge representation mechanisms. Further, different experts may enter the design scenario when required and leave once their contribution is no longer needed. Collaboration between the experts might be through stigmergic communication, but typically proceeds via the exchange of high-level messages using the common shared ontology of the design drawing. The experts are agents in the intelligent software agent sense because (a) they are autonomous with their own private beliefs and goals, (b) they communicate and interact with one another, (c) they are responsive to changes to the design and to the actions of the other experts, (d) they are pro-active in their responses to such changes, and (e) they learn from their experience of performing the design. Fig. 1 depicts a schematic view of the interactions between human agents in real-world collaborative design scenarios.

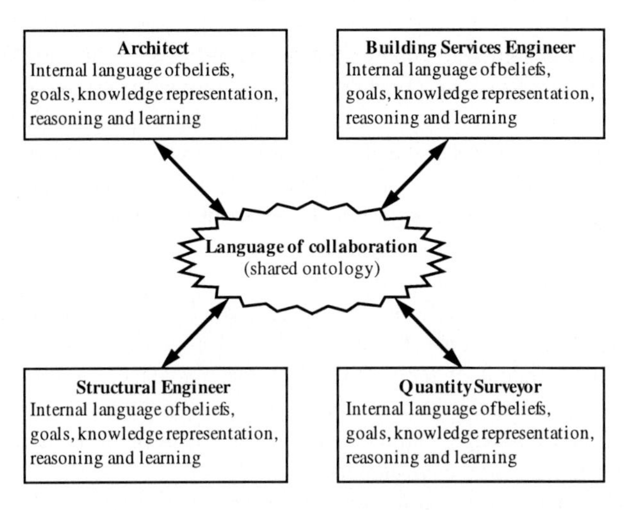

Fig. 1. Schematic view of the interactions between human agents in real-world collaborative design scenarios

It is important to note that in Fig. 1 control is decentralized, therefore coordination of the experts' behavior and resolution of conflicts are performed via negotiation between the experts. Indeed, even conflict detection is a non-trivial activity, since each expert maintains his/her own private view of the design that may be in conflict with other experts' private views or with the global shared view. In fact, for most of the design process the global view is virtual, existing only in the 'minds' of the participants. Only at the end of the design is a definitive global view agreed on.

Most current computer-based collaborative design systems fail to provide a structural and semantic symmetry-preserving mapping between the agents in the system and (human) real-world design agents. This is typified by blackboard design systems

[6]. Here, a central blackboard records the current state of the design and knowledge sources (representing different design agents) have access to and can opportunistically delete, add or modify information on this blackboard. Control of access to the blackboard and conflict resolution is managed by a blackboard control mechanism. The primary problem with this approach is the lack of flexibility caused by the centralization of control within the blackboard manager. Although centralization in itself is not a desirable feature, it can be lived with. However, to effectively resolve conflicts and progress the design, the blackboard manager needs to know in detail how and why each knowledge source forms its recommendations. This, in effect, means that (aside from architectural simplicity) the knowledge sources may as well be modules within the blackboard manager. Furthermore, with centralized control based on such intimate knowledge of each knowledge source, it is difficult for new (third party) knowledge sources to enter the design or for old ones to leave.

In summary, most current design systems provide inflexible, non-scaleable solutions, which do little to facilitate the interoperability of heterogeneous design systems required for collaborative design. The intelligent software agent approach, with its emphasis on autonomy, social ability, responsiveness and pro-activeness, is said to provide a flexible solution to this interoperability problem along with support for collaborative problem solving. In the balance of this paper, we assess these claims with respect to the design domain.

3 Requirements for Interoperability of Heterogeneous Systems

The previous section alluded to some of the basic requirements for interoperability of heterogeneous computing systems, such as an agent communication language, and a common ontology for shared domain concepts. Other issues include information discovery and coordination.

Communication

As argued earlier, for heterogeneous agents to collaborate at the knowledge level, they require an agent-independent agent communication language (ACL). A few languages have been developed to meet this need, notably the Knowledge Query and Manipulation Language (KQML) [14] and FIPA ACL [15]. Both languages specify a number of performative speech-acts [16] such as *ask*, *tell* and *reply* as legal for use in inter-agent communication, but do not specify the internal language for representing the actual content of messages. The rationale for this being that different content languages may be more appropriate in different domains; for example first-order predicate logic may be better for intercommunicating knowledge bases, whereas SQL might be better for intercommunicating databases. A number of content languages have also been developed, e.g. KIF (Knowledge Interchange Format) [8], a first-order logic language commonly used as the content language with KQML, and SL, the FIPA ACL preferred content language. Also, in a lot of applications custom content languages are typically preferred.

Ontology

Agents that communicate in a common language will still be *unable* to understand one another if they use different vocabularies for representing shared domain concepts. Ontologies (vocabularies of common concepts) for complex domains such as civil engineering design are notoriously difficult to create or maintain, and the lack of such published ontologies is currently one of the major impediment to interoperability of heterogeneous multi-agent systems. Initiatives such as STEP [15] which attempt to define and standardize product description models are important in the creation of such ontologies; however, to date, there are no large scale machine-readable ontologies for engineering. Nonetheless, for specific applications it is possible to develop a limited common ontology for the agents [9]. (The Stanford University Knowledge Sharing Laboratory website [31] hosts a number of papers addressing many issues related to creating, maintaining and porting large shared ontologies.)

Given a shared ontology for a multi-agent system application, there still remains the problem of translating between the internal language of each agent and the common ontology. Huhns et al. [17] use so-called *articulation axioms*, logical constraint expressions, to provide this mapping.

Information Discovery

The problem of information discovery at the multi-agent system level is really one of finding out about the capabilities of other agents in a society and how to contact them. These problems are typically handled using special-purpose utility agents such as *nameservers* and *facilitators*. Nameservers function as society-wide white-pages (address books) and may be arranged in hierarchies similar to Internet domain nameservers. Facilitators functions as yellow-pages, providing a look-up service for agents' abilities. Sometimes they are also used to manage the message traffic between within and between agent societies.

Coordination

In order for a group of agents to work in a coherent manner on some joint activity, they need to coordinate their actions with one another. The process of coordinating the behavior of multiple agents is currently an active area of research with many techniques in use. The main approaches can be broadly classified as organizational structuring, contracting, multi-agent planning, and negotiation.

In organizational structuring the *a priori* defined structure of the society (that is, the roles of the different agents and their relationships with one another) is exploited for coordination. The is typified by client-server systems, but also includes systems where agents engage in activities according to their roles, thus coordination is implicitly performed by assigning a role to an agent. Wiederhold [18] advanced the use of *mediators*[1] for coordinating multi-agent systems. Mediators are similar to facilitators but include the functionality of intra- and inter-society coordination. Their design and

[1] In the literature, some authors use facilitators and mediators interchangeably, while others see facilitators as 'yellow-pages' facilities only, with no coordinating role which mediators typically allude to. In this paper, we adopt the definition of facilitators as 'yellow-pages' only.

behavior is domain and application dependent, and relies on knowing the abilities of the agents they control as well as their inter-relationships. In some applications, they also perform a conflict resolution role, and in others contain the articulation axioms for translating agents' internal languages into the shared ontology.

Contracting as a coordination mechanism is typified by the classic contract-net protocol of Davis & Smith [19], where a manager agent announces a contract, receives bids from other interested contractor agents, evaluates the bids and awards the contract to a winning contractor. The simplicity of the schemes makes it one of the most widely used coordination mechanism with many variants in the literature. Some interesting alternatives to the contract-net protocol include various auction protocols such a the *english* and *dutch* auctions.

In multi-agent planning, the agents utilize classical AI planning techniques to plan their activities, resolving any foreseen conflicts. The planning normally takes one of two forms, centralized planning – in which a central agent performs the planning on behalf of the society, or decentralized planning – in which the agents exchange partial subplans, progressively elaborating the overall plan and resolving conflicts in it.

With negotiation, the agents engage in dialogue, exchanging proposals with each other, evaluating other agents' proposals and then modifying their own proposals until a state is reached when all agents are satisfied with the set of proposals. Typical negotiation mechanisms are based on game theory, on some form of planning, or on human-inspired negotiations. Nwana [20] presents a gentle introduction to the literature on coordination of multi-agent systems.

4 A Brief Review of Agent-Based Design Research

This sections presents a brief representative (but not comprehensive) review of agent-based collaborative design systems.

Cutkosky et al. [9] describe the Palo Alto Collaborative Testbed (PACT) which integrated four legacy concurrent engineering systems into a common framework. It involved thirty-one agent-based systems arranged into a hierarchy around facilitators. The agents cooperated in the design and simulation of a robotics device, reasoning about its behavior from the standpoint of four engineering disciplines (software, digital, analog electronics and dynamics). The agents communicated via KQML/KIF and used a custom ontology, built by successive refinement through dialogue between the researchers involved in the experiment. The PACT experiment was designed to research issues regarding the sharing of knowledge between heterogeneous systems and computer-aided support for negotiation and decision-making in concurrent engineering. The experiment clearly demonstrated the potential of the agent-based approach in facilitating knowledge sharing between heterogeneous systems. However, no clear results were presented on providing support for negotiation.

Shen et al. [10] present a review of recent research on agent-based systems for concurrent design and manufacturing of mechanical systems. They also describe how they have integrated a number of design, planning and scheduling systems through use of a mediator architecture. They used an extended form of KQML for communication, and the contract-net protocol as the primary coordination mechanism. While, their system allows interoperability between the design, planning and scheduling subsystems, they do not address the problem of collaboration during the design process itself.

Wellman [11] describes a computational market model for distributed configuration design. The configuration design process entails assigning parts to functions (i.e. values to attributes), within the bounds of imposed constraints so as to maximize a given utility function (performance measure). Wellman showed that if different self-interested agents were responsible for different functions, then modeling the problem as a computational market and allowing the agents' self-interest to drive the system into competitive equilibrium solves the global assignment problem (i.e. the design). He further argued that for simple examples the approach produces pareto-optimal designs quickly, but fails to do so for more complicated cases. Although the configuration design problem is not the most general form of design, and the computational market approach does not guarantee optimal designs, the approach provides a promising line of research for routine design problems that can be reformulated as configuration design. In addition, it suggests the computational market mechanism as a possible conflict resolution strategy for design. (In more sophisticated forms of design, the problem is not so much what values to assign different attributes, but determining what attributes to consider, their applicable value ranges, the desired functions for the artifact, how the functions relate to the attributes, and a reasonable performance measure for success.)

Bento et al. [12] describe a reactive agent-based approach to design founded on an extended logic and object-oriented representation of design object descriptions. They utilize object reactivity to propagate design changes and constraints. Here, the collaboration is not between design systems, but between object descriptions within a single design system. In more recent work [13], they distribute their reactive object descriptions using distributed object technology such as CORBA.

5 Experiments Using the ZEUS Agent Building Toolkit to Build Collaborative Design Systems

Having reviewed the literature on agent-based design research, in this section, we describe our experience using a multi-agent systems approach to model construction supply chain provisioning and management and collaborative building design. Our aim is to illuminate the discussion so far, to identify the potential benefits of the agent-based approach, and also to highlight some of the challenges it poses researchers developing support tools for the AEC industry. The examples are supposed to be illustrative only, to help tease out some of the benefits and problems in using an agent-based approach to support collaboration in the construction industry. As such, they are not conclusive in any sense. Both examples were developed to test an agent building toolkit we have developed called ZEUS [21]. In the following subsection, we briefly describe the ZEUS toolkit, and then proceed to describe our two experiments.

5.1 The ZEUS Agent Building Toolkit

The ZEUS agent building toolkit was developed to provide a rapid-engineering environment for developers of collaborative agent systems. The toolkit comprises a suite of Java classes which help users to develop agent-based applications by integrating and extending some predefined classes. The design philosophy behind the toolkit was to delineate domain-level problem solving abilities from agent-level functionality. Thus, the toolkit provides classes that implement generic agent functionality such as communication, co-ordination, planning, scheduling, task execution and monitoring and exception handling. Developers are expected to provide the code that implements the agents' domain-level problem solving abilities. The main components of the toolkit include an agent component library, a set of visualization tools, and an agent building environment which also includes an automatic agent code generator. The toolkit also provides utility agents such as a nameserver and a facilitator for use in knowledge discovery. Fig. 2 illustrates the architecture of agents created using the ZEUS toolkit.

- The Mailbox handles communications between the agent and other external agents.
- The Message Handler processes incoming messages from the mailbox, dispatching them to other modules of the agent.
- The Coordination Engine and reasoning system takes decisions concerning the goals the agent should be pursuing, how they should be pursued, when to abandon

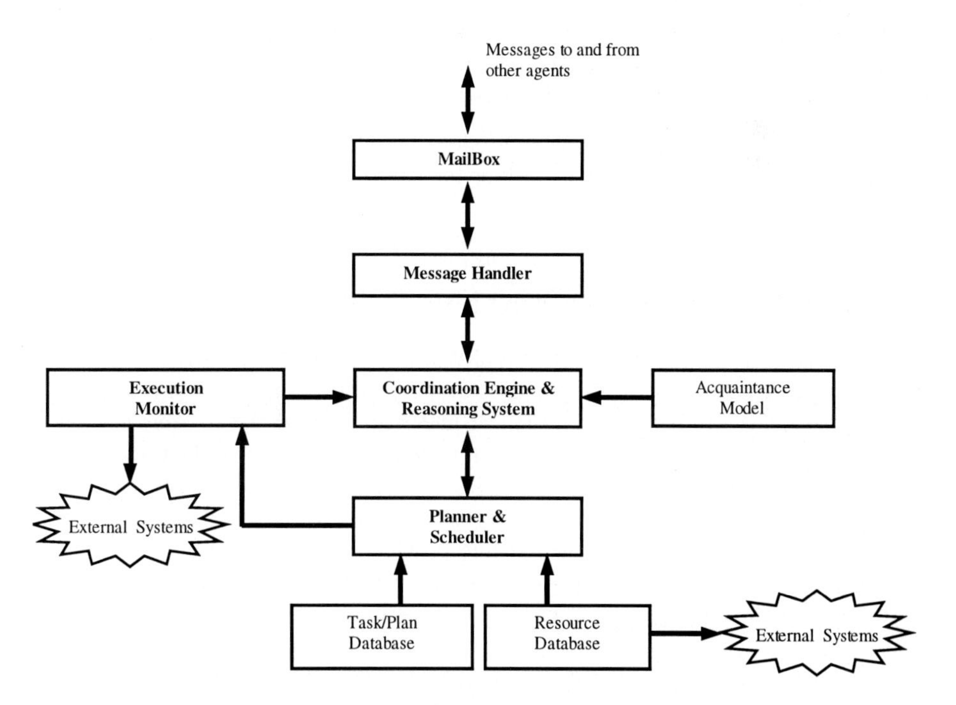

Fig. 2. Architecture of a ZEUS agent

them, etc., and coordinates the agent's overall activities. It has a database of pre-built coordination strategies including the contract-net protocol and various auction protocols. Users are also able to define custom strategies.
- The Acquaintance Model describes the agent's beliefs about the capabilities of other agents in the society, and its relationships to them.
- The Planner and Scheduler plans and schedules the agent's tasks based on decisions taken by the coordination engine and the resources and task definitions available to the agent. ZEUS agents are capable of hierarchical decentralized resource-bounded planning.
- The Resource (Fact) Database contains logical descriptions of resources available to the agent. It also provides an interface between the agent and external database systems such that the agent can query these systems about the availability of resources. Resources (facts) are specified using a frame-based representation with a very expressive language for specifying attribute-value constraints.
- The Task Database provides logical descriptions of planning operators (or tasks) known to the agent. It also contains a store of scripts which the agent uses in event-based reactivity. The scripts are described using the classic (*trigger-event*$^+$, *context, action**) triplet.
- The Execution Monitor starts, stops and monitors external systems scheduled for execution or termination by the planner. It also informs the Coordination Engine of successful and exceptional terminating conditions of the tasks it is monitoring. The services provided by these external systems are described by task specifications, with each primitive task defining an *external code stub* to the code driving or implementing the external domain function. Thus, it is possible to integrate ZEUS agents with legacy applications through the task stubs.

The ZEUS toolkit also provides, among others, an Ontology Editor for defining the shared domain ontology and a Task Editor for describing the planning operators and reaction scripts for the agent.

5.2 Application 1: Construction Supply Chain Provisioning

Various recessions since the Second World War have forced the construction industry contractors to end their historical practices of "vertical integration". Many main contractors have shed craftsmen and no longer undertake work directly. This has led to the use of labor only contracting, buying-in materials and hiring of plant. Some sections of work may be subcontracted wholesale and subcontractors may further subcontract their work. The multi-disciplinary nature of the industry and the proliferation of specialisms have aggravated the existing problems of co-ordination, communication, monitoring, and control. Clearly, there are a number of distinct disciplines required to complete a construction project, and the manner in which it is coordinated and integrated will affect the efficiency and effectiveness of the construction process. Latham's recent review of the UK industry [22] suggests that the industry's problems could be overcome by using more "collaborative" and "teamwork" approaches. The major issue facing construction clients is knowing what can be done, and by whom, in order to improve the efficiency and effectiveness of the construction procurement process.

This example currently focuses on materials management as construction materials constitute a high proportion of project costs, and the supply process is fraught with many problems that lead to program delays, overrun costs, and poor quality. These problems include late delivery, late purchase, unreliable suppliers, damaged materials, poor planning, poor co-ordination of suppliers, poor monitoring and control of project and inventory, and poor communication. The objective of material supply chain management is to obtain the right quality material, in the right quantity, from the right source, at the right time, at the right price, and in the right place. Hence the use of techniques such as *just-in-time* supply and materials requirement planning. This work is part of a major effort being undertaken to investigate the potential of agent technology to facilitate value-added decision support in such a scenario.

The material procurement process is best viewed within the general context of construction procurement. Thus, it commences with the receipt of a tender, running through the award/contract stage, onto completion and finally reconciliation and review of achievement. The key issues covered in the initial demonstration were the following:

- tasks identification and decomposition;
- supplier short listing, evaluation, and selection; and
- monitoring, co-ordination, and control of delivery and execution of tasks during construction.

The case under study, in this and the subsequent example, is a three-storey steel-framed building located on the UK M4 corridor. It reflects typical commercial low-rise office buildings in that area. The building is of a square shape, with the gross floor area approximately 1500m^2 per floor based upon 7.5 x 7.5m structural grids, and pad foundations.

Modeling the problem

The problem was modeled with over thirty agents, distributed across a network of computers, and representing the different disciplines involved in the project (e.g. project management, earthworks, concrete, steelwork, fire protection, etc.). Most disciplines were represented by more than one agent, allowing for competition and the necessary flexibility for exception handling. The capabilities of each agent were specified by defining one or more tasks it could perform. The preconditions of each task specified the resources required for the task, and its post-conditions specified the expected effects of performing the task. All tasks had an associated duration and cost, which could be functions of the resources used or produced by the task. The Project Manager agent, whose responsibility it was to oversee the entire project could perform three tasks: a complex `ManageProject` task, and two simpler `EstablishSite` and `HandOver` tasks. The `ManageProject` task was in fact a direct translation of the program of works into a task decomposition graph. The links between tasks in the graph specify both the precedence relation between the predecessor and successor tasks, and the resources produced by the predecessor task which are used by the successor task. The Project Manager agent could only perform two of the tasks in the decomposition graph, thus it had to contract out the others and coordinate their execution. Fig. 3 depicts a partial view (using the ZEUS visualizer) of the `ManageProject` task network after the Project Manager agent had contracted out portions of the work to other agents.

In order to specify the resources available to each agent and the pre- and post-conditions of tasks, we defined a small ontology for the domain. Even given the relative simplicity of the building, the ontology was complicated since it had to capture both geometric information about buildings and material specifications of their different components. Fig. 4(a) depicts a fragment of the ontology specification for building substructure.

```
ontology_item(
  Type: Substructure
  Reference: String
  Site: SiteReference
  Geometry: Rectangle
  Slabs: Slab*
  Columns: Column*
  :
)
```

```
goal(
  reference: String
  required_resources: Fact+
  desired_by: Agent
  supplied_resources: Fact*
  start_time: [Time]
  end_time: Time
  reply_time: Time
  confirm_time: Time
  cost: Cost
  invocations: Integer
  :
)
```

Fig. 4. (a) Fragment of the ontology specification for building substructure. (b) A typical goal structure

Model Simulation

The procurement process was simulated by giving the Project Manager agent the goal to achieve `handover` of the input building description by a particular date. To achieve this goal the agent scheduled the `ManageProject` task which had `handover` as one of its post-conditions. The `ManageProject` task, however, is complex, requiring a number of subtasks that the Project Manager agent could not perform. Thus, the Project Manager agent decomposed the task into its subtasks (thirteen in all) and contracted out the subtasks. During contracting the Project Manager agent used a facilitator to determine the abilities of other agents in the society. Next, following the contract-net coordination protocol, it asked potential bidders to bid for the available subtasks, evaluated the returned bids and awarded the contract to the winning agents. Bid request to bidders were goal descriptions of the form shown in Fig. 4(b). Ordering constraints in the decomposition of the `ManageProject` task (see Fig. 3), meant

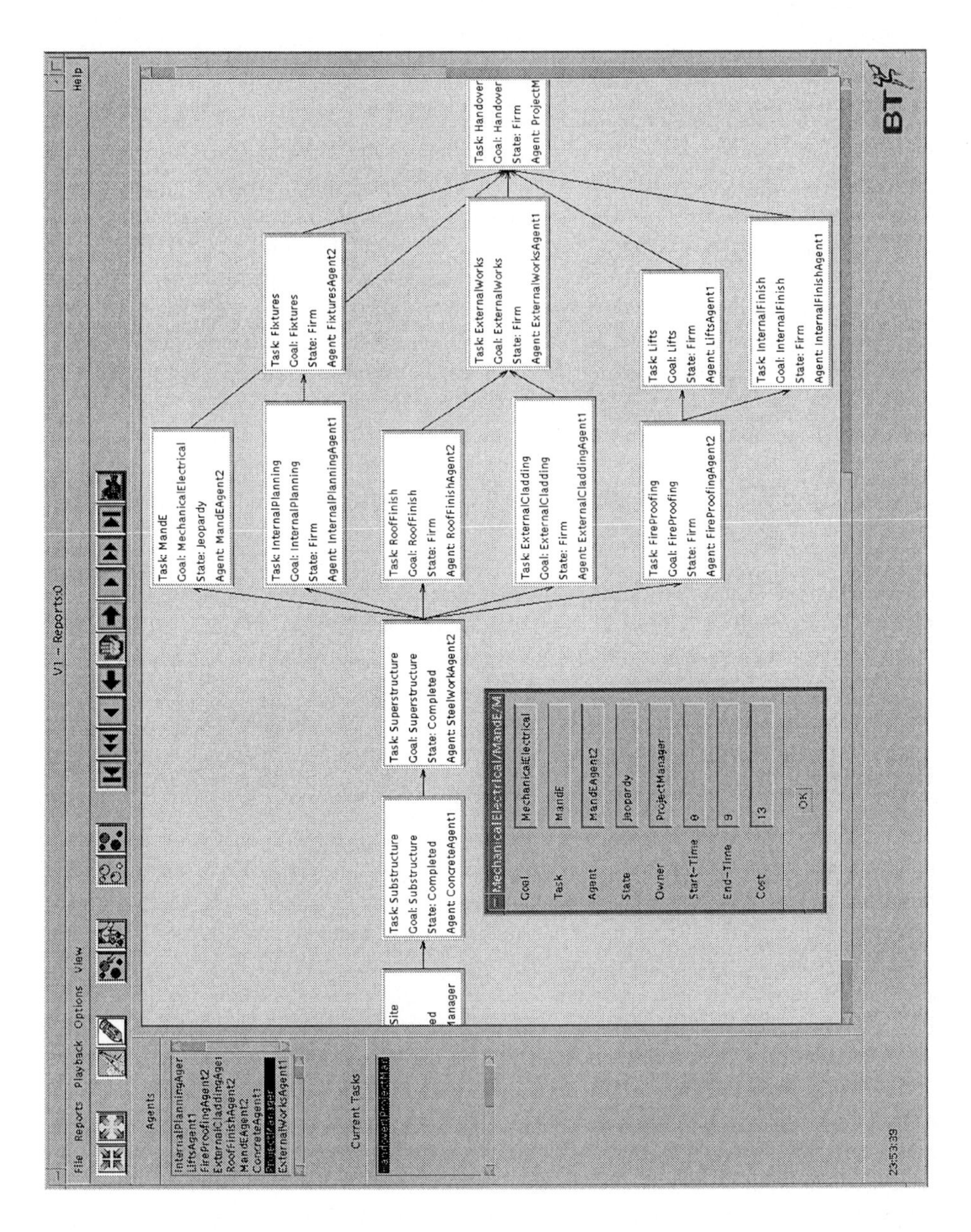

Fig. 3. A partial view of the construction supply chain after the Project Manager agent had contracted out subtasks to various agents. Most of the subtasks are either completed or on target (firmly booked) except for Mechanical and Electrical (MandE) which is in jeopardy (as shown on the popup window) because MandEAgent2 has unexpectedly run out of some resources. In the simulation, MandEAgent2 replans to reacquire the required resource

that the contracting progressed in stages, with high precedence task contracted out first.

In the simulation, the selections of winning agents for a contract were based on cost only, although other factors such as quality, relationship with the contractor, previous performance, managerial capability, etc. could have been used in the selection criteria. In other simulations, we implemented a simple reinforcement learning capability into the agents such that when contracting tasks, agents attempt to minimize the cost of the process by announcing contracts to as few potential bidders as possible. In selecting potential bidders, the contracting agent ranked the candidates by their previous performance over a time window. The final decision of how many agents to announce the contract to was determined by the level of risk failure (of the contracting process itself and the job) the contracting agent was prepared to take.

We also simulated exception management by randomly deleting resources that agents had pre-booked for use in performing a task. In such a case, the agent had to dynamically replan to achieve the task goal either by scheduling another task that did not require the lost resource, or by contracting out the task to another agent.

In summary, coordination of the procurement process was performed through a combination of distributed multi-agent planning and the contract-net protocol. The role of the agents was primarily to coordinate joint activity and to plan and schedule the procurement tasks while taking into account available resources and their associated costs.

Discussion

The first question to address in this discussion is why use an agent-based approach, that is, what does the approach give one? The key argument in favor of the agent-based approach is the drive in most companies to minimize costs and increase reliability by automating as much as is possible relatively standard procedures. A second argument is the growth of the internet, and the increasing willingness of companies to conduct business over this medium. In an article on electronic commerce, The Economist magazine [23] notes that, General Electric, a US company, currently does $1 billion a year worth of its subcontracting via the World Wide Web. This move has not only improved the fairness of the process, it has cut the length of the bidding process in some divisions to less than a half, and the company now reaches a wider base of potential subcontractors, and receives better quotes even from overseas contractors. In fact, McDonalds Restaurants UK, is spearheading a *just in time construction* initiative, whereby all its restaurants conform to a number of standard designs based on prefabricated modular units, and the construction time for new restaurants is squeezed to under one month. With such a degree of standardization the entire procurement process for such contracts could be automated with agent technology.

On the technical level, the agent-based approach effectively performs the materials procurement process and provides significant added-value over and above simple just-in-time techniques and materials requirement planning. For example, there is automatic handling of exceptions such as delivery failures by replanning. Change orders, for example design changes, equally can be handled by replanning and merging the new procurement plan with the old. Further, the agents can be made to automatically learn the performance profile of different subcontractors, which can be used to further minimize the chances of delay in the project. Finally, as a model only, the agent-based approach facilitates simulations that could be used to determine acceptable project risks prior to implementation.

The major problem faced when applying agent technology to this problem was with the domain ontology which would need to be better modeled if the approach is to scale up. In subsequent work we plan to use available parts of STEP standards to see how well they address this problem, while contributing to their development.

The supply chain provisioning case study demonstrated somewhat the potential for agents in the AEC industry. The approach was relatively successful because of the loose coupling between the task activities. Thus, planning and contracting were sufficient to coordinate the various agents' activities, to the extent that the entire process was automated. In design, there is tighter coupling between task activities, with consequently more room for conflicts. Furthermore, the design problem itself is never fully declaratively specified, nor is the solution clear even when it has been found. In the next case study we assess the agent-based approach when used to support collaborative design.

5.3 Application 2: Building Design

The problem addressed was supporting the collaboration between four experts involved in the re-design of the three-storey building described earlier, following a change of use order from the client. The experts involved in the scenario were an architect who was also the client's representative, a structural engineer, a quantity surveyor and a building services engineer. Fig. 5 shows the initial architectural layout and structural design plans prior to the change order.

Modeling the Problem

The problem was modeled using four software agents *supporting* the architect, structural engineer, quantity surveyor and building services engineer. The agents did not perform the design themselves, but provided front-ends for human designer – thus, their role was only supportive. The architectural agent had in its plan database the complex task network shown in Fig. 6 which specifies how to produce a final design, and which required collaboration with the other three agents in the scenario. Each agent also had in its plan database primitive task descriptions specifying its capabilities. For example, the structures agent's plan database contained the task `Design-Structure` which required an input `BuildingLayoutPlan` and produced as output `StructuralPlan`. The primitive task stubs (links to the external realization of tasks) created the front-ends used by the human designers, and also set up event monitors to monitor the actions of the designers.

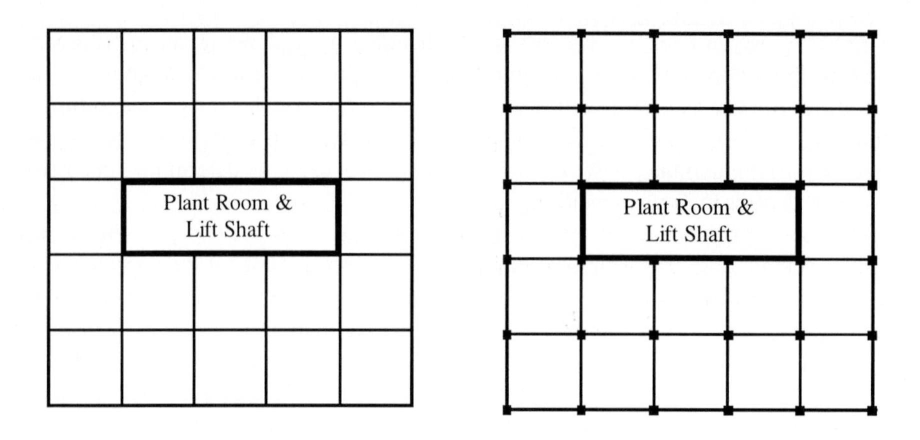

Fig. 5. The (a) architectural layout and (b) structural design plans prior to the change order from the client

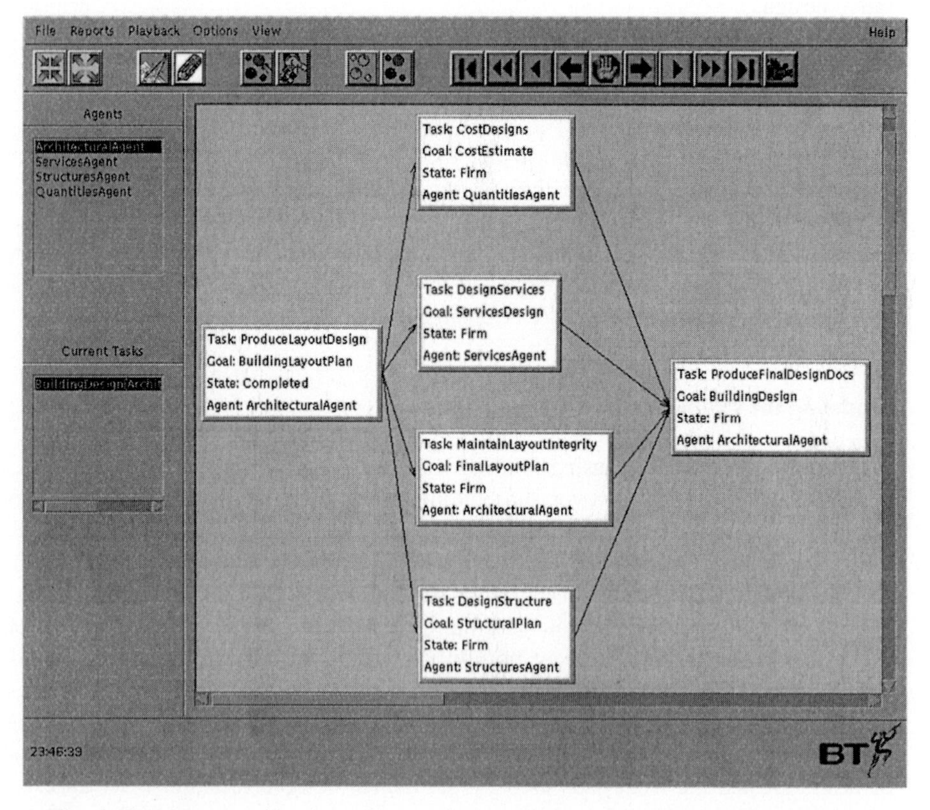

Fig. 6. The plan created by the architectural agent to manage the design of the building

Effectively, the monitors perform the collaboration support role by monitoring the evolving design and notifying changes to other concerned agents (Jennings [24] outlines cooperativity guidelines for agents engaged in a joint activity). The monitors also performed the role of articulation axioms, translating from the graphical representation used by the architect and structural engineers into the common ontology we devised for the domain. The code fragment below shows one of the monitors set up by the structures agent. The first if-block monitors changes to the graphical representation of the plan and modifies the graphics-neutral representation. The second block detects modifications to the graphics-neutral representation and notifies appropriate agents of this modification.

```
public class DesignStructure implements ZeusTask {
   public void exec() {
      ⋮
      win = new StructuresWindow();
      monitor = new GraphicsMonitor(inputArgs[0],win);
      agent.addFactMonitor(monitor,
                           FactEvent.MODIFIED_MASK);
      ⋮

public class GraphicsMonitor implements FactAdapter {
   public GraphicsMonitor(Fact initialLayout,
                          StructuresWindow win) {
   ⋮
   }
   public void factModifiedEvent(FactEvent event) {
      String architect, services, quantities;
      ⋮
      Fact f = event.getFact();
      String type = f.getType();
      if ( type.equals("GraphicalStructuralPlan") ) {
         /* updatePlan() translates the GraphicalStructuralPlan into a new
            StructuralPlan and saves the previous StructuralPlan in the design
            history */
         Fact plan = db.get("StructuralPlan");
         Fact new_plan = updatePlan(plan,f);
         db.modify(plan,new_plan);
      }
      else if ( type.equals("StructuralPlan") ) {
         architect = initialLayout.getValue("architect");
         msg = new Performative();
         msg.setType("tell");
         msg.setReceiver(architect);
         msg.setContent("structural_plan changed " + f);
         agent.sendMsg(msg);
         ⋮
```

Model Simulation

Given the goal to produce the final design, the architectural agent set up the design plan shown in Fig. 6, where it contracted in the services of the other three agents in the scenario using multi-agent planning and the contract-net protocol coordination mechanisms. During plan execution, the human architect[2] produced the initial layout plan of Fig. 5(a) which the architectural agent set as input to its `MaintainLayoutIntegrity` task, and also copied to the structures, services and quantities agents as dictated by the design process plan of Fig. 6. Next, the human structural engineer produced the structural plan of Fig. 5(b). Through the firing of the event monitors set up by the `DesignStructure` task, the structural plan is copied to the other three agents in the scenario. (The structures agent knows the identity of the services and quantities agent from the architectural agent when it confirms the contract – by then the architectural agent has decided on all the participants; and also when the architectural agent copies the `BuildingLayoutPlan` to it.) Assume for now, that the building services engineer and quantity surveyor produced their initial designs and costs respectively, which were acceptable to all the parties concerned.

Now, the architect receives a change order (for a more open plan building) from the client. He then rejects the intermediate columns proposed by the structures agent earlier, and moves the central plant room and lift shaft from the center of the building to the bottom-middle. Again, the design changes are propagated to the other agents. If the structures agent refuses to accept the rejection of the intermediate columns proposed by the architectural agent, the design deadlocks.

Resolving Design Conflicts

To resolve design conflicts such as the deadlock above, we modeled the actions of the designers as a computational market. Each design proposal by an agent had an associated cost of generating the proposal, and a subjective value the designer attaches to the proposal. The goal of each designer is to maximize its utility by generating a profit (value – cost). The design process starts with each agent given an initial endowment. In order to reject a design proposal from another agent, the rejecting agent must be willing to pay for the decision at a rate acceptable to the proposing agent, that is, a rate greater than the cost of the design to the proposing agent. Note that only the agent generating a proposal knows its cost and value. Thus, an agent rejects a design only if the sum of the cost of rejection and the cost of its own proposal is less than the expected value of its new proposal. Agents are allowed to go into overdraft during normal design changes but not when rejecting other agents' proposal. We utilized the market-based transaction protocols of ZEUS [25] to implement this conflict resolution scheme. The scheme guarantees deadlock free two-party negotiations (we have yet to assess the scheme in multi-party negotiations).

---

[2] Please note that the authors simulated the roles of all the human participants in this example, thus, no real architects, structural engineers, etc. were involved.

Discussion

In general, the agent-based approach provides at least base-level support for distributed collaborative design, by providing mechanisms for setting up the design team and communicating intermediate results between team members. However, it also raises a number of issues including the ontology problem, which was the key problem in the previous example.

The event-based model used to manage communication of design proposals does not address the questions of what to communicate, when and to whom. Currently, proposals are communicated to all team members regardless of relevance (strictly speaking, this is not an agent issue, but more of a modeling one). However, once selective notification of design changes is introduced, a distributed truth maintenance mechanism (e.g. [26]) will be required to manage the consistency of the emerging design.

Regarding conflict resolution, the market model used in the example, was limited in a number of ways. First, it is purely syntactic, failing to consider the rationale for design changes. Many authors, e.g. Pena-Mora et al. [27], have argued for the use of design rationale as a conflict resolution mechanism. While, we plan to incorporate design rationale in future work, we note that communicating one's rationale to another agent imposes extra demands on the shared communication language, and makes the assumption that the receiving agent can adequately interpret and understand the rationale. Parsons & Jennings [30] suggest an argumentation based negotiation framework that could be used to communicate design rationale during conflict resolution. The second way in which the market approach is limited results from the fact that there is no *objective* mechanism for determining the cost and value of a design proposal. Sycara [28] proposes the use of multi-attribute utility theory in evaluating tradeoffs during a negotiation process, nevertheless, one is still left with the problem of assigning costs to design processes and values to designs. However, we believe that if the cost/value assignment problem is resolved, the market model offers a powerful domain-independent conflict resolution mechanism.

A problem raised implicitly by the example, was the need for better models (to allow reasoning) of the role relationships between the members of the design team. For example, the architect, in its role as the client's representative, should be able to override any design recommendation from the other agents. However, in its role as a team member, it should only be able to negotiate settlements with the team. Barbuceanu [29] describe a deontic logic framework for ensuring agents operate according to their role relationships.

6 General Discussion

The previous sections described the general ideas underpinning the intelligent software agents approach and presented two illustrative examples of applications of the approach in the AEC industry. In this section, first, we evaluate our examples with respect to the criteria of Section 2 – "what is an agent"; and next, we summarize some of the challenges posed to the computer-assisted collaborative design research community by the agent-based approach.

In Section 2 we saw that agents in the intelligent software agents approach emphasize four attributes – autonomy, social ability, responsiveness and pro-activeness. Furthermore, in Section 3, we argued that interoperability of heterogeneous agent-based systems demanded an agent-independent communication language, a common ontology for representing shared domain concepts, and mechanisms for coordinating the problem-solving activities of societies of agents. In our two examples, the agents satisfied both the criteria for agenthood and for interoperability. For example, in the construction supply chain scenario, the agents were completely autonomous; utilized KQML (an agent-independent communication language) for communication; used a shared, custom-built domain ontology for representing communicated domain concepts; and coordinated their activities using a combination of multi-agent planning and the contract-net protocol. In addition, the agents were responsive, reacting to unexpected events that interfered unfavorably with their plans by replanning and amending their plans. In the collaborative design support example, the agents were less autonomous since human designers performed most of the design; nonetheless, they were responsive to their environment, monitoring the actions of the human designers and propagating design changes. Further, they proactively initiated and performed conflict resolution using the market model we described earlier.

Regarding the challenges posed to the collaborative design research community by the agent-based approach, our examples identified a few as the ontology problem, domain-independent conflict management, distributed truth maintenance and role modeling. Another challenge only implicitly raised in the collaborative design example, was the need for *any-time problem solving* by the agents. By this we mean that because any one agent may at any point in the collaborative design process reject one of its prior proposals, or even that of another agent, all the agents need to be able to backtrack to any point of their design history. With human designers, such any-time behavior is relatively straightforward, although it becomes less so if the human designer are replaced or augmented with legacy design support systems. In Ndumu & Nwana [3] we discuss general research and development challenges for agent-based systems.

We conclude this section by arguing that the agent-based approach provides both a *metaphor* and *framework* for reasoning about design systems. As a metaphor, the approach emphasizes the view of (agent-based) design support systems as autonomous, socially able, responsive and pro-active intentional entities. This view makes minimal assumptions about the internal behavior of the system, and forces other third-party systems to interact with such systems at the knowledge level – a much more scaleable alternative. As a framework, the approach defines an abstract structure for researching and developing heterogeneous interoperable systems. For example, it clarifies the rationale for and positions traditional AI design support applications development problems such as managing shared ontologies, conflict management, belief revision and role modeling. Note however, that the agent-based approach does not in itself provide a solution to any of these problems. It simply defines a structure for downstreaming results from various AI disciplines into applications development. For example, in the case of collaborative design, we still require design techniques such as case-based reasoning. Agents simply support such design systems, allowing interoperability and collaboration between systems.

7 Conclusions

In this paper, we have reviewed the literature on agents research and development, and outlined the potential contributions of the agent-based approach to the AEC industry. We argued that the approach facilitates interoperability between the many, diverse and heterogeneous decision-support systems in use in the industry today. Using examples from construction supply chain provisioning and collaborative building design (both implemented using our ZEUS agent-building toolkit), we explored in some detail the potential benefits of the approach as well as the challenges it poses the design community.

In summary, we argue that the agent metaphor is natural to the AEC industry, and furthermore, that it provides a natural framework for collaborative design research.

Acknowledgment

The authors work like to acknowledge the help of the members of the Intelligent Systems Research group in producing this paper. Special thanks to Hyacinth Nwana and Jaron Collis. The advice of Gilbert Owusu of LMU is also gratefully acknowledged. Thanks to Chimay Anumba for encouraging us to write this article.

References

1. Nwana, H.S.: Software Agents: An Overview. The Knowledge Engineering Review, 11(3) (1996) 205–244
2. Nwana, H.S., Ndumu, D.T: An Introduction to Agent Technology. BT Technology Journal, 14(4) (1996) 55–67
3. Ndumu, D.T., Nwana, H.S.: Research and Development Challenges for Agent-Based Systems. IEE Proceedings – Software Engineering, 144(1) (1997) 2–10
4. Jennings, N.R., Wooldridge, M.: Software Agents. IEE Review, January 1996 pp. 17–20
5. Newell, A.: The knowledge level. Artificial Intelligence, 18 (1982) 87–127
6. Morse, D.V., Hendrickson, C.: Model for Communication in Automated Interactive Engineering Design. Journal of Computing in Civil Engineering ASCE 5(1) (1991) 4–24
7. Maes, P.: Agents that reduce work and information overload. Communications of the ACM 37(7) (1994) 31–40
8. Genesereth, M.R., Ketchpel, S.P.: Software agents. Communications of the ACM 37(7) (1994) 48–53

9. Cutkosky, M.R., Englemore, R.S., Fikes, R.E., Genesereth, M.R., Gruber, T.R., Mark, W.S., Tenenbaum, J.M., Weber, J.C.: PACT: An experiment in integrating concurrent engineering systems. IEEE Computer 26(1) (1993) 28–37
10. Shen, W., Xue, D., Norrie, D.H.: An agent-based manufacturing enterprise infrastructure for distributed integrated intelligent manufacturing systems. In Proc. 3rd Int. Conf. Practical Appl. Intelligent Agents and Multi-Agent Technology, Nwana H.S. & Ndumu D.T. (eds) March 1998, pp. 533–548
11. Wellman, M.P.: A computational market model for distributed configuration design. AI EDAM 9 (1995) 125–133
12. Bento, J., Feijó, B.: A Post-Object Paradigm for Building Intelligent CAD Systems. Artificial Intelligence in Engineering, 11(3), (1997) 231–244.
13. Feijó, B., Rodarki, P., Bento, J.P., Scheer, S., Cerqueira, R.: Distributed agents supporting event-driven design processes. Artificial Intelligence in Design '98, (eds) John Gero and Fay Sudweeks, Kluwer Academic Publishers.
14. Finin, T., Labrou, Y.: KQML as an agent communication language. In Bradshaw, J.M. (ed.) Software agents. MIT Press, Cambridge, Mass. (1997) pp. 291–316.
15. STEP: The ISO STEP standards (ISO 10303): http://www.steptools.com/library/standard/
16. Searle, J.R.: Speech acts. Cambridge University Press, 1969, Cambridge MA.
17. Huhns, M.N., Singh, M.P. Ksiezyk, T. Global information management via local autonomous agents. In Readings in Agents, Huhns, M.N. , Singh, M.P., Morgan Kaufmann Publishers, Ca. 1998 pp. 36–45
18. Weiderhold, G.: Mediators in the architecture of future information systems. IEEE Computer 25(3) (1992) 38–49
19. Davis, R., Smith, R.G.: Negotiation as a metaphor for distributed problem solving. Artificial Intelligence 20 (1983) 63–109
20. Nwana, H.S., Lee, L. Jennings, N.R.: Coordination in software agent systems. BT Technology Journal, 14(4) (1996) 79–88
21. Nwana, H.S., Ndumu, D.T., Lee, L.C.: ZEUS: An advanced toolkit for engineering distributed multi-agent systems. In Proc. 3rd Int. Conf. Practical Appl. Intelligent Agents and Multi-Agent Technology, Nwana H.S. & Ndumu D.T. (eds) March 1998, pp. 377–391
22. Latham Report: Constructing the team. http://www.t-telford.co.uk/Nec/latham.html
23. Anderson C.: In search of the perfect market. The Economist, May 10th 1997.
24. Jennings, N.R: Controlling Cooperative Problem Solving in Industrial Multi-Agent Systems using Joint Intentions. Artificial Intelligence, 75 (2) (1995) 195–240
25. Collis, J.C., Lee, L. Building electronic marketplaces with the Zeus Agent Toolkit. Autonomous Agents 98: Workshop on Agent-Mediated Electronic Trading, Minneapolis May 1998.
26. Petrie, C.: The REDUX' Server. In Readings in Agents, Huhns, M.N. , Singh, M.P., Morgan Kaufmann Publishers, Ca. 1998 pp. 56–65.
27. Pena-Mora, F., Sriram, D., Logcher, R.: Design rationale for computer-supported conflict mitigation. Journal of Computing in Civil Engineering, 9(1) (1995) 57–72
28. Sycara, K.: Utility Theory in Conflict Resolution. Annals of Operations Research 12 (1988) 65–84
29. Barbuceanu, M.: How to make your agents fulfil their obligations. In Proc. 3rd Int. Conf. Practical Appl. Intelligent Agents and Multi-Agent Technology, Nwana H.S. & Ndumu D.T. (eds) March 1998, pp. 255–276
30. Parsons, S.D., Jennings, N.R.: Negotiation Through Argumentation-A Preliminary Report. In Proc. 2nd Int. Conf. on Multi-Agent Systems, (1996) Kyoto, Japan, pp. 267–274
31. KSL: http://ksl-web.stanford.edu/knowledge-sharing/papers/

A Collaborative Negotiation Methodology for Large Scale Civil Engineering and Architectural Projects

Feniosky Peña-Mora

Gilbert Winslow Assistant Professor of Information Technology and Project Management, MIT Room 1-253, Intelligent Engineering Systems Laboratory, Department of Civil and Environmental Engineering, Massachusetts Institute of Technology, Cambridge, MA 02139. E-mail: feniosky@mit.edu.

Abstract. Large-scale engineering and architectural projects are unique endeavors that require a high level of human creativity from multiple professional disciplines. In such an environment, successful collaboration is critical. Because the different participants are from different technical backgrounds, and typically come from different organizations, competitive stresses exist within project relationships. Better methodologies are needed to relieve these stresses that exist in this environment. The research presented in this paper outlines a collaborative negotiation model and methodology that successfully takes into consideration the following four issues faced during large scale civil engineering and architectural projects: (1) a collaborative-competitive environment; (2) domain-dependent information; (3) strategy-influenced outcomes; and (4) the geographical and time distribution of the negotiating individuals. The methodology explored in this paper utilizes the quantitative nature of game theory and the qualitative nature of various conflict resolution theories, and better defines the interactions of collaborating professionals to use the strengths of humans and computers alike. This research also sought to complement the development of new communication technologies with the development of a collaborative negotiation methodology that best allows professionals to take advantage of those technologies by defining the protocols of interaction during the negotiations presented in large-scale engineering systems.

1 Motivation and Problem Description

Large-scale civil engineering and architectural projects are unique endeavors, requiring a high level of human creativity and input from each participating professional. The professionals involved are brought together in what is intended to be a collaborative situation in which all the participants work together toward the goal of a successful project. Despite the team appearance, however, these individuals are typically from different organizations and differing ideological backgrounds, and must compete for limited resources and maintain balance between the project priorities and their own individual organization's priorities. Thus, the risk of conflicts occurring exists. Unmitigated conflict leads to a breakdown in team structure, the creation of less efficient solutions to technical problems, and many times, serious and crippling disputes. All of these problems may result from poor collaboration and negotiation between professionals through each phase of the project cycle. The successful avoidance of most disputes, coupled with the successful resolution of those that do occur, is critical to the maintenance of an effective large-project team. For any such project to be successful, therefore, those professionals must not only reach collaborative decisions on the particular technical issues at hand, but must also negotiate the proper allocation of limited project resources. Given this inherent strain between the collaborative need and the competitive reality, better methodologies are needed to improve the collaborative process and to create more effective, efficient, and sustainable results.

Additionally, the globalization of the design and construction industry is occurring at an unprecedented rate, bringing a distributed flavor to a process that has traditionally been handled in a face-to-face manner. As the world becomes "smaller," professionals must deal not only with the barriers that come from the differing technical backgrounds, but also with the difficulties of working across large distances for extended periods of time, with people of varying ideological and cultural backgrounds. Because the traditional face-to-face meeting environment will be less of an option, professionals must use the available communication technologies to interact and exchange ideas with their colleagues, to negotiate issues regarding their decisions, and to produce and share their intermediate and final products.

These negotiation issues that the project team faces can be summarized in four basic hurdles, each of which must be overcome for successful collaboration in the large-scale engineering domain. Failure to overcome all four of these hurdles will result in sub-optimal, inefficient, or short-lived negotiation outcomes.

1. The first hurdle in a collaborative negotiation is to explore sound solutions to the technical problem facing any given team of professionals. Negotiation in large-scale engineering and construction projects are extremely domain-dependent, as professionals are experts in each of their individual fields, and therefore bring their expertise to solve problems and to exert considerable influence on a negotiation. This collaborative nature of the working relationships among professionals must be recognized and addressed.
2. The second hurdle to overcome is one of resource allocation and objectives implementation. Most problems are not simply differences in technical opinions, but represent a distributive aspect [Raiffa, 1994]. In negotiating those problems, different team members are competing on how best they could implement objectives that provide their organizations with maximum benefit. This competitive nature of the interaction, therefore, must be taken into consideration when developing a methodology for collaborative negotiation.
3. The third hurdle is the human negotiation style and interaction problem in which professionals must find ways to communicate effectively [Ury, et al, 1988]. Different negotiators will use different strategies and tactics in their interactions.

Thus, the influence of the strategy used by negotiators must also be explored and understood in a collaborative framework.

4. Finally, the professionals involved must deal with the difficulties of working in a geographically and time-distributed environment. These conditions will challenge those accustomed to working only within the boundaries of the traditional face-to-face negotiation. The use of new communication technologies to address this issue must recognize the collaborative aspects required by the domain, and must seek to encourage sustainable (i.e., long lasting) group processes.

In order to overcome those hurdles, this research explored and tapped into the established literature of game and negotiation theories, and modified those for the large-scale engineering and construction domain. This research also sought to complement the development of new communication technologies with the development of a collaborative negotiation methodology that best allows professionals to take advantage of those technologies by defining the protocols of interaction during the negotiations presented in large-scale engineering systems.

2 The Research Approach

Given the problem statement described in Sections 1.0, this paper describes a five-prong research approach for developing a more effective collaboration and negotiation methodology. The first three are covered in detail in this paper, and the last two are work in progress.

1. Generic Negotiation Model: The first prong of the research involved the development of a generic negotiation model for the domain of large-scale engineering and construction. This model is a generalized representation of the typical parties, structure, relationship, and attributes that make up a negotiation, and can assist researchers in understanding how the developed methodology fits into the subject environment.
2. Game Theory: The second prong involved research into the existing game theoretic literature. Game theory, the study of rational behavior in situations involving interdependency [McMillan, 1992], is an established science that offers valid insight into strategic bargaining in competitive situations and serves as an important foundation for the research presented here.
3. Negotiation Theory: The third prong of the research involved an exploration of negotiation theory with respect to collaboration, conflict mitigation, and dispute resolution. These topics describe the attributes that a human agent should posses in order to most effectively assist communication between both collaborative and competitive groups.
4. Project Delivery Systems: The fourth prong of the research will examine different pure project structures and delivery systems, and will evaluate how they impact the negotiation through changes in participant roles and relationships. Furthermore,

this research prong will examine how the negotiation methodology would be adapted to accommodate such conditions.

5. Global Collaboration: The fifth and final prong of the research will investigate the ramifications that globalization will have on negotiation and collaboration across international borders. Any successful collaborative negotiation methodology will need to reflect such an understanding of cultural approaches and techniques of negotiation.

Figure 1 illustrates how the five prongs of the research fit together in the development of a better methodology for collaborative negotiation. Game theory and negotiation theory form the pillars of the collaborative negotiation research in the domain, and the generic negotiation model comes from a combination of the two disciplines. The relevant aspects of global collaboration and project delivery systems then modify that generic model, so that the final methodology reflects the unique aspects of the large-scale civil engineering and architectural project domain.

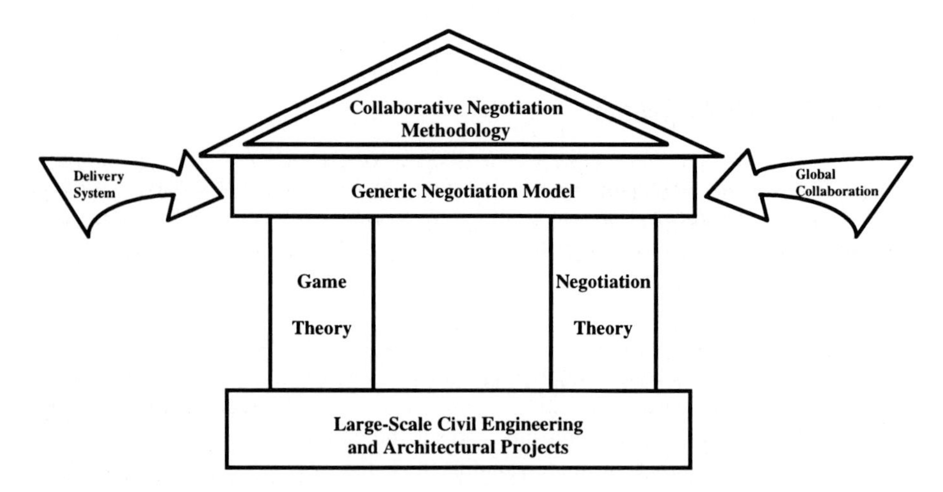

Fig. 1. Proposed components of the research

To illustrate the research efforts and methodology development, the following simplified scenario has been developed. Clearly, real world problems will be more complicated, and so will actual case studies utilized in the research. This scenario, however, can highlight some of the concepts discussed, and describe how the methodology will address them. As this fictitious negotiation moves forward, it will be revisited in each section of the paper to illustrate the importance of each research thrust. Finally, a resultant outcome will show how a particular methodology can be used to resolve the conflict to the satisfaction of the participants.

An owner, the state highway department, is beginning to implement a large-scale upgrade to a twenty-mile section of a major expressway. Included in the scope of the overall project is the reconstruction of several ramp and bridge locations. The owner has prioritized the different projects within the overall program, and has selected a designer for the first ramp section. The designer, in turn, has completed the plans and specifications. The owner, who must by law use the traditional delivery system for this type of public work, has awarded the contract for construction to the low bidder. That contractor has begun operations on site, but is forced to stop work because, while digging, an unexpected abandoned pipe and tank arrangement is discovered. After some investigation, it is concluded that the lines are old process piping for a factory that once occupied the site. The pipe is insulated with asbestos wrapping, and must be removed in order for work to proceed. A meeting is called to discuss the options, and all three parties are in attendance. Between them, these parties must decide on how to solve the technical problems associated with abatement and removal, and design issues regarding proper preparation of the subsequent soil conditions. Additionally, they must discuss the schedule implications of the delay, and deal with the additional scope and delay costs of the work from both the designer and the contractor. Work has stopped until the negotiation can be concluded.

3 Research Components

3.1 Generic Negotiation Model

Given that the design and construction domain is complex, involving multiple participants and multiple negotiation variables, it is essential to break the problem down to its basic factors. Figure 2, a preliminary representation of a generic negotiation model, illustrates the structure of a generic project in the large-scale engineering and construction domain. The model consists of five basic elements: (1) the project; (2) the participants; (3) the negotiation interaction process; (4) the negotiation assisting methodology or system; and (5) the outcome.

In a negotiation, each participant brings several attributes with them to the project domain. For any given issue, problem, or conflict encountered, the participant will have a negotiating *position* [Fisher and Ury, 1981], or an explicit set of requirements for the settlement of the negotiation. These positions can be considered the yardstick by which the initial stance and the subsequent movements can be measured. Behind positions lie the *interests* of the participants. These interests are the basic underlying needs, preferences, concerns, or goals that need to be satisfied for a successful and sustainable results [Susskind and Cruikshank, 1987]. Interests may or may not be reflected in the positions of the participant, and represent the more elusive of the two variables. These interests may range from factors that correlate well with positions, such as money or time, or factors that do not correlate, such as aesthetic value or professional reputation. Even more elusive, yet just as critical, are the *attitudes* of the participants within the negotiation. Attitudes are general descriptions of a negotiator's style, such as competitive, altruistic, or nihilistic [Darling and Mumpower, 1990]. Attitudes can also typify a particular negotiator's approach, or may be purely a situation-specific variable. An example might be a negotiator who tends to use the tactic of bluffing as part of an overall intimidation strategy.

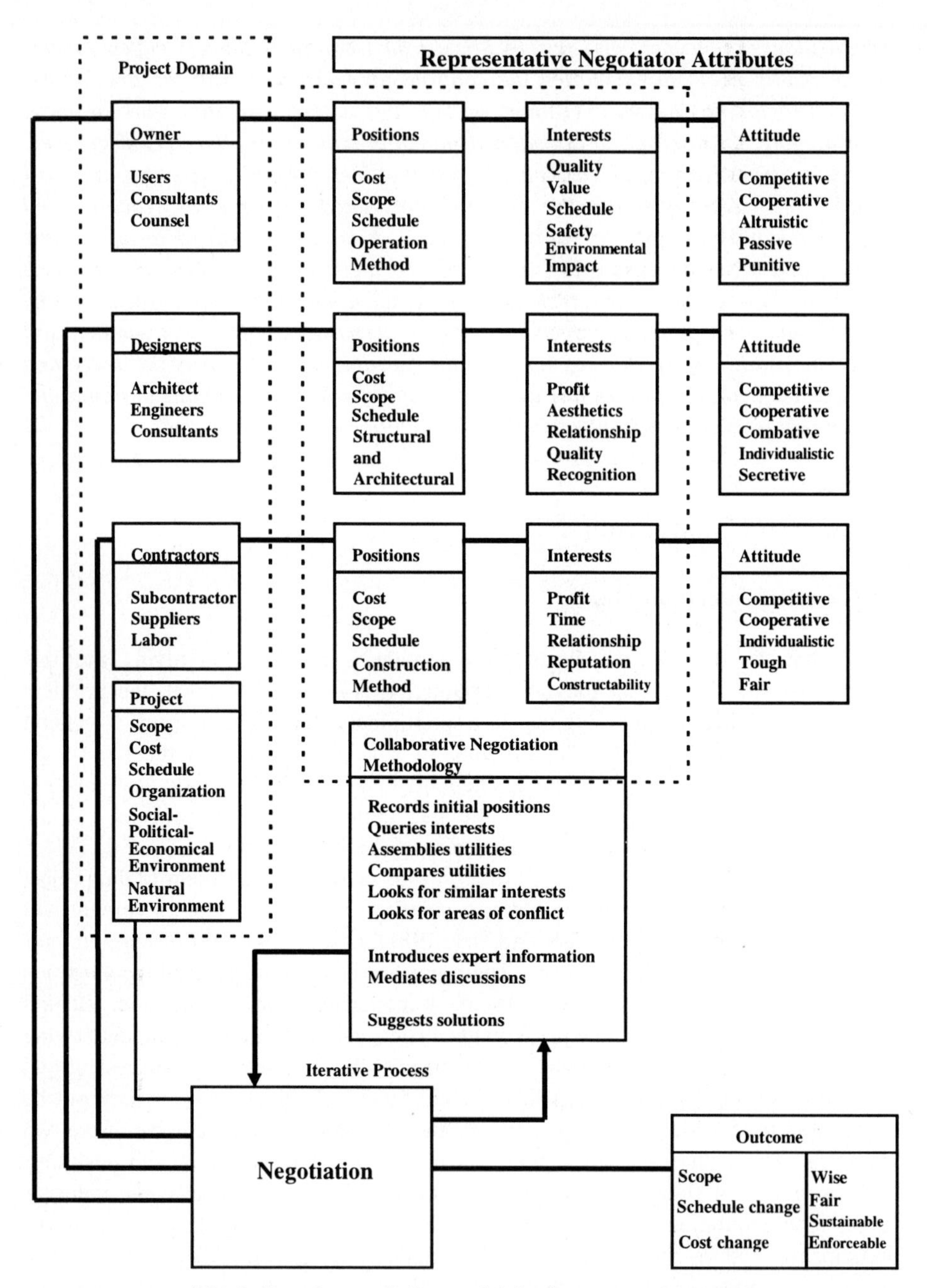

Fig. 2. Generic negotiation model for the construction domain

In the sample highway construction scenario, the contractor has entered the negotiation with a "first offer" position that he requires $100,000 for delay compensation, which he doubts that he can receive in full. In reality, he is concerned

with covering his true costs, which is only $60,000. He also has a strong interest in not performing the abatement work, as he needs to retain bonding capacity for another bid, and is nearing his aggregate limit. The designer, in the other hand, submitted an extremely reasonable estimate of $5,000 for additional design fees with the hope of gaining the monitoring fees. He knows that the monitoring fee could climb as high as $20,000, with a guaranteed profit margin. And finally, the owner is under a crucial time restriction on the project because it is the first of many subsequent projects. A delay on the first one would be disastrous in the long run, and she is willing to pay more money if both the designer and contractor can get through the problem in half the time they have requested. None of the three participants enter the negotiations with the plan to give away their true interests first. To do so would put them at a disadvantage when it got down to the hard-nosed bargaining.

As can be seen in the above example, the important element to be concerned about is how interests are satisfied, with less concern toward individual position changes. The value of an explicit expression of interest will be reflected in the ability of a potential solution to address their basic interests. In order to identify the relationships those interests have with their positions, the research looks at utility functions in game theory. In order to define the interests of participants, this research looks at negotiation theory to assist in defining which variables to use in building the utility function, as well as how to assign weights to those variables. Thus, the critical aspect of this research approach to the generic model is to define the attributes of the negotiators and the project in a manner that is useful to the negotiators themselves. This requires an iterative process that involves collecting information from the participants, manipulating it according to the interaction guidelines of the methodology, and then presenting it back to the participants so that negotiators can reach a successful goal.

3.2 Game Theory

In order to address these five areas, the focus of the research was on computer mediated collaborative negotiation for the A/E/C industry using a combination of game and negotiation theories. However, there have been extensive research on agent (machine, software) negotiation and computer supported human negotiations that also apply those two theories. Thus, these two categories were surveyed to determine how the proposed research is distinctive from and yet built on the previous studies, for more detailed information please see [Peña-Mora, and Wang, 1998].

Figure 3, 4, and 5 show how different systems in that area score with respect to the issues that are relevant on large-scale infrastructure projects. There are six axes that stand for six important issues for resolving conflicts in the large-scale civil engineering and architectural projects. These are plotted in two three-dimensional graphs to show the position of current research. The first issue is whether computers can highly assist the negotiation process. This means that computers should provide suggestions of settlements, negotiation information for the participants, preparation for the participants, and support for the process that should be followed. The second issue is how a methodology takes into consideration the collaborative-competitive nature of negotiator interactions. The third issue relates to the awareness of the methodology to the strategies used in the A/E/C industry. The fourth element is how temporary contractual arrangements are taken into consideration the methodology. The fifth issue is if the methodology allows for geographic dispersion of negotiators. Finally, the sixth issue is the ability of the methodology to account for the dynamic behavior of the participants where preferences change with time.

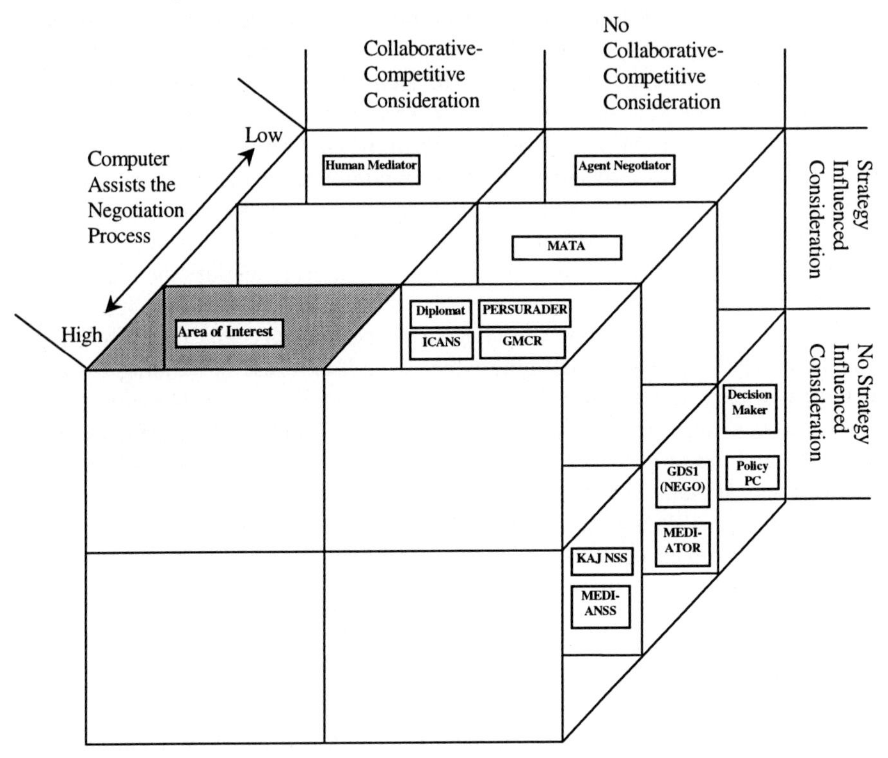

Fig. 3. Area of current research compared to other negotiation system (Part I)

From reviewing that literature, it was determined that there are some concepts in the rule of agent encounters that can be implemented in the computer supported collaborative negotiation field. First is the concept of utility and cost functions. Utility function in the agent negotiation field is used to measure the satisfaction level of an agent to the partial completion of the goal. It can be used in the computer-supported negotiation to evaluate a human being's satisfaction level. Cost functions can also be incorporated in the computer supported human negotiation field to help evaluate the cost of executing different levels of task to eliminate the uneven settlement of negotiation. The way of evaluating deceptions, which translates into strategies, can also be applied in the computer-supported negotiation. Issues of coalition and negotiation language can also become really important in human negotiation. Thus, the adoption of these concepts in the computer-supported collaborative negotiation is of great value. However, the problem is that human beings are not always rational and have constant strategies. Moreover, their strategies are more complex than the ones identified on those research efforts. As previous studies have identified, applications of the game theory to human negotiations in many ways fall short; "humans do not always appear to be rational beings, nor do they necessarily have consistent preferences over alternatives" [Rosenschein and Zlokin, 1994]. However, the work on negotiation theory in the areas of facilitation, mediation and non-binding arbitration tries to overcome some of those deficiencies. The concept of providing a process by which humans can conduct their interactions as well as generating alternatives under mutual consent bring human negotiators to think of the overall situation and not only their individual interests. Thus, the research brings that component into the methodology so that some of the power of the game theory can still be applicable on human negotiations.

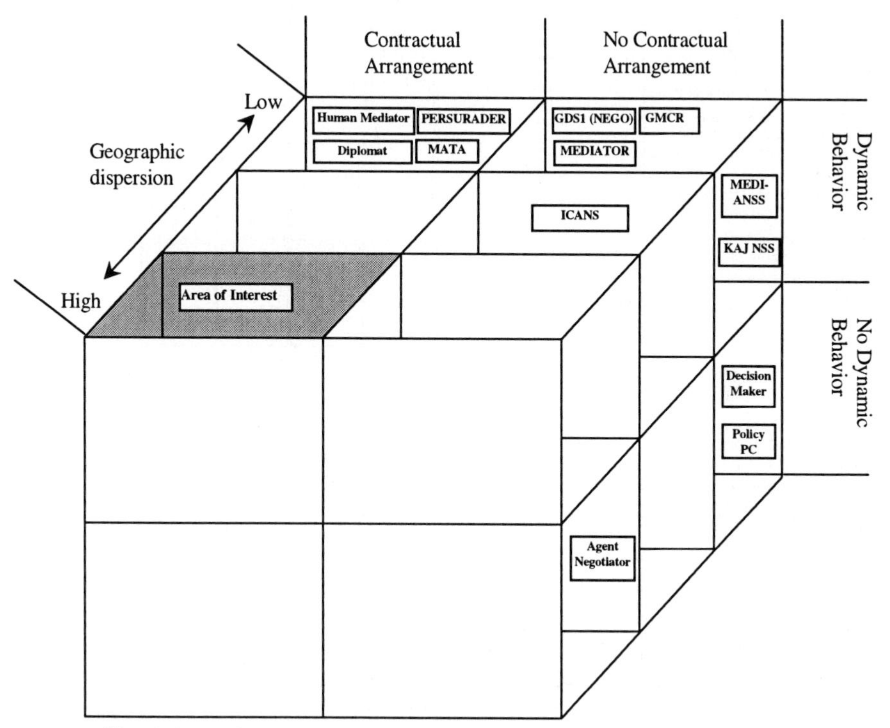

Fig. 4. Area of current research compared to other negotiation system (Part II)

Reference:

Decision Maker: Waterloo Engineering Software, 1989
GDS1: Kersten G., 1988
GMCR: Kilgour M. et. al., 1995
DIPLOMAT: Kraus and Lehmann, 1995
ICANS: Thiessen E. and Loucks D., 1992
KAJ NSS: Anson, R. Jelassi M., 1989
MATA: Cavuchi, 1997
MEDINESS: Carmel, E. and Herniter, B., 1989
MEDIATOR :Jarke, M., Jelassi, M. and Shakun M., 1987
PERSURADER: Sycara K., 1991
Policy PC: Executive Decision Services, Inc., 1988

Fig. 5. Legend and reference of compared research efforts

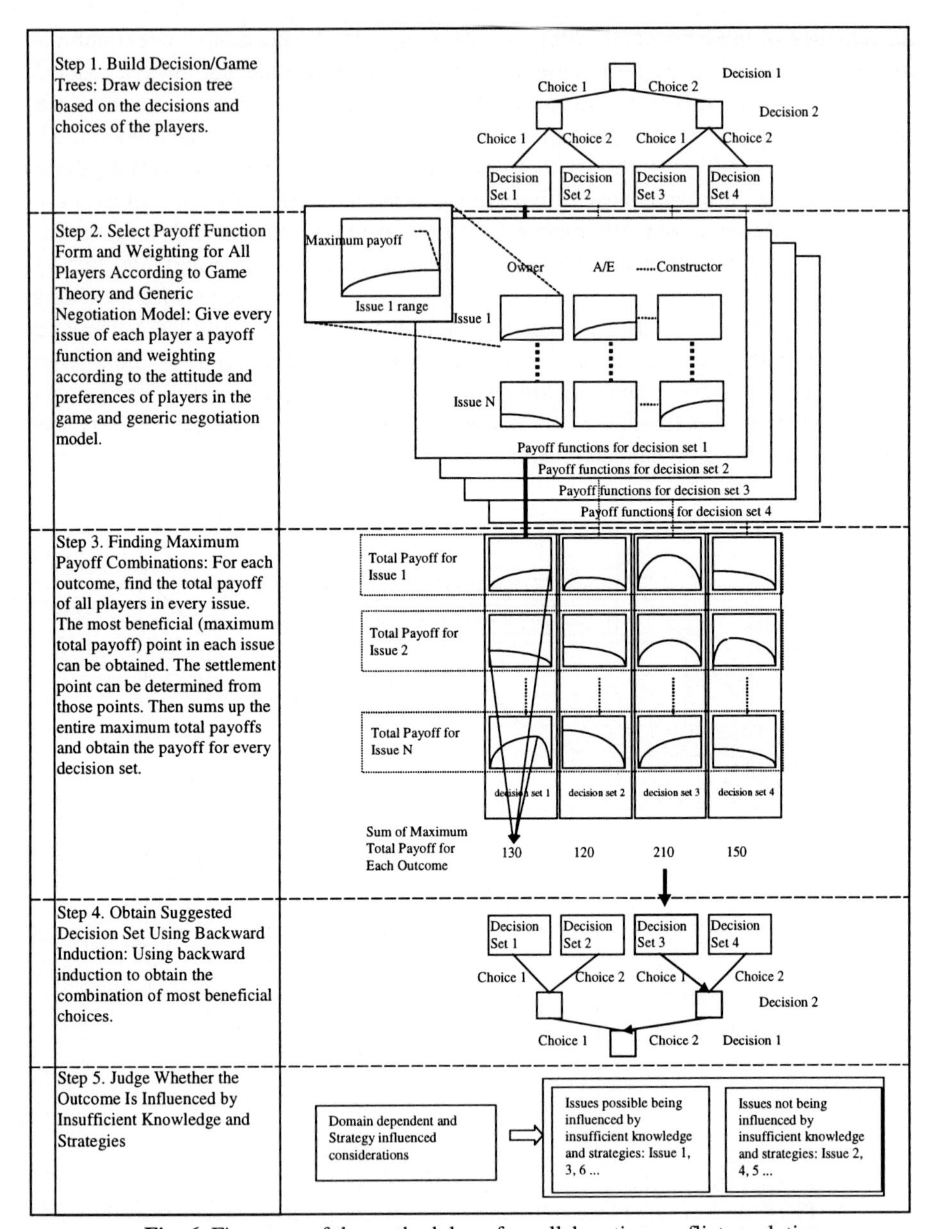

Fig. 6. Five steps of the methodology for collaborative conflict resolution

Work on the methodology for collaborative conflict negotiation resulted in a five-step approach : (1) build decision/game trees from players' decisions; (2) select payoff function form and weighting for all players; (3) finding maximum payoff combinations; (4) obtain suggested outcome using backward induction; and (5) judge whether the outcome is influenced by insufficient knowledge and strategies. Figure 6 shows the graphic illustration of the five steps. Detailed descriptions of these steps are in [Peña-Mora and Wang, 1998]. Figure 7 and 8 presents a preliminary implementation of the collaborative conflict resolution methodology stated above, where the rule of engagement has been embedded in an agent named CONVINCER [URL: http://star.mit.edu/convincer].

Looking at how the previously defined concepts relate to reality, the highway construction example represents a game that is non-zero sum, as the groups can use combinations of outcomes on different variables to enlarge the pot of rewards. To accomplish that, however, each must gain an understanding of the other participant's true interests. Utility functions could be constructed to find out which result will give the highest benefit. The owner's function places the greatest weight on the variable that represents project schedule, and less on the settlement amount variable. The designer places more weight on future monitoring (future business) fees, and less on design fees for the remedial work. The contractor places the highest emphasis on two variables, namely maximizing the amount of the delay claim, and minimizing the amount of additional work that it would need to bond. Accurately determining the variables of concern, and their individual weights, will influence the strategic approach that each takes to the bargaining table. A shared and accurate understanding by all parties of the interests of each other, with the appropriate weights, would lead to a faster and more acceptable solution.

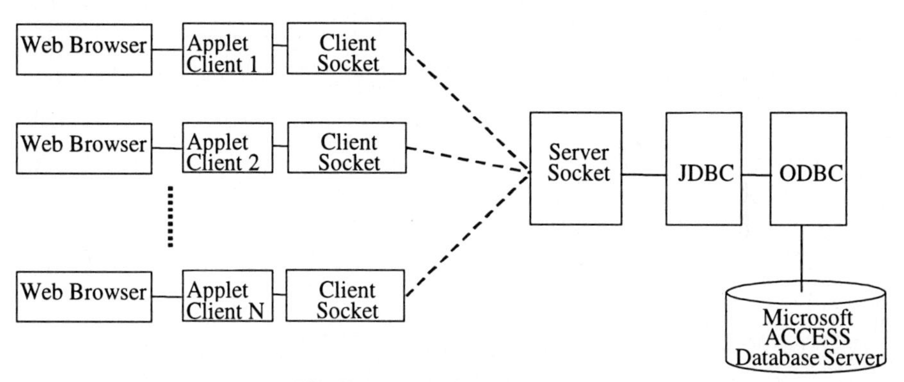

Fig. 7. CONVINCER architecture

3.3 Negotiation Theory

A successful negotiation methodology needs to address both these qualitative and quantitative natures of any negotiation. Alternative dispute resolution is a growing field fueled by volumes of behavioral research [Susskind et al., 1993]. Examples of these efforts are partnering, mediation, and arbitration [Moore, 1986]. All three are non-quantitative methods used to avoid or reduce conflict. Looking at each method from another perspective reveals that each is also an attempt to improve collaboration. Partnering is to set the stage for productive collaboration. Mediation is an attempt to improve relationships mid or late stream in a project. Arbitration is an example of trying to salvage whatever collaborative spirit might be left at the end of a dispute. The research presented in this paper applies these same principles to the large-scale engineering negotiation table, and at an early time in the process when the collaborating parties are defining their relationship and interdependence.

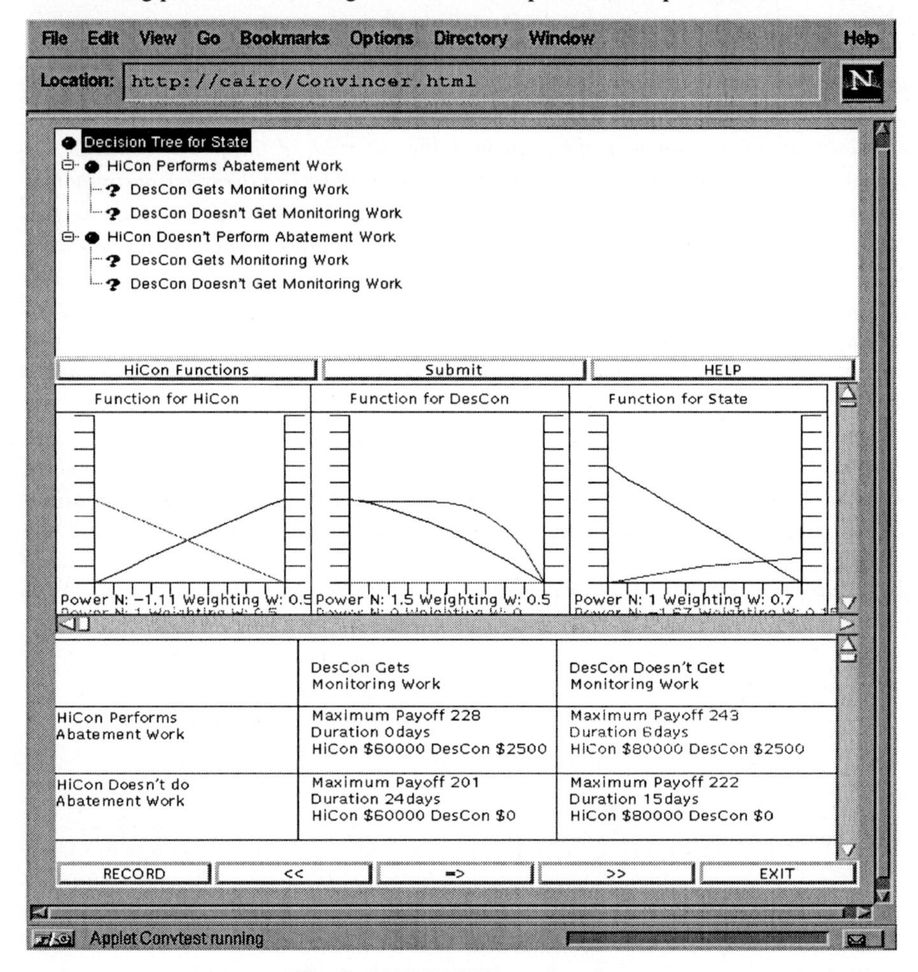

Fig. 8. CONVINCER user interface

The continuum in Figure 9 illustrates the range of options available to negotiators when they encounter a problem. The continuum can be broken down into three distinct categories: unassisted, assisted, and adjudicated. Some of the problems encountered in large-scale civil engineering projects are resolved through simple negotiation, an unassisted method in which participants do not require additional information or assistance from outside parties. As problems become more complex, however, or involve more participants, it may become necessary to move to an assisted negotiation forum [Doyle and Strauss, 1982].

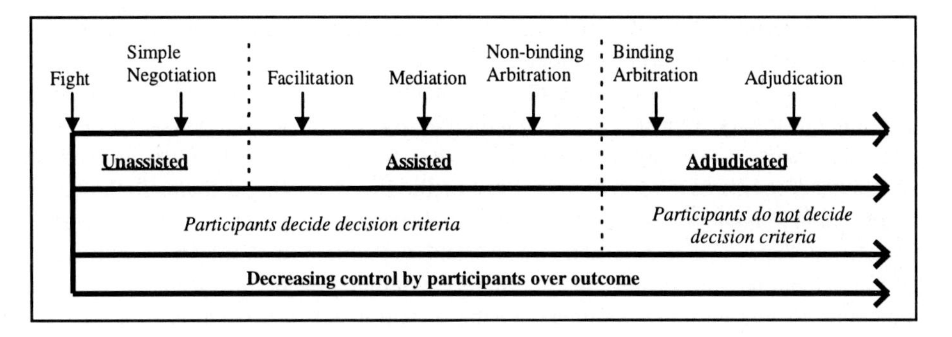

Fig. 9. Dispute resolution continuum (adapted from Susskind, 1995)

The focus of this research is on the assisted negotiation model, which contains three methods. Facilitation, the first of the assisted methods, is the intervention of a third, or additional, party into the process of the negotiation only [Doyle and Strauss, 1982]. Process is the basic system by which participants interact, and professional facilitation involves the manipulation of the group process to improve communication and to bring an increased chance of reaching a cohesive group result. Facilitation is the method that serves as the common building block of all other assisted dispute resolution methods, all of which use facilitation as the basic skill set with which to initiate conflict resolution. Thus, facilitation adds to the collaborative negotiation methodology the process by which negotiators go about resolving their differences. Mediation involves the assistance by a third party person in matters of both process and content [Moore, 1986]. Essentially, a mediator is a facilitator who crosses the line into the area of content by playing the role of catalyst, and encouraging communication between parties. Finally, like a facilitator, a mediator is not a decision maker, and is not empowered to bind the groups through any kind of judgmental result. Thus, mediation brings to the proposed collaborative negotiation methodology the relationship and interplay between process and content for negotiators to reach agreements. In non-binding arbitration, the third party is expected to render a decision based on presentations by participants [Lewicki et al., 1993]. However, since the solution can be rejected by any of the parties, the solution tends to be one of mutual agreement. Non-binding arbitration allows the collaborative negotiation methodology to suggest outcomes that enhance the benefits of all the parties involved.

By utilizing the methods prescribed by the negotiation theories, information are extracted from the participants and organized within the framework of the generic model (Section 3.1). Similarly, these theories assist in the extraction of the preference information that is critical to the construction and evolution of the utility functions needed for game theory (Section 3.2). Additionally, the concepts developed in assisted negotiation also include the process by which efficient, effective, and sustainable results are achieved. Therefore, the negotiation theory complements the game theory by presenting the process through which negotiators reach a settlement, as well as defining the information that must be extracted from the negotiators in order to reach a successful settlement.

While the game theory represents the quantitative aspects of the relationships, negotiation theory represents the qualitative aspect of negotiations. These two complementary aspects are used to reach negotiation results that involve the concept of mutual consent, which provides the best method for resolving disputes in the large-scale engineering domain. Mutual consent leaves participants with the power to decide on their own negotiation process and outcome [Susskind and Cruikshank, 1987]. Leaving this power in their hands, combined with the need to reach an outcome, fosters group commitment to the development of a solution. Groups are committed to resolving their conflict, and committed to maintaining that result once it is reached. By forcing the outcome to be a product of their own work and collaboration, participants feel ownership towards a solution, and this leads to more sustainable results [Ury et al., 1988].

Recalling the overall goal, then, of developing a methodology for assisting in a human collaboration and negotiation forum, this research uses a methodology that combines, based on the specific situation, the game theory and the three assisted conflict resolution methods of facilitator, mediator, and non-binding arbitration. As a support for facilitation, the methodology assists in the basic process of negotiation interaction, helping participants to communicate more effectively, and make explicit their interests and preferences by providing them with a set of workflow processes that encode different negotiation techniques. As a support for mediation, the methodology assists in actively bringing groups together to review informed, educated options for resolution through forum and reminder like mechanisms that encompasses both synchronous and asynchronous meeting structures. And finally, the methodology has the capability to support non-binding arbitration by developing solutions as formal decisions provided from the game theoretic analysis of the issues.

In the highway construction scenario, under a facilitation approach, the owner will realize through the information extracting process, that no matter how much she is willing to pay, the contractor does not want to perform the work. She will also discover that the contractor is willing to work double shifts to make up for the time lost for the abatement. Somewhere in this combination of time and money variables lies a range of possible results, which the facilitator can assist the participants in finding. In this scenario, the contractor agreed to work overtime to complete the work within the existing schedule, for a cost of $60,000. In return for that, the owner agreed to hire another contractor to perform the actual abatement work. To also save time, the owner elected to use the designer for monitoring services. In this example, an integrative solution was reached through proper facilitation and information exchange, and all participants agreed to move forward with the work.

3.4 Project Structures and Delivery Systems

Section 3.1 has covered the generic negotiation model, which illustrates a framework envisioned as needed for a collaborative negotiation methodology. Sections 3.2 and

3.3 addressed the use of the game theory and the negotiation theories for representing the quantitative and qualitative aspects of negotiations. In this section, yet another factor is introduced as an important driver of negotiations - the issue of project structure and delivery systems, which define the temporary formal and informal relationships among the different parties of a project, and subsequently, the negotiations within that project.

Figure 10 describes four different project delivery systems, each with very different structures. Examination of such project structures is important to this research, because the project structure dictates so many of the roles, responsibilities, and relationships that exist within the professional collaborative environment [Potter, 1995]. That information, in turn, is critical to the discovery of typical or pattern behavior by the different functional groups, and for the development of preference or utility functions that reflect not only the human negotiator, but also his or her professional alliances.

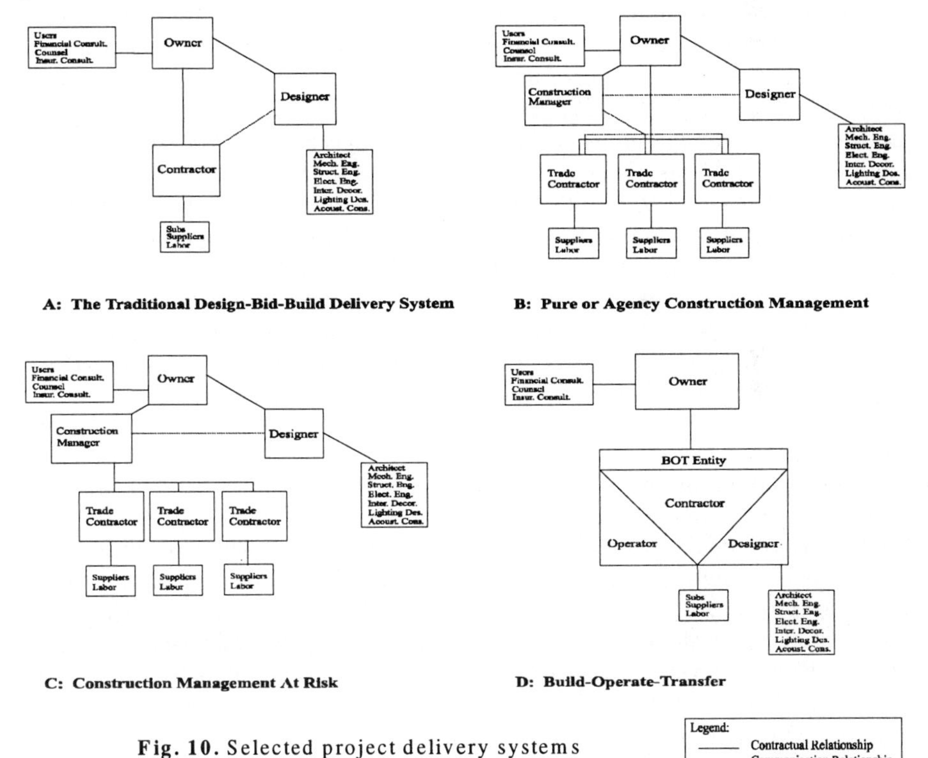

A: The Traditional Design-Bid-Build Delivery System

B: Pure or Agency Construction Management

C: Construction Management At Risk

D: Build-Operate-Transfer

Fig. 10. Selected project delivery systems

These sample project delivery system diagrams (Figure 10) are important because they illustrate the flexibility that a methodology must display in dealing with the participants of a negotiation. Depending on such a structure, each participant has a different set of interests and attitudes, and these differences greatly affect the negotiation at hand. It is critical for a methodology to be able to handle such implications, and to incorporate the kind of information that is relevant to each type of negotiation. Whereas the generic negotiation model describes participants as being independent and distinct entities, the adoption of a more descriptive and realistic project structure greatly alter the structure of the negotiation. Thus, it is critical for the development of the negotiation methodology to obtain an understanding of the different possible relationships that exist within the domain of large-scale engineering and architecture.

In the highway construction scenario, the delivery system is the traditional design-bid-build method. Under that project structure, the owner, designer, and contractor exist in a tripartite balance of power. The low bid selection process of the contractor has placed financial security high on the contractor's list of interests in any negotiation, as he is in an at-risk financial position. The designer has no tie to the contractor, and receives new contracts from the owner organizations, and therefore will tend to side with the owner in a conflict situation. Given all these relationships, the participants can be given preliminary classifications with respect to the analysis. The owner knows she has selected the low-bid contractor, and will have to address his issues of financial risk on the extension of the work. She does so by reaching a cost agreement that they can both live with. The designer under this structure has a natural alliance with the owner, and is therefore selected to act on her behalf during the monitoring phase. Had a different delivery system been utilized, for example, build-operate-transfer (BOT), the designer and the contractor would have been on the same side, placing the owner at some disadvantage with respect to this negotiation. Had the project been conducted with the contractor as a pure construction manager, the lack of financial risk would have freed up the contractor to act on the behalf of the owner. Thus, it can be seen that delivery systems when utilized can greatly affect the roles that participants play in different negotiations.

3.5 Global Collaboration

The final aspect to be taken into consideration for the development of a collaborative negotiation methodology is the issue of global collaboration. The globalization of the large-scale project domain brings into question the abilities of negotiators who are in close contact, versus those who are separated across geography and time. This aspect adds to the complexity presented by the above discussions of delivery systems, together with the way the quantitative and qualitative aspects of negotiations are represented through the game and negotiation theories.

For distributed collaboration to support the synchronous communication and coordination processes in distributed negotiation, a set of critical processes must be understood. These processes are based on the discourse mechanisms meeting participants need to assert their control over the floor in a negotiation setting, thereby coordinating group collaboration. Floor control is the critical process that affects the dynamics of group collaboration since it restricts the amount of information flow within a group interaction. Lack of floor control can greatly influence the negotiation outcome (common pitfalls of group interaction are lack of group focus - caused by distracting floor interjections, and limited exploration of negotiation space due to the monopolization of the floor by particular individuals). Hence the computer mediated communication system developed is able to recognize transitions in floor control and correct errant group dynamics to ensure effective group results and member satisfaction. Effective floor control enhances the group problem-solving process and eliminates the need for expensive facilitation services. For a more detail description, please see [Peña-Mora, Hussein and Sriram, 1996].

In order to enhance the floor transition process, there are two main issues in group dynamics that this paper addresses. The first is extending current work on dyadic (two-person interactions) turn-taking theories [Thorisson, 1995] and applying them to group floor control in a negotiation setting. The second problem addressed by the research was the derivation of a model for floor transition based on observed data and on the discourse analysis influenced by turn-taking. The model is based on a consistent description of the various states that a group will experience while exercising floor control. Mapping some of the concepts back to the actual data shows the validity of the derived model. This model then becomes the basis for allowing global collaboration in negotiations.

The discourse analysis revealed two key discourse phenomena in group negotiations. These phenomena are the focus of attention and degree of engagement. In addition, two models have been derived from this analysis to describe floor transitions in group discourse. The first model describes the state of an individual within the group. *For example, this model would characterize the designer's state in a discussion concerning the removal of the old process pipe on the sample highway construction (presumably the designer would either be speaking, mildly engaged, or listening in such a discussion).* The participant model is complemented with a model that indicates the state of the floor, which is the combined state of all participant states. This model demonstrates the extent of confusion or simultaneous disruptive conversations in the group setting.

Some of the current computer mediated communication systems (both academic and commercial) are classified in Figure 11. The figure delineates the multimedia capabilities of the systems on the y-axis. The x-axis describes the extent to which these systems support multiple participants in an interaction. Finally, the z-axis expresses the degree to which the systems allow effective control of the floor (concurrency control derived from gesture, gaze, focus, and engagement) in collaborative interaction. The core focus of the global collaboration truss of the research was on the z-axis although the existence of multimedia and multi-user support are necessary prerequisites for the work described herein but was obtained by using standard commercial components.

Thus, the research presented here is based on the hypothesis that an understanding of group interaction and the enhancing of the coordination process are critical for the adoption of computer mediated global collaborative negotiation. Thus, explicit computer coordination of group processes is necessary in an environment to support global collaborative negotiation since typical social and engagement protocols have to be provided in computer mediated communication when there is no physical interaction and complex issues have to be negotiated. To achieve that goal, the research presented integrates and builds on some of the concepts explored in electronic meeting systems, video conferencing and shared social spaces.

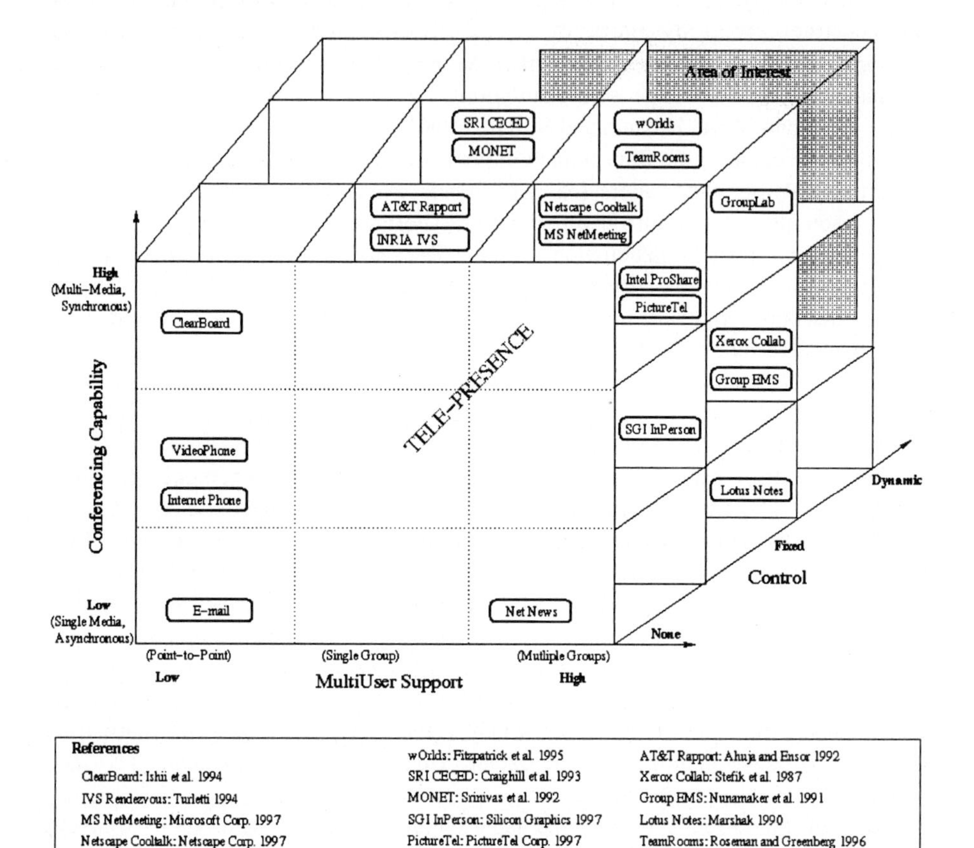

Figure 11: Area of research compared to other collaboration systems

A sample implementation of a prototype conferencing system was developed based on the criteria presented in the previous paragraph. The prototype conference system includes several extensions using commercial or off-the-net components that enable voice, visual, textual, graphical interaction, as well as communication using the World Wide Web. Finally, a scheduling interface to Primavera is included for large scale collaborations (Figure 12 and 13 show the architecture and various tools within the conferencing system).

Notice that this prototype conferencing system provides a large set of interaction tools including whiteboards, text tools, audio, video and CAD or document sharing. These tools, although useful, can be distracting to the user since they do not provide a clear focus of attention. This does not suggest that the tools be reduced, instead it is necessary to include a mechanism for identifying the focal tool of the discourse. In addition, the group aware conferencing system must support deictic referencing in both gaze and pointing. Hence, the tool has a pointing feature as well as a feature that clearly distinguishes between hearers of conversation and those to whom the conversation is addressed (as dictated by gaze in traditional meeting settings). The research also indicated that the participant should have greater flexibility in defining his/her intent to take control of the floor. An interface that provides this functionality is shown in Figure 14.

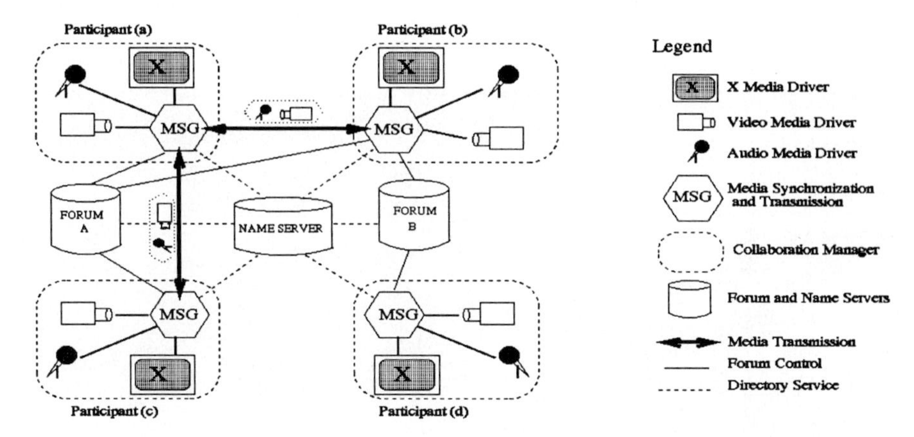

Fig. 12. System architecture

A final issue for integrating group aware conferencing to the collaborative negotiation methodology is the notion of floor control strategies (e.g. chairman controlled, brainstorming, and lecture). In regular meetings a strategy is adopted either explicitly or implicitly due to group norms or due to particular meeting room arrangements. These strategies govern floor control on the macro level; they define a style for a group meeting. The effective choice of floor control strategy can improve the fluidity of the meeting process and enhance the collaborative effort. A toolkit of strategies has been developed [Pena-Mora et al., 1996]. This toolkit represents a knowledge base that maps these strategies to various meeting situations supporting the concept of encoding negotiation methodologies from the negotiation theory to support the use of reasonable and sustainable results from the game theory. These strategies are specific to meetings or to individual agenda items. The user interface representation of two strategies and degree engagement is shown in Figure 14 and 15.

However, further research remains to be conducted in group dynamics applications in computer mediated negotiation. The data has shown that different individuals express their interest in acquiring the floor in varying ways. Hence, it is not clear whether the model developed in the preliminary research is universal. Furthermore, the states outlined in the participant model are not the only way in which the increasing engagement of a participant may be modeled. The level of segmentation of this continuum of engagement is still an open issue. Finally, the model does not account for parallel conversations, which occur often within a group discourse.

However, ``effective'' groups are not supposed to exhibit this type of behavior. Furthermore, a protocol needs to be devised to reduce participant's over- or under-stating their degree of engagement, thereby reducing individual irregularities. Finally, techniques need to be developed for detecting dysfunctional group dynamics based on a thorough discourse analysis of various negotiation situations.

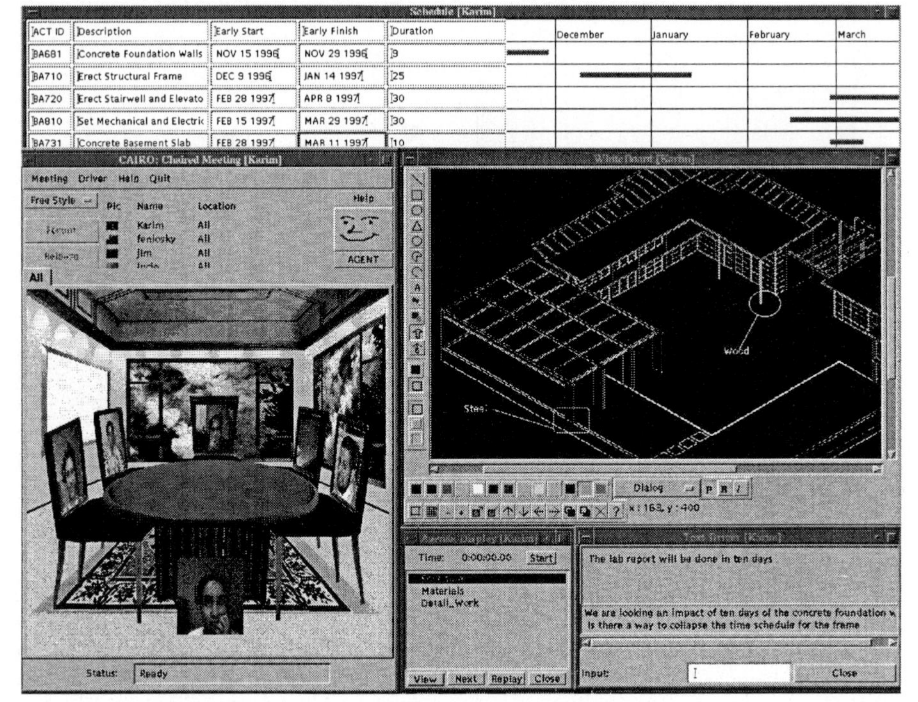

Fig. 13. Screen dump of user interface

4 Proposed Testing of the Methodology

Five educational institutions: University of Sydney (USYD), Australia; Centro de Investigacion Cientifica y de Educacion Superior de Ensenada (CICESE), Mexico; Ecole Polytechnique Federale de Lausanne (EPFL), Switzerland; Pontificia Universidad Catolica de Chile (PUC), Chile; and Massachusetts Institute of Technology (MIT), USA are seeking to develop the curriculum for a joint class taught using a computerized communication network. The project would develop an integrated research laboratory and classroom alliance aimed at improving the skills of students and the understanding of negotiation and collaboration in the subject domain. The alliance between the research laboratory and the class would be supported, augmented and complemented by the involvement of the Kajima and Shimizu Corporations of Japan, and the management staff of the Boston Central

Artery and Tunnel Project. Through the students from all five schools, the distributed collaboration concepts that form the basis of the work described in this paper will be taught, utilized, and then tested. Members of each class will develop relationships with their counterparts through the Internet and the Da Vinci collaboration system [Peña-Mora, 1996]. Within the class, the distributed collaboration concepts that form the basis of the research work will be taught, utilized, and tested. The existence of many groups, working on a multitude of simulated professional projects enables researchers to improve the methodology with each class, in a more controlled environment than in the professional arena, and in a more realistic environment than artificial laboratory simulation.

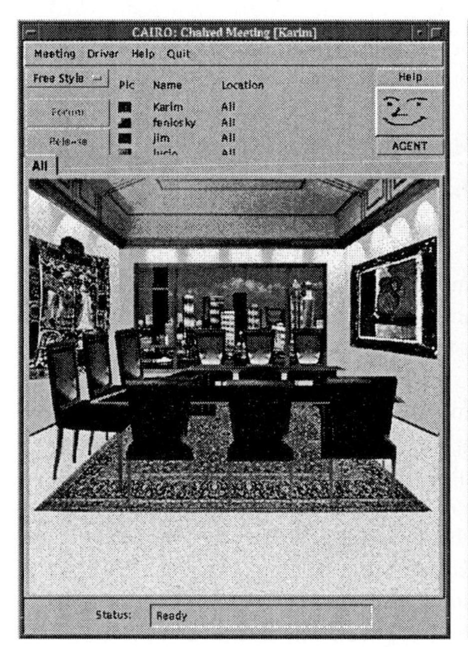

Fig. 14. Different process strategy representations

6 Summary

Large-scale engineering and architectural projects are unique endeavors that require a high level of human creativity from multiple professional disciplines. In such an environment, successful collaboration is critical. Because the different participants are from different technical backgrounds, and typically come from different organizations, competitive stresses exist within project relationships. Better methodologies are needed to relieve these stresses that exist in this environment. The methodology explored in this paper utilizes the quantitative nature of game theory and the qualitative nature of various conflict resolution theories, and attempts to better define the interactions of collaborating professionals to use the strengths of humans and computers alike.

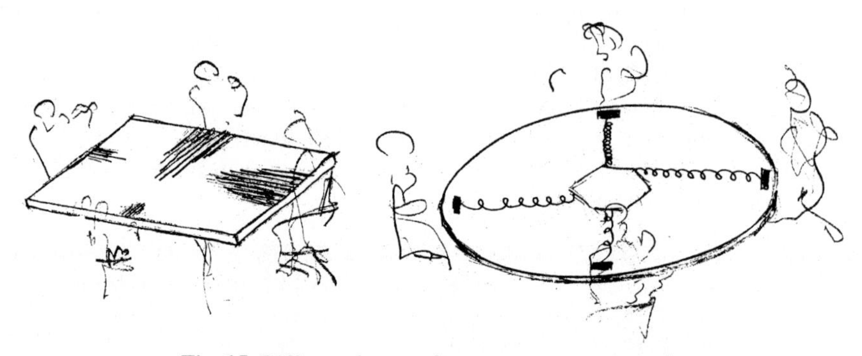

Fig. 15. Different degree of engagement representations

Substantial literature exists in each of these fields, yet no concerted approach has been undertaken to apply the integrated benefits of both together to the domain of large-scale engineering and construction. This research utilized game theory to represent the quantitative aspects of the interdependence of negotiation. The other negotiation theories were utilized to represent the qualitative aspects of effective negotiation, and help to guide those negotiations toward sustainable outcomes. In addition, this integrated methodology is developed within the context of the characteristics of large-scale engineering systems. Negotiations within that environment tend to be collaborative-competitive, domain dependent, and strategy-influenced, as well as geographically and time distributed. Because relationships within a project structure are critical to the negotiations encountered, project delivery systems are examined as a means by which negotiators can be categorized and analyzed, by role, for the purpose of understanding negotiator patterns. Finally, the research examines how global collaboration affects negotiation within the domain. It explores the issues of floor control, degree of engagement, and focus of attention to support the collaborative efforts of the negotiating individuals. Combining all of these elements the computer can increase the effectiveness of the communication and the accessibility to group preferences and information.

Acknowledgements

The author would like to acknowledge the support received from the National Science Foundation. Funding for this project comes from the Information Technology and Organization Program and the NSF-CONACyT Program, award number IRI-9630021; Structural Systems and Construction Processes Program award number CMS-9626315; Kajima Corporation, Shimizu Corporation, Intel Corporation, MIT School of Engineering, and MIT Department of Civil and Environmental Engineering. The author would also like to thank the student members of the Da Vinci Agent Society Initiative specially: Karim Hussein, James Kennedy, and Chun-Yi Wang, whose work contributed greatly to this paper.

References

1. Ahuja, S., and Ensor, J. (1992). Coordination and Control of Multimedia Conferencing, IEEE Communications Magazine, May 1992, pp. 38-43.
2. Anson R., Jelassi, M. (1989). *A Developmental Framework for Computer-Supported Conflict Resolution*, European Journal of Operations Management 1989.
3. Carmel, E. and Herniter, B. (1989), *MEDIANSS: Conceptual Design of a System for Negotiation.* Ninth Conference on Decision Support Systems.
4. Cavicchi, A. (1997), A Multi-Attribute Tradeoff Analysis for Water Resource Planning: A Case Study of the Mendoza River, MIT Master Thesis.
5. Craighill, E., Lang, R., Fong, M., and Skinner K. (1993). CECED: A System for Informal Multimedia Collaboration, SRI International, Menlo Park, CA.
6. Darling, T. and Mumpower, J. (1990), Modeling Cognitive Influences On The Dynamics Of Negotiations, *Proceedings of the Twenty-Third Annual International Conference on System Sciences*, IEEE Computer Society Press.
7. Doyle, M and Straus, D. (1982), *How To Make Meetings Work.* Jove Books, New York.
8. Executive Decision Services Inc. (1988), P.O. Box 9102,Albany NY 12209-0102. *Policy PC Reference Manual*, 2nd edition .
9. Fisher, R. and Ury, W. (1981), *Getting To Yes*, Penguin Books, New York.
10. Fitzpatrick, G., Kaplan, S., and Mansfield, T. (1996). Physical Spaces, Virtual Places and Social Worlds: A Study of Work in the Virtual. In *Conference on Computer-Supported Cooperative Work,* pages 334-343.
11. Intel Proshare (1996). Intel Corp., User's Guide: Intel Proshare Conferencing Products.
12. Ishii, H., Kobayashi, M., and Arita, K. (1994). Interactive Design of Seamless Collaboration Media, Communications of the ACM, 37:84-92
13. Jarke, M., Jelassi, M. and Shakum M.(1987). *Mediator: Towards a Negotiation Support System*, European Journal of Operational Research (31) 1987.
14. Kersten G. (1988), *A Procedure for Negotiating Efficient and Non-Efficient Compromises*, Decision Support Systems, (4) 1988.
15. Kilgour, D., Fang, L., and Hipel, K. (1995), *GMCR in Negotiations*, Negotiation Journal, April 1995.
16. Kraus, S. and Lehmann. D. (1995), *Designing and Building a Negotiating Automated Agent*, Computational Intelligence, 11(1):132-171.
17. Lewicki, R., Litterer, J., Saunders, D., Minton, J., (1993), *Negotiation*, Richard D. Irwin.
18. Marshak, D. (1990). Lotus Notes: A Platform for Developing Workgroup Applications. In *Patricia Seybold's Office Computing Report.* July 1990
19. McMillan, J. (1992), *Games Strategies and Managers.* Oxford University Press.
20. Microsoft NetMeeting (1997). Microsoft Corporation, NetMeeting Home, http://www.microsoft.com/netmeeting.
21. Moore, C.W. (1986), *The Mediation Process: Practical Strategies for Resolving Conflict.* Jossey Bass, San Francisco.
22. Murnighan, J. and Conlon, D. (1991). The Dynamics of Intense Work Groups. *Administrative Science Quarterly,* 36(2).
23. Netscape CoolTalk (1997). Netscape Corporation,Welcome to Netscape Navigator Gold Release 3.0, http://home.netscape.com/eng/mozilla/3.0/relnotes/unix-3.0Gold.html.
24. Nunamaker, F., Dennis, A., Yalacich, J., Vogel, D., and George, J. (1991). Electronic Meeting Systems to Support Group Work. *Communications of the ACM,* 34(7):40-61.

25. Peña-Mora, F., (1996), Da Vinci Initiative: Computer-Supported Negotiation Across Space and Time for the Life-Cycle Development of Sustainable Large-Integrated Engineering Systems in a Collaborative-Competitive, Domain-Dependent, and Strategy-Influenced Environment, *MIT Intelligent Engineering Systems Laboratory Technical Report Number IESL 96-03.*
26. Peña-Mora, F., Hussein, K., and Sriram, D. (1996), "CAIRO: A system for Facilitating Communication in a Distributed Collaborative Engineering Environment," *Computers In Industry*, Elsevier Science Publishers, B.V. (North Holland).
27. Peña-Mora, F. and Wang, C., "Computer-Supported Collaborative Negotiation Methodology", *ASCE Journal of Computing in Civil Engineering*, Vol.2, No. 12, pp. 64-81, April 1998
28. PictureTel (1997). PictureTel Corporation, PictureTel – Products, http://www.picturetel.com/products.htm.
29. Potter, A., (1995), Delivery System of Choice, *Construction Business Review, November/December.*
30. Raiffa, H. (1994), *The Art and Science of Negotiation.* Harvard University Press, Cambridge.
31. Roseman, M. and Greenberg, S. (1996b). Teamrooms: Network Places for Collaboration. In *Conference on Computer-Supported Cooperative Work,* pages 325-333.
32. Rosenschein, J. and Zlotkin, G. (1994), *Rules of Encounter,* MIT Press, Cambridge.
33. SGI InPerson (1997). Silicon Graphics Corporation, InPerson 2.2 Product Guide, http://www.sgi.com/Products/software/InPerson/index.html.
34. Srinivas, K., Reddy, R., Babadi, A., Kamana, S., Kumar, V., and Dai, Z. (1992). MONET: A Multimedia System for Conferencing and Application Sharing in Distributed Systems, Concurrent Engineering Research Center, West Virginia University, Working Paper # CERC-TR-RN-91-009
35. Stefik, M., Foster, G., Bobrow, D., Kahn, K., Lanning, S., and Suchman, L. (1987). Beyond the Chalkboard: Computer Support for Collaboration and Problem Solving in Meetings. *Transactions of the ACM,* 30(1):32-47.
36. Susskind, L., Babbitt, E. and Segal, P., (1993), When ADR Becomes the Law: A Review of Federal Practice, *Negotiation Journal*, Vol. 9, No. 1.
37. Susskind, L. (1995), *Notes Taken During Instructional Presentation*, Massachusetts Institute of Technology.
38. Susskind, L and Cruikshank, J. (1987), *Breaking the Impasse.* Basic Books.
39. Sycara, K. (1991), *Problem Restructuring in Negotiation*, Management Science, Vol. 37 No. 10.
40. Thiessen, E. and Loucks, D. (1992), *Computer Assisted Negotiation of Multi-Objective Water Resource Conflicts*, Water Resources Bull. 28(1).
41. Thorisson, K. R. (1995). Computational Characteristics of Multimodal Dialogue, In *AAAI Fall Symposium on Embodied Language and Action,* Massachusetts Institute of Technology.
42. Turletti, T. (1994). The INRIA Videoconferencing System (IVS), ConneXions – The Interoperability Journal, 8(10):20-24.
43. Ury, W., Brett. J and Goldberg, S. (1988), *Getting Disputes Resolved.* Jossey Bass, San Francisco.
44. Waterloo Engineering Software (1989), Univ. of Waterloo, Waterloo, Ontario, *Decision Maker: User Manual.*

An Investigation into the Integration of Neural Networks with the Structured Genetic Algorithm to Aid Conceptual Design

Rafiq, M. Y., and Williams, C.,
School of Civil & Structural Engineering, University of Plymouth, UK.
mrafiq@plymouth.ac.uk and cwilliams@plymouth.ac.uk

Abstract. Genetic Algorithms (GAs) and structured Genetic Algorithms (sGAs) are powerful tools for modelling some of the activities related to the conceptual stage of the design process. Artificial Neural Networks (ANNs) are Artificial Intelligence (AI) tools which can learn and generalise from examples and experience to produce meaningful solutions to problems even when input data is fuzzy, discontinuous or is incomplete. Human creativity, intuition and expertise can be combined and incorporated when training ANNs. Research has shown that the ANN can be a powerful tool for modelling some of the activities of the conceptual stage of the design process. The current paper investigates possibilities of integrating the sGA and the ANN in the context of a decision support tool to assist designers.

Keywords: Genetic Algorithms, structured Genetic Algorithms, Artificial Neural Networks, Artificial Intelligence, Conceptual Design, Optimisation, Integration.

1.0 Introduction

The conceptual stage of the design process is one of the most imaginative stages of the design process in which human creativity, intuition, and successful past experience play an important role. This stage of the design process is identified with a high degree of uncertainty concerning the design information and lack of clarity of the design brief. For the above reasons activities related to this stage of the design process requires a high level of expertise in every discipline involved in the process.

Some of the attributes of the tasks related to the conceptual stage of the design process can be summarised as:

- Analysis of many dimensions of the problem in the search for possible solutions.
- Synthesis of a number of possible solutions within the framework of constraints and requirements specified in the design brief.
- Critical evaluation of alternative solutions.
- Decision making - selection of a design option which is the best ‘fit for purpose’.

Activities described above are highly non-linear and non-algorithmic by nature. There are no pre defined rules for formulating design solutions. At present, design solutions developed mainly rely on heuristics and the successful past experience. It is therefore very difficult to model the design activities related to this stage of the design process.

Recent research has shown that structured Genetic Algorithms (sGA) have proved to be a powerful tool for modelling some of the activities of the conceptual stage of the design process [1-6].

A major disadvantage of the sGA is the rapid growth of the objective function, when the problem hierarchy is developed (see Figure 1).

Artificial Neural Networks (ANNs) are amongst the AI tools in which the human creativity, intuition and past experience can be incorporated in the network training process. Like human experts, ANNs learn from experience and examples. Research has shown that it is possible to model some of the activities of the conceptual stage of the design process using ANNs [7]

The main objective of this paper is to investigate the possibility of integrating the GA with the ANN to:

- Solve the problem associated with the rapid growth of the sGA objective function.
- To model some of the activities of the conceptual stage of the design process using ANNs.

Results of pilot studies related to above issues will be reported in this paper.

2.0 Genetic Algorithms and the Conceptual Design

The Genetic Algorithm (GA) is an adaptive search and optimisation technique which is based on the theory of biological evolution in which characteristics of parents are transmitted to their offspring by means of genes which lead to the evolution of organisms. GAs employ an artificial version of natural selection and use artificial genetic structures to solve problems [8,9]. They are robust algorithms capable of traversing a large, complex multi-dimensional search space to obtain a design solution. GAs rapidly identify discrete regions within a large search space in which a concentrated search may yield an optimum solution.

The GA is a powerful search and optimisation technique. Research has shown that GAs can very successfully handle multi-dimensional and multiple-criteria problems in a variety of engineering disciplines [10-12].

The 'structured GA' (sGA) developed by Dasgupta and MacGregor [13] is a mechanism allowing a multi-level chromosome structure which leads to multiple simultaneous changes. In the sGA chromosome structure, genes within a single chromosome string can either be 'active' or 'inactive'. This activation process determines the direction of the further search within the process.

The distinct multi-level chromosome structure of the sGA makes it is possible to encode and evaluate a number of design alternatives in a single sGA chromosome string concurrently. Within the sGA chromosome higher level genes (discrete parameters) act as switches to activate or deactivate the lower level genes (continuous parameters) and according to the status of these switches, various design alternatives are generated concurrently within the sGA population. This characteristic of the sGA makes it a powerful tool for synthesis, evaluation and selection of alternative designs at the conceptual stage of the design process [14].

Representing the problem using the sGA structure allows more variety and an extensive search within the available time limits would be possible. As a result of this, a much wider region of the global search space is sampled to identify a variety of potential design alternatives which are often ignored within conventional practice due to time limitations. In a simple GA each branch of the problem covered by the sGA would have to be presented as a separate problem for evaluation and optimisation purposes.

2.1 Representation of a Simplified Building Design Problem in the sGA

A typical design domain, for a building, as shown in Figure 1, represents the hierarchy which includes only part of the major elements of a building. Of course other material types such as masonry, timber, etc. and various compatible floor systems should be included for a complete system. The main objective of the research was to use a sGA to investigate the possibility of including as many as possible different alternative solutions (diverse regions of the design search space) which include a variety of construction materials, concurrently. The advantage of this approach is that it would enable the designer to investigate a wide range of alternative designs, and a number of different construction types, at the conceptual stage of the design process. The sGA has proved to be a powerful tool for accomplishing this task [5, 14,15].

Figure 1 includes two different set of parameters, discrete parameters (frame type, various floor system types, etc.) and continuous parameters (in-situ concrete floor and parameters related to this floor system, etc.). Representing this diversity was perfectly possible within the sGA chromosome structure. The whole structure of Figure 1 was included in a single sGA chromosome string.

Figure 2 shows a genotype representation of the problem presented in Figure 1 encoded into the sGA chromosomes. The GA chromosome is generally encoded by a binary number format (a string of 0's and 1's) or more efficiently in a real number format [14,15]. Figure 2 shows a symbolic representation of the genotype for clarity.

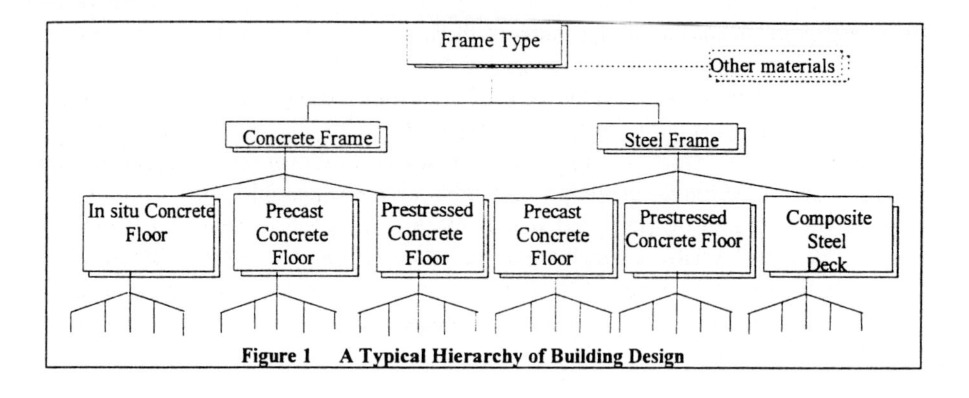

Figure 1 A Typical Hierarchy of Building Design

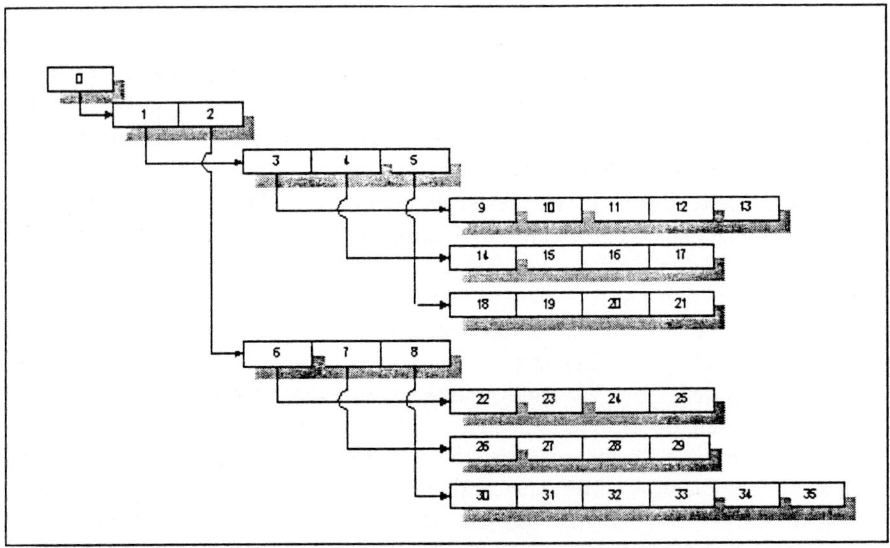

Figure 2 Genotype representation of building hierarchy in sGA

A typical sGA string for this problem is symbolically shown in Figure 3.

Discrete Parameters					Continuous Parameters					
0	1	2	3	4-8	9-13	14-17	18-21	22-25	26-29	30-35

Active Parts   Inactive Parts

Figure 3 Typical sGA chromosome for the building hierarchy

2.2 Advantages and disadvantages of the sGA

The sGA is a very powerful tool for the concurrent evaluation of various design alternatives. One of the main problem with the sGA is that it uses much redundancy, only 'active' genes within the chromosome string (a small part of the chromosome representing a single design set) are used at any given time. Other genes within the chromosome (which constitute a large part of the string) remain redundant until the higher level switches activate these at a later time. Dasgupta and MacGregor [13] argue that the redundancy characteristic of the sGA matches the 'biological genetic material'. They argue that 'a large percentage of higher organisms is *junk* (over 80% of human DNA has no apparent function)'.

Figure 3 shows the active and inactive parts of the sGA chromosome.

The sGA has been successfully implemented in high-level design studies by Roberts and Wade [15] and Parmee [5,16] to tackle complex real-world problems. Advantages of using the sGA can be summarised as:

- Concurrent representation, synthesis and evaluation of many conceptual design alternatives.
- Capable of sampling a wider range of design search space, to locate potential design solutions with variety.
- Very powerful for use at the conceptual stage of the design process (locating various local optima), where conventional software fails to provide any assistance.
- Useful for the integration of the design activities.
- Some of the shortcomings of the sGA can be summarised as:
- Crossover between differing configurations [16] (e.g. crossover between a high fitness steel frame with a high fitness concrete frame). This may result in a premature convergence or sometimes disruption to good designs discovered at early generations.
- A high degree of redundancy within the sGA parameter string (see Figure 3); as a result of this a high proportion of crossover and mutation could be ineffective as they may be within a region of the string which is not active (e.g. in the inactive region of the chromosome, representing the steel frame, while the concrete frame is active).
- Complexity of the problem when encoding and decoding the string.
- Rapid and explosive growth of the objective function size. This is particularly true when the problem hierarchy is developed (see Figure 1).

To resolve some of the above problems the following solutions are proposed.

- To solve the problem with crossover between differing configurations, alternative methodologies have been developed. Variable mutation probability approach and the GAANT algorithm [16], which combines the Ant Colony metaphor and the simple GA have been successful in resolving some of the above problems. In the GAANT algorithm the crossover can only occur between members of the same class and the length of the string is much shorter as it is restricted to only a single design set.

- For more efficient representation and to reduce the length of the sGA chromosome string, it would be easier to use an integer or real number representation scheme. Details of this scheme with relevant examples can be found in [14,15]. This will greatly resolve the complexity associated with encoding and decoding.
- To avoid explosive growth of the objective function, the Neural Network was integrated with the GA in which the NN replaces part of the conventional objective function of the GA for fitness evaluation. Discussion into this process is the main theme of the current paper.

3.0 The Neural Network

Artificial Neural Networks are interconnected neurones, each with some kind of internal activation functions (sigmoid or linear) that allows the network to mimic the activities of the human brain (to some extent). During the training process the network learns from experience and examples presented to it.

Over the past decade artificial neural networks have been widely used for many applications. This includes a number of applications in the field of structural analysis and design [19]. Among the more widely adapted neural network architecture is the back-propagation network.

Neural network can be used as a function approximation tool to replace lengthy procedural algorithms. This paper investigates the possibilities of using artificial neural networks to replace part of the objective function of the GA for fitness evaluation process.

Selection of the number of training patterns for the training neutral network is an extremely important issue. If the number of training patterns is too little, then the gap between data presented will be too large. In this case generalisation will be too difficult and hence the result of the training will not be satisfactory and the network response to unknown patterns will be poor. On the other hand, if too many training patterns are presented to the network, the network would train too closely and it will learn details. Once again the generalisation would become too difficult and the network response to previously unseen patterns will be poor. The issue of patterns selection has been addressed in a separate paper [7].
Some of the major attributes of artificial neural networks can be listed as:

- Neural networks can learn and generalise from examples to produce meaningful solutions to problems even when input data is fuzzy, discontinuous or is incomplete.
- Neural networks are able to adapt solutions over time and to compensate for changing circumstances.
- Data presented for training neural networks can be theoretical data, experimental data, empirical data based on good and reliable past experience or a combination of these. Training data can be evaluated, verified or modified by the human expert to inject human intelligence and intuition.

These criteria make ANNs a promising candidate for modelling activities of the conceptual stage of the design process.

3.1 Artificial Neural Network for Modelling Reinforced Concrete Flanged Beams

To demonstrate the usefulness of NNs an example of reinforced concrete flanged beam was used.

To train the artificial neural network, parameter ranges for RC flanged beam were carefully selected to cover the majority of the practical designs. An initial investigation was also conducted to evaluate the level of influence of various parameters which affect the practical design of the reinforced concrete flanged beams.

Training data for the neural network was generated by using a procedural program. A linear optimisation process was used to obtain near to optimum solutions for the reinforced concrete flanged beam design parameters. The back-propagation network used in this study consisted of 4 input neurones (applied load, beam span, flange width to web width ratio and depth to web width ratio) , 12 hidden neurones and 4 output neurones (optimum depth, steel weight, shear requirements and deflection requirements) as shown in Figure 4. The output from the neural network was used to obtain the overall cost of each design. This information was used to evaluate the fitness of each alternative design. As cost models , generally, depend on many factors, it was not included in the network training.

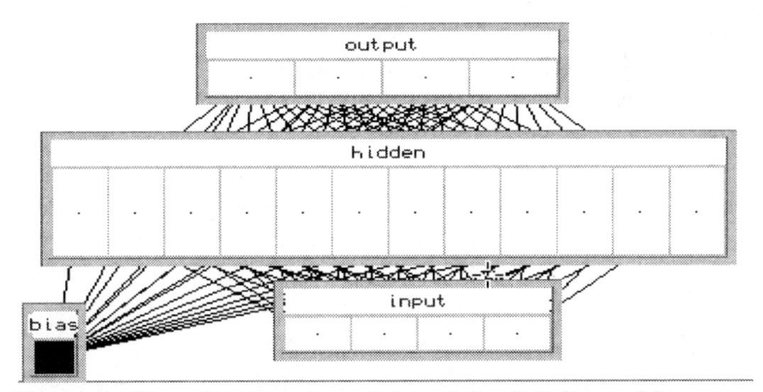

Figure 4 Neural network model used for RC flanged beams

The beam depth is the most important parameter which influences many other design parameters and the final cost. For the calculation of the optimum beam depth the minimum cost criterion, serviceability requirements (deflection requirements) and buildibility requirements (maximum percentage of reinforcement within the section) were considered.

A selection strategy was adopted to select data for the NN training to minimise the complications associated with too many or too little data which severely affect the performance of the NN. A total of 41 beams with near to optimum designs were selected from a total possible range of 980 beams (details of data selection has been reported in [7]). Comparison of the target and the neural network predictions for the optimum beam depth are shown in Figure 5.

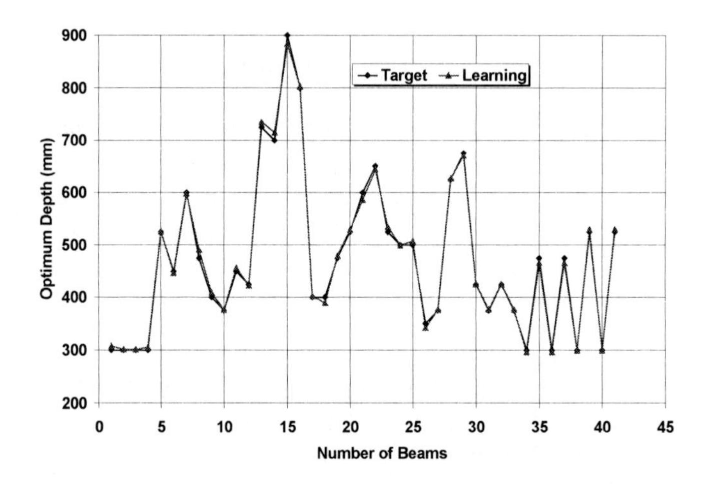

Figure 5 Neural Network training results

To test the performance of the trained NN, a set 108 beams were selected for testing the network. Most of these patterns were new to the neural network. Figure 6 shows the result of this test. Maximum error resulted from testing 108 beams was 9.6% which is acceptable for all practical purposes.

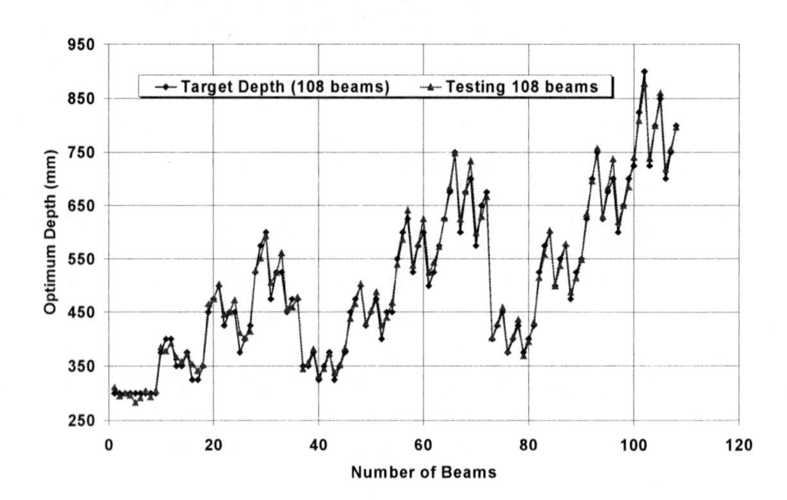

Figure 6 Result of test on the trained Neural Network

From Figures 5 and 6 it is clear that the neural network has successfully learned patterns presented to it and has found a generalised relationships which can respond to unseen data satisfactorily.

Output information resulted from the trained NN was designed to produce initial sizes and material quantities. This level of information is sufficient for the conceptual design purposes.

4.0 Integration of the Neural Network with the Genetic Algorithm

The following processes are identified with genetic Algorithms:

- Initialising an initial population of a given size
- Fitness evaluation
- Selection of pairs to be included for mating
- Reproduction by means of crossover and mutation

In GAs, fitness evaluation is problem specific and some times can consist of many long operations. For example for optimum design of a reinforced concrete beam, the objective function of the GA may contain routines to satisfy:

- Bending and shear reinforcement requirements
- Serviceability requirements (e.g. deflection check, etc.)
- Buildibility requirements (e.g. concrete cover, bar spacing and amount of steel reinforcements in the section etc.)

The beam depth is the most important parameter which influences many other design parameters and the final cost. To obtain an optimum depth for the beam a linear optimisation program can be used either within the objective function of the GA or independently, which can be a repetitive long process for the GA applications. Alternatively beam depth can be encoded as part of the GA string for optimisation purposes. In the latter case longer string, larger population and greater number of generations may be required to achieve an optimum design.

The above process covers only one element of the whole structure. It is obvious that if the whole building and all possible combination of construction material within the design hierarchy, part of which is shown in Figure 1, are represented in the sGA string, the objective function of such systems can easily get out of control.

To avoid this explosive growth of the objective function a trained Neural Network can be used to as part of the objective function of the GA to replace lengthy procedural algorithms. The output from the NN would be used to evaluate the fitness of the design alternatives within the GA or the sGA population.

Results of pilot studies are discussed in the following section.

4.1 An Example of the Integration of the NN and the GA

To investigate the possibility of integrating the NN with the GA, a rectangular plate representing a floor plan or a symmetrical part of the floor plan of a building was studied. Figure 7 shows a flow chart of the process.

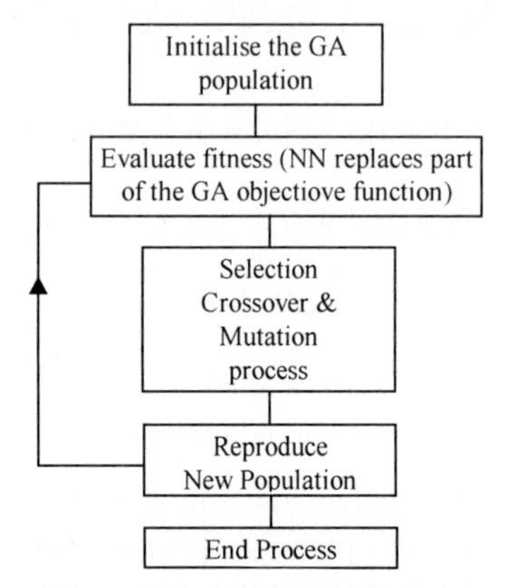

Figure 7 Flowchart of the GA process

For illustration a 48m by 12m plate was considered. Dimensions of the building footprint are generally, determined by the availability of the land and are therefore known. For simplicity only an in-situ reinforced concrete floor system was considered. To further simplify the problem, for this study only reinforced concrete flanged beams were included in the fitness evaluation.

Other essential information supplied as input to the GA were: floor design load, upper and lower limits for grids in both x and y directions and material costs.

The objective of this exercise was to obtain an optimum grid layout for the floor plan. For this study minimum material quantities were used as a fitness indicator.

The output from the GA were: grids in x and y directions, beam depth to beam web width ratio and beam flange width to beam web width ratio. A real number representation scheme was adopted for a more efficient representation.

The output from the GA was used to determine the number and locations of main and secondary beams. The information obtained from the GA along with the floor design loading were then passed to the NN to obtain material quantities. This information was used to calculate the cost of the floor and hence the fitness of each design alternative. The NN process included within the GA objective function is only the 'feed forward' part of the NN.

Final outcome of the GA, for the above example, is shown in Figure 8 and the GA average population fitness is shown in Figure 9. The format of the GA output was designed to be similar to bill of quantities which gives material quantities, floor grid layout and details of structural components dimensions.

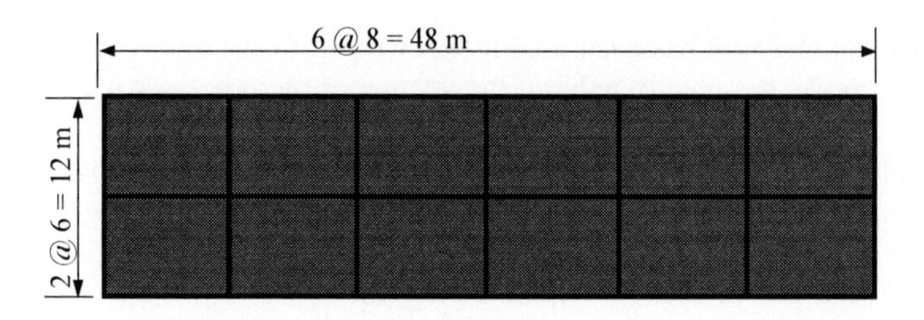

Figure 8 Optimum grid layout generated by the GA

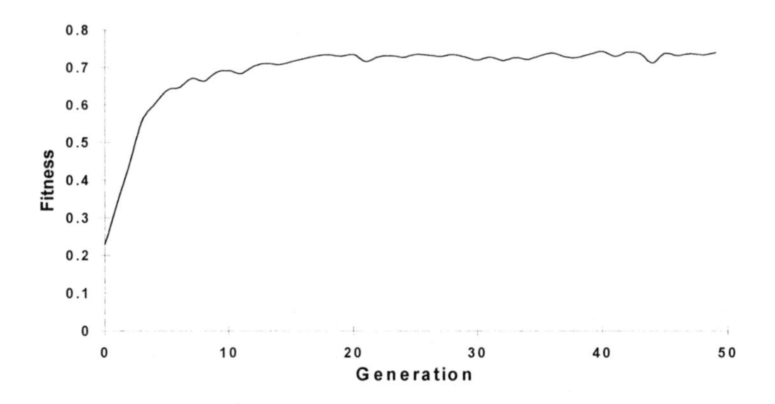

Figure 9 The GA average population fitness over 50 generations for a population size of 100

5.0 Conclusions

The structured Genetic Algorithm is a very powerful tool for synthesis, evaluation and selection of various design alternatives. It allows many design alternatives with different construction material to be processed, concurrently. This makes it a powerful tool for modelling the activities of the conceptual stage of the design process.

It is clear that activities related to the conceptual stage of the design process, generally, rely on human intuition, creativity, past experiences which have been very difficult to model.

Neural Networks are AI tools which have flexibility and allows the human creativity and intuitions to be incorporated in its learning process. This characteristic of NNs makes them a promising tool for use at the conceptual stage of the design process.

One of the drawbacks of the sGA is the rapid growth of the objective function. The paper has investigated the possibility of the integration of the NN with the GA. A trained Neural Network has been used to replace part of the objective function of the GA. This process has greatly reduced the volume of the processes undertaken by the GA.

Trained Neural Networks can also be used independently as a powerful decision support tool at the conceptual stage of the design process to assist designers. Examples of the use of NNs are initial member sizing, material and cost calculations for tender purposes. It can also be used for checking designs.

The paper has addressed the possibility of integrating the GA with the NN and has put forward suggestions for solving the problems associated with the explosive growth of the sGA objective function. Although current studies have concentrated on a very small part of the whole design domain, it has been shown that this process greatly simplifies the GA process and it should be possible to extend this to other areas with a wider scope.

References

1. Mathews, J. D., and Rafiq, M. Y., "Adoptive Search for Decision Support in the Preliminary Design of Structural Systems", *First International Conference on Adaptive Computing in Engineering Design and Control*, Plymouth, 1994.
2. Mathews, J. D., Rafiq, M. Y., and Parmee, I., "Adaptive search to assist in the conceptual design of concrete buildings", *Civil-Comp95, Sixth International Conference on Civil and Structural Engineering Computing,* Cambridge, ED B. H. V., Topping, Civil Comp Press Edinburgh, pp 179-187, August 1995.
3. Bullock, G. N. B., Denham, M. J., Parmee I. C., and Wade, J. G., "Developments in the use of the genetic algorithms in engineering design", *Design Studies*, Vol. 16, No. 4, October 1995, pp 507-524.
4. Rafiq, M. Y., and Mathews, J. D., "An Integrated Approach to Structural Design of Building Using Genetic Algorithms", *2nd World Conference on Integrated Design and Proces Technology*, IDPT-Vol. 3, Eds Esat, I. I., Ventiali, F., Rasty, J., Gransberg, D. D., and Ertas, A., pp 84-90, 1996.
5. Parmee, I. C., "High Level Decision-Support for Engineering Design Using the Genetic Algorithm and Complementary Techniques", in, *Proceedings of Applied Decision Technologies*, Brunel Conference Centre, London, 1995.
6. Rafiq, M. Y., " Optimum Building Concept Generation and Integration of Design Process Using a Structured Genetic Algorithm", *Proceeding of the World-wide ECCE Symposium, in the Practice of Building and Civil Engineering*, pp 117-121, September 1997.
7. Rafiq, M. Y., and Easterbrook, D. J., "Artificial Neural Networks for Modelling Some of the Activities of the Conceptual Stage of the Design Process", submitted to the ASCE Computing Congress, 1998.
8. Goldberg, D.E., "*Genetic Algorithms in Search, Optimisation, and Machine Learning",* Addison -Wesley, Reading, MA, 1989.
9. Davis L., "*Handbook of Genetic Algorithms",* Van Nostrand Reinhold, New York, 1991
10. Jenkins, W. M., "Structural Optimisation With the Genetic Algorithms", *The Structural Engineer,* Vol. 69, No. 24, pp 418-422, 1991.
11. Rafiq, M. Y. "Genetic Algorithms in Optimum Design, Capacity check and Detailing of Reinforced Concrete Columns", *Proc. Opti 95: 4th Int. Conf. on Computer Aided Optimum Design of Structures*, pp 161-169, 1995.

12. Grierson, D. E. and Pak, W. H., “Optimal Sizing, Geometrical and Topological Design using a Genetic Algorithm”, *Structural Optimization*, Spring-Verlag, Vol. 6, pp 115-159, 1993.
13. Dasgupta, D., and MacGregor, R., 'Structured genetic algorithms', Research report IKBS-2-19, University of Strathclyde, Glasgow, Scotland, 1991
14. Rafiq, M. Y., "A Design Support Tool for Optimum Building Concept Generation Using a structured Genetic Algorithm", accepted for publication in the *International Journal of Computer-Integrated Design and Construction.*
15. Roberts, A., Wade, G, “Optimisation of Finite Word length Filters Using a Genetic Algorithm”, *Proc. Adaptive Computing in Engineering Design and Control*, Plymouth UK Parmee IC editor, 37-43, 1994.
16 .Parmee, I. C., “The Development of a Dual-Agent Strategy for Efficient Search Across Whole System Engineering Design Hierarchy”, Forth International Conference on Parallel Problem Solving from Nature, Berlin, Germany, September 22-27, 1996.
17. Eberhart, R. C., and Dobbins, R. W., *“Neural Network PC Tools”*, Academic, San-Diego, 1990.
18. Caudill, M., “Neural Network Primer”, *AI Expert,* a series of articles from December 1987 to May 1989
19. Hajela, P. and Berke, L., “Neurobiological Computational Models in Structural Analysis and Design ”, Computers and Structures, Vol. 41, No. 4, pp 657-667, 1991.

Finding the Right Model for Bridge Diagnosis

B.Raphael and I.Smith

Institute of Structural Engineering and Mechanics (ISS-IMAC)
EPFL-Federal Institute of Technology
CH-1015 Lausanne
Switzerland

Abstract. This paper describes a hybrid reasoning system for complex diagnostic tasks in structural engineering. This project combines results from research into compositional modelling with model reuse for improving the quality of diagnosis through a systematic consideration of feasible models for explaining observations. This leads to more accurate predictions of behaviour and as a result, improved structural management.

KEYWORDS: Diagnosis, case-based reasoning, model-based reasoning, compositional modelling, structural monitoring.

1. Introduction

Our research is motivated by the observation that traditional engineering diagnostic assessments often cannot systematically cover all relevant modelling possibilities for explaining observations. There are a large number of assumptions involved in formulating models for full-scale civil engineering structures (see section 2). Different models are associated with different assumptions. Some assumptions cannot be justified in certain situations, while others lead to models that are far too complex to be used in practice. Only a few models are capable of adequately explaining a particular diagnostic observation. Computer assistance for selecting an appropriate complete model improves the quality of diagnoses, predictions and subsequent monitoring, maintenance and replacement decisions.

We have developed a hybrid reasoning system that combines compositional modelling with model reuse in order to improve the quality of diagnosis. The system assists the engineer in formulating complete models by:

- composing model fragments
- reusing complete models that are stored in a case base
- using a combination of the above

Our representation of model fragments ensures that assumptions are made explicit and that their implications are evaluated during model formulation.

2. Complexity of diagnostic tasks

The complexity of diagnostic tasks in structural engineering is often caused by a large number of possible models for interpreting behaviour. For example, a bridge can be modelled through making different assumptions about the way it functions and how it reacts to environmental effects. In order to illustrate modelling possibilities, a bridge in service in Switzerland is used as an example. **More than a million models** can be formulated for this bridge.

2.1 Lutrive bridge

This example uses the Lutrive bridge in Switzerland. The North and South Lutrive bridges are two parallel twin bridges (Figure 1). Each bridge supports two lanes of the Swiss national highway RN9 between Lausanne and Vevey. Built by the cantilever method with central hinges, the two bridges are gently curved (radius = 1000 m) and each bridge is approximately 395 m long with four spans. The two bridges have the same cross section consisting of a box girder of variable height (from 2.5 m to 8.5 m) and two slightly dissymmetric cantilevers which are meant to reduce the effects of torsion in the curved bridges.

Figure 1: **Lutrive bridge**

2.2 Assumptions involved in modelling

Different models can be formulated for this bridge through making different assumptions. Assumptions are related to material properties, geometry and structural loading. Typically, a simple or minimal model is initially formulated. This model is then modified to accommodate variations in assumptions. Some assumptions are fundamental to the creation of this minimal model while others are made for practical reasons, for example, to interpret data at reasonable computational cost. Assumptions are also made when there is a lack of knowledge of the behaviour of the structure.

The following table, Table 1, is a list of modelling assumptions possible for the Lutrive bridge. Characteristics of assumptions are indicated in the table. That is,

- whether they are fundamental to the creation of a minimal model
- whether they affect the accuracy of the model
- whether they are made because of a lack of knowledge of behaviour. This characteristic can be subdivided according to whether or not clarification is possible through inspection of the bridge.

Table 1: Modelling assumptions for the Lutrive Bridge

	Assumptions	Fundamental	Accuracy	Lack of knowledge	
				Verifiable	Difficult to verify
1.	Base model: Bernoulli-beam hypothesis, linear elastic behaviour, beams continuous over supports, (Figure 2a).	X		X	
2.	Twin column support (Figure 2b).		X		
3.	Varying moment of inertia of the beam cross section. (The depth of the beam varies from 2.5 m. to 8.5 m.).		X		
4.	No relaxation in prestressing force.	X			X
5.	Stiffness of columns considered in bending (Figure 2c).		X		
6.	Rigid beam-column connection.		X		X
7.	Cracks at the supports (Figure 2d).	X		X	
8.	Rotational springs assumed due to cracks within the span.	X		X	
9.	Curvature of the beam (in the plan).		X		

	Assumptions	Fundamental	Accuracy	Lack of knowledge	
				Verifiable	Difficult to verify
10.	Deep beam hypothesis (Bernoulli-beam hypothesis not valid).		X		
11.	Load carrying capacity of the deck considered (Figure 2e).		X		
12.	Ideal pin rollers at the supports.		X		X
13.	Initial displacement due to shrinkage.	X		X	
14.	Young's modulus of the concrete assumed not to have changed since the time of construction.				X
15.	Ideal hinges at mid-spans. (Hinges do not transmit any moment at all).		X		X
16.	Non-linear material.		X		
17.	Geometric nonlinearity.		X		
18.	Support settlement.	X		X	
19.	Ideal point loads.	X			
20.	Distributed wheel loads (figure 2f) instead of ideal point loads.		X		
21.	Distribution of repeated wheel loads assumed (since traffic on the bridge cannot be controlled during measurements).			X	
22.	Constant temperature gradient assumed throughout the length of the bridge.	X			
23.	Temperature difference between the top and bottom chords of the beam is constant.			X	
24.	Linear temperature variation along the depth of the beam (Variation due to the hollow box section not considered).				X

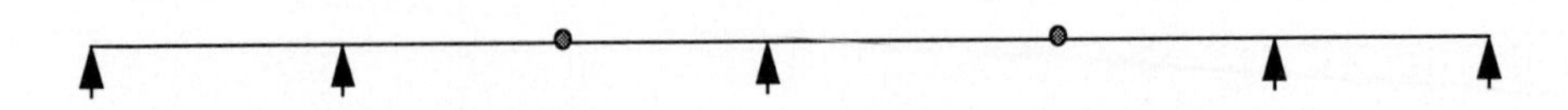

Figure 2a: **The simplest means of modelling the Lutrive bridge is by assuming an ideal four span continuous beam with constant cross-section, elastic material, satisfying the Bernoulli-beam hypothesis.**

Figure 2b: **To make the model closer to reality, the interior supports in figure 2a can be replaced by twin supports separated by 12 m.**

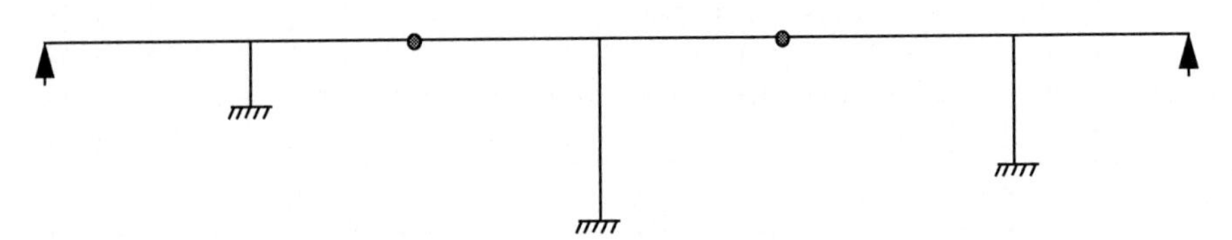

Figure 2c**: In order to study the effect of the stiffness of the columns on the behaviour of the bridge, a frame model can be assumed.**

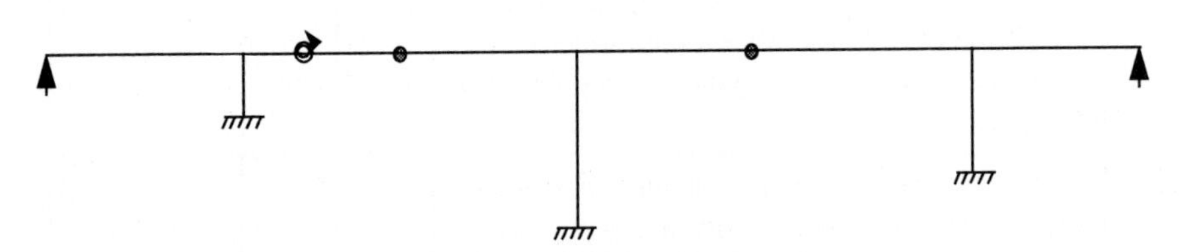

Figure 2d**: Support cracks could considerably reduce the stiffness of the continuous beam. They can be modelled as rotational springs.**

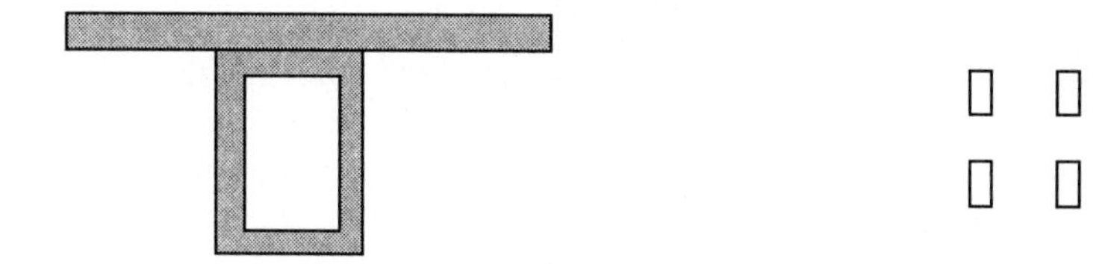

Figure 2e: **Load carrying capacity of the deck adds to the stiffness of the beam. A certain percentage of the width of the deck can be assumed to be taking part in the bending.**

Figure 2f: **It will be more accurate to model vehicular traffic on the bridge as distributed loads than ideal point loads.**

2.3 Discussion

Twenty four assumptions related to the material property, geometric property and loads have been listed for a real bridge. (This is not an exhaustive list and there could be many other assumptions). Approximately twenty are independent assumptions. Different combinations of assumptions create more than a **million** different models. Several sub-choices are possible under each of these assumptions. (For example, whether the relaxation in prestressing force is 5%, 10% or something else). If these sub-choices are also taken into account the number of modelling possibilities increases further. In practice, engineers make reasonable assumptions based on their intuition and experience and consider only a few models, as it is not possible to consider many options manually. However with computer support it becomes possible to consider more alternatives and thereby improve the quality of diagnosis.

3. Current techniques

Traditional means of diagnosis involve the use of a single model. The disadvantages of this method are the following:

- There are hidden assumptions in the model. It is often not ascertained whether the assumptions can be justified. Since the assumptions are not made explicit, their effects may be overlooked.
- Results may not be accurate because other possible means of diagnosis (using other models) may be more appropriate.

In spite of these drawbacks, engineers continue to use single models because there is no support for managing multiple models. Manually working with all the assumptions and models is extremely difficult. Hence, engineers often make simplifying assumptions to create a minimal model and verify only a few of them after diagnosis.

Computer systems have been developed to perform diagnostic tasks in many domains [1,2,3,4]. Usually, these systems employ one or more of the following three different approaches:

- Rule-based reasoning,
- Model-based reasoning (MBR) [3,5],
- Case-based reasoning (CBR) [1].

Drawbacks of current systems using these techniques include:

- Difficulty in maintaining rules, cases and knowledge bases
- Implicit assumptions in models (and rules)
- Implicit knowledge in cases
- Lack of support for dealing with a large number of assumptions
- Multiple models are not allowed
- Difficulties associated with combining cases and case adaptation
- Infeasibility of scaling up the system to accommodate full-scale practical problems
- Inability to provide explanations (other than problem solving traces)

A summary of the current methods used for structural diagnosis is given in Figure 3.

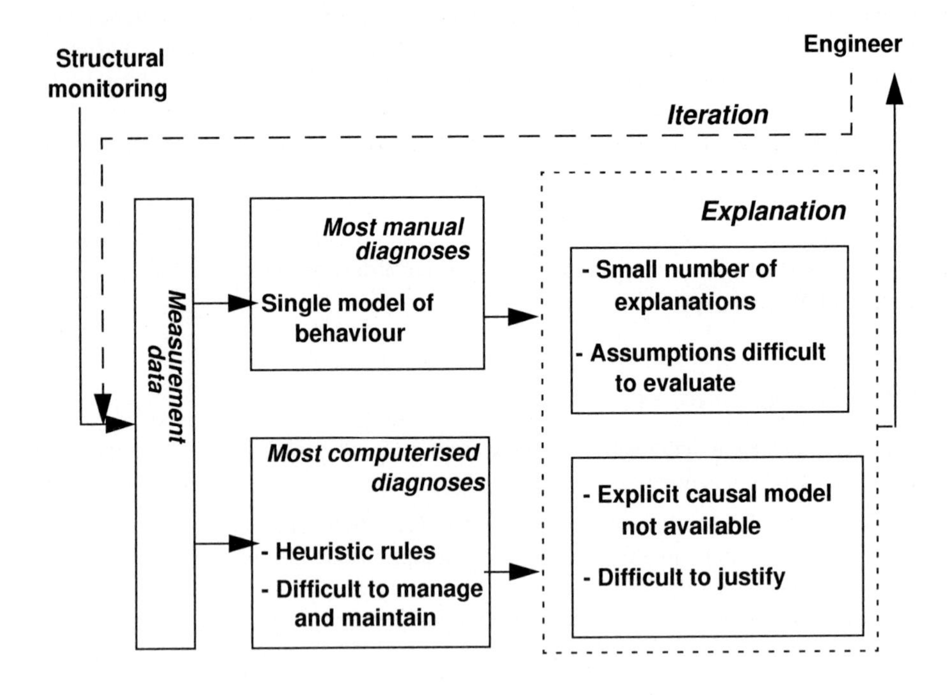

Figure 3: **Current methods for structural diagnoses**

Two diagnostic systems that have been applied to full-scale problems in structural engineering are DAMSAFE [3] and IGOR [4]. DAMSAFE is an expert system tool for the monitoring of dams. In this system, the behaviour of dams is represented as a network of processes. Features extracted from monitoring data are mapped onto processes (that describe behavioural states) by means of rules defined by experts. IGOR is a decision support system that helps technicians seismically assess buildings. This system uses multiple models of different levels of abstraction and is able to generate causal explanations that are tailored to different types of users. However, in both systems the influence of assumptions in the formulation of multiple models is not explicitly considered.

Compositional modelling [6] has been used for diagnostic tasks for physical devices. This research focuses on the need to accommodate multiple models and the necessity to make assumptions explicit. Compositional modelling is a framework for constructing adequate device models by composing model fragments selected from a model fragment library. Model fragments partially describe components and physical phenomena. Model construction is a search problem where the goal is to select a model from the space of possible models defined by the model fragment library while satisfying all modelling constraints. This research concentrates on providing causal explanations for the behaviour of small devices. Its application to complex diagnostic tasks in structural engineering has never previously been assessed.

4. A hybrid reasoning system for diagnosis

We combine the techniques of compositional modelling and model reuse (Figure 4). The idea of a hybrid reasoning system is not new [2,7,8]. However, some of the hybrid reasoning systems [7, 8] use CBR only to speed up the reasoning process. In our system, the role of cases is different; cases are employed to reduce problem complexity due to the large number of possible combinations of model fragments. We do not use cases to capture heuristics for diagnosis. Since our emphasis is on performing model-based diagnosis using data obtained from measurement techniques [9], our cases contain models. Models have also been used for the adaptation of cases [2]. In our system, model identification is the goal and cases assist in creating appropriate models. In this way a set of appropriate models becomes available for providing causal explanations to the user.

4.1 System design

The design of the system is shown in Figure 4. Observed behaviour (measurement data) is interpreted using models which are identified by means of two techniques, model composition and model reuse. The first technique uses model fragments stored in a library whereas the second technique uses complete models stored in a case base.

The criterion for selection of the technique has to account for three situations, **S1, S2** and **S3**. In situation **S1**, cases that have an exact match with the current situation exist in the case-base. Model reuse is appropriate in such a situation. In situation **S2**, none of the cases have any similarity with the current situation. Compositional modelling is

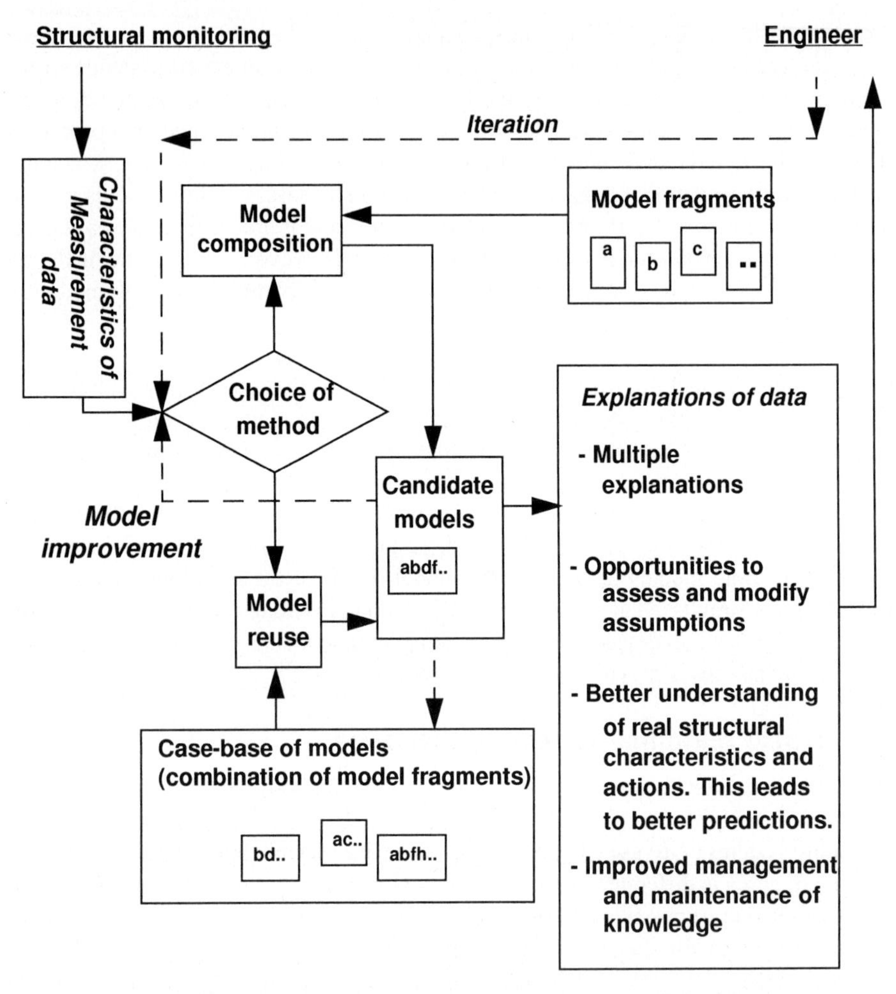

Figure 4: Diagnosis using model composition and model reuse

used here. In situation **S3**, there are some similarities between the current situation and those in the case-base. The following three methods are appropriate for such a situation:
S3.1: A case is selected and model composition is used to adapt it.
S3.2: Model composition is activated and guided by common attributes of selected cases.
S3.3: Parts of relevant cases are combined using methods of model composition.

We have begun with the approach that corresponds most closely to current engineering practice. Since model reuse is often most efficient, cases that have a reasonable degree of similarity with the current diagnostic task are selected (Situation **S1**). In the event of failure to retrieve relevant cases, the technique of model composition is used (Situation **S2**). When cases are partially relevant, users have the opportunity to replace inappropriate model fragments with more relevant fragments using the model composition module (Situation **S3.1**). Engineers are also able to go back and modify assumptions (and models) after browsing through the interpretations. Subsequent prototypes will develop the second and third methods in situation **S3**.

4.2 Defining model fragments

There are several ways to define model fragments for the example given in section 2.1. The approach we have followed is to have different sets of model fragments for modelling structural aspects such as, geometry, material, cross-section properties, support conditions and loads. In order to create a complete model capable of interpreting measurement data, a model fragment has to be selected from each set.

Some examples of model fragments belonging to these sets are shown in the following table:

Table 2: Sample model fragments

	Name of Fragment	Set	Assumptions	Key attributes
1.	Geom-1	Geometry	Continuous beams satisfying Bernoulli-beam hypothesis (Assumption-1), Ideal hinges (Assumption-15)	Nodes, Beams, Connectivity, Hinges
2.	Material-1	Material	Young's modulus of concrete not changed (Assumption-14), Linear material (Opposite of Assumption-16)	Young's modulus, Poisson's ratio
3.	Section-1	Cross-section	Constant cross-section (Opposite of Assumption-3)	Moment of inertia, area
4.	Support-1	Support-conditions	Ideal pin rollers (Assumption-12)	Position of supports, Type of supports
5.	Load-1	Structural-loading	Distributed wheel loads (Assumption-20)	Position of loads, Magnitude

By combining all the fragments in this table, a complete model consisting of a four span continuous beam with distributed wheel loads is created. The behaviour of this model can be deduced by standard methods of structural analysis. More generally, the composition of model fragments is a complicated process which might involve addition, deletion and modification of several attributes of individual fragments.

The system described in this paper received preliminary approval by engineers responsible for bridge evaluation in Switzerland. We are currently implementing the system in order to proceed with full scale validation and testing. Scientific challenges include adequate support for the choice of method for model selection, retrieving relevant models for reuse and adapting models when there is no exact match in the case base.

5. Conclusions

A large number of modelling possibilities exist for explaining behaviour of structures in service. Different models are associated with different sets of assumptions. Choosing the right model for comparison with observations require systematic evaluations of the implications of assumptions. Our project combines compositional modelling with model reuse for improving the quality of interpretations. More effective evaluation and prediction of structural behaviour leads to better decision making for repair, retrofit, allowable load modifications and ultimately, structural dismantlement.

Acknowledgements
We would like to thank S.Vurpillot for assistance with the example used in this paper, as well as to R. Stalker and K. Shea for useful discussions. Finally, we would like to thank Logitech SA and Silicon Graphics Incorporated for supporting this research.

References

[1] J. L. Kolodner, “Case-based Reasoning”, *Morgan Kaufmann*, San Mateo, CA, 1993.

[2] M.L.Maher and D.M.Zhang, “CADSYN; A case-based design process model”, *AI EDAM*, vol. 7(2), pp. 97-110, 1993.

[3] P. Salvaneschi, M. Cadei, and M. Lazzari, “Applying AI to structural safety monitoring and evaluation”, *IEEE Expert/Intelligent Systems and Their Applications*, 11, No. 4, August, 1996.

[4] P.Salvaneschi, M. Cadei and M. Lazzari, “A causal modelling framework for the simulation oand explanation of the behaviour of structures”, *Artificial Intelligence in Engineering*, vol. 11, pp. 205-216, (1997).

[5] M. J. Chantler, G. M. Coghill, Q. Shen, R. R. Leitch, “Selecting tools and techniques for model-based diagnosis”, *Artificial Intelligence in Engineering*, vol. 12, pp. 81-98, 1998.

[6] B. Falkenhainer, K. D. Forbus, “Compositional modelling: Finding the right model for the job”, *Artificial Intelligence*, vol. 51, pp. 95-143, 1991.

[7] M. Someren, J. Surma, and P. Torasso, "A utility-based approach to learning in a mixed case-based and model-based architecture", *Proceedings, Second International Conference on Case-Based Reasoning (ICCBR-97)*, Rhode Island, July 25-27, pp. 477-489, 1997.

[8] D. J. Macchion, D. Phuoc Vo, "A hybrid knowledge-based system for technical diagnosis learning and assistance", *EWCBR-93*, p.301, 1993.

[9] D. Inaudi, *Fibre optic sensor network for the monitoring of civil engineering structures*, EPFL Thesis, 1997.

Knowledge-Based Assistants in Collaborative Engineering

W. M. Kim Roddis

Associate Professor
Civil and Environmental Engineering
University of Kansas
Lawrence, Kansas, USA
roddis@ukans.edu

Abstract. Collaborative engineering requires the exchange and use of design information in a multidisciplinary team with time and space separations among the designers over the course of the design process. A challenge to effective collaboration is the delivery of existing knowledge to the time and place required. Upstream communication is necessary when participants later in the sequence, such as the construction team, possess information needed by those earlier in the sequence, such as the design team. For separate entities sequentially performing design and construction, knowledge-based assistants can provide such essential information as relative costs of construction choices when the construction partner is not yet identified. For a single entity team performing design-build, knowledge-based assistants can provide consistent and thorough implementation of design choices considering cross-disciplinary and construction issues. Steel building design/fabrication/ construction serves as an example to provide specific context for discussion of the use of knowledge-based assistants with emphasis on upstream communication of design information. Cost savings are used to measure the value of such knowledge-based assistants. Knowledge-based assistants encapsulating cross-disciplinary knowledge can also be used to assist in design situations encountered on a semi-regular basis by a design organization but infrequently by an individual designer. Cases and rules in combination can give designers guidance based on pooled organization experience. Such use of knowledge-based assistants is briefly illustrated for 1) steel bridge fabrication error resolution and 2) repair of fatigue damage found for in-service bridges.

1. Knowledge delivery

The practicing engineer has a voracious need for information and the expertise to apply it. This need has grown due to the increasing sophistication, complexity, and sheer amount of extant engineering knowledge. Although the knowledge exists, there are problems in finding the best source at the right time. A fundamental objective of knowledge-based systems (KBS) technology is to deliver existing knowledge at the time and place needed.

As stated in the call for papers for this conference, transfer of the results of artificial intelligence research, including KBS, into structural engineering applications has been slow. The fragmentation of the architecture, engineering, and construction (AEC) industry is not only one of the persistent barriers to innovation in practice, but is also a perennial motivation for pursuing knowledge-based solutions. Advances in product modelling currently provide greatly improved means of overcoming the barriers of incomplete information and constantly changing contexts. Lower hardware costs and availability of robust software on inexpensive platforms make possible the incorporation of KBS into organizations with modest computing resources. For some applications, practical value can be found for fairly small systems that can be expanded incrementally. This paper looks at some of the opportunities for successfully applying knowledge-based assistants in practice. The objective of this paper is not to look at topics for further research, but rather to focus on some of the established results ready for near term incorporation into standard design office computing tools. Much benefit would be gained by practicing engineers if more widespread technology transfer were made to automate routine design subtasks and checking of design details.

2. Collaborative engineering

The engineering design process is inherently a multidisciplinary activity requiring cooperation. Although a large engineering problem is often decomposed into a number of specialized tasks to manage the complexity of the project, each sub-task relies on information and feedback from other areas of specialization to solve its problem. Thus, the success of a multidisciplinary design effort depends heavily on the ability of the individual specialists to communicate and share information with each other while working concurrently.

Upstream communication of information is necessary when participants later in the sequence possess information needed by those earlier in the sequence. The deleterious consequences of separation of design of an artifact from production of that artifact is well known in manufacturing industries. The need for feedback of production information to engineering design is even more critical for the AEC industry since most constructed facilities are produced as singly realized structures. Without the ability to refine and improve a design in a production cycle of multiple items, the need for transfer of knowledge from parties later in the building process forward to those earlier in process is thus even more important. In particular, there is substantial need to provide feedback concerning constructability during the early stages of design (Fischer 1991).

Much of the AEC industry has temporal and organizational separation of design and construction phases. Many times, the construction team is not brought under the contract umbrella until after the majority of the building design is complete. This is significant because the design team's ability to influence the overall project cost is greatest in the early stages of design and decreases rapidly after that (ASCE 1990). By the time the preliminary design is completed, much of the project cost has been determined. Without the advice of a constructor, the designer must rely on past experience to assure that the design is not difficult to construct. The designer in isolation frequently has insufficient information on the relative costs of material, fabrication, and construction tradeoffs to fully evaluate various design alternatives. By including construction input early in the design phase, project costs decrease, schedules are shortened, and safety levels increase on the job site (Zimmerman 1995).

Fortunately, the value of sharing information among parties in the AEC industry (Fischer et al. 1994) is becoming widely recognized in practice as demonstrated by the increasing use of product models and electronic exchange of these models between project phases and participants (Novetski 1996). The availability of these product models provides an opportunity to implement knowledge-based assistants to leverage the data contained in the existing models and provide substantial benefits in improved quality and reduced costs of completed structures. For separate entities sequentially performing design and construction, a KBS can provide such essential information as relative costs of fabrication and construction choices when the construction partner is not yet identified. For a single entity team performing design-build, a KBS can provide consistent and thorough implementation of design choices considering cross-disciplinary and construction issues.

3. SteelTeam: design/detailing/fabrication/construction

The steel building industry is used in this section to provide a specific context for discussion of the use of knowledge-based assistants with emphasis on upstream communication of design information. Decisions made at each stage in the steel building design/detailing/fabrication/ construction process depend on information held by parties responsible for following phases. This provides many opportunities for applying value engineering and improving steel economy by providing better information concerning later stages, such as fabrication, to those performing earlier phases, such as design (Ricker 1992). Indeed, much progress has been made in Europe on integrating steel building engineering (Crowley 1997, Hannus 1996). The American steel building industry remains highly fragmented, with parties to the overall process organizationally separated. The specific parties responsible for detailing, fabrication, and construction are frequently not yet identified at the time of design. Assuming the constraints of organizational and possibly temporal separation among parties to an engineering process requiring collaboration, a steel design environment called SteelTeam (Pasley 1996, Ernst and Roddis, 1994) was implemented in a research setting. The following subsections discuss: 1) the data model and knowledge base components of SteelTeam; 2) testing of SteelTeam on simulated and actual structures to quantify benefits; and 3) establishing the feasibility of transferring the KBS technology to practice using an existing engineering data management system to provide the product model.

3.1 SteelTeam: integrating CADD with KBS

Many KBS addressing various aspects of structural building design have been developed in research settings. Fundamental work addressed preliminary framing of multistory buildings (Maher and Fenves 1985). Systems with direct relevance to SteelTeam includes the following. The Designer Fabricator Interpreter (DFI) (Werkman and Hillman 1989) acts as an intelligent interface between designers, fabricators, and constructors by offering advice and criticism of beam-to-column connections. FLEX (Floor Layout Expert) is a prototype engineering design system that addresses the task of designing and arranging the layout of steel floor framing systems (Morse and Hendrickson 1991). FLEX takes into consideration the mechanical equipment placement, ease of construction, and structural efficiency of the design. The Construction Knowledge Expert (COKE) (Fischer 1993) uses a KBS to provide constructability input during the preliminary design phase of cast-in-place concrete. COKE uses the CIFECAD system (Fischer 1991) to obtain the building model. Tizani et al. (Tizani et al. 1995) present a system with a costing module that determines relative cost of design alternatives. This cost is given in terms of time required to carry out all fabrication operations, total weight, and total surface area of the structure. The Automated Design Aid (ADA) (Moore and Tunnicliffe 1994) is intended to act as an advisor on the constructability of designs that have been produced using CAD software.

Many of these systems address constructability concerns at the preliminary design phase as the SteelTeam system does. However, none of these systems address the overall steel building design in terms of constructability, ease of fabrication, and other concerns that the parties to the design process have. SteelTeam addresses the detailed design phase and provides knowledge-based assistance during later phases, such as the shop drawing review.

SteelTeam, a CADD environment integrated with a KBS (Pasley 1996, Ernst and Roddis 1994), improves communication during the steel building design and construction process in two specific ways. First, it defines and uses an object-based building data model that provides a mechanism for downstream communication from designers to the parties involved with detailing, fabrication, and construction. Second, SteelTeam provides a KBS used by the designer as a decision support tool representing upstream communication of the concerns of the detailer, fabricator, and constructor. SteelTeam was developed on a 90 MHz Pentium with 16 MBytes of RAM. SteelTeam makes use of AutoCAD for its Graphical User Interface and is implemented in AutoLISP.

A product data model is used by SteelTeam to represent the steel building design and provide for further integration through the use of translators that allow the data model to be linked to third-party structural analysis, structural design, detailing, and connection design software. The International Standards Organization (ISO) is developing STEP (Standard for the Exchange of Product model data), formally known as ISO 10303 (Crowley 1997), as an international standard for the computer representation and exchange of product data. Use of the STEP standards in the steel building industry is being undertaken as part of the Pan European CIMsteel project (Crowley 1997). The CIMsteel project was started in 1985 to improve the effectiveness and efficiency of the European steel construction industry through the integration of design codes and the design, detailing, fabrication, and construction of steel buildings. One of the most advanced areas of the CIMsteel project is in the area of product modeling and data exchange. Although the robustness of the STEP standard is attractive, STEP was still under development at the time SteelTeam was developed and in particular did not yet address steel connections. For this reason, the product model in the SteelTeam system was chosen to resemble simpler models that have been used in research.

CIFECAD (Fischer 1991) is a symbolic design built as an extension to AutoCAD. Reinforced concrete building components are drawn as objects with associated information describing composition, structure, dimensions, orientations, etc. A building can be designed as a collection of building components, and the resulting geometric and symbolic descriptions are then available to other systems (Winstanley et al. 1993). CIFECAD is used to provide a greater level of detail on building designs than conventional CAD representations. The CIFECAD system shares many characteristics with the SteelTeam building model. An object-oriented data structure is used to describe the building in terms of design objects such as beams, columns, joists, connections, and braces. Design objects are divided into classes that are organized into a class hierarchy. The SteelTeam building data model is made up of three components. First, the AutoCAD drawing database contains information about the design objects stored as extended entity data. Second, additional information about the building design and building specific preferences is stored in lists as variables inside AutoCAD. Third, the object list containing the instances of the attributes of each object that is part of the steel building design. The object list includes instances of beams, columns, bracing, joists, and connections. When the data model is stored between invocations of the SteelTeam program, the three components listed above are stored as the AutoCAD drawing file and six other supporting files.

These six accompanying files are: 1) the load case file providing end reactions for each load case for each member, 2) the floor file associating AutoCAD layers and floor elevations, 3) the construction preferences file containing contractor preferred defaults for issues such as erection bracing, work heights, and bolting practices, 4) the fabricator preferences file specifying relative cost data and comparative ranking of connection types, 5) the mill preferences file setting mills likely to be used for supply along with milling and transportation restrictions, and 6) the section preferences file listing structural shapes preferred for use even at a cost penalty, for example when an inventory of particular shapes is on hand.

A knowledge-based shell is provided within SteelTeam. This rule based shell uses a backward chaining inference engine implemented in AutoLISP. The KBS accesses the building data model to provide advice to the user on how to improve the overall economy of the steel building. The data model provides the KBS with access to both geometric data, such as member location, and non-geometric data, such as member type.

To optimize the overall economy of the steel building, the designer needs to be aware of where money is spent on a project. In steel building construction, project cost is normally divided fairly equally among material, fabrication and erection (Ricker 1992). One of the focuses of the KBS is to make the designer aware of how designer choices influence the overall economy of the structure, not only material costs. The rules are designed to supplement, and in some cases take the place of, feedback that would normally be received from the detailer, fabricator, or constructor. The rules came from various publications, the steel design codes and specifications, conference presentations, and the detailer and fabricator representatives on the SteelTeam advisory panel (Pasley 1996).

The KBS is able to make recommendations based on six areas of concern: design issues, fabrication issues, constructability issues, relative costing, availability, and completeness. For example, design includes setting the optimum bay ratio and bay area; fabrication includes fitting connections within the available beam depth; constructability includes keeping column splice heights within 5 ft. (1.5 m) of the top of steel to avoid the use of scaffolding or extra rigging to make the connection; relative costing includes cost comparison between a heavier, clean column and a lighter column that requires stiffeners and doubler plates; availability includes section sizes rolled at which steel mills; and completeness includes provision for non-typical connection details. The type of problems addressed would generally be characterized as routine problems. The currently implemented knowledge base consists of approximately 150 rules. The coverage of the domain is not exhaustive and is intended to be expanded through practical use.

3.2 SteelTeam: quantification of knowledge-based assistant benefits

It is difficult to motivate the design office to spend money in their operation to improve and upgrade communication when it saves cost in someone else's work. This is the case in the American steel building industry because the design, fabrication, and construction are each performed by separate entities. However, clearly showing reduction in overall project cost gives the designer a competitive advantage when the owner chooses the design team.

Two approaches quantify the effect of SteelTeam on the cost of the structural steel frame. The first approach uses a typical building assumed to trigger all applicable SteelTeam advice directly dealing with value engineering. This approach establishes an optimistic measure of SteelTeam's performance, giving a potential cost savings at the upper end of that possible using the current SteelTeam implementation. The second approach uses two actual buildings, one excluding connections and one including connections. Since much of SteelTeam's cost benefits are associated with design-fabrication collaboration and hence heavily influenced by connection costs, the actual building excluding connections establishes a pessimistic measure of SteelTeam's performance, giving a potential cost savings at the lower end of that possible using the current SteelTeam implementation. The actual building including connections establishes the most realistic measure, giving a potential cost savings expected initially in practice. As the currently implemented SteelTeam knowledge base is quite small and intended to be expanded through practical use, all three performance measures are conservative.

Building	Potential Savings	Approximate Steel Frame Cost	Savings/ Cost
"Typical" 20-story with connections	$184,000	$921,000	20.0%
Building A (bank) without; connections	$4,330	$212,000	2.0%
Building B (refinery) with connections	$56,000	$1,050,000	5.4%

Table 1. Summary of Savings from the Use of the SteelTeam System

The potential savings for the three structures is summarized in Table 1. The two percent savings from Building A sets an overly pessimistic lower bound for expected initial SteelTeam performance due to the absence of detailed connection information in the building model and lack of quantified cost benefits for most of the improvements identified. The twenty percent savings for the "typical" 20-story building represents an upper bound for expected SteelTeam performance due to the generous assumptions that were used to determine potential savings. The five percent savings from Building B sets a conservatively measured level of savings that would be expected to be easily realized in initial practice. In addition, the presence of the connection design information, member end reactions, and a more complete set of project data results in a more efficient steel building design process, benefits from the use of SteelTeam that are not directly measured in terms of cost in this study.

3.3 SteelTeam: application to existing engineering data management systems

In practice, it is unlikely that a design firm would completely replace its current design methods and processes in favor of a system such as SteelTeam, regardless of its potential to lower overall costs and to shorten schedules. A more realistic scenario is that the SteelTeam system would be integrated into existing systems already used by designers. In order to evaluate the viability of combining SteelTeam with engineering data management systems in practice, the POWRTRAK engineering data management system used at Black and Veatch was considered as a candidate for integrating with SteelTeam. POWRTRAK is a concurrent engineering system. At its heart is a centralized database that increases communication between engineering disciplines, and reduces repetitive data entry. Several engineering application areas add information to the database and use information supplied by other application areas. The interface between application areas is strictly through the information that is shared in the common database. The eight functional application areas of POWERTRAK are: Project Scheduling System, Project Cost Analysis System, Computer Automated Engineering, Engineering Design, 3-D Modeling, Drawing Control, Construction Control, and Procurement Control.

Combining the SteelTeam system with a concurrent engineering system such as POWRTRAK creates a synergy in which the best features of each system help to strengthen the weaker parts of the other. The strength of SteelTeam is its KBS, described in section 3.1 The KBS helps designers make informed decisions that effect fabrication and construction costs by reviewing design data early in the process. It is weakest when it comes to modeling data and managing changes to the data. On the other hand, a concurrent engineering system like POWRTRAK inherently has the ability to store, manage, and report engineering data unambiguously by keeping information in a single, centralized location. Substantial value can be added to this system by integrating a decision support tool like SteelTeam that can review the large amounts of data gathered during design for issues related to fabrication and construction. POWRTRAK applies this idea to one aspect of construction already. It has the capability to review the design data for "space control", reporting interferences between components designed and modeled by different engineering disciplines. Integrating the engineering data model of POWRTRAK with the SteelTeam KBS has the potential to make each system work better. However, this hybrid system requires some feasibility analysis and evaluation of the key issues of platform portability, data model compatibility, and value added justification.

The operating environments of SteelTeam and POWERTRAK are different on several different levels. The SteelTeam system architecture is based on a KBS that uses a CAD system for its working memory. The KBS was developed in and works with the customization tools (primarily AutoLISP) that the CAD system makes available. POWRTRAK has a distributed process architecture in which many independent modules use a central database. It is a client/server architecture in which the modules (as clients) access a relational database (the server). Next, the operating scope of SteelTeam is completely within and dependent upon the AutoCAD program. AutoCAD must be started and running in order to use SteelTeam. POWRTRAK, on the other hand, is a system of many individual programs that run in various OS/hardware environments: Windows on a PC, UNIX on a Sun workstation, or VMS on a Digital VAX mainframe. Their common characteristic is that they access a relational database and share data through that interface. Because SteelTeam is dependent upon the AutoCAD environment, its options for interfacing with other systems is limited to the programming interfaces that AutoCAD provides. SteelTeam, as it currently exists, would not be able to be integrated into POWRTRAK as simply another application module. This is because a fundamental basis of POWRTRAK is that all persistent data is stored and shared in the relational database. SteelTeam is based on storing its persistent data in an AutoCAD drawing file (and auxiliary support files).

SteelTeam can be adapted to work on data other that the building data model that it currently uses. This can make it a much more effective and flexible tool. Conceptually, the building data model used by SteelTeam is very similar to the approach taken by POWRTRAK. Both use object-oriented principles to define abstract design objects that are the fundamental elements of the system. The data models of SteelTeam and POWRTRAK differ only in their implementations. SteelTeam defines a class hierarchy of data structure types that represent designed elements of a structure such as beams, columns, and connections. Its building data model is a collection of instances of these data structure types. These instances are implemented by graphical entities with attached data in an AutoCAD drawing and auxiliary support files. POWRTRAK defines a hierarchy of component types that represent designed components of a power plant facility, like beams, columns, pipe, and equipment. Its data model is a collection of instances of these component types. The instances of component types are implemented as records in several tables in a relational database.

The value of combining the SteelTeam KBS with the existing data model of POWRTRAK is powerful. The SteelTeam KBS provides a way to formalize knowledge about avoiding potential downstream problems and make it available during the early phases of design. It can be applied using an agent-based approach in which designers would enlist the help of an agent to review and report problems in the current state of the design. Some potentially useful issues to check for are: (1) constructability and construction costs, (2) ease of fabrication and fabrication costs, (3) completeness of design information, (4) design efficiency considerations. The result is an effective decision support system that provides crucial information that has a significant impact on the overall schedule, cost, quality, and safety of a project.

4. Leveraging Product Data Models

The above example looked in some detail at the integration of a product model and KBS for steel building design/fabrication/construction to conservatively quantify benefits directly realizable from KBS application with emphasis on upstream communication of design information. In addition to the feedback of detailing/fabrication/construction knowledge to the designer, the KBS provides other uses of the data available in the product model, such as: automation of routine tasks, consistency of design, and collaborative communication.

Automation of routine tasks, such as connection review during the shop drawing checking process, can be done using the KBS to implement a knowledge-based assistant to compare each shop drawing specified connection's capacity to the forces required by analysis. This task requires access to relevant data from the engineering and shop drawings as well as the structural analysis and connection capacity software, and was demonstrated in the SteelTeam system using the internal product model and KBS as well as third-party analysis and connection software linked to SteelTeam by file exchange and file translators.

Consistent and thorough implementation of design choices can be done using the KBS to implement a knowledge-based assistant that reviews the design documents for conformance to design standards formalized and represented in an appropriate knowledge base (Ernst and Roddis 1994). Specific design checking issues enumerated within the context of the SteelTeam and POWRTRAK discussion above are constructability, fabricability, design completeness, and design efficiency. This KBS application enables the engineer to focus on the higher level design issues and transfers the burden of tedious, repetitious checking for conformance to computer tools.

An important requirement for concurrent engineering is that when a change is made to the design, other participants in the design process that are affected by the change are notified. Again, the KBS can be used to facilitate this information sharing by implementing knowledge-based assistants to track changes by each user and identify items to be brought to the attention of another user. For example, when a flat plate concrete floor system is selected by the structural engineer , a rule would become active that would flag all changes by any design discipline that required floor penetrations in the slab near the columns to be brought to the attention of the structural engineer automatically.

The upstream communication advice, automation of routine tasks, consistency of design, and collaborative communication roles all lend themselves to an agent based view (Genesereth and Ketchpel 1994). Each can be applied using an agent based approach in which designers would enlist the help of an agent to review and report problems in the current state of the design. The SteelTeam example above essentially uses the knowledge base as an advisory agent to provide valuable feedback during the early design stages. This feedback allows the designer to consider the concerns, represented in the knowledge base, of each of the parties to the design process. By providing this information that may be missing due to asynchronous design, savings are realized in the project cost by avoiding downstream problems that can lead to costly changes and re-work.

The current expanded use in practice of routine electronic data exchange and shared product data models is finally moving the AEC industry beyond "islands of automation." To reiterate, the availability of product data models provides an opportunity to implement knowledge-based assistants to leverage the data contained in the existing models and provide substantial benefits in improved quality and reduced costs of completed structures.

5. Collection and sharing of design experience

The above discussion illuminates the potential for KBS to make substantial contributions to structural engineering applications in the near term by combining the data access and integration capability of product data models with the knowledge representation and inferencing capability of the KBS. A different potential near team application of AI and KBS to structures is in areas where similar design situations are encountered on a semi-regular basis by a design organization but infrequently by an individual designer. In such areas, cases and rules in combination give designers valuable guidance. Case based reasoning can be used as a tool for contributing to and sharing in the benefits from a pool of past design experience. In addition, documentation and exchange of information on failed designs could be used by designers to prevent the re-occurrence of such errors, improving the quality of the product. A KBS can be used to search for and retrieve previous designs for adaptation to the current problem. There are domains in structural engineering where a relative modest knowledge base of rules and cases can be useful in practice. Following the theme of delivery of knowledge across the design/detailing/fabrication/construction separations, knowledge-based assistants encapsulating fabrication knowledge can also be used to assist in redesign necessitated by changes downstream. An example from practice is resolution of steel bridge fabrication errors.

5.1 Fabrication Indexed eXamples and Solutions (FIXS)

In bridge engineering, errors occur during the manufacture of elements of the steel bridge assembly at the mill, shop, and field stages. These errors can cause deleterious effects on the performance of a bridge if they are not repaired properly. Errors also cause delays in the bridge fabrication and construction process. These errors need to be recognized and corrected properly and efficiently according to each individual situation. A rule-based expert system, the Bridge Fabrication Error Solution Expert System (BFX) was developed and deployed (Melhem et al. 1996). This system examines fabrication errors of steel bridge members when detected in the plant and recommends corrective action. The expert system is performing well at the Kansas Department of Transportation (KDOT), but it has some restrictions in performance and expandability. To investigate the usefulness of a case based approach, CB-BFX was developed (Roddis and Bocox 1997). The major performance strength of the rule based approach was that an incorrect repair was never recommended. The major weakness of the rule based approach was that for a substantial fraction (one-third) of the time, no solution repair was found so no guidance was provided. The major performance strength of the case based approach was that guidance in the form of a relevant previous case was provided 80% of the time. The major weakness of the case based approach was the need for engineering judgment on the part of the user to adapt the retrieved repair to be suitable to the current use.

In light of this experience with the rule and case based approaches, the desirability of a combined system is apparent. It is also desirable to generalize the approach to make it useable by other State DOTs. Regional standardization of the approach used to resolve fabrication errors would expedite bridge fabrication and would be expected to reduce fabrication costs. Use of a common approach by DOTs would reduce uncertainty for fabricators, reducing their costs and in turn reducing costs to DOTs. In addition, documentation and exchange of information on fabrication errors could be used by fabricators to prevent the occurrence of such errors in the first place, leading to a reduced number of errors, improving the quality of bridge member fabrication, directly benefiting both fabricators and DOTs.

For these reasons, a new system taking a combined rule and case based approach is under development. This system, named FIXS (Fabrication Indexed eXamples and Solutions), emphasizes the sharing of design experience rather than the expert system technology. The American steel bridge industry supports continued development of computer software cataloging repair methods for mill, shop, and field errors. This application again demonstrates the value of formalizing design/fabrication interface knowledge. It also demonstrates the usefulness in practice of even modest sized knowledge bases (347 rules, 121 cases).

5.2 Consultant Reasoning About Cracking Knowledge (CRACK)

Much of the American steel highway bridge population is entering a service life phase where fatigue becomes of increasing concern. In addition, many design details commonly used in construction of much of the interstate system are now known to be prone to fatigue damage. This is a second example of an area where similar design situations are encountered on a semi-regular basis by a design organization but infrequently by an individual designer. A knowledge-based tool for analysis of fatigue and fracture in steel bridges was developed for research use, but was not transferred to practice. CRACK (Consultant Reasoning About Cracking Knowledge) used rules, model based reasoning, and qualitative simulation (Roddis and Martin 1992). Current interest by DOT's in evaluation and repair of fatigue prone details has initiated development of KBS for this problem domain. This KBS also uses a combined rule and case based approach. The cases provide not only recommendations of r fixes to consider but also examples of options to avoid.

6. Conclusion

KBS technology has the potential to make substantial contributions to structural engineering applications in the near term by combining the data access and integration capability of product data models with the knowledge representation and inferencing capability of the KBS. Testing of an implemented knowledge base for design/detailing/fabrication/construction indicates that it is realistic to expect at least a 5% savings on in-place steel frame costs in practice. Leveraging product data models, knowledge-based assistants can be used to enhance upstream communication of design information, automate routine tasks, enhance design consistency, and facilitate collaborative communication. Such KBS applications enable the engineer to focus on the higher level design issues and transfer the burden of tedious, repetitious checking for conformance to computer tools. Opportunity is also ripe for application of AI and KBS to structures in areas where similar design situations are encountered on a semi-regular basis by a design organization but infrequently by an individual designer. In such areas, cases and rules in combination give designers valuable guidance. KBS has the proven ability to deliver existing knowledge to the time and place required. There are many current opportunities to make substantial gains in quality and effectiveness of design by applying KBS to structural engineering.

Acknowledgement

Part of this work was supported by the National Science Foundation under NSF Grant MSS-9221977. Additional support was provided by the Kansas Department of Transportation, the Mid-America Transportation Center, and the National Steel Bridge Alliance.

References

ASCE. (1990). Quality in the Constructed Project; A Guide for Owners, Designers and Contractors, American Society of Civil Engineers, New York, NY.

Crowley, A. J. (1997). "The CIMsteel Project: Homepage," www.leeds.ac.uk/civil/research/cae/cicmsteel/cimsteel.htm, 23 April 1997.

Ernst, J. and Roddis, W. M. K. (1994). "Checking of CAD Drawings for Fabrication Issues," Analysis and Computation Proceedings, Eleventh Conference, ASCE, 248-253.

Fischer, M. A. (1991). Constructability Input to Preliminary Design of Reinforced Concrete Structures, 64, Center for Integrated Facility Engineering (CIFE), Civil Engineering, Stanford University.

Fischer, M. A. (1993). "Linking CAD and expert systems for constructability reasoning." Proceedings of the 5th International Conference on Computing in Civil and Building Engineering – V-ICCBE, Anaheim, CA, 1563-1570.

Fischer, M., Froese, T., and Phan, D. (1994). "How do integration and data models add value to a project?" 1st Congress on Computing in Civil Engineering, Washington, DC, 992-997.

Genesereth, M. R., and Ketchpel, S. P. (1994). "Software agents." Communications of the ACM, 37(7), 48-53.

Hannus, M. (1996). "RATAS-IT in Finnish Construction," www.vtt.fi/cic/ratas/index.html, 16 May 1996.

Maher, M. L., and Fenves, S. J. (1985), HI-RISE: Knowledge-based expert system for the preliminary structural design of high rise buildings, in J. S. Gero (ed.), Knowledge Engineering in Computer-Aided Design, North Holland, Amsterdam.

Melhem, H. G., Roddis, W. M. K., Nagaraja, S., and Hess, M. R. (1996). "Knowledge acquisition and engineering for a steel bridge fabrication expert system." Journal of Computing in Civil Engineering, ASCE, 10(3), 248-256.

Moore, D., and Tunnicliffe, A. (1994). "An Automated Design Aid (ADA) for constructability." 1st Congress on Computing in Civil Engineering, Washington, DC, 1584-1591.

Morse, D. V., and Hendrickson, C. (1991). "Model for communication in automated interactive engineering design." Journal of Computing in Civil Engineering, 5(1), 4-24.

Novetski, B. J. (1996). "Computer integrated design: the real payoff." A/E/C Systems Computer Solutions, 5(6), 35-39.

Pasley, G. P. (1996). SteelTeam - Creating a Collaborative Design Environment for the Steel Building Industry, Doctoral Dissertation, Department of Civil and Environmental Engineering, University of Kansas.

Ricker, D. T. (1992). "Value Engineering and Steel Economy." Modern Steel Construction, 22-26.

Roddis, W. M. K., and Hess, M. R. (1997). "Case-based approach for steel bridge fabrication errors." Journal of Computing in Civil Engineering, ASCE, 11(2), 84-91.

Roddis, W. M. K., and Martin, J. L. (1992). "Qualitative Reasoning about Steel Bridge Fatigue and Fracture," IEEE Expert, IEEE Computer Society, 7(4), 41-48.

Tizani, W. M. K., Nethercot, D. A., and Smith, N. J. (1995). "Decision support for the fabrication-led design of tubular trusses." International Association for Bridge and Structural Engineering, Bergamo, Italy, 117-120.

Thornton, W. A. (1995). "Connections: Art, Science and Information in the Quest for Economy and Safety." National Steel Construction Conference, San Antonio, Texas, 1-1 1-22.

Werkman, K. J., and Hillman, D. J. (1989). "Designer fabricator interpreter system: sharing perspectives between cooperating agents to evaluate alternate connection configurations." Workshop on Distributed Artificial Intelligence, Eastsound, Washington, 95-110.

Winstanley, G., Chacon, M. A., and Levitt, R. E. (1993). "An integrated project planning environment." Intelligent Systems Engineering, 2(2), 91-106.

Zimmerman, W. (1995). "Steel Erection Awareness: An Erector's View." National Steel Construction Conference, San Antonio, Texas, 38-1 38-6.

CAD Modelling in Multidisciplinary Design Domains

M. A. ROSENMAN, J. S. GERO

Key Centre of Design Computing
Department of Architectural and Design Science, University of Sydney
NSW 2006 Australia
{mike, john}@arch.usyd.edu.au

Abstract. In a multidisciplinary design environment, such as the architecture, engineering and construction (AEC) domain, the various designers will have their own views, concepts and representations of design objects, making communication in a CAD environment a complex task. This paper demonstrates that by taking into consideration the concepts of function and purpose such multiple views and representations can be accomodated. The representation of the functional properties of design objects and their purpose is the underlying basis for the formation of different representations and the coordination of these representations. The paper puts forward definitions for function and purpose which allow for the representation of these properties of a design object and for interdisciplinary communication and integration in a CAD environment.

1 Introduction

Large scale design projects involve many different disciplines each with their own area of concern and expertise. At various stages of the design designers from different disciplines will represent an abstraction (a model) of the current design according to their views. These different models will initially be incomplete and inconsistent but through collaboration they will undergo changes as inconsistencies are removed and details are added and eventually a consistent representation emerges.

While currently paper-based representations are the conventional method used for representation, it is being realised that the complexity of large-scale design projects can only be adequately handled by a systems integration and automation approach and that computer-aided design (CAD) is the vehicle for providing this integrated information processing (1). However, the use of CAD systems for representing design objects brings into focus the aspects of explicit/implicit representations and especially the requirement of different views and representations of the same design object by different design disciplines.

In order to make CAD modelling useful to designers in a multidisciplinary collaborative environment, such as the Architecture, Engineering and Construction (AEC) domain, each designer's view and representation must be accommodated and integrated within a comprehensive representation of the design under concern. This paper argues that a multiple view approach is essential for any meaningful representation in a multidisciplinary environment. Since views and representations depend upon a functional context, i.e. a particular set of functional concerns, the representation and application of functional properties is an essential aspect of any successful collaborative CAD modelling.

2 Multidisciplinary Design Domains

2.1 Concerns and Concepts

The AEC domain typifies a multidisciplinary design domain. In the AEC design environment, many disciplines are involved, each dealing with a specialized aspect of the building design and each with its own concepts and interpretations of the object (the building). The fragmentation of the design and construction disciplines in the AEC domain is due to the specialization of each discipline according to functional concerns.

Architects are mainly concerned with providing sufficient, efficient and aesthetic spatial environments for a given set of activities. They are thus concerned with concepts such as spatial sufficiency, spatial organization, comfort, aesthetics, weatherproofness, rooms, storeys, facades, floors, walls, etc. Structural engineers, on the other hand, are concerned with providing stability by resisting or transmitting forces and moments. They are concerned with concepts such as gravity/lateral loads, support, bending, shear, deformations, beams, columns, shear walls, etc. Contractors, on the other hand, are concerned with the constructability of a design and hence with the relationships between the physical elements and the operations and sequence of operations required to construct the building. That is, they are concerned with concepts such as availability, composability, time and place, stability, walls, windows, beams, pipes, etc. Some aspects are the concern of more than one discipline, e.g. stability is the concern of both the structural engineer and the contractor.

2.2 Collaboration between the Disciplines

Paper-based representations, in the form of line drawings, have been the conventional method used for representation and communication of information between designers. Each discipline represents its model in its own set of drawings (blueprints) where each such set of drawings represents that discipline's model of the building using that discipline's set of representation conventions. Any inconsistencies between the various models are corrected by marking the appropriate drawings and sending them back to the appropriate discipline. This process usually goes through several iterations. The result is a number of sets of drawings, one per discipline, where, although each set represents the building using a different model, the comprehensive representation is consistent. There is no attempt to integrate the various sets of drawings into one drawing.

The advantages of using CAD as a modelling tool for systems automation and integration and for communication between distributed members of a project have been extensively presented (1, 2]. The method which has generally been accepted, as the means of enforcing consistent representation and interpretation is the construction of a single unified model of the design object under consideration (3-6). The argument put forward in this paper is that this single model is incapable of representing the different views and models of the different disciplines and that the traditional paper approach, actually represents a necessary approach which has to be dealt with in any electronic communication medium such as CAD. This multiple model approach needs to ensure that consistency is achieved and maintained throughout the various representations.

3 Multiple Views and Models

3.1 Multiple Views

We are concerned with the perception, conception and representation of design objects by different design participants. We build a conceptual model of an object based on that view, i.e. a representation, and manipulate that representation when we communicate. In a design context, the view that a person takes depends on the functional concerns of that person. Given a design object, such as a building, there are many views that we may take, leading to different conceptual interpretations. For example, a building may be viewed as a set of activities that take place in it; as a set of spaces; as sculptural form; as an environment modifier or shelter provider; as a set of force resisting elements; as a configuration of physical elements; etc. A building is all of these, and more.

3.2 Multiple Models

A model of an object is a representation of that object resulting from a particular view taken. For each different views of a building there will be a corresponding model, Figure 1.

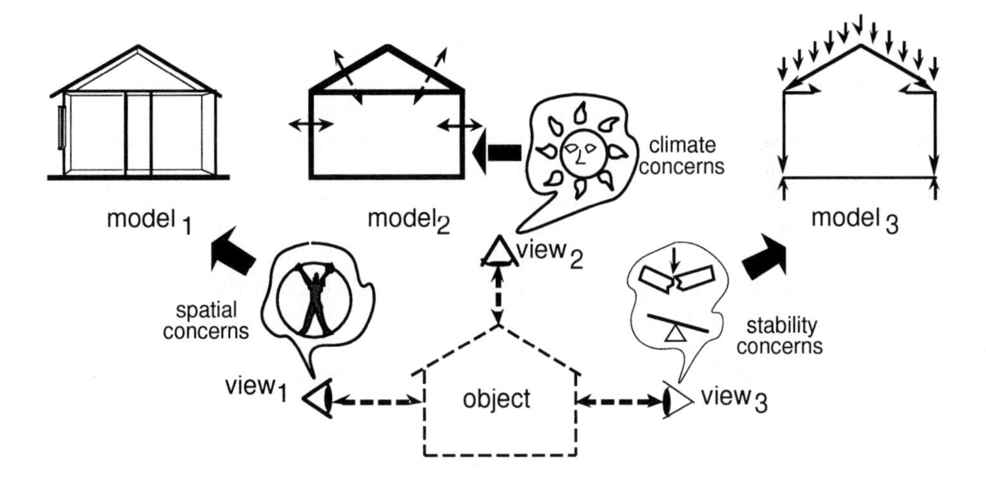

Fig. 1. Multiple views and models

Depending on the view taken, certain properties and descriptions of the object become relevant. The sound insulating properties of a wall are not relevant to a structural engineer's description of that wall. In fact, many walls may not be relevant at all to a structural engineer if they do not either contribute directly to the stability of a building or indirectly by providing a substantial load. The architects will model certain elements such as floors, walls, doors and windows. For the architects, these elements are associated with the spatial and environmental qualities with which they are concerned.

Structural engineers, however, see the walls and floors as elements capable of bearing loads and resisting forces and moments. Both models must coexist since the two designers will have different uses for their models. For example, the structural engineers will need to carry out calculations based on their model while the architects may need to ascribe different properties to their separate wall elements,. The engineers may modify some of the properties assigned to these element by the architect and may add some new elements, such as beams and columns. The addition of such new elements may affect the architect's model (and vice versa). Any such decisions taken by the engineer must be conveyed to the architect by making changes in the architect's model as appropriate. It will be shown that such changes in another discipline's model can be done when the change affects a function which is the concern of that discipline.

3.3 Representing Multiple Models

There exists considerable work using a single model approach based on the construction of a model from 'primitive' elements from which multiple interpretations are derived (7-10). This approach is analogous to the formation of views in database management systems. However, it is argued that, this approach is insufficient since the 'primitive' elements themselves are subject to the views taken by the different viewers and hence different primitive models are constructed by each such viewer (11, 12). Since the basic description of an object differs from viewer to viewer, each viewer may represent an object with different elements and different composition hierarchies. For example, while architects may model walls on different floors as separate elements, bounding various rooms, the structural engineers may model only a single shear wall. So that, not only is the interpretation of the meaning of a design object different from one viewer to another but, also the description of the structure of the object differs. This approach is similar to that taken by Nederveen and Tolman (13, 14]) and Pierra (15). There exists no single unified model nor even a single set of unique elements but rather different descriptions of the same elements and different subsets of these descriptions in different models. Each model must be consistent vis-a-vis the object being described.

Since the various models constructed by the various disciplines are representations of elemental models as seen through views based on functional contexts the representation of functional properties of design objects is the underlying basis for the formation of different concepts.

4 Purpose, Function, Behaviour and Structure

4.1 Definitions

The essential factor in a description of any design object allowing for the formation of multiple interpretations is a description of its functional properties in addition to its structural properties. There have been various attempts at defining the concepts of purpose, function, behaviour and structure, among them the following, (16-24). Notwithstanding such a large number of work, there is still confusion especially as regards the relations between the concepts of purpose and function and the concepts of function and behaviour. For example, Umeda et al. (19) state that 'to support X' is a representation of function, since it is 'to do 'something whereas 'A is supporting B' is a description of behaviour.

The definitions that will be used here are those put forward in [25) where the concepts are explained in detail.

purpose:	is the reason why an artefact exists or what it is intended for;
function:	is what is performed by an artefact; what it does (intended or not);
behaviour:	is the manner in which an artefact acts under specified conditions;
structure:	is what constitutes an artefact (or defines its constitution).

Purpose is a human-related concept associated with human needs [25]. A clock's purpose is to tell the time where 'telling time' is a human-related concept. A beam's purpose may be to provide an unencumbered space below. The acceptance that the function of an artefact is what it does, rather than what it should do, results in the recognition that artefacts perform many functions, only some of which were intended. Motor cars belch out exhaust fumes, they clog up streets, they make noise. None of these functions were intended but occur. There is a close relation between function and behaviour, behaviour being the mechanism by which function is achieved. Thus, a beam supports a load (function) by bending and transmitting the load to supports (behaviour). Objects occupy space (function) by exhibiting volume (behaviour) as a result of their shape and dimensions (structure).

4.2 Purpose as Intended Function

Many design problems are removed from direct human needs as they are subproblems of some larger design need. Nevertheless, within their particular sphere of concern, they are treated as design problems in their own right. Thus the designs of gears, amplifiers, trusses, etc., are treated by their designers as encapsulated design problems. As such, purposes are assigned to the design of these artefacts which are removed from human socio-cultural needs and hence take on a more technical aspect. Such purposes are, in effect, intended functions. They describe the functions that an artefact should achieve, e.g. transformation of torque, transformation of a signal, transfer of loads, etc. To an architect the purpose of an element may be to provide an unencumbered uniform space under it, to the structural engineer the intended function or surrogate purpose is to transfer some set of loads in a given way. Thus the design of some artefact which is a design problem to some designer but where the artefact will only be used as a component in some larger system is assigned a purpose in terms of its intended functions.

4.3 Representing Purpose and Function in CAD Systems

The representation of functional properties becomes the essential factors in a representation schema for modelling in a multidisciplinary collaborative environment. The current practice in CAD systems is to represent merely the structure properties of an object, usually only the graphical representation. It is not always possible to infer functional information from a structural description. For example, one cannot determine that a wall is loadbearing from topological relations alone. The recognition that graphical properties, while important, are not the only properties that need be described in an object's representation forms the underlying basis of the STEP effort for electronic data exchange of product information (26).

5 Concepts and Descriptions

Design prototypes describe classes of design elements and include a categorization of purpose, function, behaviour and structure properties (27]. In a fragmented environment, such as AEC, each discipline has its own set of design prototypes with its own concepts, terminology and visual representations which are not necessarily shared between the disciplines. Specific examples of design prototypes, i.e. instances, are described using the design prototype schema and by instantiating all relevant properties to specific values and form that discipline's model.

However, to provide integration between the concepts of the different disciplines, generic concepts which contain properties common to the disciplines are necessary (13, 15). Any element of any discipline's model will inherit these properties and thus be subject to being part of other disciplines' models. Figure 2 shows an example of generic concepts for walls and columns and the architects' and structural engineers' concepts, which are views of the generic concepts. Not all the behaviour and structure properties are shown.

The generic concepts do not include any purpose, since purpose is strictly a view-based concept but include those functions which occur whenever an element of that type occurs. For example, the generic wall concept includes three functions which occur when a wall exists but are actually side-products to the existence of a wall. That is, these are things which the wall does regardless of intention. The 'provides_load' function is marked as optional since it depends on the actual composition of the wall and the structural engineers's decision. However, any wall element created by the architect for space separation and environment controlling purpose which inherits a possible provide_load function from the generic concept produces a corresponding element in the structural engineers's model. The structural engineers may then assign any of the two optional functions and purpose to this element. Conversely, any wall element created by the structural engineers for a support purpose will inherit the 'occupies_space' and 'affects_visual_envmnt' functions from the generic model and hence produce a corresponding element in the architect's model.

Once a designer's view has been expressed as a set of concerned functions, all elements, whose functions contribute to those functions defined for that view will become part of that designer's model even if created by another designer. This will be so even if the functions were not intended by that other designer as described above. Finding the relationships between contributing functions is not a simple text match but may have to be carried out through various levels of abstraction (28).

6 Functional Modelling in Collaborative CAD Modelling

6.1 Assignment of Purpose and Function

There are two main ways in which the modelling of functional properties, can help in communication between the different disciplines as they collaborate to achieve the intentions of each designer as well as consistency in the description of the artefact under consideration.

1. assigning purpose to define intentions, i.e. intended functions;
2. assigning functions to elements and relating those functions to the concerns of the various designers.

In the first case, an intended function, a purpose, assigned to an element by a designer will result in an indication that the existence of that element is contingent on that purpose. Thus, the element cannot be modified in a way that will impair the intended function. For example, the assignment of a lateral force-resisting or a loadbearing function to a wall by a structural engineer should now prevent the architect from removing that wall.

In the second case, elements will, by their existence, carry out certain functions which will be associated with their conceptual description, e.g. in a design prototype. So that even if a designer does not assign a particular intended function to an element, this function will still be assigned to the element by default. This function may not be of concern to that particular designer but may be of concern to other designers. For example, the structural engineer may add a column in a space to carry out some intended support function. However, one of the unintended yet existing functions of columns is that they occupy space. This function of space occupation is of concern to the architect and as a result, that description of the column which relates to the space occupation function will now form part of the architect's model. That is, the column will appear in the architect's model.

Thus, it is through the concepts of function and purpose, assigned to design objects, that information is transmitted, allowing for the coordination of the overall decision-making effort.

6.2 Relationship between Models

Elements in the different models which are related must be related explicitly through explicit relationships. For example a floor element in the architect's model and a slab element in the structural engineer's model, which refer to essentially the same physical element, must be related by a relationship such as a *same_as* relationship. This 'same_as' relationship specifies that the structural properties of the 'two' elements are the same. Another relationship is the *part_of* relationship. This is not the same as the common understanding of 'part_of' which in actuality is more accurately labelled as *component_of*. Thus, a bicycle wheel is a 'component_of' a bicycle whereas a designated part of a concrete slab is 'part_of' that slab. The 'part_of' relationship allows inheritance of properties whereas the 'component_of' relationship does not. The 'part_of' relationship allows elements of one designer's models to be 'part_of' an element of another designer's model thus allowing for consistency of properties such as dimensions and material. Other constraining relationships need also be stated, as for example, that the height of a wall element in the architect's model is related to the depth of a beam element in the structural engineer's model.

7 A Building Example

Below is set out a simplified example of a collaborative CAD session between different disciplines. Firstly, the architect models some concept for part of a horizontal slab-type office building. The wall, floor and roof are represented as lines since their material and thickness are as yet undecided. Some of the dimensions also are not fixed. Figure 3 shows the architects' first conceptual graphical representation. Figure 4 shows a simplified description of some of the objects in the architect's first model. Only some elements and attributes are shown. Behaviour attributes are not shown.

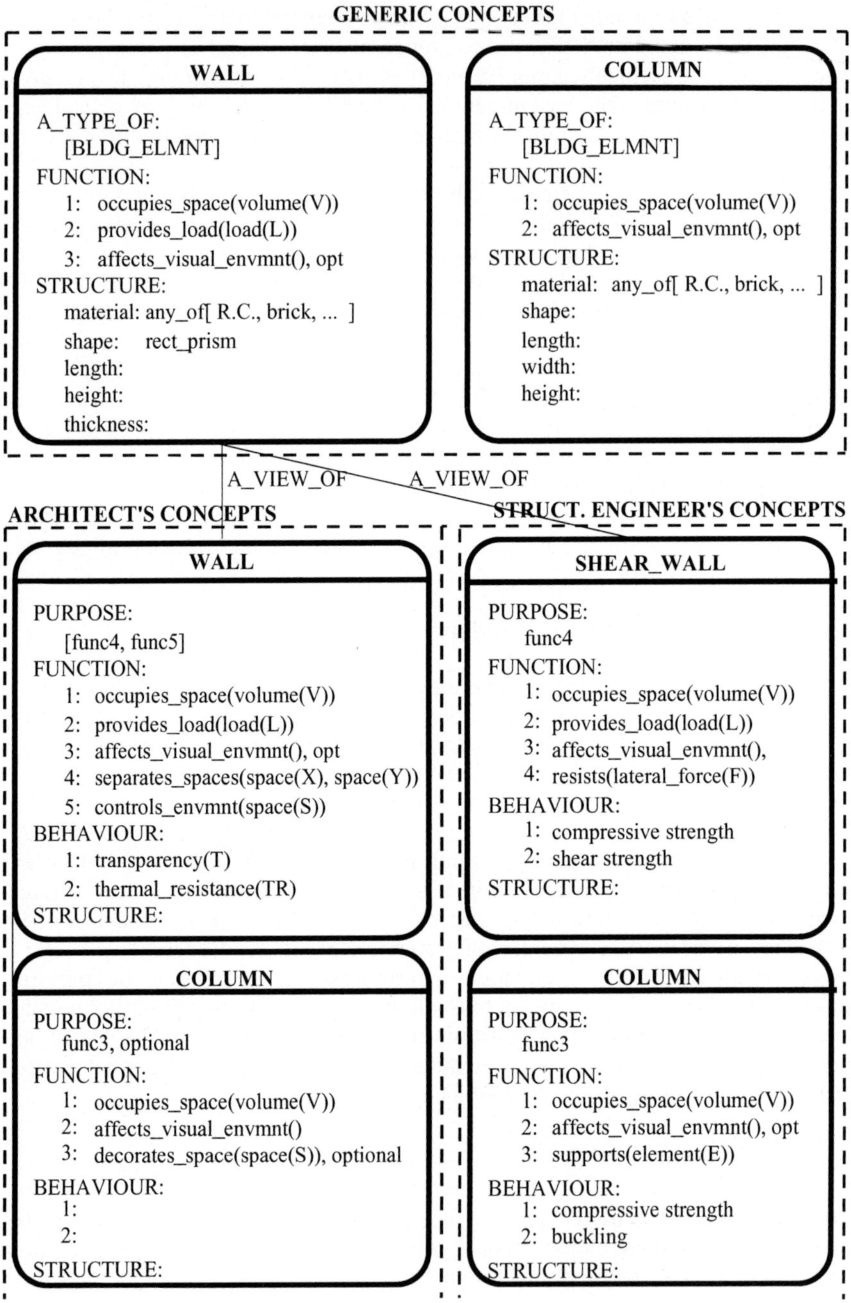

GENERIC CONCEPTS

WALL

A_TYPE_OF:
[BLDG_ELMNT]
FUNCTION:
1: occupies_space(volume(V))
2: provides_load(load(L))
3: affects_visual_envmnt(), opt
STRUCTURE:
material: any_of[R.C., brick, ...]
shape: rect_prism
length:
height:
thickness:

COLUMN

A_TYPE_OF:
[BLDG_ELMNT]
FUNCTION:
1: occupies_space(volume(V))
2: affects_visual_envmnt(), opt
STRUCTURE:
material: any_of[R.C., brick, ...]
shape:
length:
width:
height:

A_VIEW_OF A_VIEW_OF

ARCHITECT'S CONCEPTS

WALL

PURPOSE:
[func4, func5]
FUNCTION:
1: occupies_space(volume(V))
2: provides_load(load(L))
3: affects_visual_envmnt(), opt
4: separates_spaces(space(X), space(Y))
5: controls_envmnt(space(S))
BEHAVIOUR:
1: transparency(T)
2: thermal_resistance(TR)
STRUCTURE:

COLUMN

PURPOSE:
func3, optional
FUNCTION:
1: occupies_space(volume(V))
2: affects_visual_envmnt()
3: decorates_space(space(S)), optional
BEHAVIOUR:
1:
2:
STRUCTURE:

STRUCT. ENGINEER'S CONCEPTS

SHEAR_WALL

PURPOSE:
func4
FUNCTION:
1: occupies_space(volume(V))
2: provides_load(load(L))
3: affects_visual_envmnt(),
4: resists(lateral_force(F))
BEHAVIOUR:
1: compressive strength
2: shear strength
STRUCTURE:

COLUMN

PURPOSE:
func3
FUNCTION:
1: occupies_space(volume(V))
2: affects_visual_envmnt(), opt
3: supports(element(E))
BEHAVIOUR:
1: compressive strength
2: buckling
STRUCTURE:

Fig. 2. Generic and discipline concepts

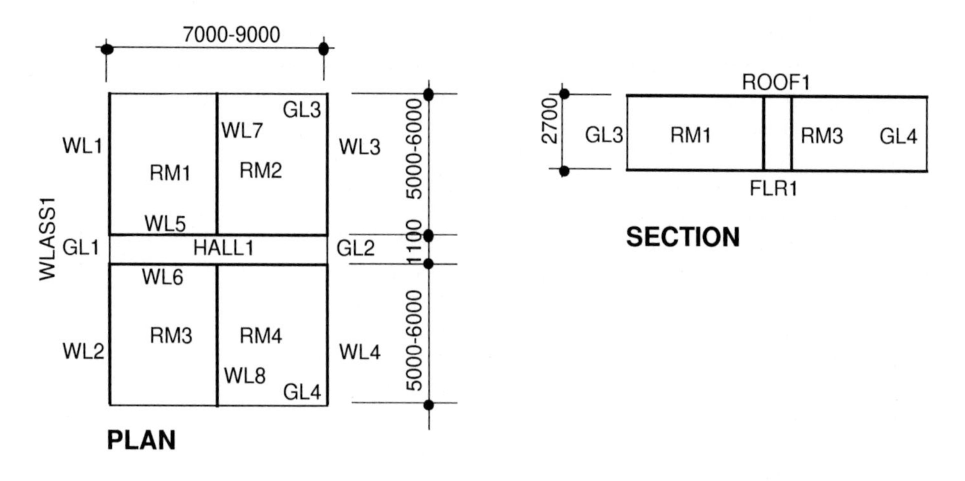

Fig. 3. Graphical representation of architect's first model.

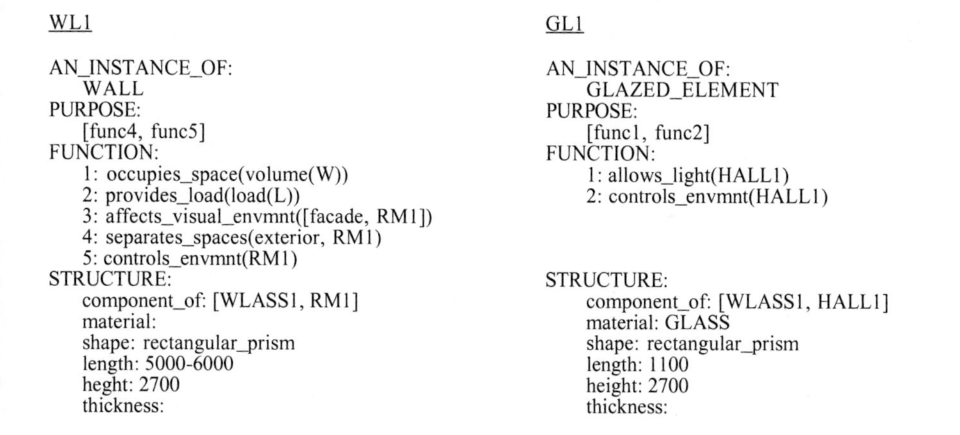

WL1

AN_INSTANCE_OF:
WALL
PURPOSE:
[func4, func5]
FUNCTION:
1: occupies_space(volume(W))
2: provides_load(load(L))
3: affects_visual_envmnt([facade, RM1])
4: separates_spaces(exterior, RM1)
5: controls_envmnt(RM1)
STRUCTURE:
component_of: [WLASS1, RM1]
material:
shape: rectangular_prism
length: 5000-6000
heght: 2700
thickness:

GL1

AN_INSTANCE_OF:
GLAZED_ELEMENT
PURPOSE:
[func1, func2]
FUNCTION:
1: allows_light(HALL1)
2: controls_envmnt(HALL1)
STRUCTURE:
component_of: [WLASS1, HALL1]
material: GLASS
shape: rectangular_prism
length: 1100
height: 2700
thickness:

Fig. 4. Non-graphical representation of architect's first model

The structural engineer examines the architect's first model and notes that walls WL7 and WL8 have intended functions of providing flexibility to the respective room spaces. As such the structural engineer proposes the scheme shown in Figures 5 and 6. The structural engineer decides that the transverse walls should act as shear walls as well as supporting the roof slab. The two walls, WL1 and WL2 of the architects are effectively considered as one wall and the height of the opening for GL1 is reduced from full storey height to 2100 mm, thus creating one wall assembly element. The wall opening has no constructive structural purpose (actually it has a degrading effect on the function of the shear wall assembly), but is to provide space for GL1, an architectural purpose, assumed by the structural engineer. This will ensure that the element OPN1 will form part of the architect's model. Further, the relation that the wall SW1 has parts WL1 and WL2 will cause those elements to inherit the material and thickness properties of SW1. Beams BM1 and BM2 are added and two of their functions (inherited from the generic concept) is that they occupy space and affect the visual environment, ensuring that they appear in the architect's model and indicate required changes in the height of the glazed elements. The structural engineer attempts to take into account the architect's intention of a flexible location for walls WL7 and WL8 by orienting the columns with their long dimensions along the hall walls. This provides a constrained flexibility for the location of walls WL7 and WL8. Again the inherited functions for the columns will cause them to be part of the architect's model.

The architects now discover that new elements and inconsistencies exist. They accept the need for the shear walls and beams and the reduction of the height of GL1, GL2, GL3 and GL4, and modify their model accordingly, Figure 7.

The wall instance, WL1 has values of properties, e.g. material, thickness, inherited from SWL1 resulting from the relationship 'WL1 part_of SWL1'. From the functions of the columns regarding space occupation and visual effects, column instances are created in the architect's model, since these are functions with which the architect is concerned, as defined in the architect's view.The architects may not accept the columns since these interfere with the rooms. However, the architects cannot remove the columns since their purpose as supporting SLAB2 has been assigned by the structural engineer. They must notify the structural engineer that this solution is unacceptable. A method of electronic annotation is provided for in the VisionManager system (29). The structural engineer may either decide on a new system, such as providing for beams above walls WL5 and WL7 or may argue that the columns are necessary. If the beam solution is chosen, the HVAC engineers may subsequently notify the architects that they need penetration for their ducts. The negotiations continue through several stages of development and modelling until all participants are satisfied and consistency is reached.

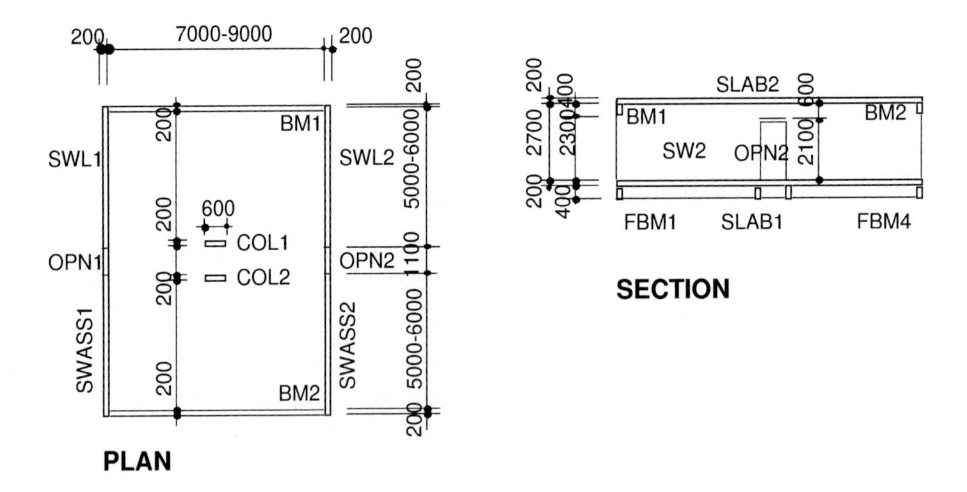

Fig. 5. Graphical representation of structural engineer's model

SWL1

AN_INSTANCE_OF:
[SHEAR_WALL, LDBEARING_WALL]
PURPOSE:
[func1, func2]
FUNCTION:
1: occupies_space(volume(V))
2: provides_load(load(L))
3: affects_visual_envmnt()
4: resist(lateral_force(F))
5: supports(SLAB2)
STRUCTURE:
component_of: [BLDG1]
parts: [WL1, WL2]
material: R.C.
shape: rectangular_prism
length: 11500-13500
height: 2700
thickness: 200

OPN1

AN_INSTANCE_OF:
WALL_OPENING
PURPOSE:
[func1, func2]
FUNCTION:
1: creates_hole(SW1)
2: provides_space(GL1)
3: reduces_strength(SW1)

STRUCTURE:
width: 1100
height: 2100
thickness: same_as(SWL1)

BM1

AN_INSTANCE_OF:
BEAM
PURPOSE:
[func3, func4]
FUNCTION:
1: occupies_space(volume(V))
2: affects_visual_envmnt([RM1, RM2])
3: supports(SLAB2)

STRUCTURE:
component_of: [BLDG1]
material: R.C.
shape: rectangular_prism
length: 8000-9000
depth: 400
thickness: 200

COL1

AN_INSTANCE_OF:
COLUMN
PURPOSE:
[func3]
FUNCTION:
1: occupies_space(volume(V))
2: affects_visual_envmnt([RM1,RM2, HALL1)
3: support(SLAB2)
4: transfers_force(F, SWL1, SWL2)
STRUCTURE:
component_of: [BLDG1]
material: R.C.
shape: rectangular_prism
length: 600
height: 2700
width: 200

Fig. 6. Non-graphical representation of structural engineer's model

WL1

AN_INSTANCE_OF:
WALL
PURPOSE:
[func4, func5]
FUNCTION:
1: occupies_space(volume(6.21))
2: provides_load(load(15.5))
3: affects_visual_envmnt([facade, RM1])
4: separates_spaces(exterior, RM1)
5: controls_envmnt(RM1
STRUCTURE:
component_of: [WLASS1, RM1]
part_of: [SWL1]
material: R.C.
shape: rectangular_prism
length: 5200-6200
height: 2700
thickness: 200

COL1

AN_INSTANCE_OF:
COLUMN
PURPOSE:
[func4, func5]
FUNCTION:
1: occupies_space(volume(0.32))
2: affects_visual_envmnt([RM1,RM2,HALL1])
STRUCTURE:
same_as: COL1.SE
component_of: STOREY1
material: R.C.
shape: rectangular_prism
length: 600
width: 200
height: 2700

Fig. 7. Revised architect's model

8 Summary

The paper has shown how the concepts of function and purpose are essential in collaborative CAD modelling between different design disciplines to allow the representation of the different viewpoints and to provide the necessary coordination and consistency. The essential factors are the explicit representation of functional properties of design objects where function is defined as any effect resulting from the behaviour of an object and purpose as intended function. The above simplified example showed how these concepts can be used in such a multiview approach.

Work to date has already demonstrated the potential for CAD systems to allow the modelling of different views through the linking of graphic and non-graphic databases using a graphic and a relational database and an interface command language (28).

Further research and development is required to investigate the degree of automation or the nature of the notification required or possible. Should an architect be allowed to alter the structural engineer's model and vice versa? Can such permission be authorised? Alternatively, should any notifications regarding dissatisfaction or suggested changes be limited to a bulletin board-type notification as in VisionManager (29).

Acknowledgments

This work is supported by the Australian Research Council Large Grants, A89601961.

References

1. Madison: Conference papers. 1st Int. Symposium Building Systems Automation-Integration, June 2-8, Madison, Wisconsin. Dept. of Eng. Professional Development, College of Engineering, University of Wisconsin-Madison/ Extension, Madison (1991).
2. Howell, I.: The need for interoperability in the construction industry. In: INCIT 96 Proceedings. The Institution of Engineers, ACT, Australia (1996) 43-48 .
3. Bjork, B-C.: RATAS: A proposed Finnish building product model. Studies in Environmental Research No. T6. Helsinki University of Technology, Otaneimi, Finland (1987).
4. Bjork, B-C.: Basic structure of a proposed building product model. CAD (1989) **21**(2): 71-77.
5. Gielingh, W.F.: General AEC Reference Model (GARM). ISO TC184/SC4/WG1 Document N329 (1989).
6. Nederveen, S.V., Plokker, W. and Rombouts, W.: A building data modelling exercise using the GARM approach. COMBINE Report (working draft) (1991).
7. Howard, H.C., Abdalla, J.A. and Phan, D.H.: Primitive-composite approach for structural data modelling. Journal of Computing in Civil Engineering (1992) **6**(1):19-40.
8. Amor, R.W. and Hosking, J.G.: Multi-disciplinary views for integrated and concurrent design. In: Mathur, K.S., Betts, M.P. and Tham, K.W. (eds.): Management of Information Technology for Construction. World Scientific, Singapore (1993) 255-267.
9. Clayton, M.J., Fruchter, R., Krawinkler, H. and Teicholz, P.: Interpretation objects for multi-disciplinary design. In: Gero, J.S. and Sudweeks, F. (eds.): Artificial Intelligence in Design '94. Kluwer Academic Publishers, Dordrecht, Netherlands (1994) 573-590.
10. MacKellar, B.K. and Peckham, J.: Specifying multiple representations of design objects in SORAC. In: Gero, J.S. and Sudweeks, F. (eds.): Artificial Intelligence in Design '94. Kluwer Academic Publishers, Dordrecht, Netherlands (1994) 555-572.

11. Rosenman, M.A., Gero, J.S. and Hwang, Y-S.: Representation of multiple concepts of a design objects based on multiple functions. In: Mathur, K.S., Betts, M.P. and Tham, K.W. (eds.): Management of Information Technology for Construction. World Scientific, Singapore (1993) 239-254.
12. Rosenman, M.A. and Gero J.S.: Modelling multiple views of design objects in a collaborative CAD environment. CAD Special Issue on AI in Design (1996) **28**(3): 207-216.
13. Nederveen, S.V.: View integration in building design. In: Mathur, K.S., Betts, M.P. and Tham, K.W. (eds.): Management of Information Technology for Construction. World Scientific, Singapore (1993) 209-221.
14. Nederveen, G.A. van and Tolman, F.P.: Modelling multiple views on buildings. Automation in Construction (1992). **1**:215-224.
15. Pierra, G.: A multiple perspective object oriented model for engineering design. In: Zhang, X. (ed.): New Advances in Computer Aided Design & Computer Graphics. International Academic Publishers, Beijing, China (1993) 368-373.
16. Bobrow, D.G.: Qualitative reasoning about physical systems: an introduction. Artificial Intelligence (1984) **24**:1-5.
17. Sembugamoorthy, V. and Chandrasekaran, B.: Functional representation of devices and compilation of diagnostic problem-solving systems. In: Kolodner, J. and Riesbeck, C. (eds.): Experience, Reasoning and Memory. Lawrence Erlbaum, Hillsdale, NJ (1986) 47-73.
18. Pahl, G, and Beitz, W.: Engineering Design: A Systematic Approach. Springer-Verlag (1988).
19. Umeda, Y., Takeda, H., Tomiyama, T. and Yoshikawa, H.: Function, behaviour, and structure. In Gero, J.S. (ed.): Applications of Artificial Engineering in Engineering V, Vol 1: Design. Computational Mechanics Publications, Southampton, (1990) 177-193.
20. Goel, A.K.: Representation of design functions in experience-based design. In: Brown, D.C., Waldron, M.and Yoshikawa, H. (eds): Intelligent Computer Aided Design, North-Holland, Amsterdam (1991) 283-308.
21. Hundal, M.S.: Conceptual design of technical systems. In: Proceedings of the 1991 NSF Design and Manufacturing Systems Conference. Society of Manufacturing Engineers, Michigan, (1991) 1041-49.
22. Johnson, A.L.: Designing by functions. Design Studies (1991) **12**(1): 51-57.
23. Gero, J.S., Tham, K.W. and Lee, H.S.: Behaviour: A link between function and structure in design. In: Brown, D.C., Waldron, M. and Yoshikawa, H. (eds): Intelligent Computer Aided Design, North-Holland, Amsterdam (1992) 193-225.
24. Sturges, R.H.: A computational model for conceptual design based on function logic. In: Gero, J.S. (ed.): Artificial Intelligence in Design '92. Kluwer Academic, Dordrecht, (1992) 757-772.
25. Rosenman, M.A. and Gero, J.S.: Purpose and function in design: from the socio-cultural to the techno-physical. Design Studies (1998) (to appear).
26. STEP: Part 1: Overview and fundamental principles, Draft N14. ISO TC 184/SC4/ WG6 (1991).
27. Gero, J.S.: Design prototypes: A knowledge representation schema for design. AI Magazine, (1990) **11**(4):26-36.
28. Hwang, Y.S.: Design Semantics and CAD Databases. PhD Thesis, Department of Architectural and Design Science, University of Sydney, Sydney. (1994) (unpublished).
29. Fruchter, R., Reiner, K., Leifer, L. and Toye, G.: VisionManager: A computer environment for design evolution capture. In: Gero, J.S. and Sudweeks, F. (eds.): Artificial Intelligence in Design '96. Kluwer Academic, Dordrecht, The Netherlands (1996) 505-524.

A Family of Software Components to Deliver Solutions for the Interpretation of Monitoring Data

Paolo Salvaneschi, Marco Lazzari

ISMES - via Pastrengo, 9 - 24068 Seriate (Bergamo) ITALY
Tel: +39 35 307 337 Fax: +39 35 302 999
email: {psalvaneschi, mlazzari}@ismes.it

Abstract. In this paper we summarize the results of our efforts, during the last ten years, to apply artificial intelligence techniques to the interpretation of engineering monitoring data. These efforts led us to create a set of software components which can be adapted to develop *real-world* applications to face specific customers' requirements, and to deploy several systems for the on-line interpretation of data coming from automatic monitoring systems of large dams, monuments, and landslides.

1. Introduction and History

Ten years ago our group started applying AI techniques to improve the ability of automatic systems to provide engineering interpretations of data streams coming from monitored structures. In the following years some applications were developed and are now in service. They range from monitoring of dams to monuments, and finally the same set of concepts and software components was applied to landslides. During the various developments, we moved, with some success, from applied research to the delivery of industrial solutions for our clients.

This led us to acquire knowledge and experience and to create a set of software components collectively called MISTRAL which can be adapted to develop a particular solution for a specific client.

The aim of this paper is to provide an overview of the existing set of components, recall the conceptual structure and the behavior of the system[1] and give quantitative data and qualitative comments about the results gained until now.

---

[1] we do not mean to deal here with the details of our interpretation systems, but we provide references to other publications that are more concerned with modeling techniques and implementation issues

2. Software Components, Solutions and Enabling Technologies

The MISTRAL system is packaged in a way that is possible to deliver *solutions* for specific problems. A so called *solution* for specific business needs is composed of:

- a context of application providing a set of requirements which can be supported;
- a collection of software components which can be used as building blocks to be adapted and integrated into a software product;
- a technological environment (hardware and system software environment, common software architecture, standard communication mechanisms) which can enable the integration of the existing components;
- the ability to deliver adaptation and integration services based on a specific know how both of software technology and engineering solutions.

The context of application is the management of the safety of structures. Data about structural behavior are collected through tests and visual inspections, while automatic instrumentation and data acquisition systems are used for real time monitoring. The interpretation of such data is not easy owing to different factors, such as the large amount of data, the uncertainty and incompleteness of information, the need for engineering judgement, knowledge of the particular structure, experience of the behavior of structures in general and a background of general engineering knowledge.

According to this context, the aim of a MISTRAL-based solution is to help safety managers, engineers and authorities to deal with safety problems of structures in the following two ways:

- by filtering false alarms;
- by supporting the automatic detection and early warning of dangerous situations.

Both aspects are valuable. The former is more related to the reduction of costs associated with human interpretation required even in case of false alarms, while the latter deals with the improvement of the safety level of the structure.

From the organizational point of view, two situations may be interesting for the use of such a kind of systems.

The first one is the interpretation of data coming from structures of particular relevance or criticality, such as a large structure or a structure which manifested some problem.

The second is the case where a significant amount of structures (e.g. a set of dams of a region) are operated by a central office which collects all the data and monitors the status.

Specific requirements are satisfied adapting and integrating software components taken from an existing set. They need to be tailored both in the sense that not all the components may be of interest (e.g. a specific interpretation technique is chosen) and that the chosen components have to be adapted (e.g. to the configuration of the existing sensors).

In the following we provide a list of the components that we have already developed and are suitable to be used for any new instance of MISTRAL:

- *communication with a data acquisition system*
  Manages the data communication from the data acquisition system to MISTRAL; calls the monitoring system and receives the data gathered during the last acquisition (normal real-time procedure) or collected while MISTRAL was, for some reason, not active. It can work via serial connection or over a network.
- *data base of measurements and sensors and associated management functions*
  Three kinds of database of measurements and interpretations are available: a static data base of test cases, a dynamic database collecting all the data related to the monitoring system (measurements, evaluations, explanations), an archive of past *reference* situations, used by case-based reasoning procedures (explained in the following).
- *empirical and model-based interpretation*
  Essentially, the interpretation is a process of evidential reasoning, which transforms data states into states of the physical system and interprets them in terms of alarm states. The interpretation may be based on three types of knowledge [1]:

  1. Empirical knowledge. States of sensors (derived from comparisons between measurements and thresholds) may be combined through sets of rules. The rules may be applied to groups of instruments of the same type providing a global index having the meaning of an anxiety status based on empirical past experiences. Rules may be also applied to sensors of different type belonging to the same part of the physical system (e.g. a dam block). This may suggest the existence of dangerous phenomena in that part of the physical system.
  2. Qualitative models. Qualitative relationships between measured variables may be evaluated to provide evidences for physical processes (e.g. excessive rotation of a dam block)
  3. Quantitative models. Quantitative models (e.g. relations between cause and effect variables based on statistical processing of past data or deterministic modelling approaches) may be used, in co-operation with thresholds, to provide states of sensors (e.g. measuring effect variables).

- *case-based interpretation*
  Analogical reasoning is used to retrieve, given the qualitative description of the state of a structure, the closest-matching cases stored in a case base, which can help safety managers to interpret the current situation. Usually, situations stored in the reference database are those which were more deeply investigated by experts, because of some singularities, and are enriched with experts' comments; therefore, at run-time, the analogical reasoning allows to understand whether the current situation is similar to a past one, and the comments of the latter can address the investigation of the current one. This enables to record and manage special behaviors that cannot adequately managed by explicit models.
  Two different approaches were adopted for developing such tools. The first approach is based on numerical/symbolic processing: several metrics were defined to compute the analogy of a couple of cases. These metrics span from

simple norms within the hyperspace defined by the state indexes, to more complex functions also taking into account the gravity of the cases, in a fashion similar to the law of gravitation. In such way, instead of computing the distance between two situations, their reciprocal attraction is evaluated. The similarity between the situations is then defined by checking their attraction against an adequate threshold system. This approach led to the development of a tool (DéjàViewer) which encompasses these different metrics. The second approach is based on the application of neural networks to implement a sort of associative memory that allows to compute the similarity of a couple of cases [2].

- *empirical interpretation based on neural networks*
  Within MISTRAL, deep knowledge is usually codified by numerical algorithms, and qualitative reasoning is implemented by symbolic processing; when heuristic knowledge is involved, the shallow knowledge about the structure and the instrumentation is managed through empirical rules based on the alarm state of single instruments, taking into account their reliability and significance. Such rules are derived from the analysis of a set of exemplary cases, which allow to identify weights to be given to the parameters used by the rules.
  This process, time consuming and boring for the experts, can be adequately dealt with neural nets. Therefore, neural nets can be developed to perform the empirical evaluation of data in order to achieve the same results as the symbolic processors previously used, but with reduced development and tuning effort [3].
- *explanation*
  The result of the interpretation is the identification of the current state of the structure. From the trace of execution, using knowledge about the behavior of the monitored structure and the instruments, the explanation module generates natural language messages. They describe the current state of the structure and the deductions of the system.
- *statistical processing*
  Statistical analysis of data may be performed both to provide evidences for the evaluation process (the measure of a plumb line is constantly increasing), and for periodical synthesis of past evaluations (weekly reports).
- *charting*
  It is possible to select a situation from the data base and show on the screen its graphic representation.
- *reporting*
  Reports may be extracted from the data base providing the required detail for various types of users (e.g. safety managers or local authorities)
- *GIS based interaction*
  A window of the system may interfaces a GIS, to exploit its cartographic facilities and offer a powerful representation tool to *navigate* through data and exploit standard features of GISs, such as zooming functions to highlight sub-areas to be analysed. This may be useful when the monitoring instruments are spread over a large area, as for environmental monitoring (we have exploited a GIS for monitoring landslides).

- *man/machine interface*
  The window-based interface draws on the screen graphical representations of the physical system and its composing objects (e.g. the monument and its chosen structural components) as well as the sensors on the structure and the interpretation objects (e.g. physical processes). The interface displays the objects using a colour scale based on the objects' state; natural language explanations of the analysis are shown on the screen too. Interactors are available to give users more refined information by focusing on interesting details. Via the interface the user can also activate the other components such as access the internal data base.

The existing set of components may be used as a basis to integrate them into a product only if they are based on a common technological and architectural environment. The reference technologies for our projects are: PC Windows, relational DBMS (e.g. Access) interfaced through SQL and ODBC, GIS Mapinfo, communication mechanisms based on OLE, interpretation kernel based on Prolog, numerical processing in C, interface in Visual Basic.

The choice of Prolog enables to clearly separate data and procedures within the code; this feature allows to quickly reconfigure the system, whenever the addition or removal of objects to be evaluated (e.g. new monitoring instruments) or functions (e.g. new evaluation rules) is required. Moreover, Prolog proved very useful for the development of rule-based evaluation sub-modules and for the generation of natural language explanations.

Using the standard interfaces such as OLE and ODBC, a reference architecture has been established where the components may be plugged in. Essentially the system is composed of three layers communicating through standard mechanisms: the man machine interface, the layer of components providing types of processing of data and the data layer.

3. Delivered Solutions

Below we report the list of the solutions delivered till now, including the name of the structure, the organization in charge, the year of installation and the number of sensors interpreted by the system; more details can be found in [4] for applications to dams, [5] for application to monuments and [6] for landslides.

System	Operational from	N. of sensors
Ridracoli Dam (Consorzio acque Forlì e Ravenna)	1992	46
Pavia Cathedral and six Towers (Lombardia Reg. Auth.)	1994	120
Pieve di Cadore Dam (ENEL)	1996	54
Cancano Dam (AEM)	1997	68
Valtellina landslides (Lombardia Reg. Auth.)	1996	250
Valgrosina Dam (AEM)	in progress	

The first version of MISTRAL has been installed since October 1992 on a PC linked to the monitoring system of the Ridracoli dam and is currently in use. After that, new applications were delivered for the Pieve di Cadore dam (operated by ENEL) and two dams operated by the AEM company (the application for the second of them is in progress). This last case is interesting because, while in the previous cases the system is interpreting the data from a single dam and is installed at the warden house near the dam, in this last case the system is installed at a remote control site and is designed to progressively host the interpretation of data coming from many dams. Moreover, the same software is loaded on a portable computer used by inspection people and may be linked to the acquisition system on site to get data on the fly and provide interpretations.

Another application was delivered for the management of the safety of the Cathedral of Pavia and of six towers in the same town.

On March 17, 1989 the Civic Tower of Pavia collapsed. After this event, the Italian Department of Civil Defence required the installation of an automatic monitoring system with the ability to automatically provide engineering interpretations, linked via radio to a control centre, located at the University of Pavia.

The instrumentation installed on the Cathedral and on the towers allows to acquire the most important measurements on each monument, such as opening/closure of significant cracks, displacements, stresses and also cause variables, such as air temperature, solar radiation, groundwater level. In case of anomalies the system calls by phone the local authority and ISMES. The system is installed in Pavia and is operational from the beginning of 1994.

Finally, the same approach has been applied to the interpretation of data coming from monitoring of landslides. In the summer of 1987 a ruinous flood affected large areas of the Valtellina (Northern Italy) and caused many landslides. On July 28 a large mass of rock, estimated to be 34 million m^3, suddenly moved down towards the Adda valley, destroyed some villages with casualties. As a consequence, the regional authorities and the Italian Department of Civil Protection appointed ISMES to develop a hydrogeological monitoring net (about 1000 sensors).

The net includes devices such as: surface strain-gauges; topographical and geodetic benchmarks; inclinometers; settlement-meters; cracking gauges, rain gauges, snow gauges and thermometers.

Currently, the data of the most significant instruments of the net (about 250) are processed by an instance of MISTRAL (called Eydenet), that supports real time data interpretation and analysis of data concerning the stability of the slopes affected by the landslides. The system has been operational since October 1996 at the *Monitoring Centre for the Control of Valtellina*, set up by the Regione Lombardia at Mossini, near Sondrio; it is operated by a senior geologist and a team of engineers of the Regione Lombardia.

4. System Overview and Underlying Concepts

The interpretation system is linked (normally through a serial link) to the acquisition system and gets data periodically. Every time a new set of data is available, it is stored into the data base (acting as a circular buffer). The new set in then interpreted and the data are transformed into sets of object's states. Objects may be instruments or families of instruments, physical active processes, objects composing the physical system. They may assume one the possible states: *unknown*, *normal*, *anomalous*. We use different levels of anomaly, depending on the specific instance of MISTRAL and on the specific object to assess.

The basic concept underlying the interpretation is that the available information may be exploited using and integrating many different reasoning paths. The rationale is that we have to deal with incomplete information and knowledge, and expert engineers use various types of knowledge, such as empirical associations and causal models, organising them through reasoning processes related to the specific tasks to be performed. More details about the conceptual structure of MISTRAL may be found in [1].

The user can access the results of the processing through a *window-based interface* (fig. 1). The interface draws on the screen graphical representations of the objects which have been assessed (instruments, structural elements) and displays them using a colour scale based on the state of the object. Interactors are available to get more refined information about the dam state, by focusing on interesting details (e.g. current readings recorded by the instruments, current state of instruments and parts composing the structure and the explanations of the current state). Through the interface the user can also activate functions (e.g. reporting or natural language explanation of the system deductions) and access the internal data bases.

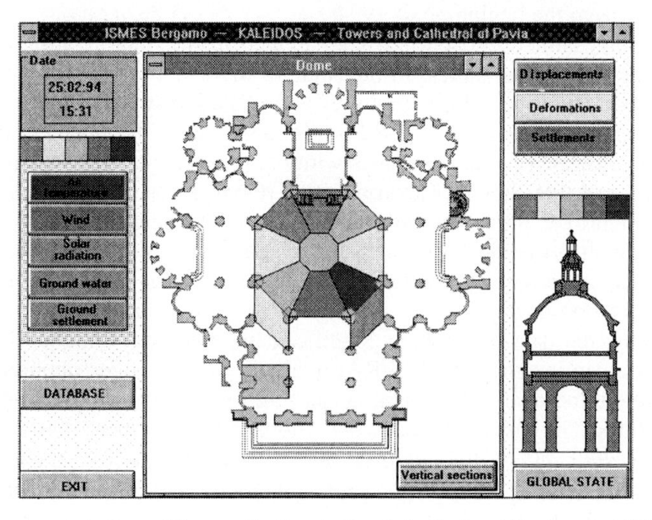

Fig. 1. The user interface of the application to the Cathedral and the towers of Pavia

5. Quantitative Results: Report from a Specific Case

While planning the development of MISTRAL, we have decided to concentrate our efforts on the achievement of three main quality objectives:

1. ease of use of the system,
2. reliability,
3. filtering capabilities.

With reference to the firsts aim, we believe that often very sophisticated A.I. prototypes fail and are not accepted by users because their functionalities are difficult to use. A.I. researchers tend to consider the *intelligence* of the system their main target, while we think that our target is the intelligence of the system's behavior, and this quality strongly depends, for interactive applications, on the *communication skills* of the system, that is on its ease of use.

Therefore, we have sometime *sacrificed* clever functions which appeared too hard to be used or, even better, we have delivered different releases of the same application, whose complexity, both of reasoning and of use, was progressively higher, so that the users could get accustomed to the system in its simpler version, before facing more complex functionalities.

As an example, the system developed for the dam at Ridracoli was initially delivered in a version for a restricted number of instruments and with a limited set of interpretation procedures, mainly based on the validation procedures of the acquisition system, which were already well known by the users; we concentrated our attention on the man/machine interaction, and quickly get the system used by the safety managers. Subsequently, we delivered three other releases of the system, which incorporated more sophisticated numerical procedures, evaluation of trends, syntheses of the evaluation results, graphics, interpretation algorithms on a larger set of instruments, and their improvements were easily accepted by users on the ground of their previous experience with the system.

With reference to the second topic - that is the reliability of the system - we are rather satisfied by MISTRAL's behavior: in six years of continuous operation, from the first installation at Ridracoli to the latest release at our fifth site (an improvement of Eydenet), we had to fix bugs or tune procedures no more than five times, and extraordinary maintenance interventions are required only when accidental crashes of the system, the acquisition devices or the computer nets occur.

We believe that our development process, based on the ISO 9000 standard, is the main responsible of such achievement, since it provides a very strong and clear reference framework for deriving correct code from specifications, and facilitates all those project reviews which highlight possible defects of the production process.

Eventually, the third, essential goal was related to the interpretation and filtering capabilities of the system.

In these years the different *incarnations* of MISTRAL have evaluated several millions of measurements; both users and our safety experts have analysed off-line the system's behavior, and their qualitative judgement is that MISTRAL works correctly.

More specifically, it never missed any significant situation, while it never created unjustified alarms; it filtered *correctly* several accidental warnings of the acquisition systems; moreover, it pointed out some situations that were significantly different from what expected, since the physical structures under evaluation were *operated* in situations violating the assumptions of MISTRAL's models.

Let us have a look at some figures, which describe an interesting sequence of events that were monitored by MISTRAL on one of our dams. First of all, let us stress that *nothing dangerous* ever happened in this period, nor the figures show any failure of the socio-technical apparatus that is in charge of the safety management: our main concern about these figures is related to the fact that MISTRAL, even tested with situations that it is not trained to manage, can provide useful suggestions to safety managers.

We have analyzed a set of 684 acquisitions gathered in 1997 by some dozens of instruments, for a total of about 37,000 measurements. This instance of MISTRAL derives the global state from row measures through empirical and model-based reasoning, taking into account numerical and statistical processing; case-based reasoning is not applied, since we were not requested by our customer to incorporate this functionality. We have evaluated the global state of the structure according to the values shown in the following table:

Level	Global state
1	normal
2	normal - some local warnings
3	normal - some anomalies
4	anomalous - warning
5	very anomalous - alarm

With reference to those values, MISTRAL produced the following evaluations of the global state of the dam:

Level	# of situations	% of situations
1	426	62
2	230	34
3	20	3
4	8	1
5	0	0

These figures are very different from those usually produced by MISTRAL, which are much more concentrated on the first level of the scale.

But an off-line analysis of the data pointed out that the warnings were due to an extraordinary combination of events (higher level of the basin, abnormal air temperatures), which were outside the limits of MISTRAL's models. In fact, the result of MISTRAL's evaluations was, in those situations, a reminder for the dam managers of the unusual, delicate, but not severe operating conditions.

Finally, if we extract from the whole data set only the measurements gathered in the period when the dam was operated in *normal* conditions, that is according to the limits we considered for tuning MISTRAL's models, we achieve the following results:

Level	# of situations	% of situations
1	416	96
2	13	3
3	3	1
4	0	0
5	0	0

A deeper examination of the state indexes highlighted that MISTRAL classified as normal situations (Level 1) about 200 acquisitions that presented accidental over-threshold values. According to our safety experts, such *filtering* was proper and correct; as a result, the dam managers were not requested to react to those situations, with a sensible reduction of management efforts.

On the other hand, when MISTRAL highlighted some unusual situations, the off-line analysis proved that its judgement was correct, and those situations, although not dangerous, were suitable to be carefully taken into consideration.

As a combined result of these behaviors, we have achieved that the dam managers are not distracted by false alarms, and can concentrate their attention and efforts on really significant situations. Moreover, the reduced number of *stimuli* avoids that safety managers discard alarms, since they are too much to be dealt with.

6. Impact on the Organization

The above chapter has shown some quantitative data to exemplify the results that may be achieved adding such types of programs to an organization managing the safety of complex systems.

More generally we can draw the following considerations about the impact on the existing organization (for a discussion of the topic, see also [7]):

1. The diagnostic program must fit the existing information systems in terms of existing machines, flows of information and human roles. A key way for the success is to understand the existing organization and how some requirement may be satisfied adding a new A.I. based component. The fundamental requirement to justify the cost of the system is the filtering of false alarms. This performance may be quantified and used to support the installation with respect to the management. The advantage is as relevant as the number of sensors to be interpreted increases. Obviously the availability of an automatic interpretation device increases the safety level providing a continuous, distributed and on line expertise.
2. An interesting aspect is the collection and (partial) accurate modelling of a corpus of knowledge, which was formerly belonging to a limited number of people. Even in the mind of that experts that knowledge was not organised in a clear conceptual structure able to be communicated. This knowledge is now an explicit property of the company and may be distributed and sold.
3. Finally, a last comment is related to the interaction between the user and the diagnostic program and the issue of responsibility. The system is used on a regular basis and, at the end, the decision to trust in the system or call the responsible people and the experts, is in the hands of the users. From this point of view, the transparency of the system and the ability to justify decisions and support them with levels of detailed data are of key relevance. Nevertheless, we cannot forget that confidence is placed in a machine, that is in its developers. This means that the interpretation system is a critical application which must be specified, designed, built and validated carefully. In fact, even if decisions and responsibility lie in the hands of human beings, the elements for decisions are in the co-operative system made by human beings and machines.

7. Future Developments

The suit of programs presented in this paper is under continuous development and improvement at ISMES, since it is still used to set up new packages tailored on customers' requirements.

Currently, we are concerned with improvements to the man/machine interface, for allowing easier communications through user and system; to the generation of reports from the database, for providing syntheses of MISTRAL's interpretations over a period of time; to the graphic capabilities (time series diagrams).

Moreover, we are designing new interpretation mechanisms for identifying physical processes affecting the structures under examination (e.g. rotation of a dam block). Currently we use quantitative and qualitative models of such processes, but we wish to experiment their identification via the definition of well-known reference situations to be checked against current measures by means of analogical reasoning.

References

1. Salvaneschi, P., Cadei, M., Lazzari, M.: The Application of AI to Structural Safety Monitoring and Evaluation. IEEE Expert 11(4) (1996) 24-34
2. Lazzari, M., Salvaneschi, P., Brembilla, L.: Looking for analogies in structural safety management through connectionist associative memories. IEEE International Workshop on Neural Networks for Identification, Control, Robotics, and Signal/Image Processing (NICROSP '96), Venezia, Italy, (1996) 392-400
3. Brembilla, L., Lazzari, M., Salvaneschi, P.: Structural monitoring through neural nets. Second Workshop of the European Group for Structural Engineering Applications of Artificial Intelligence (EGSEAAI '95), Bergamo, Italy (1995) 91-92
4. Lazzari, M., Salvaneschi, P., Ruggeri, G., Mazzà, G.: Information Systems for Dam Safety: Evolution through Artificial Intelligence. Engineering Intelligent Systems 6(1) (1998) 57-63
5. Lancini, S., Lazzari, M., Masera, A., Salvaneschi, P.: The diagnosis of Ancient Monuments' Behaviour through Expert Interpretation Software Systems. Structural Engineering International 7(4) (1997) 288-291
6. Lazzari, M., Salvaneschi, P.: Integrating Geographic Information Systems and Artificial Intelligence for Landslide Hazard Monitoring. Cahiers du Centre Europeen de Geodinamique et de Seismologie. Walferdange, Luxembourg (to appear)
7. Garrett, J.H., Smith, I.F.C.: AI Applications in Structural/Construction Engineering. IEEE Expert 11(3) (1996) 20-22

AI Methods in Concurrent Engineering

Raimar J. Scherer

Dresden University of Technology, Faculty of Civil Engineering
Mommsenstraße 13, 01069 Dresden, Germany
scherer@cib.bau.tu-dresden.de

Abstract. An AI-based hierarchical framework for concurrent engineering is suggested. It is based on the research carried out on product modelling, interoperability by mapping and matching, activity control, workflow management and information logistics in virtual enterprises and the research done in AI application to conceptual structural design and product information systems for the electronic market. This research has resulted in several prototypes in the past four years. By extrapolating the results achieved in the single prototype areas, this framework is expected to reduce the lead time of briefing, design and constructs to the range of 50 %. The paper is focused on the cognitive architecture in which the AI methods are embedded and the information logistics part with the newly suggested electronic board. The framework has to support object-oriented as well as object-centred, document-centred, process-centred and production-centred working environments with different working cultures. This needs multiple presentation of data, data transformation, data condensation and dedicated AI methods.

Introduction

Real cognitive engineering tasks, whether for design or for project management, are too complex to be formalised by one single method or modelling paradigm. Approaches that are able to combine multiple AI methods are considered more suitable. Thus, an intelligent tool should be considered as the sum of its cognitive architecture and the algorithms implemented in this architecture.

Intelligent tool = cognitive architecture + basic AI algorithms

This reveals that a dedicated problem has to be analysed according to two criteria: the general cognitive aspects determining the cognitive architecture as the upper layer, and the specific cognitive steps mapped on selected AI algorithms as the basic layer. The intelligent tool becomes a system, if a user is incorporated as a part of the system. Adaptability of the tool to the user's preferences, intention and behaviour is the most important aspect here. This needs AI methods.

Intelligent system = human + intelligent tool

Tasks are parts of activities and activities are parts of processes. Therefore intelligent tools and the dedicated users have to be co-ordinated by integration and communication, which again need AI methods. Because of the distributed and concurrent nature of most building construction processes, this is first of all a logistic problem, i.e. the problem of information logistics.

Intelligent process = information logistics + intelligent systems

The information logistics architecture consists of several intelligent agents based on AI methods, which have to generalise and condense the individual incoming information in order to partition, archive, retrieve and distribute the right information to the right person at the right time. It is not only a simple through-put task with one or several new addressees, but it demands cognitive management work, which needs a responsible human - the project manager. The information logistics system should support the manager by taking over routine management and logistic tasks and prepare decision making.

Concurrent engineering system = project manager + intelligent processes

In a virtual enterprise we can identify four main processes: the management process, the business process, the technical process and the commerce process. The horizontal co-ordination and integration of these four processes as well as the vertical co-ordination and integration of the various activities down to each single resulting individual task together with the appropriate integration of the various individuals involved is not only an information technology task but demands also the extensive use of AI methods with very specific configuration requirements. Therefore, an architectural approach for intelligent hybrid AI tools is first introduced and later on in this paper a scenario of a concurrent engineering system is discussed, which is organized according to the above given four processes and based on the hierarchical components of a concurrent engineering system.

Intelligent Hybrid Tools

Due to the complexity of real-world problems intelligent tools should be considered as the combination of different AI methods in a cognitive architecture. For the purpose of adapting AI-technology to AEC we propose a process that starts with the adaptation of basic AI-algorithms to the engineering domain, combines these in generic cognitive architectures, and transfers these architectures in real-world prototypes. The necessary domain-dependent transformations of the basic AI algorithms and architectures can, to a wide degree, be generalised for knowledge based systems. The reasons for this are the specific, but common characteristics of building and construction problems:

- Building and construction usually deal with one-of-a-kind products but in almost all cases each product is a member of a product family. Thus the need for fast evaluation of alternatives and re-design is very important.
- In many parts of building and construction design numerical computation is essential and there are in existence a large set of mathematically based methods.

- In all areas of building and construction the design process relies, to a high degree, on rules of thumb.
- Project specifications are, in general, not fully given in detail. They often remain on a general level and are vague and conflicting in the details.

As a result of the adaptation, a set of basic domain specific algorithms can be obtained and can serve as **conceptual building blocks** for a variety of intelligent tools.

Cognitive Architectures

Cognitive architectures have been the subject of active research in cognitive science and Artificial Intelligence. A broad set of architectures has been invented, either based on requirements for general intelligence and problem-solving behaviour [1,2,3,4] or with respect to autonomous robot and vehicle behaviour [5,6]. Hybrid architectures can combine the developed basic algorithms as building blocks, depending on the associated cognitive abilities necessary for their corresponding problem class (fig. 1). If the intelligent tool is configured as a toolbox [7] the combination of the AI building blocks needs specific civil engineering knowledge from the practical domain of the application, provided either by the provider of the intelligent tool or by the user.

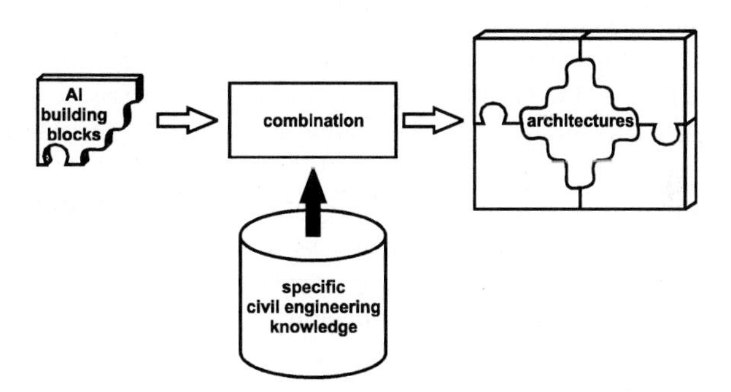

Figure 1: Combination of AI building blocks by cognitive architectures

Depending on abstract and generic classifications of the problems in terms of *strategic, tactic and reactive reasoning* [16], cognitive architectures for knowledge based systems in building and construction can be distinctly characterised as follows [8]:

Reflex behaviour. A system that acts on the basis of connected input/output patterns. This behaviour is sufficient for design tasks that do not require anticipation or consideration of past experience.

Utility-based behaviour. Apart from pure inference capability, it might also be desirable for a system to judge the value of information, the actions and the sensitivity of decisions to small changes in the assessments made. This is an architectural characteristic that is suited for design systems frequently involving a high-level interaction with the user.

Planning behaviour. In problem solving, a system makes guesses by choosing actions and evaluating resulting states. Since this leads to huge branching factors and thus is inapplicable for some domains, often a direct linkage between actions and states and discarding independent parts of the world' is necessary. This is achieved with planning methods.

Decision-theoretic behaviour. A system that makes decisions by choosing the action from a set of considered actions that has the best expected outcome. In that sense, this behaviour is an integration of probability theory and utility theory.

Learning capability. A system that is able to improve itself in order to increase its problem solving performance. This is desirable behaviour in domains that are difficult to formalise in a priori knowledge, because either the knowledge is too data-state dependent or the knowledge to be provided is too user-dependent, i.e. the tool behaviour has to adapt to the user's preference.

Communication. A system that can communicate about its state and knowledge in a formalised way to other agents or to the user.

The definition of the methodology for assigning and implementing intelligent tools with dedicated cognitive architectures involves three main steps:

1. Define problem categories.

   Design and construction problems are to be generalised in common categories that are comparable in respect to the necessary cognitive skills and behaviour. This requires, as a very first step, to define a suitable granularity on which problems and sub-problems are to be characterised.

2. Develop suitable cognitive architectures for the identified categories.

   It is also necessary to categorise cognitive behaviour and architectures, which can be matched with problem categories. Starting point here are behaviour facets.

3. Define a formalised approach for adapting a selected architecture to a given problem.

   Cognitive architectures and behaviours facets, in general, will not be reusable in an identical way in different problems, although these problems are classified in the same category. In order to obtain best performance, domain-specific adaptation of the generic building blocks for a dedicated application will be necessary.

Areas of application should be key areas with a great amount of routine cognitive labour, where pure mathematical tools fail because of imprecise information, e.g. of the necessary parameter values, or where information is only available as rules of thumb such as in preliminary design, uncertain material parameters in geotechnical engineering or uncertain cost values.

Intelligent Systems for Concurrent Engineering Support

Concurrent engineering technology has been the subject of intensive research in recent years [10,11,12]. The main objectives of concurrent engineering methodology are to support design for value and considerably reduce the lead time by overlapping briefing, design, bidding and product and supplier selection (fig. 2). Alongside the emerging information technology (IT) methods for integration and communication, electronic commerce can add an additional important dimension to the concurrent engineering system if we are able to integrate detailed technical product and cost information available on the electronic market in the briefing and design phase tools.

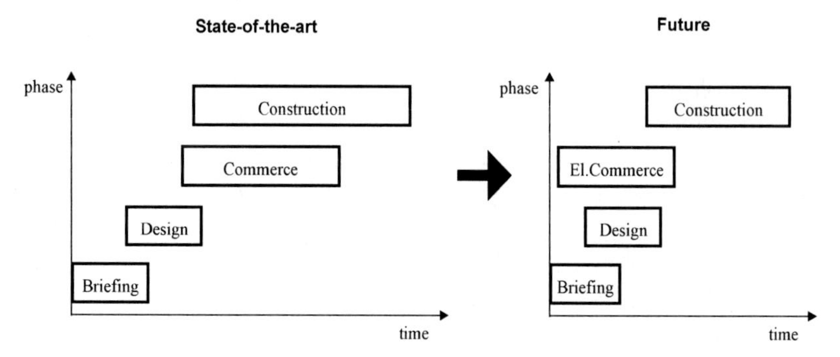

Figure 2: Reduced lead time through IT based concurrent engineering

Vision

We can imagine the following visionary scenario. The early consulting session between the investor and the architect-salesman will be based on virtual reality, deriving benefit from typified building blocks attached with functional requirements and cost values which can be assembled in a VR environment to already provide the investor at the very beginning with an understandable impression and reliably cost values. Cost intensive and architecturally or functionally critical building elements can be figured out and downloaded from suppliers' catalogue servers for fast and precise alternative conceptual design studies. Therefore consequences and impacts on the design and investment costs can be discussed at the very beginning with high precision.

The virtual design team including a cost consultant will be set up by Internet bidding in order to find the best team. The virtual design team will work in a concurrent engineering manner using the electronic market place to obtain the technical information about building components and services, whilst the cost designer will control the costs and organise the bidding, negotiations and contractual agreements with the suppliers via the Internet electronic market place.

In parallel, the investor will be kept informed by the architect-salesman about the progress of the design and can for example interact via video conference for critical parts with the virtual design team. Thus, in the future, the design process will simultaneously be driven by the investors requirements, the technical knowledge provided by the design team and the products and services offered on the electronic market without any time delay.

Approach

An information logistics system called *co-ordination board* is developed for the information integration of the concurrent processes (fig. 3). There the information is not simply collected and distributed, but it is also condensed, merged, re-classified and transformed into active objects serving as reactive and pro-active agents to support the project manager and trigger follow-up actions.

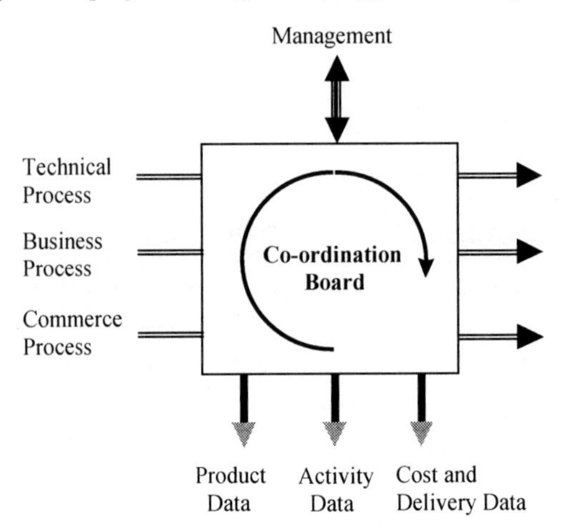

Figure 3: Cross section of the processes to be co-ordinated and the resulting data issues

The client-server-agent architecture is centred around the Internet/Intranet-based co-ordination board, which can operate on top of a middleware layer (e.g. a CORBA implementation) and an advanced information logistics layer [19]. The co-ordination board is structured into two layers – a wrapper layer and a classification layer. The objective of the wrapper layer should be to enhance the basic technical data with meta data describing the context, the status, permitted variations, constraints and dependencies, permitted notational transformations and interoperability information. The objective of the classification layer is to condense and classify dynamically the wrapped design data into high-level objects with the help of the object-centred approach based on description logic [14].

The system will be driven by the investor's requirements for the technical process, and the availability of the required products on the market. The goal is to get both these two sets of requirements to match. The connection of the technical process and electronic commerce results in several electronic market links in the following design phases:

Briefing phase:	–	Fast and concise catalogue look-up (download of technical information and standard prices)
	–	Selection design team
Design phase	–	Detailed catalogue look-up (download of 'intelligent' technical components)
	–	Cost information on services and components (pre-bidding and pre-negotiation)
Tendering:	–	Bidding and negotiation for services and components.

The Information Logistics Process

The information logistics system has to link together data, information and intention. The project manager must be able to control the process and therefore the status of the activities has to be transparent to all project participants in a presentation corresponding to their dedicated views and responsibility in the project. As a result, two major objectives can be identified:

Firstly, a communication architecture is needed that carries out the compression and presentation of information for all participants across the technical design, business and commerce domains. This defines the **data-oriented view**.

Secondly, the data manipulation and communication must be coupled with the management process in order to monitor the execution of the planned project workflow and adapt it to the current alternatives. For this purpose, software-agents shall automatically trigger small routine activities, whereas responsibility-dependent tasks are left to human actors. This defines the **activity process-oriented view**.

Both objectives demand interoperability of different degrees and a high level of semantics in order to identify the necessary activities. We claim that besides the sharing of data and objects also a common understanding and interpretation must be supplied in order to allow co-operative work. From the viewpoint of ontologies [15], a theory for common object representations for the co-ordination board has to be derived. We call this **co-ordinational interoperability** because it should provide the co-ordinated access and interpretation of semantic objects. The key point here is that integration can not be reduced to common data structures and transportation of data across networks. High level interoperability is the sharing of interpretable and useful information between multiple partners. This implies that, in general, information must be condensed when moving from the specific context of an actor to differing contexts of other actors.

Below co-ordinational interoperability, there are the **platform interoperability** and the **notational interoperability**. Platform interoperability aims at integrating heterogeneous computer hard- and software and notational interoperability is the effort to standardise the syntax notation and complete it by mapping procedures as suggested in STEP [17].

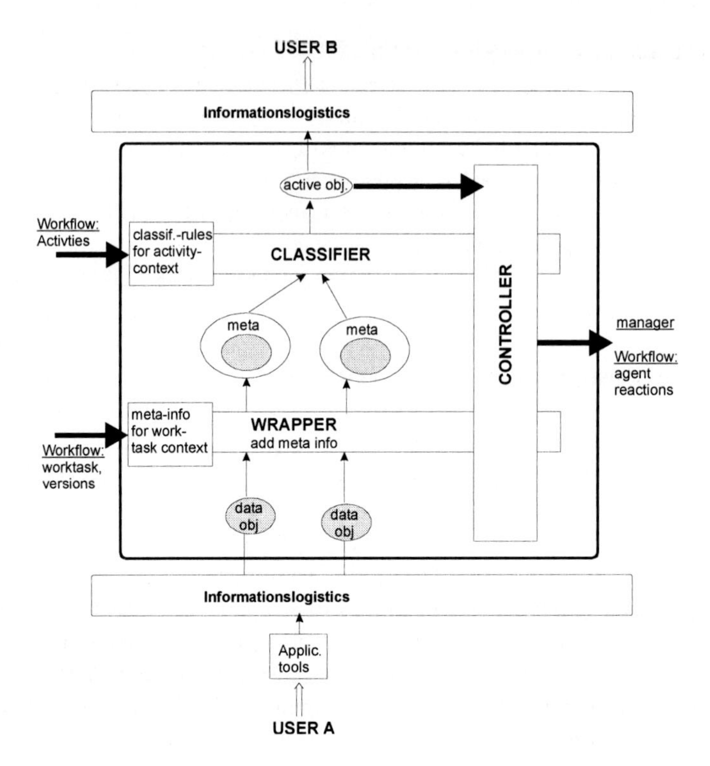

Figure 4: Architecture of the co-ordination board

Co-ordinational and notational interoperability

In order to realise co-ordination and notation interoperability, we have introduced the concept of an electronic co-ordination board. This approach will be discussed from a data and process-oriented view. The data-oriented view is shown in fig. 4 in the data transportation and compression process from the bottom to the top and the process-oriented view is given by the direction of workflow control from left to right.

Data view. In respect to the data view, the co-ordination board shall provide the interoperability mechanisms to compress data objects resulting from arbitrary software tools to common-understandable objects on a high semantic level. This compression is a data transformation process that requires data transportation and transformation on dedicated levels of interoperability. We define this transformation as a two-step process modelled by the wrapper and the classification layer.

The *wrapper layer* can be categorised as an advanced means to achieve notational interoperability. The basic functionality is to wrap the native result data of the software tools in a commonly parseable description - the meta information. Since the objective is information compression for decision support it is not necessary to map all elements of the result data to a common neutral description, instead they are selectively extracted. This is done with the help of *a priori* definitions of formats, tools, keywords and data structures configured in the wrapper layer and controlled by the context of the actual worktask. The wrapped result data and associated meta information serve as input to the classification layer.

The *classification layer* can be interpreted as the level on which co-ordinational interoperability is provided. There, native data are reduced to their basic, common understandable, characteristics. This means classification from the object-oriented point of view. In contrast to a traditional class-centred approach, the process of categorising information on the classification layer *after* the instantiation of the corresponding objects requires an *object-centred approach* [14], which allows dynamic classification and re-classification based on the actual properties of an object instance applying methods of description logic. The resulting classified *information objects* should be accessible by various client software. The information they contain will be on the one side the native result data and on the other side background knowledge that is inherited from the activity context through classification. These information objects serve as a high-level representation of project and domain knowledge to support decision making.

Activity process view. Some of the classified information objects act as agents themselves. Their reactive behaviours modify the planned workflow in response to actual data and process information on the board (fig. 5). Thus the board will also serve for activity control, i.e. project management. The pre-planned work-flow serves as input configuration data on both levels of the board.

The context of a worktask allows characterisations of task, role, actors and software tools. These characterisations serve as a valuable information source for the wrapper in order to derive meta information. Thus the configuration of the wrapper layer needs input data from activity control on the level of worktasks.

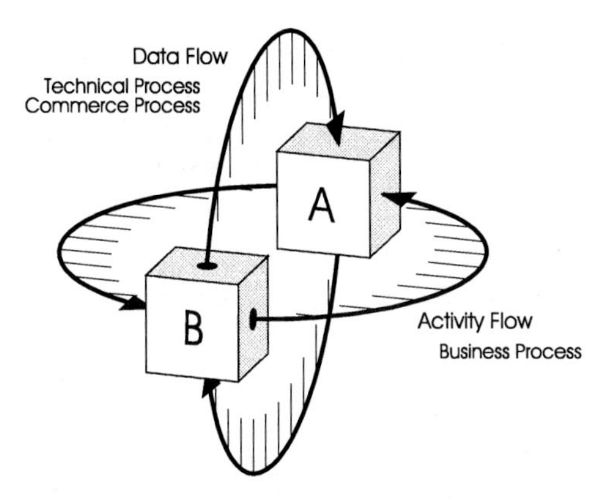

Figure 5: Interaction cycle of activities (A) controlled by the project manager and co-ordination board (B) for concurrent engineering.

On the classification layer, the received data shall be compressed to objects on a high semantic level. In order to derive a meaningful classification in a given project context, the classification rules must be adjusted to the context of the actual project activities, inherent goals and intentions. In the terminology of workflow modelling, the level that considers goals and intentions is the level of activities with corresponding *process templates* configured in the business process.

The objects that result from the classification are in the data-view condensed information objects representing the status of product data for the project participants. With the methods, which the objects inherit when they are classified and in consequence evolve to active objects, reactions to object configurations on the board are implemented. These influence the activity model and support decision-making of the project manager.

Platform Interoperability

For the interoperability of distributed system components, we have developed in the ESPRIT project ToCEE a mechanism called Uniform Project Resource Locator (UPRL) [18]. The main extensions of UPRL are summarized in table 1. They allow project-wide object addressing and extends the standard WWW addressing technique towards co-ordinated concurrent access to shared object-oriented models, client applications to manipulate server-side documents and objects, based on a formal interface definition [13, 19].

Resource Identification	URL	UPRL
Server	www server	UPRL object server
Client	browser, robot	browser, robot, helper application
Response content	HTML, multi media	HTML, multi media, objects
Access methods	browsing, full text search	browsing, full text search, object queries
Spec. of request semantics	-	EXPRESS-C language
Semantic reflection	-	Concept registry
Semantic multi-server integration	-	Concept registry + Server Interoperability Protocol (SIOP)

Table 1: Comparison of URL and UPRL

Interoperability of helper applications and servers is achieved by mapping object addresses to URLs. The basic mechanism URL is extended for addressing project relevant objects and includes modelling of (inheritable) object behaviour, access authentication (role dependent visibility of objects and object behaviour), parallel execution of object behaviour, transparent access to *all* meta data.

All objects of the environment can be addressed by UPRLs of a dedicated Information Logistics server (fig. 6) although they may be physically located and managed by other servers. Systems can retrieve and manipulate data of a server by sending requests to the server across the network and specifying an object address and name at run-time. Models can be physically distributed across several servers, based on a Server Interoperability Protocol specification (SIOP). All requests to UPRL objects can be co-ordinated by a common request broker corresponding to the concepts of the CORBA technology.

For UPRL servers, additional client applications, such as CAD systems, can be included on the client side, and which can actively manipulate server side documents and objects. These client applications can be extended with a respective middleware interface. It can be implemented either on the basis of a Java class library developed in the ToCEE project, or, for other target implementation languages, by using HTTP libraries.

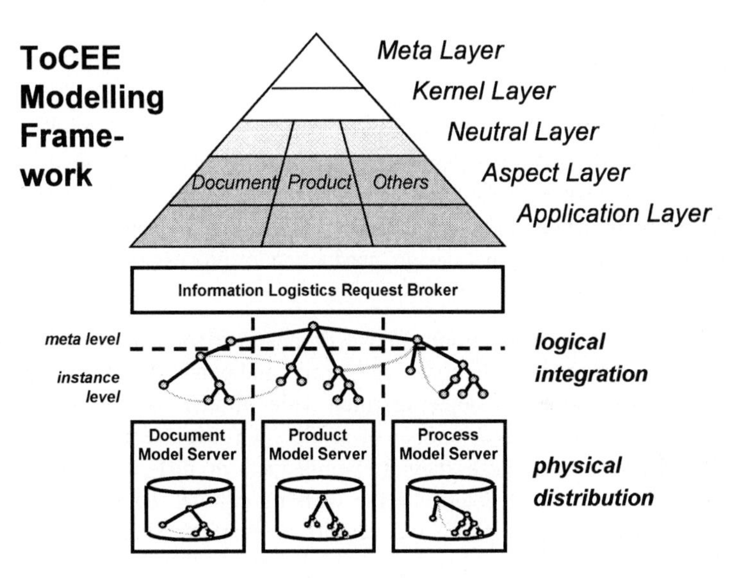

Figure 6: Partitioning models across several servers [19,27]

The Business Process

The part of the business process we are interested in is the co-ordination of teamwork for concurrent engineering and electronic commerce. The work of the members of the virtual enterprise is organised in terms of worktasks (fig. 7), which are globally identifiable and linked to roles, required input, expected or delivered results (documents, product model views, product data objects) and project time schedules.

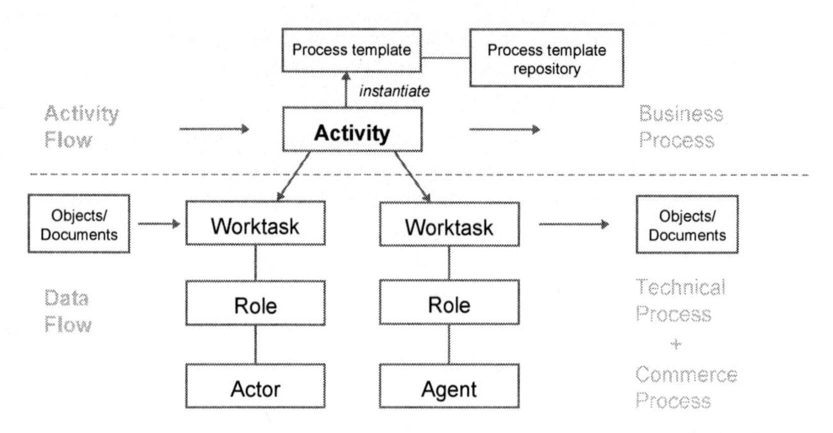

Figure 7: Business Process Model

Models are needed which support the users during the selection of correct activity input versions and the notification about which activities have to be performed when, by whom and for which reason. The data are needed according to the preferences of working either design, production, or process-oriented.

Role and Actor Model

The collaborative work environment we propose supports the activities with background knowledge about the roles, obligations and contractual relationships of the different players. The object-oriented role model is used to enrich communication with a *semantic* model of senders and receivers. The roles may have multiple aspects, because they are embedded in different environments (fig. 8). These environments determine the working culture and the kind of data represented. The role model includes devolution of responsibility, authentication, user preferences, tools to be used and notification services. The generic role model has to be adaptive to the project-specific configuration of the roles. The instantiation of a consistent role model is a non-trivial data transformation [20] even if it is based on an object-oriented contract model of the specific project. The system supports both multiple actors per role and multiple roles per actor. For the latter, we have introduced a technique which we call "organizational role abstraction" in order to address different functions of an organization not only by actors, but also by roles, e.g. HVAC expert of company C. Organizational abstraction permits to encapsulate the details of the execution of an activity from other participants.

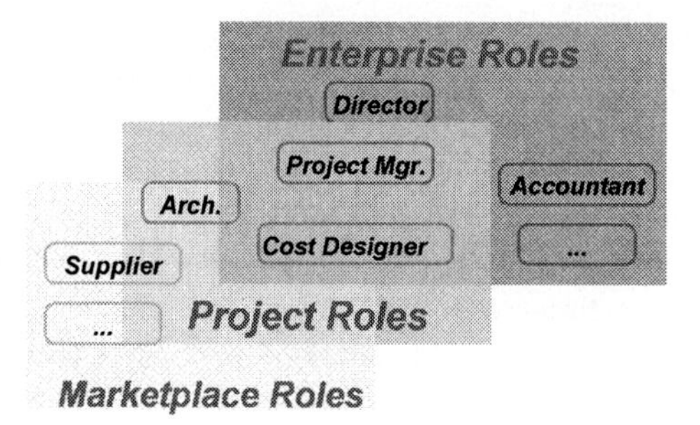

Figure 8: Role model

Activities and worktasks

It should be possible for a project manager to define and monitor project communication on the basis of responsibilities. His model is therefore role-process-centred. In our approach a project manager describes the activities of the roles in terms of *worktasks* and an activity model is built up from those atomic worktasks (fig. 7). Such an activity driven architecture extends the different communication activities of actors with an explicit semantic level, which keeps track of the causality of communication, by grouping atomic worktasks to activities and checking their relationships to predefined process templates, which are defined by a project manager and based on overall enterprise strategies. Data management is required to handle and archive the correct versions of documents in a shared document management system linked as views to the Product Model.

Monitoring role and actor interactions with Software Agents

Transparent and predictable knowledge about expected activities is needed to provide actors with the best available information when they schedule their activities. We investigate knowledge-based approaches, where dependencies are represented as constraints, so that we can detect the status of worktasks, e.g. scheduled, in execution, pending or finished by constraint propagation. Agents will trigger reactions to modify or respectively correct the planned workflow. Examples of such agents are specific simple software demons that carry out version control, verify if a given worktask is successfully completed, detect worktasks which take an unexpectually long time or identify worktasks which are blocked due to missing input.

Implementation of the Activity Model

For the integration of the activity control model special interfaces to the co-ordination board and the two processes controlled by worktasks, the technical and the electronic commerce process, have to be developed. They are closely based on AI methods. For instance, worktasks define events on the co-ordination board and vice versa. A filtering agent has to be developed which extracts those events on the co-ordination board, which are candidates for worktasks which update the activity model, monitored by a project manager. This should be based on a generic software agent. The responsibility for including agents remains in the hands of the project manager or the persons to whom he grants rights to include such agents.

The interface to the Technical Process includes, for instance, an interface to the top level semantic categories of the design product, such as "building", "building part" "wall", "column", which allows it to partition large sets of worktasks into partial models and define advanced retrieval services, e.g. "all worktasks related to windows", and a workflow aware transaction management, e.g. locking, rollbacks of the product model server. The interface to Cost Control should allow the transfer of explicit cost information as input and output of worktasks and management of the relationship to a *resource model,* e.g. by updating accounts.

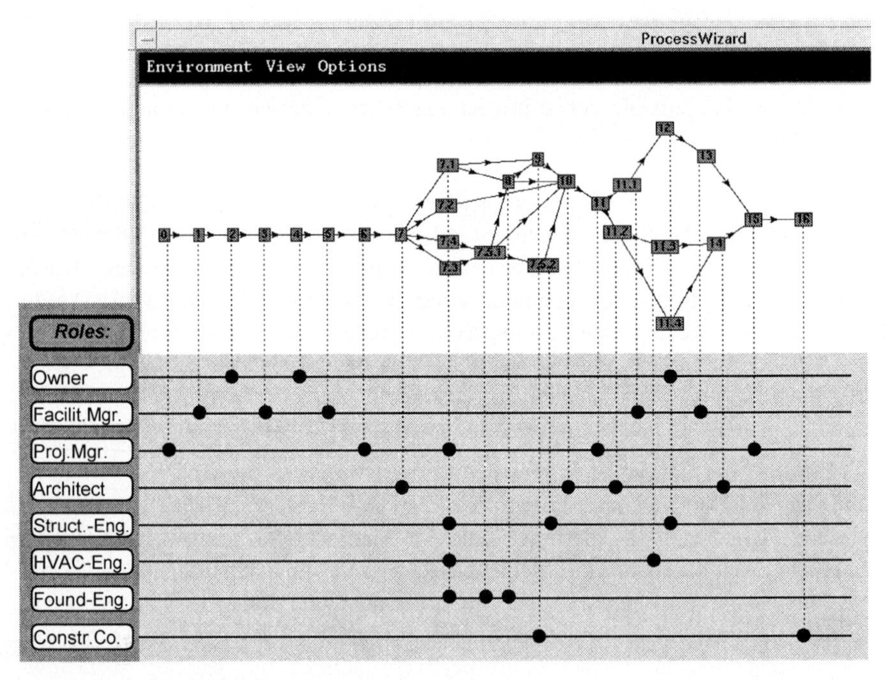

Figure 9: Process Wizard tool with worktask window and with an attached diagram detailing the dependencies and actor roles.

In the ToCEE project, such a business process environment, where the interfaces are based on formal interface definition language like CORBA-IDL, is already conceptually developed [19] and a prototype implementation called *Process Wizard* is on the way. In figure 9, an example is shown with the main activities of the ToCEE demonstration scenario. Each worktask of a user role is modelled as a node in the process network and the dependencies between them are represented as arrows. Advanced modelling techniques, such as conditional execution of worktasks and recursion, are supported by the modelling tool. In this context a process definition methodology was developed to achieve a parametric description of worktask pattern, based on *process templates*. The environment has a layered process architecture, built up from:

requests and *responses* as atomic process events, as provided by the distributed middleware (request broker),

data *transactions*, triggered by requests, as the atomic operations on product model data, but where not all consistency rules for model data are performed,

orders as units of inter-personal communication, creating worktasks for the project actors, related to aggregations of documents and transactions, e.g. "check_consistency" and

process templates as the most general level of reusable *a priori* process knowledge, modelling reusable patterns of human decisions and worktask dependencies.

The following process templates are provided [19]:

- Atomic templates, which define *orders* for worktasks
- Composed templates

They permit recursions, loops and reusable template modules. Based on these different possibilities to combine worktasks, process templates are generated, containing parametric descriptions of sets of worktasks.

For each project, a set of process templates can be maintained in a project specific process template repository. After the selection of a template, the process management tool creates the actual worktasks through an intermediate transactional layer, which can be used for concurrent access to the process model. The main dependencies between object classes of the process model are shown in fig. 10. During this overall process, the process management tool continuously updates the worklists for the different users, which contain exactly those worktasks which are relevant for one user showing date, status, data repository ID and dependencies on other worktasks. If a user finishes a worktask, he assigns his results to the process management tool which analyses possible follow-up activities for other users.

However several extensions are still necessary. For instance, the *first* operation level (fig. 10) needs extensions to ensure the high security and reliability needs of electronic commerce and business transactions. The *second* level has to be extended to support both technical product semantics and business semantics in an integrated manner. Document management can be included as an extension of classical transactions by also supporting binary large objects (BLOBs) as basic data types, maintaining all documents created in the system. The authorship and authorisation of documents can be transparently managed. The *third* level has to be extended towards cost aware activities, by explicitly modelling cost attributes and evaluating them along all interaction paths. On the *fourth* level, a new resource model is needed, with relationships to aggregated costs and account updating.

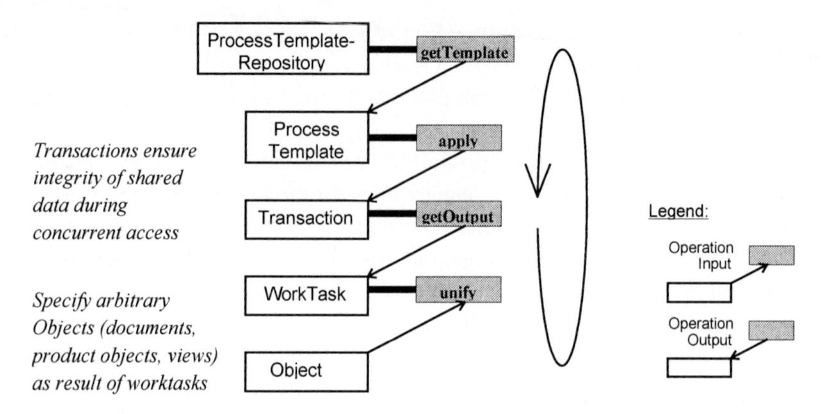

Figure 10: Basic interaction between process related object classes

The Technical Process

We envisage the technical process as driven by the requirements of the investor, the availability of technical products on the electronic market, the know-how of the design team and the sophistication of the tools. In general, the technical process will be integrated in the co-ordination board via the work tasks of the business process (see fig. 7).

Each individual technical work task will thus represent one distinct part of the consecutive changes of the state of the evolving product data. The goal of each task is to provide a feasible technical solution whereby the required functionality and behaviour of the designed building are achieved in a measurable way. The actor responsible for a certain specific task will typically work with the product data representing his specific view. He needs tools and services that can support the reliable access to a common product data repository, including view transformation methods, version management and recognition of conflicts with other designers. Since the technical solutions most often include a great variety of externally supplied products, it is important to ensure a close interaction with the electronic commerce activities, i.e. search on the electronic market, finding, negotiating, purchasing and supplying the right products and product data.

Such a holistic approach should enable:

- early consideration of the investor's specific requirements, i.e. including the selection of specific product components already at the briefing phase,
- step by step satisfaction of the requirements by means of gradual refinement of critical control parameters, such as cost and supply time, by means of a gradual reduction of initially broad ranges for the values of these parameters modelled through *fuzzy attributed objects* to concretely instantiated entities,
- continuous awareness of the attractive alternative products offered on the electronic market, in order to incorporate them into the design.

We supported the technical process by a hierarchically structured product data model of the building (fig. 6), based on STEP [17] and alternatively on the IFC model [21]. The product data will be stored in a shared distributed database and accessed through a high-level interface based on SDAI [17]. The anticipated overall architecture of the technical process system and its relation to the overall workflow is shown in fig. 11. The Product Data Server details are given in [22].

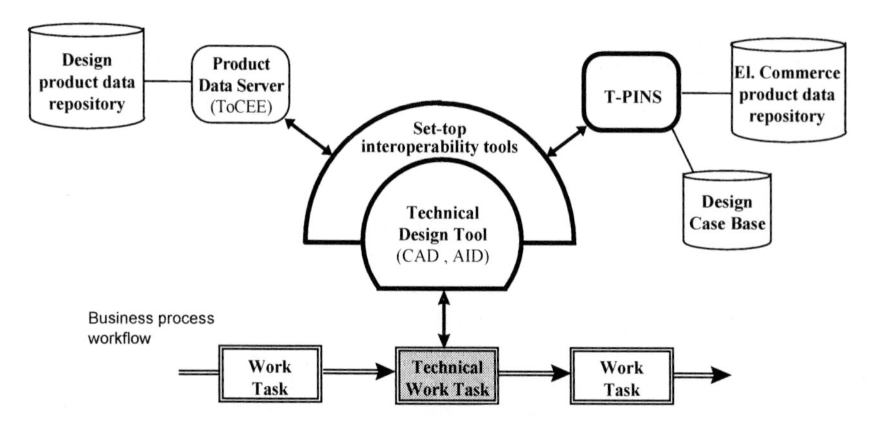

Figure 11: Architecture of the IT components for the technical process

Interoperability Methods

In order to achieve the interoperability on the technical data level, methods are needed that can enable the transformation of the data from one designer's view to another, at the same time ensuring proper management of all locally made changes and of the resulting product data versions. Accordingly mapping and matching tools based on STEP have been developed in the ESPRIT project COMBI [23] and were upgraded in the ESPRIT project ToCEE [22]. In order to achieve flexibility operational interoperability, semantic interoperability and functional interoperability must be taken into account.

```
ENTITY TC_Model  SUBTYPE OF (TC_IfcRoot);
  contents : TC_ModelContents;
  underlying_schemas : SET [1:?] OF TC_ModelSchema;
  accessRights : OPTIONAL SET [1:?] OF TC_AccessRight;
  ...
 OPERATIONS
  Map (VAR TargetRefs : LIST OF TC_ModelSchema);
  Match (compVersion : INTEGER;
         VAR ChangedContent : TC_InfoContainer);
  CheckConsistency (OPTIONAL Targets : LIST OF TC_Model;
                    VAR SyncRequest  : BOOLEAN;
                    VAR Result       : BOOLEAN);
  Commit (VAR Result : BOOLEAN);
  GetProductObjects (VAR ProdObjRefs : SET OF TC_Product);
  ...
END_ENTITY;
```

Figure 12: Part of the TC_Model entity specification in EXPRESS-C

Operational interoperability is part of the Information Logistics process and sub-structured into co-ordinational, platform and notational interoperability. On this level the operations involving semantic and functional interoperability are already formally defined, although they have to be performed by dedicated interoperability tools. An example for such definition is the concept TC_Model (fig. 12), representing a whole model instantiation, as implemented in the ToCEE project.

Semantic interoperability is defined as the ability of the conceptual model schemata to share common concepts of the different technical data models. It involves two complementary approaches:

- *Static model harmonisation*, i.e. use of the inherent features of the modelling paradigm like generalisation/specialisation at system design time,
- *Model mapping*, i.e. use of specifications and methods for the transformation of the modelling objects of one schema to another at run-time, where model harmonisation is alone insufficient.

Model mapping is a one-directional process that fully or partially transforms the classes and instances contained in a source model to new target model. The result of a mapping operation can be a modified schema (useful at system design time), or a new context (instantiation) of the target model (needed at run-time). For the formal specification of the inter-model equivalences an appropriate *mapping language* is needed, in order to avoid hard-coding of the model transformations for each particular case. As a result of extensive research efforts, a wide range of languages has been developed for such high-level specifications in recent years, e.g. EXPRESS-M, EXPRESS-V, EXPRESS-X, VML, KIF [24,25]. However, none of these languages is especially designed to support a modelling framework, such as the IFC project model [21], which strongly relies on the use of a common lean kernel model. Therefore, it can be more beneficial to use instead a more specialised language, such as CSML [26], developed in the COMBI project, which is able to support the kernel model approach and thus seems to exhibit better run-time performance compared to the other methods.

The mapping process we developed encompasses the following steps (fig. 13):

- parsing, including the analysis of the source and the target models and the mapping specifications, and the generation of the needed meta data structures,
- analysis of the instances of the source model and their representation on the common ontology level,
- expansion of the instances and relations of the source model into the target,
- reduction of the generated new target model instances to eliminate redundancy.

Functional interoperability is defined as the capacity to support at run-time the data modifications in the actual populated data. These features of the model management services are needed when one model has to be updated and checked against the constraints defined in its underlying schema, or when changes in one model have to be propagated and checked against the constraints of one or more other discipline-specific aspect models. This can be done with the help of two intelligent tools: model matching and consistency checking.

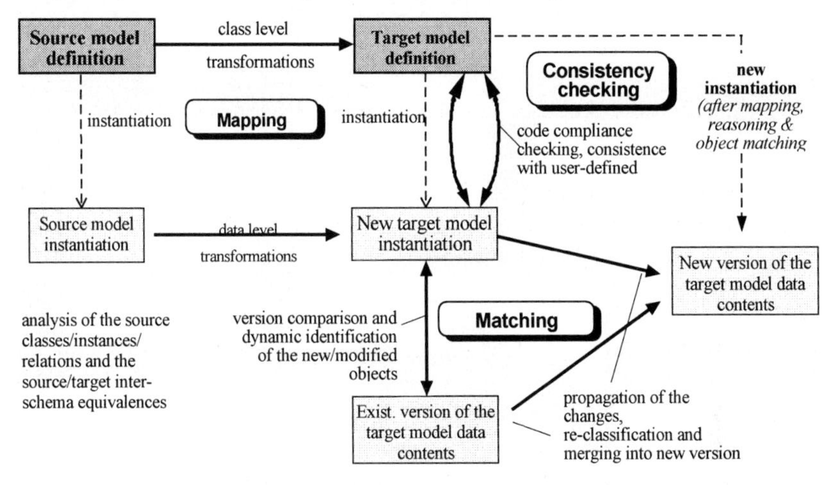

Figure 13: Principal schema of the data model transformations using the interoperability tools for model mapping, matching and consistency checking

Model matching involves a context-dependent analysis of the newly obtained target model data, including comparison with older target model versions. The matching operations are thus applied only on the target model schema. In our approach they make use of knowledge-based rules for dynamic object identification and classification, which work on specially introduced meta-level object representations [27]

The matching process we are developing encompasses the steps (fig. 13):

- identification of the changed instances in the target model,
- classification of the changed instances w.r.t. the concepts of the ontology by using the subsumption algorithm of the description logic approach,
- propagation of the changes over the whole data structure of the target model, if necessary repeatedly applying the two steps above.

The objective of *consistency checking* is to prove the validity of a new model version on the basis of 'after-add' rules applied to the new target model context. Such rules would normally not be included in the model specification itself, and can address issues such as compliance to certain codes or regulations, user requirements etc. Hence, this type of *local consistency checking* operations will be based both on the object-oriented definition of the target model classes and on a complementary rule-based extension of the model itself, represented in a separate specification schema. We are currently investigating this approach in the framework of the ToCEE project. In COMBI we have already established a prototype called PROMINENT [9] for semantic and some functional interoperability based on a three-layer product model, the kernel, the domain-dependent aspect and the application-dependent application models. The snapshot in fig. 14 visualizes three aspect models of the COMBI demonstration building.

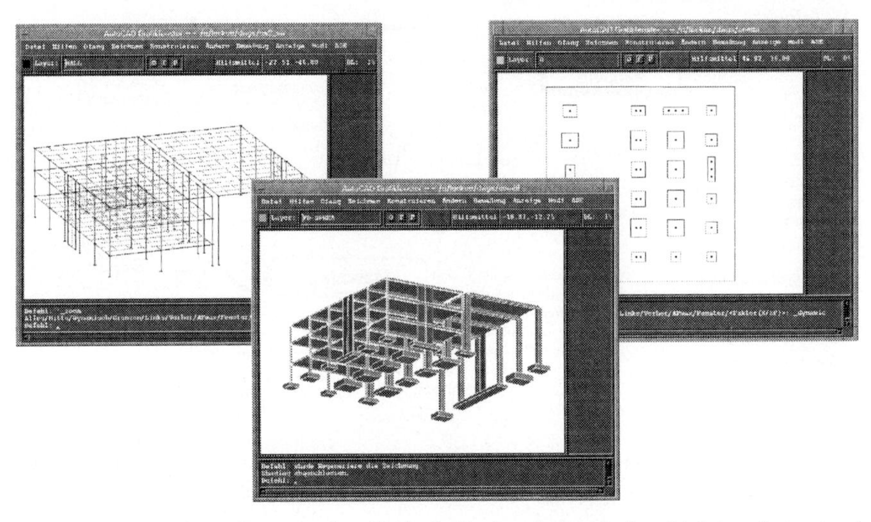

Figure 14: Integration of structural analysis (left), foundation design (right) and structural system design (middle) by semantic and functional interoperability

Technical Product Information System (T-PINS)

The efficient search for appropriate products on the electronic market needs a dedicated Technical Product Information System, which should be integrated in the design tools. Such a system should act as an intelligent advisory system supporting the designer's cognitive work on several levels [28].

On the first level the input will not be provided by pure data, but on a higher knowledge level, i.e. by design criteria complemented with allowable ranges for certain control parameters. This is comparable to shopping, when we want first to be advised by a skilled person's judgement about the product criteria. At this level no instantiated product data are needed, but the search object is enriched by knowledge through *dynamic classification.*

On the second level the input will be comprised of incomplete, i.e. only partially attributed, product data, representing a feasible, but not yet fully detailed design solution. Here the product selection will be done by comparing the available suppliers' products with the specified design data with the help of a *case-based reasoning* method and context knowledge extracted from a design environment database which will be prototyped for certain typical cases.

On the third level the real product components will be considered. The search can be carried out with a standard Internet *product broker system*, such as the GLENET system [29] This layer has to deal with different ontological problems, e.g. the presence of synonyms (different semantics for one and the same named concept) and homonyms (same semantics for differently named concepts). Such problems can be tackled with the interoperability methods mentioned before.

Technical Design Tools

The design tools used in the technical process can be general-purpose tools, such as CAD systems or analysis software, and tools or technical agents built on the basis of AI methods.

The added value from the use of intelligent technical tools for the technical design process is already proved by the prototype of a design assistant for preliminary structural design, developed within the COMBI project [16]. The structural design of a standard office building was reduced to 10 % of the normal design time. This tool is able to support the designer in his synthesis tasks and not just in analysis tasks. It is based on the methodology of intelligent systems as described in chapters 1 and 2,. Its layered cognitive architecture is:

The strategic level. Decisions of a general nature are made and the abstraction level is high. Reasoning on this level heavily involves projection and anticipation. In the system architecture, strategy is managed by a planning component.

The tactic level. The degree of freedom is limited by corresponding strategic decisions. The used models are increasingly detailed. In the system architecture a concept of so-called ''tools" is used to model tactic design actions that relate to the definition of a focused set of known design parameters.

The reactive level. On this level instances and models are more detailed. Actions can be interpreted as reactions to other actions or conflicts detected. The main AI method for this is constraint propagation

Technical Agents built on AI methods can tackle some routine design tasks in order to reduce the cognitive load of the designers. They do not have to be part of a design tool, but one may imagine that they are downloaded from the internet. Such a category of routine tasks is the continuos checking of the consistency of the actual design data. Agents could autonomously check for conflicting design decisions and inform the project participants when possibly critical design states are detected. At present, agent development is focus on the domain of geometry, and later on should be extended on code checking and functional criteria. This task gets immediately more complicated when intervals of values corresponding to contributions of different designers have to be considered.

The Electronic Commerce Process

Several electronic commerce links with the technical process are needed, at the same time distinguishing between technical look-ups and legally pre-binding and binding commercial interactions. The links are provided on different design phases as described before. The necessary tools are briefly described.

The looking up and downloading of technical information as well as standard prices can be handled with the T-PINS system as described in the previous chapter. Bidding, tendering, negotiation and procurement contracts can be handled by a Commercial Product Information System - **C-PINS**, and a Supplier Product INformation System - **S-PINS**. The envisioned overall architecture of the IT components for support of the electronic commerce process and their relation to the overall workflow are shown in fig. 15. In anticipation of the fact that many advanced IT applications for cost calculation and electronic commerce exist and more are emerging on the software market, the research focus in the context of concurrent engineering methodology should be based strongly on the necessary interoperability methods. The generic interoperability tools that must be developed in order to service such costing control applications can use basically the same IT methods as the respective tools for the technical process.

For data integration, the STEP- or IFC-based building object model has to be extended with specific high-level "electronic commerce" objects in order treat the electronic commerce process on the same semantic level as the technical and the business processes. The electronic commerce objects can be accessed through an electronic commerce product data repository, which can also be used by the Technical Product Information System T-PINS.

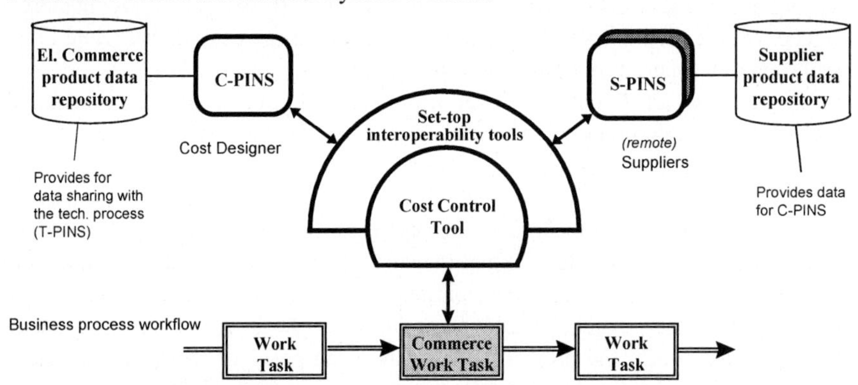

Figure 15: Overview of the components of the Electronic Commerce process

The **Commercial Product Information System (C-PINS)** for the support of Internet-enabled tendering, bidding, negotiation and procurement must be closely linked to theT-PINS system. It should consist of a set of a *tendering module* based on SGML/HTML-enhanced tender texts , a *bidding* module, a *negotiation module* which enables confidential price discussions with suppliers and a *procurement* module which is supported by an object-oriented contract model, such as the one developed in the ToCEE project. This model can be used to formalise the contract issues in order to allow rigorous support of contractual responsibilities by the system and to facilitate the preparation of contracts with the suppliers. In addition, the procurement module can provide a sample library of standard forms for typical contract parts which will be selected automatically by an agent, triggered through status changes of the "contract" objects. Filtering, sorting and evaluation of the electronic commerce data can be provided by generic AI-based agents.

The **Supplier Product Information System (S-PINS)** provides the specific additional functionality needed from the point of view of the suppliers. It should include: a module facilitating the promotion of the suppliers' products on the electronic market, a module for bidding offers, and a product data interface (STEP). The module for promoting the supplier's products should be based on STEP/SGML-based forms and provides interactive aids for publishing product information (in the form of texts, tables, technical drawings, photographic images etc.), and for presenting the supplier's profile (qualifications, portfolio) on the WWW.

Conclusion

Information technology will strongly encourage the construction industry to migrate from current document-centred fast tracking strategies to a concurrent engineering way of working. Classic information technology methods, such as data integration, electronic communication, Internet and Intranet technologies and workflow methods are only one substantial part of a concurrent engineering environment. Another, yet no less important part, comprises the AI tools. However these tools will appear in the overall framework not as readily observable large components, but rather in a very fine granular manner, distributed over the whole system. AI methods will appear in an atomic manner and - as it was demonstrated - will have to be built-in in cognitive architectures, demanding specific configurations for each dedicated task. This means that each applied AI method will need a lot of configuration and application-dependent effort to exhibit its full power, but on the other side the measurable benefit of each single built-in AI method might be almost negligible on the macro scale of a concurrent engineering system. Only when the sum of the whole AI impact on the overall system is measured, their important role in the design of complex and powerful IT environments can be understood.

The objectives in IT research and development have already shifted from the investigation of client-server systems to client-server-agent systems, where agent development subsumes the application of AI methods. However, this is mostly attempted on platform level. What is still missing is the shift from the man-machine interface to the man-machine-agent interface, which means that an intelligent system = human + intelligent tool and intelligent tool = a cognitive architecture + configurated AI methods.

Acknowledgements

The support of the Commission of the European Community for the ESPRIT project contracts No. 6609 COMBI (10/93-12/95) and No. 20587 ToCEE (01/96-12/98) is gratefully acknowledged. My gratitude goes especially to my research assistant Peter Katranuschkov and to my PhD students Rainer Wasserfuhr and Markus Hauser for their contributions to this research work, the results of which are described in this article.

References

[1] Laird J.: *Preface of the special section on integrated cognitive architectures.* SIGART Bulletin, 2, 1991.

[2] Mitchell M., J. Allen, P. Chalasani, J. Cheng, O. Etzioni, M. Ringuette, and J.C. Schlimmer: *Theo: A framework for self-improving systems.* In K. VanLehn, (ed), Architectures for Intelligence. Lawrence Erlbaum Associates, Hillsdale, NJ, 1991.

[3] Forbus D. and D. Gentner. *Similarity-based cognitive architecture.* SIGART Bulletin, 2, 1991.

[4] Carbonell, J.C., C.A. Knoblock and S. Minton. *Prodigy: An integrated architecture for planning and learning.* In K. Van Lehn, (ed), Architectures for Intelligence. Lawrence Erlbaum Associates, Hillsdale, NJ, 1991.

[5] Brooks A.: *How to build complete creatures rather than isolated cognitive simulators.* In K. Van Lehn, (ed), Architectures for Intelligence. Lawrence Erlbaum Associates, Hillsdale, NJ, 1991.

[6] Maes P.: *The agent network architecture (ANA).* SIGART Bulletin, 2, 1991.

[7] Hauser M. and R.J.Scherer: *Automatic knowledge acquisition in the reinforcement design domain.* In Choi C.-K., C.-B. Yun, H.-G. Kwak (eds.), Proc. 8th Int. Conf. on Computing and Building Engineering, pp. 1407 - 1412, Seoul, Korea, August, 1997.

[8] Russel I. S. and P. Norvig: *Artificial intelligence - a modern approach.* Prentice Hall, New Jersey, 1995.

[9] Scherer R. J. and P. Katranuschkov: *Integrated product model centred design in a virtual design office.* to appear in: Proc. of Information Technology for Balanced Automation Systems in Manufacturing, Prague, August 1998.

[10] Kusiak A.: *Concurrent engineering: automation, tools and techniques.* John Wiley & Sons, 1993.

[11] Prasad B.: *Concurrent engineering fundamentals (integrated product and process organization).* Prentice-Hall, Englewood Cliffs, NJ, 1996.

[12] Huovila P., L. Koskela and M. Lautanala: *Fast or concurrent - the art of getting construction improved.* Proc. 2nd Int. Workshop on Lean Construction, Santiago, Chile, 1994.

[13] Wasserfuhr, R. and R.J. Scherer:: *Information management in the concurrent design process.* Proc. Int. Colloquium IKM'97, Weimar, February 1997.

[14] Hakim, M.: *Modelling evolving information about engineering design products.* PhD thesis, Dept. of Civil engineering, Carnegie Mellon University, 1993.

[15] Gruber T.R.: *Towards principles for the design of ontologies used for knowledge-sharing.* In Guarino N. and R.Poli (eds): Formal Ontology in Conceptual Analysis and Knowledge Representation. Kluwer Academic Publ., Deventer, Netherlands, 1993.

[16] Hauser M., Scherer R.J.: *Application of intelligent CAD paradigms to preliminary structural design*, Artificial Intelligence in Engineering 11 (Special Issue: Structural Engineering Applications of Artificial Intelligence), pp. 217 - 229, Oxford, 1997.

[17] ISO 10303-1 IS: *Product Data Representation and Exchange - Part 1: Overview and fundamental principles.* ISO TC 184/SC4, Geneva, 1994.

[18] Scherer R. J.: *Overview of requirements and vision of ToCEE.* Public Annual Report 1996, EU-ESPRIT project No. 20587 ToCEE, TU Dresden, Germany, Feb. 1997.

[19] Wasserfuhr R. and R. J. Scherer: *Process models for information logistics in the concurrent building life cycle.* In K.S. Pawar (ed): Proc. 4th Int. Conf. on Concurrent Enterprising, Nottingham, Oct.1997.

[20] Scherer R.J.: *Legal framework for a virtual enterprise in the building industry.* In K.S. Pawar (ed): Proc. 4th Int. Conf. on Concurrent Enterprising, Nottingham, Oct. 1997.

[21] IFC Release 1.5 Final Version: *IFC object model for AEC projects.* IAI Publ., Washington DC., Sept. 1997.

[22] Hyvärinen J., P. Katranuschkov and R.J. Scherer: ToCEE: *Concepts for the product model and interoperability management tools.* Deliverable F2-1, EU-ESPRIT project No. 20587 ToCEE, TU Dresden, July, 1997.

[23] Katranuschkov P.: *COMBI: Integrated Product Model.* In Scherer R.J. (ed) Proc. 1st European Conf. on Product and Process Modelling in the Building Industry, Balkema Publ, Rotterdam, Netherlands, 1995.

[24] Verhoef M., T. Liebich and R. Amor: *A multi-paradigm mapping method survey.* Proc. CIB W78-TG10 Workshop on Modelling of Buildings through their Life-Cycle, pp. 233-247, Stanford University, CA., 1995.

[25] Genesereth M. and R. Fikes. *Knowledge interchange format 3.0, Reference manual.* Tech Report Logic-92-1, Comp. Science Department, Stanford University, CA. 1992.

[26] Katranuschkov P. and R.J. Scherer: *Schema mapping and object matching: a STEP-based approach to engineering data management in open integration environments.* Proc. CIB-W78 Workshop, Bled, Slovenia, June 1996.

[27] Katranuschkov P. and R.J. Scherer: *Framework for interoperability of building product models in collaborative work environments.* In Choi C.-K., C.-B. Yun and H.-G. Kwak (eds) Proc. 7th Int. Conf. on Computing in Civil and Building Engineering, pp. 627-632. Seoul, Korea, August 1997.

[28] Scherer R.J.: *A Product Information System with an Adaptive Classification Structure.* In J. Gausemeier (ed.) Proc. Int. Symposium on Global Engineering Networking, Antwerpen, Part I, pp. 69 - 78, Paderborn, 1997.

[29] Rethfeld U.: *GEN vision an major concepts.* In Proc. 1st European Workshop on Global Engineering Networking, Paderborn, February, 1996.

A New Collaborative Design Environment for Engineers and Architects

Gerhard Schmitt

Chair for Architecture and CAAD,
Swiss Federal Institute of Technology
ETH Zürich, Switzerland
e mail: schmitt@arch.ethz.ch

Abstract. Civil Engineering and Architecture are disciplines that highly depend on collaborative work and design. We introduce a design and collaboration system that makes use of conventional and AI communication techniques. The system addresses questions relating to authorship, design ownership, design procedures, alternative design methods, and the use of large, distributed data bases for a multitude of users to support the above activities. Three practical examples form the center of the description: Phase(X), Multiplying Time, and the Delft Experiment.

1 Introduction

The technical working environment and infrastructure in advanced architecture, engineering and construction (AEC) firms has changed dramatically since the middle of the 1990ies. Computer supported communication and collaboration are no longer mere possibilities, but, given the will and know-how of the participating partners, a reality. What was first achieved at universities and large AEC firms such as Norman Forster and Partners, is now in the reach of any small and medium firm with access to the Internet. In order to be successful, this type of cooperation requires new design and communication methods. To prove the point, we will describe three related projects and the associated methods.

Phase(X) is the name of the teaching environment for an elective CAAD course at ETH Zürich, in which more than 100 students participate in the learning and application of new design media. In ten phases, they move from using two-dimensional drawing tools towards the application of rendering, animation, and more-dimensional modeling software. The developers of Phase(X) had previous experience with collaborative design studios and collaborative work with engineers. Rather than following the approach of other collaborative design studios (Wojtowics, 1995; Chiu, 1997), they decided to place the project data base in the center and to strengthen asynchronous cooperation over synchronous design actions. By opening the resulting data base to the Internet and by displaying alternatives, the system becomes accessible also to non-designers and allows a new type of participatory design. The necessity for these aspects has been recognized (Fukuda et al., 1998; Lee et al., 1998).

Whereas Phase(X) dealt with abstract principles, Multiplying Time focused on an architectural design project. Multiplying Time was a collaborative studio between architecture students at ETH Zürich (first semester), the University of Hong Kong (advanced) and the University of Seattle (advanced). In five phases, using the same type of database as the Phase(X) environment, students started from a simple design brief and moved towards complete design descriptions. The time difference between Hong Kong, Zürich, and Seattle - on the average eight hours each - was used to expand one working day period to a twenty four hour period, thus expanding one week of exercises to three weeks of design. Each day, there were video conferences between students and faculty at the places involved. The system was similar to Phase(X), in that after each phase only the best design was chosen by the students for further development. The potential for transcontinental engineering and architecture cooperation became obvious.

The Delft Experiment built on a previous design exercise, Multiplying Time. Students in Delft started where students from Hong Kong, Zürich and Seattle had stopped. In four phases they continued to develop the design with different software and rendering programs.

2 Computer Supported Collaborative Design

Computer supported collaborative design (CSCD) is a special type of computer supported collaborative work (CSCW). Working together in an electronically linked team is a most useful addition to working individually. It can foster synergies and lead towards a goal much faster, because the combined knowledge of the partners can easily result in a breakthrough. CSCW teams form spontaneously, if common interest, spatial distances and flexible working conditions exist. If this is not the case, collaborative work must be agreed on in advance. CSCW is possible only in networked computer environments. The largest and most accessible network today is the Internet, through which partners distributed world wide can organize collaboration. Communication over the Internet is standard in scientific computing and is gaining ground in engineering and architectural firms that operate internationally or with remote partners.

CSCW supports different types of communication, each of which has specific technical prerequisites. Depending on the task and on how many people are cooperating, a CSCW environment needs e-mail, video, audio, a common drawing platform (white board), direct written communication (talk), file transfer, and the shared use of programs (application sharing). The combined use of those instruments defines the main difference to the video telephone or picture telephone, that has been known since the 1930s (Schmitt, 1996).

Video and audio contact are essential for communication through the network. To achieve good quality even with low bandwidth, techniques are used to compress the image on the side of the sender and decompress it on the side of the receiver. To be able to send audible information through the same line and at the same time as visual information, the sound also needs to be compressed.

The white board supports sketching directly on a shared document or to comment on the work of others. The talk window is useful for quickly transmitting written information, especially for establishing audio or video connections.

During a CSCW session, the exchange of data and files based on the file transfer protocol (FTP), is fundamental. FTP is increasingly superseded by direct exchange through more user friendly Internet applications. It is also possible to take remote control of another computer to simplify work on the same project. With full remote control over another machine in a different place, it is possible to leave delicate configuration, design or change operations on the shared, virtual design model to the most experienced person in the team.

CSCW has been established as a possible form of education. Experiments of CSCW in education date back to the early 1990ies, when virtual design studios evolved in the United States, Europe, and Asia (Lee, 1998, Wojtowics, 1995). Typical applications are the transmission of lectures and exercises; enabling the communication between students and professors; the follow-up on experiments or examining models in distant laboratories; the collaborative work on design problems; and finally the presentation and critique of a project through the network.

For successful use of CSCW in teaching, it is important that a gradient or differential of information and knowledge exists between the parties involved: Cooperation is not something that simply happens, but it occurs when there is a true desire to learn from somebody or to give information to somebody. This information differential - and curiosity - foster collaboration from the beginning.

3 Elements of a New Collaborative Design Environment

Our evaluation of past collaborative architectural and engineering design experiments resulted in the development of alternatives to the conventional approach. Three major differences of the resulting Phase(X) approach to conventional CSCW are:

- The center of attention is not the group of collaborating people, but the common project. It is supported by the data base and appears as the same, virtual project to all participants. Spatial separation of the design partners objectivates collaboration and focuses on the project, rather than on idiosyncratic behavior.
- Authorship in extensive collaborative projects is often an issue of conflict. We therefore propose to bundle individual authorship into a common design, and to trace the individual authorship if needed, using the data base.
- We do not only encourage to look at other people's work, but enforce it by requesting that each design phase must be based on the previous phase of another person or team.

4 Collaborative Learning: Phase(X)

Phase(X) is a design Principia course at ETH Zürich using the computer as a medium. Its purpose is to pose and explore fundamental questions concerning design ideas, modeling and authorship. Phase(X) expands the idea of the paperless studio by building more-dimensional computer models, by networking the designs and by focusing on abstract concepts such as Types & Instances (Madrazo, 1996). Adding modeling instruments such as Sculptor (Kurmann et al., 1997) offered students additional opportunities to explore new design approaches, based on playful interaction with design objects (http://space.arch.ethz.ch/ws96/).

Phase(X) treats authorship in a way that is only possible in a networked, cooperative design environment. After each phase, students do not proceed with their own design but continue with a solution they carefully select from the results of their colleagues. After the completion of their work in each phase, they place it into a MSQL data base. A graphical representation of the work becomes visible immediately after it has been submitted. Based on this representation, students must select the best project they can find in the data base. This way, a continuous evolution of the design is guaranteed. The probability, that only the best designs will be further developed, is high. Authorship is not a question, as the contribution of all people involved in the design in terms of time and model is recorded in the data base (http://space.arch.ethz.ch/ws97/).

Phase(1) starts with two-dimensional compositions on an empty grid, placed into the Phase(X) data base after completion. The result is immediately visible in a browser window (see figure 2, bottom). In Phase(2) and in all following phases, students check out a design as the basis for the following phase. They progressively refine the objects in the following design stages. The final results are complex objects with shared authorship that can be traced back to the contributing authors and co-authors using the *thread* function. Two important views of the process and the products developed: InWorld and OutWorld.

InWorld describes the perspective of the participant from within the structure of the experiment (see figure 2). The interface places observer and designer into an introverted position. They can only see the direct vicinity of the design: its parent and its children. The system presents itself as a genetic tree. Navigation from branch to branch sheds some light on the system structure. InWorld is the plane on which design ideas are developed and stored. Phase(X) objectively keeps the memory of the individual designs and makes them available in real time. Out of this develops the OutWorld.

OutWorld is the name for the presentations that evolve from the entire data set (see figure 1). Based on these overviews, cross comparisons emerge, assumptions and theses can be explored. OutWorld replaces the sequential view of the InWorld with a parallel view. The interface produces the presentation - consisting of lines and surfaces in two- and three-dimensional space - in real time. The different overview presentations give partial objectivity to the OutWorld. The observer influences the views by choosing parameters. The OutWorld view proved to be the only meaningful way to present and compare design data from more than 700 student projects.

Phase(X) introduces the notion of memes to design. We see memes as an analogy to genes (Dawkins, 1976) that contain crucial information for the replication and development of organisms. We assume that a design contains memes with different qualities: they may be strong, so a design is chosen by many others for further development in the next phase; they may be strong and sustainable, so they influence not only the next but also the following design stages. These qualities can be interactively explored in Phase(X)^2, the follow-up to Phase(X).

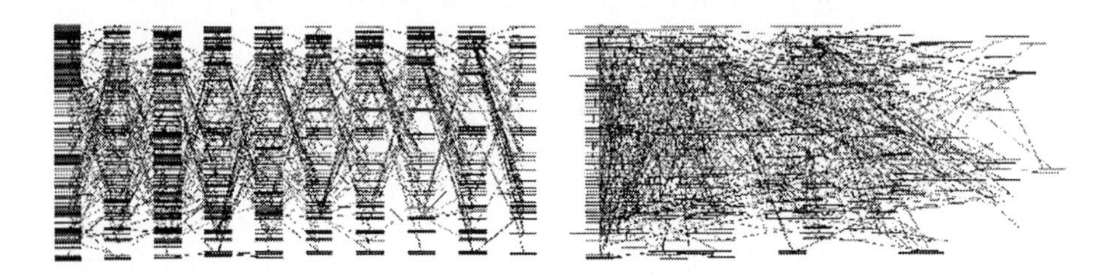

Fig. 1. Two OutWorld views of the same database. Projects ordered by authors and connections (left), authors and connections and time (right).

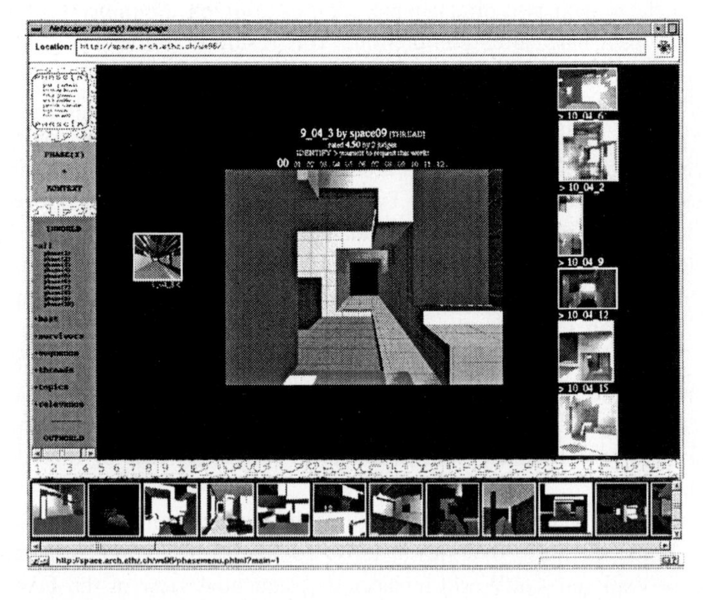

Fig. 2. Phase(X) InWorld designer and user interface. Top level menu frame on the left. Main window with ancestor project (left), focus of interest (center) and follow-up projects (right). Scrollable project database frame at the bottom. F. Wenz, F. Gramazio, U. Hirschberg, P. Sibenaler, C. Besomi, B. Tunçer, 1996.

5 Design Teaching: Phase(X)^2

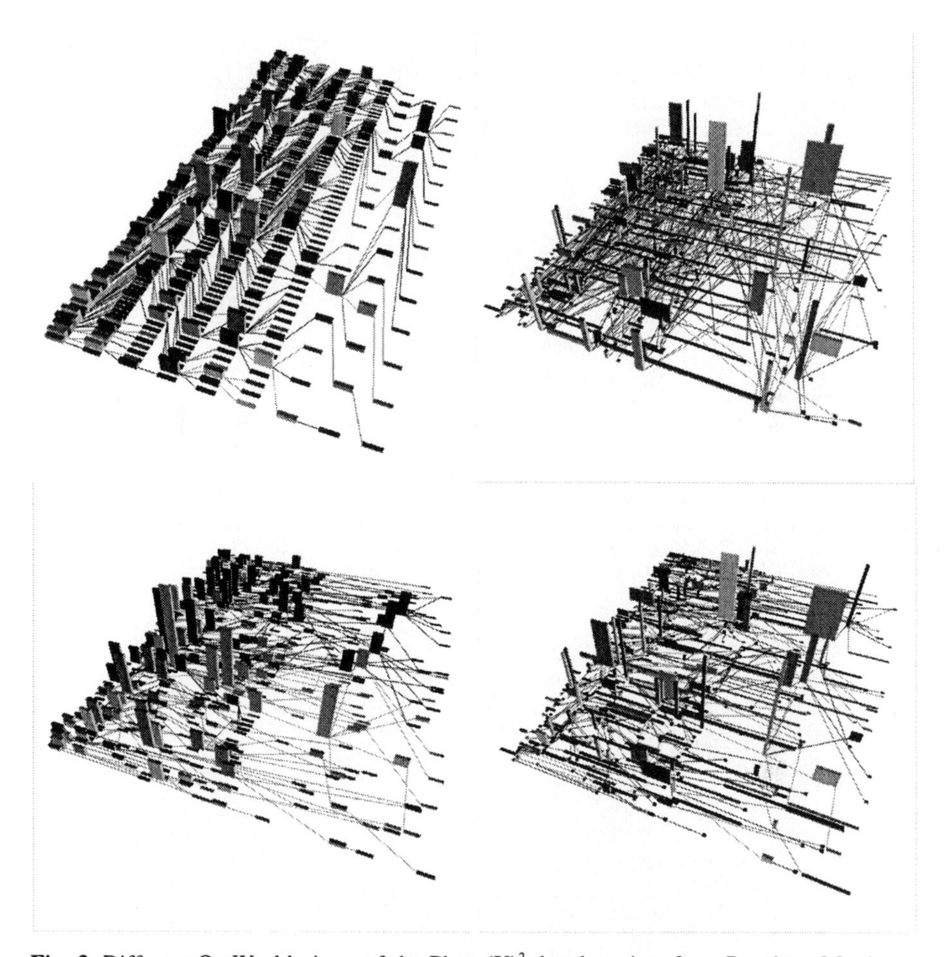

Fig. 3. Different OutWorld views of the Phase(X)^2 data base interface. Results of 9 phases before the end of the project. This interactive three-dimensional and multi-colored data view allows for the display of more individual properties than the two-dimensional OutWorld displays. Phases always develop from left to right. Height of volumes expresses performance criteria rated by the system. The rating by students and instructors was much more balanced.

Phase(X)^2 is the sequel to Phase(X), with many improvements in the interface and the data base. Whereas Phase(X) used a modified AutoCAD as modeling program, Phase(X)^2 is based on MicroStation. The procedure of designing is similar: Phase(1) - Composition in the plane; Phase(2) - Objects in the plane; Phase(3) - Positive and negative volumes; Phase(4) - Rotation in space and movement; Phase(5) - Free form surfaces - structure and cover; Phase(6) - Design vocabulary; Phase(7) - Parametric solids; Phase(8) - Self-similar structures and fractals; Phase(9) - Light and space (http://space.arch.ethz.ch/ws97/).

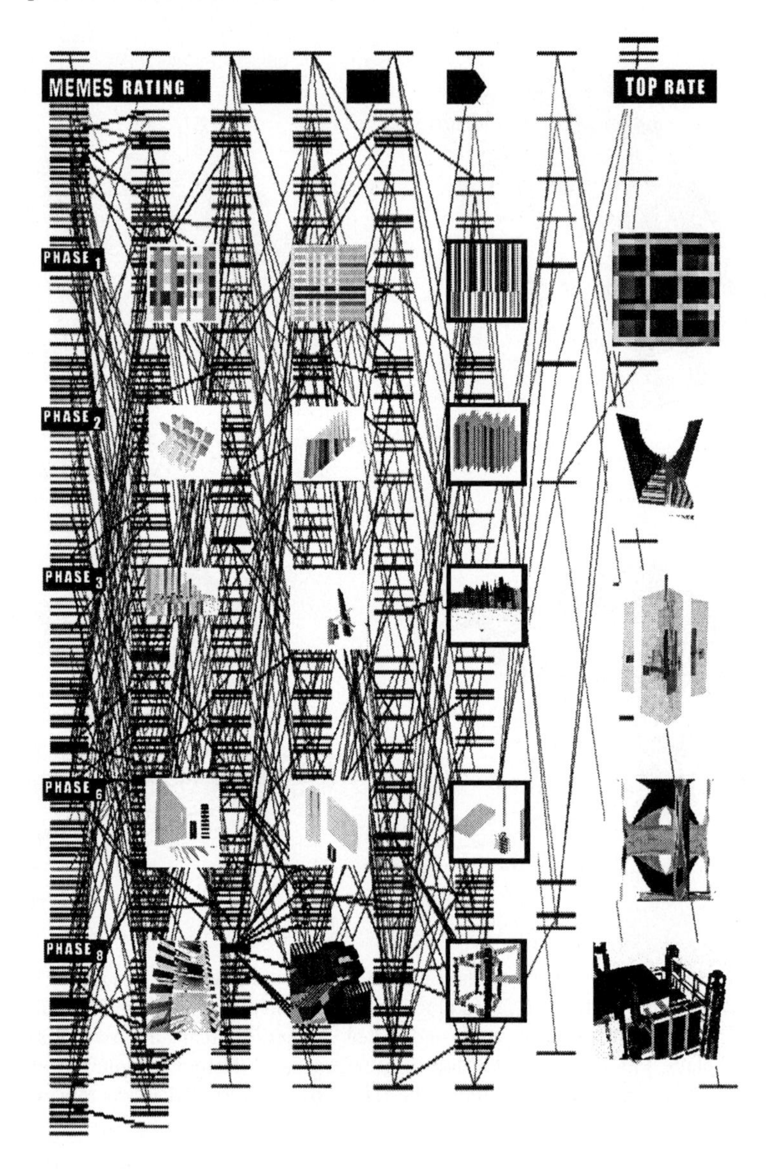

Fig. 4. Memes and quality: Note that the projects rated best in quality by students and faculty (right column) do not necessarily have the most follow-ups and memes (left columns).

A major difference to Phase(X) is that the OutWorld - the data base overview (see figure 3) - can be seen immediately and concurrently with the InWorld - the individual projects. In addition, new views on the data base are enabled: Users can choose to see the most recent entries in each phase; they can look at the most relevant designs in terms of memes and followers; they can rate designs according to their criteria and see the overall ratings of each project; they can observe the time it took to generate a specific design; they can explore the data base according to certain key words. They can also have a quick overview of the work of an individual designer.

Together, InWorld and OutWorld form an environment that could not be created or exist without the computer. More important, working in the framework of Phase(X) leads students to better results in less time. Judging from the students' and the teachers' response, this new way of teaching and learning cooperative design is attractive. The fact that no individual ownership of a design is possible did not pose a problem to anyone. Perhaps this is due to the abstract nature of the exercises and the differences between the university environment and professional practice. Yet the people we asked could imagine working in practice under similar conditions. Therefore, Phase(X) might be an indication to a possible future AEC working environment. To test this hypothesis, we conducted the Multiplying Time experiment. It introduces a real design problem, along with some working conditions closer to architectural practice. The experiment used a common modeling tool with special provisions for collaborative design.

6 A Collaborative Design Instrument: Sculptor

Sculptor is a program developed by David Kurmann at ETH Zürich to support the early conceptual phase of object and architectural design (Kurmann et al., 1997, http://caad.arch.ethz.ch/~kurmann/sculptor). It allows intuitive interaction with a virtual model and is based on known concepts and mechanisms of spatial composition and recognition.

- Sculptor provides objects, models and worlds, whose attributes and behavior - such as form, geometry, color, material, and movement - can be specified interactively in addition to modeling constraints such as collision and gravity. Objects, groups of objects and worlds can be changed by scaling, translating, condensing, and exploding.
- Sculptor offers the opportunity to model with spatial elements, or voids. Such negative volumes that create a void when intersected with a solid, can be manipulated and moved in the same manner as solids. Solids and voids have the same data structure (see figure 5, right). The interactive real time intersection of positive and negative volumes supports the direct composition of spaces.
- A third special type of objects in Sculptor are rooms, consisting of a solid and a contained space. Rooms also have the advantage that a single additional attribute, the purpose of the room, places necessary information into digital models, information that later can be interpreted by intelligent agents.
- Sculptor has been integrated with the constraint and case-based architectural floor planning software IDIOM (Smith 1996). An abstract graph of a floor plan can be generated, representing different rooms and their connections. The user can proceed with further refinement and experimentation with this model. Sculptor and IDIOM collaborate on one computer screen. All changes in the two-dimensional IDIOM

interface are reflected directly in the three dimensional Sculptor interface and vice versa.

Distributed Sculptor (see figure 5, left) adds the possibility of Computer Supported Collaborative Design (CSCD) to Sculptor (Dave 1995). Different users on locally distributed computers share a 3D-model. Every user can have an individual point of view. The user holding the pen or modeling device is allowed to make changes, while the others may only observe. Each collaboration site enables observation of current sessions and allows users to join or create a new distributed cooperation session.

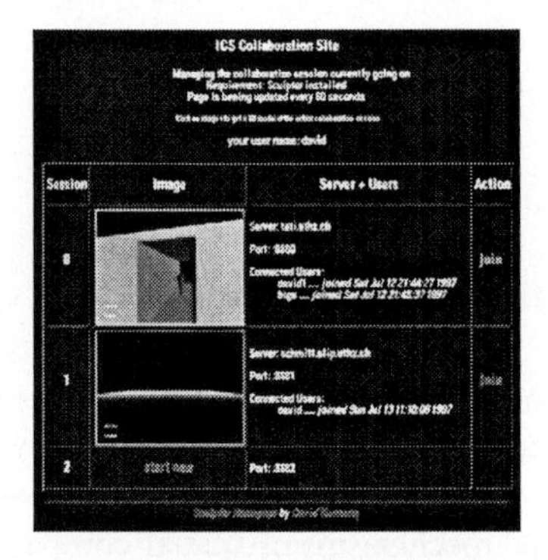

Fig. 5. Left: Individual views of a collaborative Sculptor session, using distributed Sculptor, appear in model and text windows. Right: The interface for designing with Sculptor from the inside out. David Kurmann, 1998.

Sculptor can export models into rendering programs such as Radiance (see figure 6) and into Virtual Reality Modeling Language (VRML) representations, so that they can be accessed over the World Wide Web. There are plans to enhance the CSCW through a Collaboration Agent, that helps to keep the model consistent and solves conflicts when different users want to access and modify the model simultaneously (Lashkari 1994).

Fig. 6. Sculptor models rendered with Radiance in the Phase(X) course at ETH Zürich. Mark Frey (left) and Christoph Loppacher (right), 1996.

7 The Virtual Design Studio (VDS): Multiplying Time

With increasing globalization and specialization in the design and building industry, collaboration between partners in remote locations becomes crucial. Ideally, all of them could work on a building design at any place, simultaneously together (synchronously) or separately (asynchronously), while the latest state of the design would always be available to all team members. They could collaborate on a shared object and no information would thus be lost in transfer of files. Similar programs are available for other, probably simpler applications: shared data bases or shared calendar and address files, through which several people organize their appointments and contacts.

The Multiplying time project comes close to this goal. It allows at the same time the continuous work on a design or a set of designs through different time zones around the world (http://space.arch.ethz.ch/VDS_97/). The task was to design a house for a painter and a writer on an island west of Seattle, USA. Three partners from ETH Zürich (Urs Hirschberg), the University of Hong Kong (Prof. Branko Kolarevic), and the University of Seattle (Prof. Brian Johnson) agreed on the common design project for one week in three different time zones, thus multiplying one week into three working weeks (see figure 7). The interactive program Sculptor was installed in all three locations to enable synchronous and asynchronous design. Data exchange was enabled with a data base directly connected to the Internet, similar to that of Phase(X).

On the morning of the first day, students in Hong Kong started with the design. At the end of their 8 hour working day, they placed the results in the common data base that could be seen by all partners through the browser interface. Students from Zürich began 8 hours later and could thus base their decisions on the results achieved by their Hong Kong partners. After 8 hours, they also placed their designs in the common data base, so that the participants from Seattle were able to explore the designs from Zürich and Hong Kong by the time they started to work. In addition, video conferences took place about every 8 hours, during which students could share and explain their ideas. The setup thus created an intense global think-tank, operating 24 hours a day.

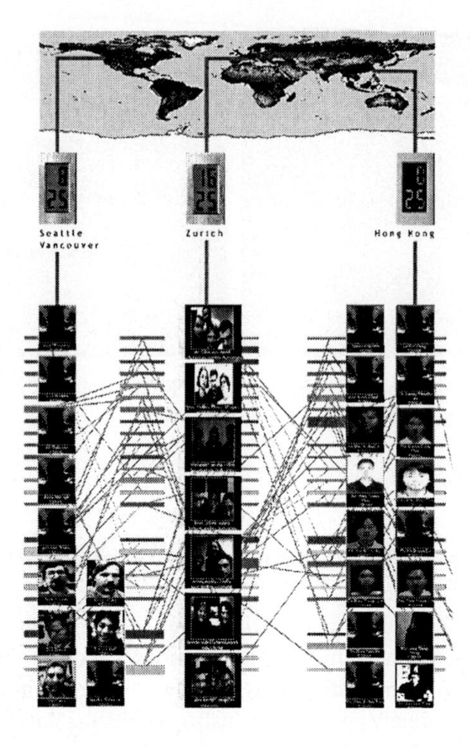

Fig. 7. The Multiplying Time setup and participants. The individual designs of the first phases are shown underneath with the connections between the student projects. Malgorzata Bugajski, 1998.

Every day a new phase was introduced along with a new design issue. In each phase, students could select a design to develop further from any of the three locations. On the last day, a video conference between all three locations took place for the evaluation of the final design proposals. Authors and critics discussed the individual designs and observed the design threads. Students from the three locations noticed that, although they had not known each other before, they found a common language to communicate. The basis for this language were the modeling program and the individual designs.

8 The Delft Experiment

As a follow-up to the Multiplying Time experiment, a group of students at the Technical University of Delft continued where the previous exercise had ended. The Delft students had access to all stages of the previous design phases through the data base. In three additional phases, they developed more refined solutions.

The question was, whether the principle would function with different software (AutoCAD and 3D Studio instead of Sculptor), different hardware (HP PCs instead of

Silicon Graphics workstations), and different operating systems (Windows NT and UNIX instead of UNIX alone). The answer to each of the questions was a surprisingly clear *yes*. The interface programmed in PhP, using Netscape Navigator, proved very stable, as did the data base. The connection between the physical place of the data base at ETH Zürich and the Technical University of Delft was superb (up to 900 kBps), so that all means of communication could be used without the interruptions we had to cope with in the connections to Seattle and to Hong Kong.

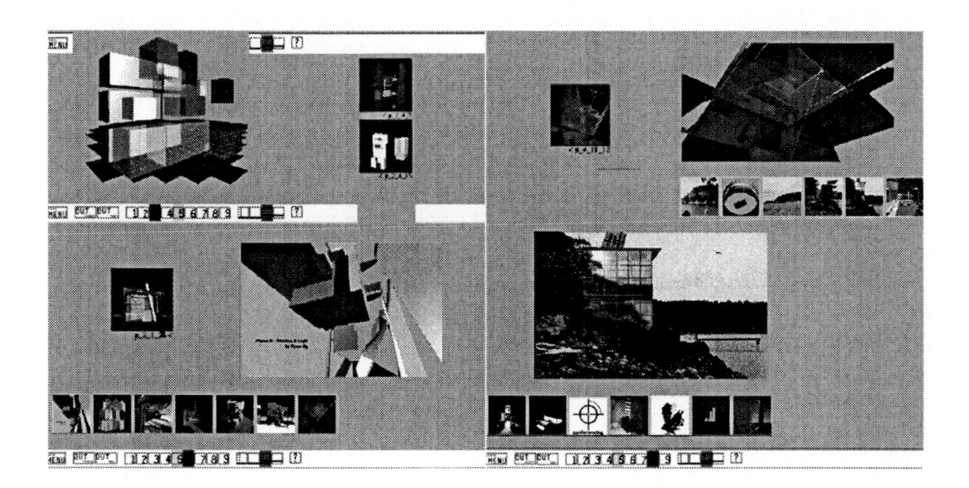

Fig. 8. Snapshots from the Multiplying Time studio. The images show the development from conceptual models (left) to more refined projects (right) that evolved over 5 days in three time zones. Left: Phase 1 Dualities, by S. Margaris, S. Lemmerzah, T. Musy - ETH Zürich; Phase 3 Light & Shadow, by Siu Hong Ryan and Chi Kit Benson Hong Kong. Right: Phase 8 Situation, A. Amin, P. de Ruiter- Delft. Collected by Malgorzata Bugajski, 1998.

The Delft students had to interrupt their work at night and found it unchanged the next morning which made this part of the experiment less dynamic. It also showed that it is preferable to work shorter hours and to fully concentrate, rather than to work long hours and lose efficiency towards the end of the cycle. Finally, the superior interface and short learning time (about 30 minutes) of Sculptor were demonstrated: All those who had no previous experience with AutoCAD and 3D Studio were severely handicapped in the Delft experiment.

The Multiplying Time and the Delft experiments demonstrate that it is possible to work from a common data base, taking advantage of different time zones and special capabilities of particular sites: Seattle provided the site, Hong Kong the first design models, Zürich the modeling program, Delft special rendering techniques. The resulting designs are of shared authorship, but the individual contributions are clearly identifiable, along with the evolvement of the design.

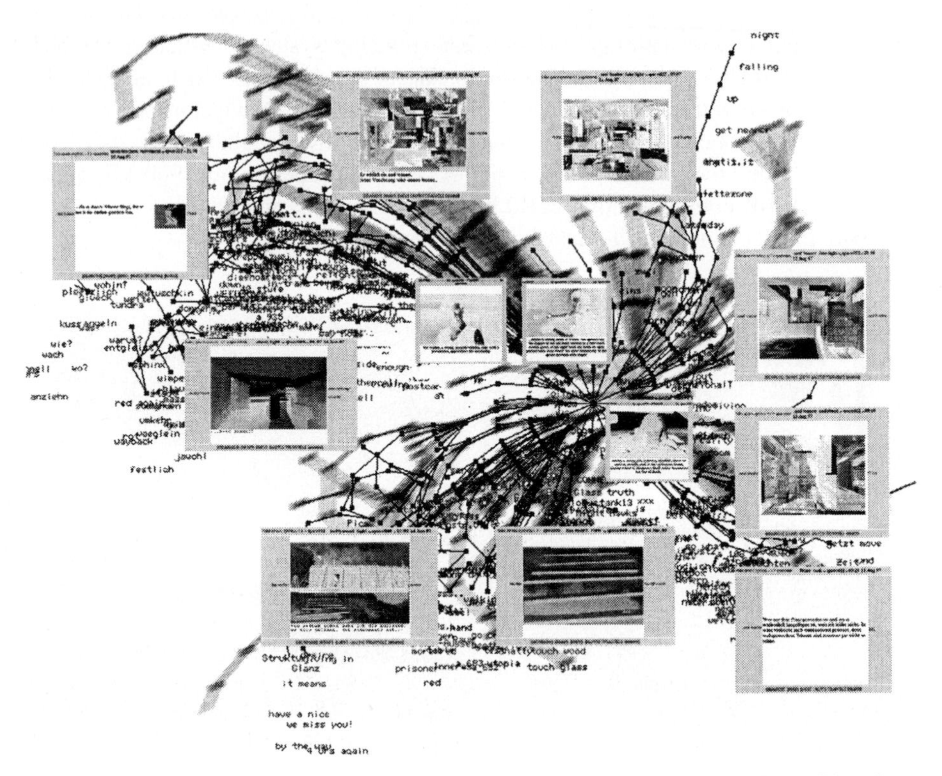

Figure 9. Data base representation of fake.space, the follow-up course to Phase(X), where the design project starts from the center. Branches grow and shrink according to the number of designs attached to each node (http://space.arch.ethz.ch/ss97/).

9 Conclusions

The development of the Phase(X) design environment, resulted in a new collaboration method and in a breakthrough of productivity and quality. The students had learned to communicate and cooperate in a networked environment, design time was reduced (roughly by one week per semester), and the ratings of the individual results were higher.

The experiments show, that by solving the question of authorship in displaying each person's contribution in the data base and in different visualization schemes, there is much less negative competition between designers. In such an environment, designers are able and willing to choose and continue to develop the best solutions, rather than continuing on their own - sometimes weaker - solutions. Through the use of software agents, communication between the individual groups and in the data base can be further improved. Software agents in the form of representatives, presenters, and moderators, have been developed in a related research project (Stouffs et al, 1998, http://caad.arch.ethz.ch/research/IuK) and will be available for further AI applications to engineering and architectural design studios.

We have strong indications from practice that the Phase(X) approach could bring an improvement to project development and to project management as well. The scale-up to commercial use is expected within the next two years.

Acknowledgments

I want to thank Florian Wenz, Cristina Besomi, Fabio Gramazio, Urs Hirschberg, Patrick Sibenaler, Bige Tunçer and Malgorzata Bugajski for their innovative and enthusiastic work in the Phase(X), fake.space and Multiplying Time projects.

References

1. Chiu, M.-L.; 1997: Representations and Communication Channels in Collaborative Architectural Design, in Gero, J. and Sudweeks, F. (eds.), *Formal Aspects of Collaborative CAD*, IFIP, pp. 77-96
2. Dave B., 1995: Towards Distributed Computer-aided Design Environments, in: proceedings *CAAD Futures 1995*, National University of Singapore, Singapore
3. Dawkins R., 1976: *The Selfish Gene*, Oxford University Press, New York
4. Fukuda, T., Nagahama, R. and Sasada, T., 1998: Collaboration Support System for Public Design, in: proceedings *CAADRIA '98,* Morozumi & Iki Laboratory, Kumamoto University, Japan, pp. 299-307
5. Kurmann D, Elte N. and Engeli, M., 1997: Real-Time Modelling with Architectural Space, *in* R. Junge (ed.), *CAAD Futures 1997*, Kluwer Academic Publishers, pp. 809-819
6. Lashkari Y., Metral M., Maes P., 1994: Collaborative Interface Agents, MIT Media Laboratory, Accepted to AAAI '94, Cambridge
7. Lee, E., Woo, S., Shiosaka, Y. and Sasada, T., 1998: Alternative Design Comparative System in Collaborative Design, in: proceedings *CAADRIA '98,* Morozumi & Iki Laboratory, Kumamoto University, Japan, pp. 327-335
8. Lee, S., Mitchell, W. J., Naka, R., Morozumi, M., Yamaguchi, S.: 1998, The Kumamoto-Kyoto-MIT Collaborative Project: A Case Study of the Design Studio of the Future, Proceedings of *Collaborative Buildings 1998*, Darmstadt, February 21-22, 1998
9. Madrazo L.: 1996, Typen & Variationen - Types and Instances, *in* G. Schmitt, *Architektur mit dem Computer*, Vieweg Verlag Wiesbaden, pp. 126-138
10. Schmitt, G.: 1996, *Architektur mit dem Computer*, Vieweg Verlag, Wiesbaden
11. Smith I., 1996: Model-Based Design using Intelligent Objects, in: proceedings *Closing Conference Priority Programme Informatics Research*, Jean-Michel Grossenbacher (Ed.), Swiss National Science Foundation, pp. 69-70
12. Stouffs R., Tunçer B. and Schmitt G.: 1998, Supports for information and communication in a collaborative building project, Proceedings of *Artificial Intelligence in Design '98*, Lisbon, Portugal, July 20-23, forthcoming
13. Wojtowics, J. (ed.), 1995: *Virtual Design Studio*, Hong Kong University Press

Intelligent Structures: A New Direction in Structural Control

K. Shea and I. Smith

Institute of Structural Engineering and Mechanics (ISS-IMAC)
EPFL-Federal Institute of Technology
CH-1015 Lausanne
Switzerland

Abstract. Since structural control of civil structures was first proposed in 1972, research and applications have focused on enhancing safety of structures under extreme conditions. This paper introduces a new direction in structural control. Computational systems and explicitly defined knowledge are used to improve serviceability and maintenance of civil structures. The objectives of such structures, called intelligent structures, are to maintain and improve structural performance by recognizing changes in behaviors and loads, adapting to meet goals, and using past events to improve future performance. To meet these objectives, synergies from research in structural control, computational methods and monitoring technology are exploited. This paper presents a review of current research and technology in structural control to provide a background on which characteristics of intelligent structures are proposed. A computational framework based on intelligent control methodology is then presented that combines reasoning from explicit knowledge, search, and learning to illustrate capabilities of intelligently controlled structures. A system based on this framework is described that uses the specific combination of case-based reasoning and simulated annealing search. An application of the system to tensegrity structures is given. Computational control systems that use this framework are stimulating the design and construction of innovative structures and thus are extending the possibilities for structural engineers.

1. Introduction

Until now, structural control has focused on active control of structures to enhance safety during earthquakes and high winds. While maintaining serviceability was mentioned as a goal of structural control in original concept descriptions [1,2], it has not been a focus point and thus there has been nearly no investigation in this area. Since the majority of work in structural control has been carried out in the US and Japan, where earthquakes are a primary concern, maintaining serviceability is viewed only as a secondary goal for active control research [3]. However, in areas where earthquakes are not a primary concern, active control of structures provides a means of continuously controlling performance of complex structural systems to maintain serviceability in changing environments. The ultimate goal of intelligent structures is to maintain and improve structural performance by recognizing changes in behaviors and loads, adapting the structure to meet goals, and using past events to improve future performance. Controlling performance will ultimately lead to reducing maintenance and longer service-lives.

Intelligent structures are a natural extension of structural monitoring. While monitoring is often limited to determining reserve capacity in existing structures, intelligent structures use this information to adapt. Structures with high sensitivity to loading and where there is much uncertainty about in-service loading are targeted as prime applications. Examples are tensegrity and cable dome structures. Currently, such structures are over designed to compensate for uncertainties. The resulting excess translates to increased material and fabrication cost as well as transportation and reassembly costs in cases of reusable structures. Adapting a structure requires determination of when, how and how much to adapt. In order to create a robust control system for a range of structures and performance goals, artificial intelligence techniques are used in addition to conventional control.

In this paper, a review of the state-of-art of related fields is first presented as a basis to evaluate a new approach to active structural control. Target applications are then given to motivate characteristics of intelligent structures. Next, a framework for computational control of complex structural systems is presented and applied to controlling a unique type of tensegrity structure. The generic nature of the framework makes it applicable to a wide range of structures.

2. Background

Intelligent structures research builds on work already carried out in the areas of: (1) structural engineering, (2) structural control, (3) intelligent control, and (4) applications of artificial intelligence methods to monitoring, diagnosis and maintenance (Figure 1). Active control of civil structures was first introduced by Yao [2] as a means of extending the limits on heights of tall buildings by actively controlling structures during high winds. Since the original concept, structural control has been used in civil structures to counteract extreme environmental conditions such as earthquakes and winds as well as in space structures to control undesirable vibrations and deformations.

The general structural control problem is to create a system capable of applying forces to a structure under excitation such that the structural response is improved for safety and serviceability purposes. Structural control systems developed for this problem can be categorized as passive, active or hybrid. Passive control systems use an actuator, for instance a tuned mass damper, that directly reacts to an excitation resulting in the exertion of a force on the structure. Active control systems use sensors and actuators to activate the application of a force on a structure [4]. Hybrid systems use a combination of passive and active control, for example a tuned-mass damper combined with an actuator. To date, structural control systems have been installed in 25 buildings and towers where the most common system used is a hybrid mass damper [5]. An application of active control to bridges under dynamic loading is shown in [6].

Extensions of active control systems include adaptive control systems where feedback is used to adjust controller parameters. Difficulties with these systems are instabilities due to disturbances and unmodeled dynamics [4, p.917]. In structural engineering, intelligent control is targeted for use in situations where conventional controller design is difficult and where success is limited, such as when there is qualitative, uncertain, and incomplete information. Currently, applications of intelligent control systems to structures are limited to controllers that use neural networks [7,8,9] and fuzzy techniques [10,11]. For example, an artificial neural network has been used to control variation in the tension of a tendon attached to a bridge to reduce excessive deformations under dynamic loading [12]. In this paper, these applications do not meet the definition of intelligent structures that will be proposed in Section 3 since they are limited to using statistical learning techniques to update control models.

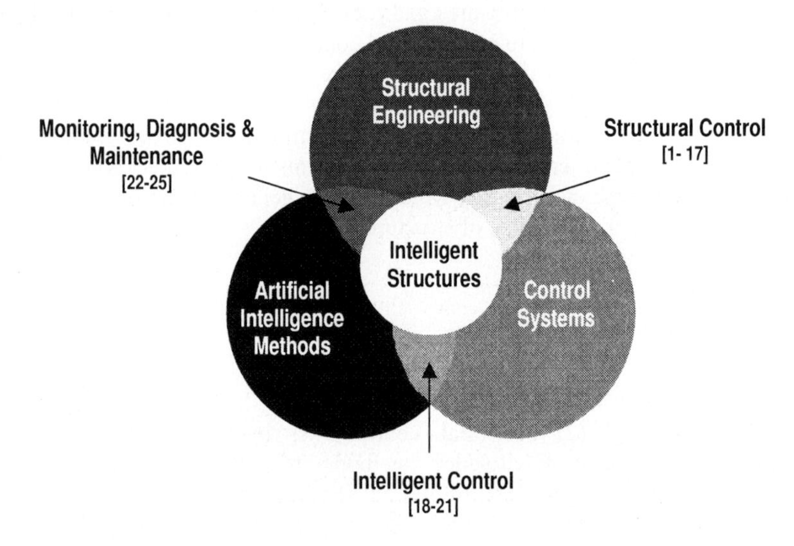

Fig. 1. Examples of related work on the interfaces between control systems, artificial intelligence and structural engineering research

To date, research into active control of civil structures does not include continuous adaptive control of geometry. However, for space applications, adaptive structural geometry has led to the development of deployable truss structures [13] where adaptive geometry is integral to the design purpose. Deployable structures are commonly very flexible and require active shape control to counteract unwanted deflections [14,15] and to enable autonomous deployment [16]. Adaptive trusses with pneumatic struts have also been used to control vibrations of structures in space [8]. [17] suggests applying work done in adaptive space structures to civil structures but no such structures are known to exist. A drawback of current methods for shape control is a strong integration of structural geometry in the control technique such that the resulting method is applicable only to a specific structure.

While structural control has focused on the design of individual controllers, intelligent control addresses global control of complex systems in uncertain environments. An autonomous intelligent controller can be broken down into three levels (Figure 2): the execution level (traditional control), the coordination level (coordinating multiple controllers), and the management level (planning control strategies) [18]. Most current structural control applications are limited to the execution level. Learning can occur at any level in the hierarchy, for example neural network controllers mentioned previously learn at the execution level. While conventional control methods are used for the execution level of an intelligent controller, computational methods are used in the coordination and management levels [18]. Applications of this hierarchy can be found in [19,20]. One method presented for the coordination of two controllers uses a combination of a knowledge-base and a neural network predicator to modify the input to the local controllers [21]. Intelligent control methodology, as defined in [18], has not been applied to the control of structures.

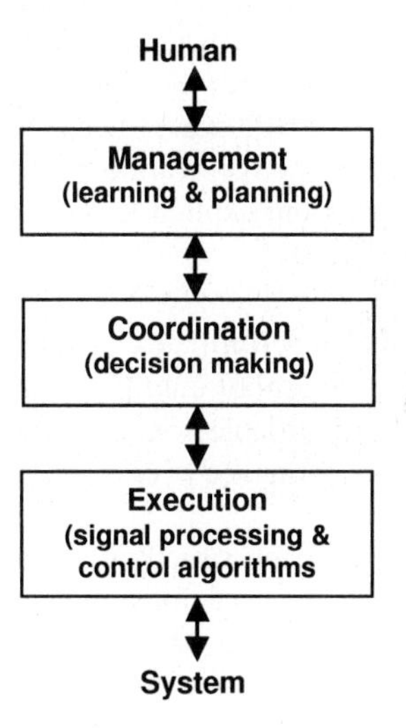

Fig. 2. Intelligent Autonomous Controller [18]

Artificial intelligence techniques used in structural engineering include applications to monitoring, diagnosis and maintenance tasks. Example applications are safety management [22], explanation of structural behavior [23], and monitoring and diagnosis [24,25]. For civil structures, artificial intelligence has provided advances in structural monitoring by enabling more meaningful data interpretation and now will be used to extend structural control capabilities.

3. Intelligent Structures

The term "intelligent structures" has been used to describe a wide range of structural engineering technologies ranging from structures with embedded active materials to structures with artificial neural network controllers. This paper introduces a new definition of intelligent structures providing a new direction in structural control that focuses on long-term performance-based control of civil structures. Target applications for intelligent control are:

- structures with high sensitivity to behavior and unknown service conditions (e.g. lightweight structures),
- movable structures (e.g. draw bridges and stadiums with roofs that open),
- structures subject to fatigue loading (e.g. bridges),
- structures with changing environmental conditions (e.g. climatic conditions),
- structures where maintenance costs are greater than initial cost, that is life cycle performance is very important (e.g. bridges), and
- structures where aptitude in service is difficult to evaluate.

To meet the challenges of controlling structures in such cases, an intelligent structure has two attributes: (1) active features, and (2) a computational control system that combines reasoning from explicit knowledge with learning and planning. Functions of intelligent structures are:

- adaptation of structural geometry to improve performance,
- control for multiple, modifiable performance goals,
- autonomous and continuous control of multiple, coupled structural subsystems,
- reasoning from explicit and modifiable domain knowledge, and
- improvement of structural performance over time (learning and planning).

Comparing current technology for structural control with intelligent structures, Figure 3 illustrates the capabilities of intelligent structures over current active structures. These capabilities transform structures from static objects to interactive machines. Intelligent structures interact with human controllers as well as the environment by autonomously adapting. The interaction between human controllers, computer controllers and structures is shown in Figure 4.

Adapting structural geometry, or other active features, usually involves the control of coupled structural subsystems. Current structural control systems do not consider coupling among controllers since they are generally limited to controlling a single actuator or multiple independent and individually controlled actuators. For example, in the vibration control of a cable-stayed bridge, each control cable has a separate controller such that the tension is adjusted individually but relative to the structure. The methods used require the interaction between the cable and the structure to be weak. Conversely, in intelligent structures, strong interactions are intentionally designed between active features and behavioral properties.

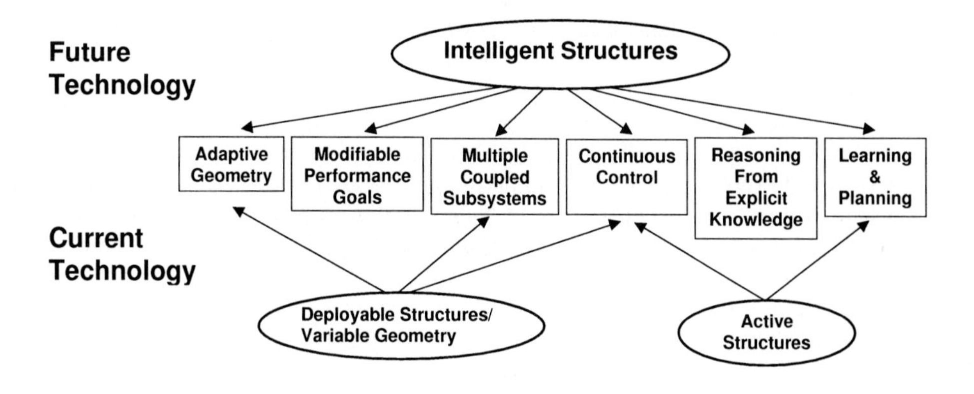

Fig. 3. Characteristics of and computational methods for intelligent structures compared with current technology applied to complex structures

Most often, active structures are limited to control of a single performance objective, for example vibrations caused by earthquakes, wind, or space conditions. Taking a broader view of structural performance goals to consider goals such as maintaining stress and deformation ranges results in improvements in the health of structures and consequently, extended service life. To create interactive structures these goals must be modifiable by either a human controller or through autonomous learning and planning. Such modifications cause the structure to react differently to future environment changes and thus improve performance.

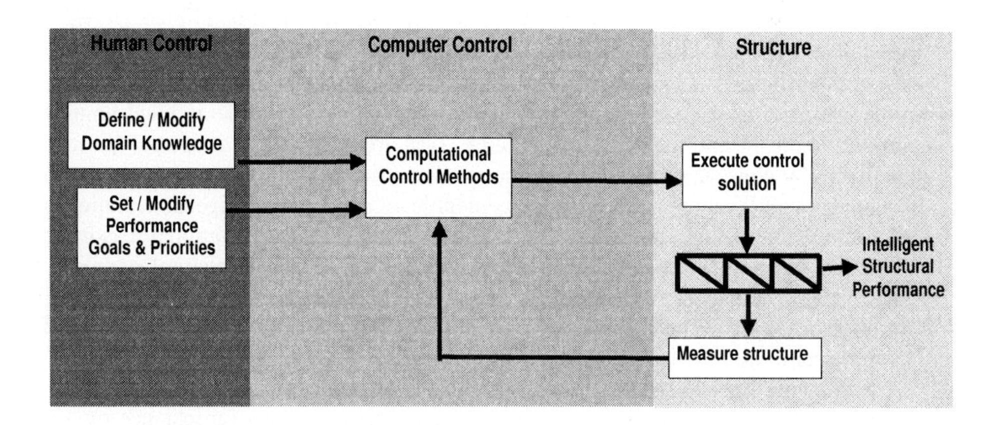

Fig. 4. Interaction between human controller, computer controller and structure for active intelligent control

A key attribute of intelligent structures is the existence of explicit modifiable knowledge used for reasoning about control actions. Explicit knowledge includes geometric and behavior models, performance goals, as well as control rules and cases.

Separation between domain specific knowledge and control methods is beneficial for creating a system that is effective for a number of performance goals and a wide range of applications.

The ultimate goal of intelligent structures is to improve structural performance throughout service life, thus requiring learning and planning. Learning involves the use of past events to improve future performance and can occur by either generalizing past events or recognizing past events and extrapolating previous solutions to produce improved control actions. Planning, for intelligent structures, involves finding control actions to satisfy goals that are based on the prediction of future events [26]. Learning is used to improve behavior models and control solutions by adapting known solutions while planning is used to set priorities among performance goals based on the prediction of future events.

Characteristics such as an adaptive geometry, a broader range of modifiable performance goals, explicit knowledge representation and control of coupled subsystems enhance current capabilities of structural control and lead to innovative active structures. Additionally, reasoning, learning and planning in a complex environment to maintain and improve structural performance extends structural life. To achieve these objectives, new methods for structural control must be developed. Since the objectives of intelligent structures are different from current control objectives it is unclear whether control techniques can be extended to intelligent structures. For this reason a computational approach to active control of civil structures is introduced.

4. Computational Control for Intelligent Structures

Although methods exist for several individual objectives, their interconnection has not been sufficiently investigated for the construction of successful intelligent structures. Due to the complexities of the control objectives, a hierarchical control approach found in intelligent control (Figure 2) is applied here to structural control. This hierarchy consists of three levels: the execution level, the coordination level and the management level [18]. At the management level artificial intelligence techniques for learning and planning are used. At the coordination level computational reasoning and search techniques are employed to direct the adaptation of structural geometry with coupled behavioral affects. At the execution level, existing conventional controllers are used. The actions of local controllers modify the shape, behavior and thus performance of the structure.

A framework for computational control of intelligent structures is shown in Figure 5. The input to the system consists of a set of performance goals, for instance ranges of acceptable stresses, and explicit domain specific knowledge, which includes geometric and behavioral structural models. The major tasks of the system are to: (1) compare the current structural performance to the desired performance, (2) reason about the goals not satisfied (diagnosis) and find a set of control actions that improve the performance (search), (3) execute the set of actions proposed in step 2, (4) measure the behavior of the new structural state, and (5) use the current event as an instance for learning and planning. This process iterates continuously over the life of the structure. Since the aim of intelligent structures is to adapt over periods of days and weeks rather than seconds, computational control is a feasible alternative. The desired output of the system is intelligent structural performance. Many combinations of reasoning, search and learning methods can be used within this framework.

The coordination level of control consists of comparing, reasoning and search tasks that provide flexibility in the structural features used for control and the performance goals considered. The first task of coordination is to determine if control must be applied thus entailing a comparison between the current performance of the structure and the defined set of performance goals. From this comparison, a set of unachieved goals can be identified. Comparing the current state to the desired performance also provides a means of assessing the success of previously applied control solutions, which can be used in learning and planning modules described later.

Controlling coupled features, such as adaptive structural geometry requires reasoning and search techniques. Techniques can be combined to simulate the effects of simultaneously executed control actions and find a set of control actions that best improves system performance. An example of control actions is position movements of controllers located on a structure used to change the behavior and thus performance of the structural system. Formulated as an optimization problem, the variables are the possible control actions, for instance ranges of feasible movements, and the objective function is determined from the performance goals and their priorities. An optimization method is then used to determine an optimally directed set of control actions, relative to the performance goals. This set of control actions is used as input for controllers located on the structure such that the configuration and thus behavior of the structure is modified. The behavior of the new configuration is evaluated again and compared with performance goals to determine the success of the control solution applied.

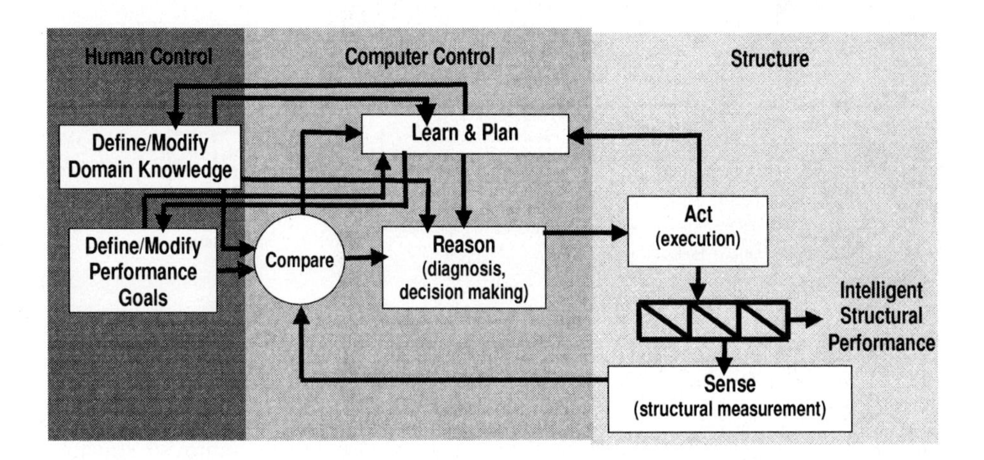

Fig. 5. Framework for intelligent computational control of complex structures

While the system described up until this point senses and reacts to changing environments, the larger goal of intelligent structures is to improve the applied control solutions based on knowledge of past experiences. Using specific methods for learning produces the computational control system shown in Figure 6. This system combines case-based reasoning and simulated annealing for solution generation with artificial neural networks and case-updating as means for learning. At the management level, three aspects of learning and planning are used: (1) improving modeled domain knowledge based on performance, (2) modifying priorities among goals to efficiently maintain performance under changing conditions, and (3) learning control solutions for specific events. An illustration of incorporated learning aspects is now described.

The first aspect of learning involves improving behavior models since the approximation of structural behavior is an integral part of simulation and evaluation of control actions within the optimization process. Multiple behavior models exist in the domain knowledge corresponding to different environmental conditions and structural states. Also, due to different performance criteria, different behavior models are used to achieve desired computation speed versus accuracy compromises. Determination of the most appropriate model to employ is made in the reasoning module. To account for large discrepancies between theory and real structural behavior, an artificial neural network is used to modify parameters in theoretical models in order to improve accuracy.

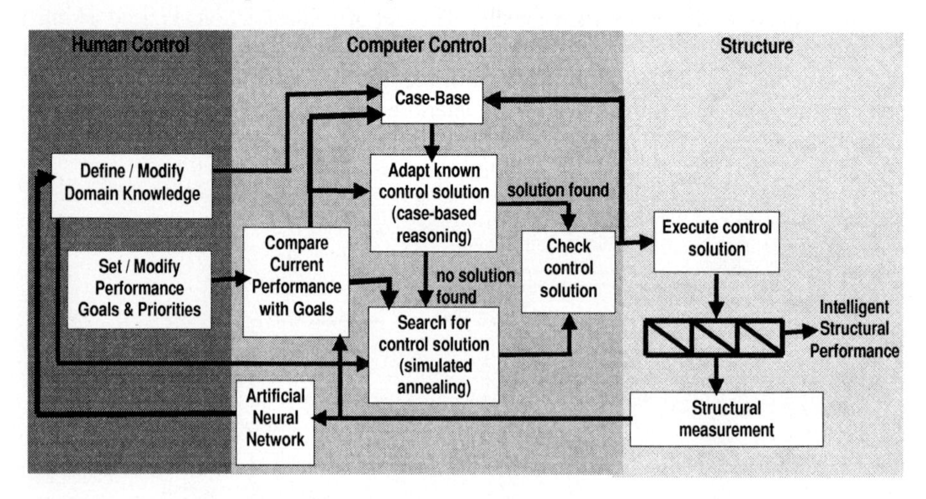

Fig. 6. A System for Intelligent Computational Control

The second aspect of learning included is automated modification of priorities among performance goals according to past events leading to improved structural performance. In observing the performance of a structure, a human controller may modify the set of performance goals by removing goals, introducing new goals and modifying priorities among goals. Since the tradeoffs among goals are difficult to predict and since they directly influence the control solution determined by the optimization, learning is used to adjust the priorities among goals automatically based on past performance. For instance, on detecting increasing wind loads, a goal to decrease the stress in parts of the structure to compensate for winds would be placed at a higher priority than a goal to use as little control motion as possible.

The final aspect of learning involves the use of a case-base to classify past events for reuse in similar current events. While employing a global optimization technique, such as simulated annealing, enables optimally directed control, intensive computational effort is required. Therefore it is advantageous to store past events and their successful control solutions such that solutions can be retrieved and adapted to new situations. For example, given an applied load, position of the controllers on the structure, and set of performance goals and priorities, a control solution that has been successful in the past can be applied again in the same situation. While human controllers can enter cases based on past experience, the case-base can also be automatically updated over the service-life of the structure.

5. Application: Intelligent Tensegrity Structures

Tensegrity (tensile-integrity) structures are lightweight, reusable structures and therefore offer attractive solutions to a growing number of structural requirements. Since tensegrity structures are sensitive to small changes in external loading, active control is necessary to make them suitable for practical use. The system presented (Figure 6) will be applied to control tensegrity roof structures being designed by Passer and Pedretti, SA for exhibition at the Expo.01 in Switzerland in 2001. The design consists of coupled active features that require continuous control in order to maintain and improve system performance. A roof system comprised of tensegrity (tensile-integrity) modules each of which consists of a self-stressing system of compression struts and tensile wires where the stress distribution in the module is controlled by a central pneumatic joint is shown in Figure 7. The roof structure is then formed from an intricate connection of modules that is then covered with a roof fabric. The behavior of the entire structure is coupled since the movement of one central joint affects the stress distribution in that module as well as the adjacent modules. The control problem is then to determine the positions of the hydraulic joints that best distribute the stress in the structural system in a changing environment. The primary change in loading on the structure is anticipated to be due to slow changes in the behavior of the covering fabric. The initial performance goals are to maintain the desired stress distribution in the structure such that the covering fabric remains in tension using the least amount of control motion possible as well as keeping the system as robust as possible to react to new changes in loading. As can be seen, this problem is highly coupled, the structural behavior is non-linear and somewhat unknown, and the structure will be placed in a dynamic environment. System development is underway.

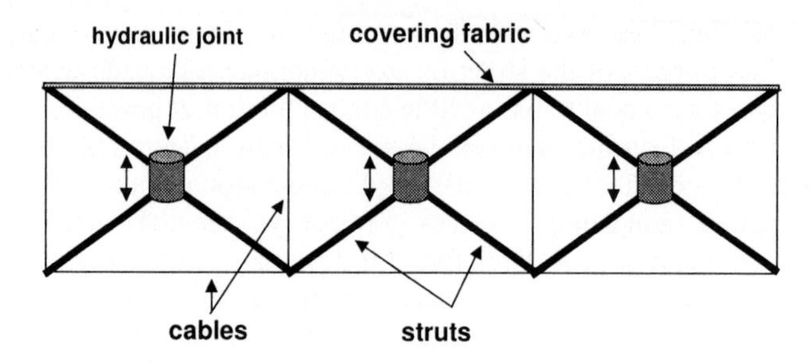

Fig. 7. Controlling stress distribution in tensegrity modules with central hydraulic joints. The actual modules are three-dimensional.

6. Conclusions

Structures are often thought of as large static objects that are over-designed to withstand extreme environmental conditions. With the incorporation of active control, a structure becomes a dynamic object capable of interacting with a complex environment. As recent climatic events have shown (for example, pylon collapses in Canada), the environments where civil engineering structures exist are increasingly difficult to predict. Therefore, opportunities for taking advantage of structural control systems using computational structural control are increasing. The creation of structural control systems based on the computational framework presented, which uses knowledge representation and learning, will enable the design and construction of intelligent structures. Just as structural control was introduced to extend the limits of high-rise buildings, intelligent computational control will create new possibilities for innovative, active structures.

Acknowledgements

The authors would like to thank the Swiss National Science Foundation (FNS), the Commission for Technology and Innovation (CTI), Logitech SA, and Silicon Graphics Incorporated for supporting this work. We would like to thank Passera & Pedretti SA, Lugano for their collaboration as well as L. Pflug, D. Inaudi, R. Longchamp, D. Gillet, B. Raphael and R. Stalker for their discussions of this work.

References

1. Zuk, W. (1968), "Kinetic Architecture", *Civil Engineering*, ASCE, December, pp 62-62.
2. Yao, J.T.P. (1972), "Concept of Structural Control", ASCE Journal of the Structural Division, 98:1567-1574.

3. Housner, G.W., T.T. Soong, and S.F.Masri (1994), "Second Generation of Active Structural Control in Civil Engineering", *First World Conference on Structural Control*, 3-5 August, Los Angeles, California, USA, vol 1, pp.Panel-3 – Panel-18.
4. Housner, G.W., et al. (1997), "Structural Control: Past, Present, and Future", *Journal of Engineering Mechanics*, 123(9).
5. Spencer, B.F., Jr. and M.K. Sain (1997), "Controlling Buildings: A New Frontier in Feedback", *IEEE Control Systems*, December, pp. 19-34.
6. Hirsch, G.H. (1991), "Active Control of Bridges", *Intelligent Structures-monitoring and Control*, Elsevier, pp.308-319.
7. Rehak and Garrett (1991), "Neural Computing for Intelligent Structural Systems", *Intelligent Structures Monitoring and Control*, Ed. Y.K. Wen, Elsevier, pp. 147-161.
8. Yen, G.G. (1994), "Reconfigurable Learning Control in Large Space Structures", *IEEE Transactions on Control Systems Technology*, 2(4):362-370.
9. Ghaboussi, J. and A. Joghataie (1995), "Active Control of Structures Using Neural Networks", *Journal of Engineering Mechanics*, ASCE, 121(4):555-567.
10. Faravelli, L. and T. Yao (1996), "Use of Adaptive Network Fuzzy Control of Civil Structures", *Microcomputers in Civil Engineering*, 11:67-76.
11. Casciati, F., Faravelli, L. and T. Yao (1996), "Control of nonlinear structures Using the Fuzzy Control Approach", *Nonlinear Dynamics*, 11:171-187.
12. Zagar, Z. and D. Delic (1993), "Intelligent Computer Integrated Structures: A New Generation of Structures",, M.R. Beheshti and K. Zreik, eds., Elsevier Science B. V., North-Holland, pp.371-378.
13. Kawaguchi, M. (1989), "Space Structures with Changing Geometries", *IASS-Congress*, September, pp 33-45.
14. Okubo, H. and N. Komatsu (1996), "Tendon Control System for Active Shape Control of Flexible Space Structures", *Journal of Intelligent Material Systems and Structures*, 7: 470-475.
15. Tabata, M. and M.C. Natori, (1996), "Active Shape Control of a Deployable Space Antenna Reflector", *Journal of Intelligent Material Systems and Structures*, 7:235-240.
16. Huang, S., Natori, M.C., and K. Miura (1994), "Motion Control of Free-Floating Variable Geometry Truss Part I: Kinematics", *Journal of Guidance, Control and Dynamics*, 19(4):756-763.
17. Wada, B.K. and Das, S. (1991), "Application of Adaptive Structures Concepts to Civil Structures", *Intelligent Structures-monitoring and Control*, Elsevier, pp.195-217.
18. Passino, K.M. (1996), "Toward Bridging the Perceived Gap Between Conventional and Intelligent Control", Chpt. 1, in *Intelligent Control Systems,* M.M. Gupta and N.K. Sinha, eds., IEEE Press, pp. 3-27.
19. Gupta, M.M. and N.K. Sinha (1996), *Intelligent Control Systems*, IEEE Press.
20. Antsaklis, P.J. and K.M. Passino (1993), *An Introduction to Intelligent and Autonomous Control*, Kluwer Academic Publishers.
21. Cui, X., and K.G. Shin (1996), "Intelligent Coordination of Multiple Systems with Neural Networks", Chpt. 9, in *Intelligent Control Systems,* M.M. Gupta and N.K. Sinha, eds., IEEE Press, pp. 206-233.
22. Comerford, J.B. et al. (1993), "Causal Models and Knowledge Integration in System Monitoring", IABSE, pp.331-338.
23. Salvaneschi, P., Cadei, M., and M. Lazzari (1997), "A Causal Modeling Framework for the Simulation and Explanation of the Behavior of Structures", Artificial Intelligence in Engineering, 11: 205-216.

24. Goodier, A. and S. Matthews (1996), "Knowledge Based Systems Applied to Real-Time Structural Monitoring", *Information Processing in Civil and Structural Engineering Design*, pp 263-270.
25. Lazzari, M., Salvaneschi, M. and M. Cadei (1996), "Applying AI to Structural Safety Monitoring and Evaluation", *AI in Civil and Structural Engineering, IEEE Expert, Intelligent Systems and their Applications*, August, pp. 24-34.
26. Dean, T.L. and M.P. Wellman (1991), *Planning and Control*, Morgan Kaufman.

Integration of Expert Systems in a Structural Design Office

Aldo Cauvin [1], Rinaldo Passera [2], Giuseppe Stagnitto [3]

[1] Professor of Structural Engineering University of Pavia
Pavia,Italy

[2] Structural Engineer Passera & Pedretti
Lugano,Switzerland

[3] Structural Engineer
Pavia,Italy

Abstract. The purpose of this paper is to address the following questions:

- Can expert systems be integrated into design procedures of a structural design office at reasonable cost, in a short period of time and in a way that can be accepted by the practising engineer?
- How usefulness can be the adoption of suitable expert systems?
- What procedures are most suited to reach these goals?

For this purpose, the experience accumulated by the authors in recent years is exposed from the point of view of the structural engineer rather than from the one of the AI scientist. The key to the success seems to lie in the following points:

- An adequate data base must be collected concerning existing structures according to a suitable format .
- Knowledge should be extracted from this data base using as far as possible an automatic self instructing procedure based on neural networks
- The inference procedure should as far as possible imitate the way of reasoning of the structural engineer.
- The system should be integrated in a CAD environment.

1 Introduction

The applications of Artificial Intelligence to structural design are now extensively studied in the Academic environment [1,2,3,4]. It is now recognised that they can help in all the phases of the design process where non-algorithmic and mainly qualitative decisions need be made. They are therefore particularly useful in the preliminary phases of design when shapes and materials need be chosen. However the practical applications of these procedures have been quite limited in structural design offices. In fact, many of the procedures which have been proposed look and sometimes are quite complicated and, most of all, require a change in mentality which cannot easily be achieved in the professional world. For these reasons, a collaboration program has been organized between the Department of Structural Mechanics of the University of Pavia and the structural design firm of Passera and Pedretti. The basic aim was to identify the most suitable structure for a design expert system which could be easily implemented in the office in a reasonable time, at reasonable cost and without interfering with the normal activity of the office.

The first applications used expert system shells and procedures previously adopted for different purposes in different fields (medical, geological), mainly diagnostic. These shells utilize deductive inference engines which are not particularly suited for fuzzy logic processes like design, which must lead to non univocal, often ambiguous answers. It became therefore apparent, from the beginning of this research, that no "commercial" expert system shell was suited for this purpose and special purpose code had to be developed. Better results were obtained using "model oriented" abductive[1] processes[8], in which knowledge is represented in form of "models", that is rules where the premise always contains the cause of the event and the consequence is represented by the conclusion. Using this approach an abductive process needs not be transformed in an equivalent deductive one, introducing the so called "certainty factors". As a result the knowledge base can be updated and modified with ease without any need to update also at every change the certainty factors. In the case of design the "causes" are the design choices and the "consequences" are the performances (expected behaviours) that we can expect from the choice which was made. However it was soon discovered that building the knowledge base extracting it from human experts was a slow, difficult and time consuming task.

2 Basic Requirements of a Simple and Effective Expert System Devoted to Design

It would therefore be extremely useful to identify a procedure which permits building of the knowledge base, deriving it for the "rough" data, which to some extent, can be found in the technical literature and in the available data bases concerning existing structures. The solution can to some extent be found using an extremely simple "neural network" As shown in fig.1, a knowledge base composed by models can be represented by a net of units and connections which can be interpreted as a "neural network" characterized by two layers:

- The units of the first layer represent all the possible design choices
- The units of the second layer represent all the possible performances (expected behaviours) corresponding to all the possible design choices.

Each unit in the first layer is connected with each unit in the second layer (although only some connections are active),while units in the same layer are not connected among them. This is a "classical" neural net called "Perceptron", widely used in pattern recognition works [1]. Here however its use is completely different. The connections between the two layers can be represented by a boolean matrix where 1 represents a connection between a design choice and a performance (and therefore the existence of a rule or "model" which expresses the fact that that performance is one of the consequences of that choice), while 0 denotes no connection (see again fig. 1 where the boolean matrix corresponding to the net of fig.1 is represented.) That

---

[1]A logical process is "abductive" when the conclusion and the rule are known and the premise need be derived

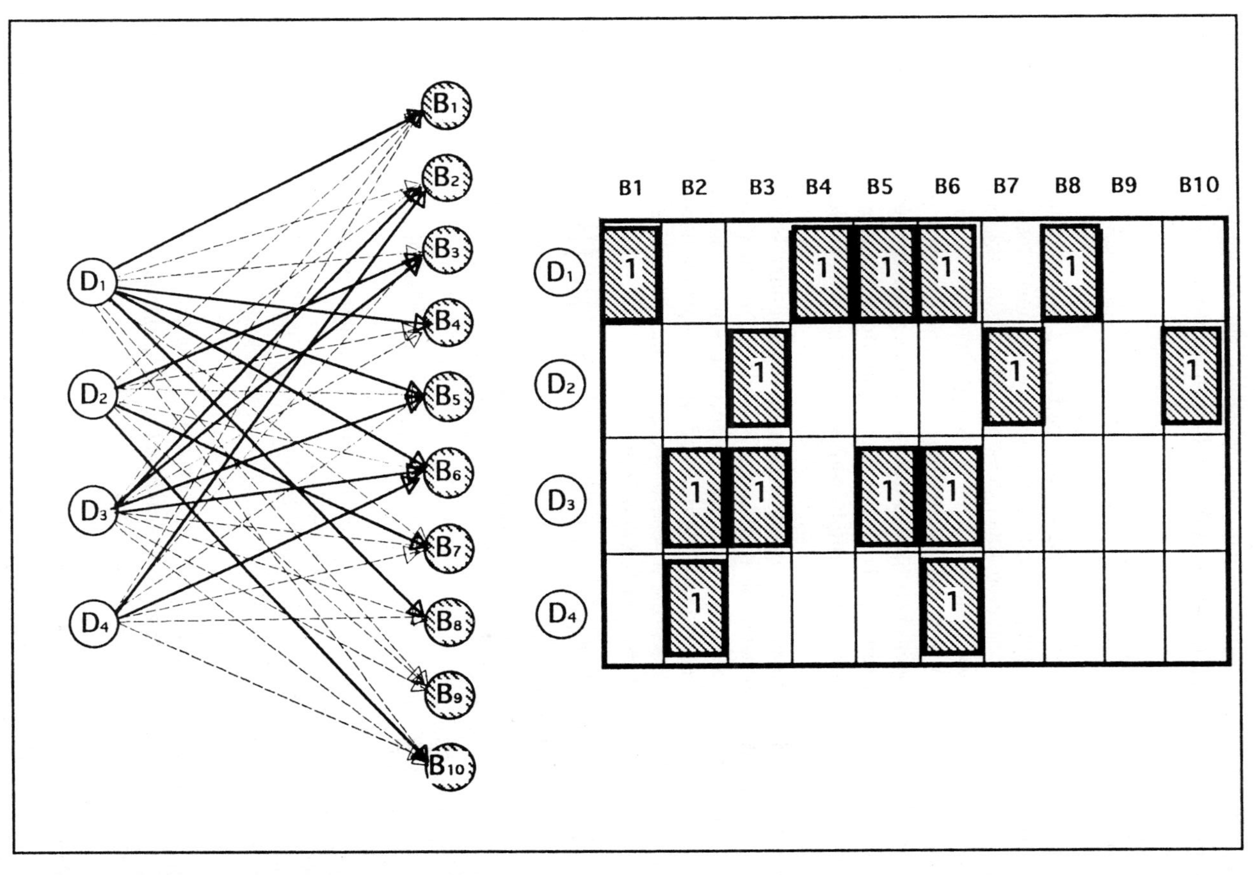

Fig.1-Boolean matrix of the Neural Net

matrix therefore synthesises and univocally defines the knowledge base. Now the question is:how can this matrix be built and,if it is the case, modified?

3 "Instruction "of the Net

The more rational way to build the knowledge base is by careful examination of an adequate number of existing structures. We already mentioned the necessity to build a database containing information about these structures [5,6]. By careful examination of these information the boundary conditions and therefore the required behaviours according to which the structure was designed should be detected. In other words **we use in this case a procedure which is inverse to that of design: instead of making a design choice according to given boundary conditions, given a choice we detect the boundary conditions according to which the choice was made.**

Once for an adequate number of structures the connections between choices and behaviours have been detected, it becomes possible to "instruct" the net so that the knowledge base can be gradually accumulated in it. For this instruction procedure, once the structural types and possible required behaviours have been individuated and classified, it is not necessary to use a specialist, the procedure being essentially mechanical. We could proceed in this way (see fig.2):

- At the beginning of the procedure the boolean matrix $[\beta]$ is empty (every β_{ij}=0). Its dimensions are given by the number of possible design choices(number of rows) and by the maximum assumed number of expected behaviours.
- Sheets are prepared for every example containing a given structural choice and the connected behaviours.
- Counters are associated to every design choice (first layer) and to each connection.
- Every time a sheet is considered the counters of the corresponding choice i (N_i) and connections (M_{ij}) are incremented by one.
- The ratio $R_{ij}=N_i/M_{ij}$ is computed for every connection.
- Whenever R_{ij} becomes greater than a given threshold T(let's say 0.8 for example) β_{ij} is set=1 (connection on)
- Whenever $R_{ij}<T$ b_{ij} is kept (or set again) =0 (connection off)
- The "instruction" of the net is terminated when an adequate number of examples (more than 100) has been processed in this way.

The meaning of the threshold is obvious: when an adequate number of examples has been considered and more than 80% of a given structural type have been found associated with a given behaviour, it is reasonable to assume that the same behaviour has contributed in an essential way to determine that choice. It is therefore reasonable, as required behaviours must match performances (expected behaviours) to adopt a rule connecting choice and behaviour.

The choice of the threshold must be made carefully, according to the distribution of available data. It is not therefore an "a priori" choice. We could say that the threshold

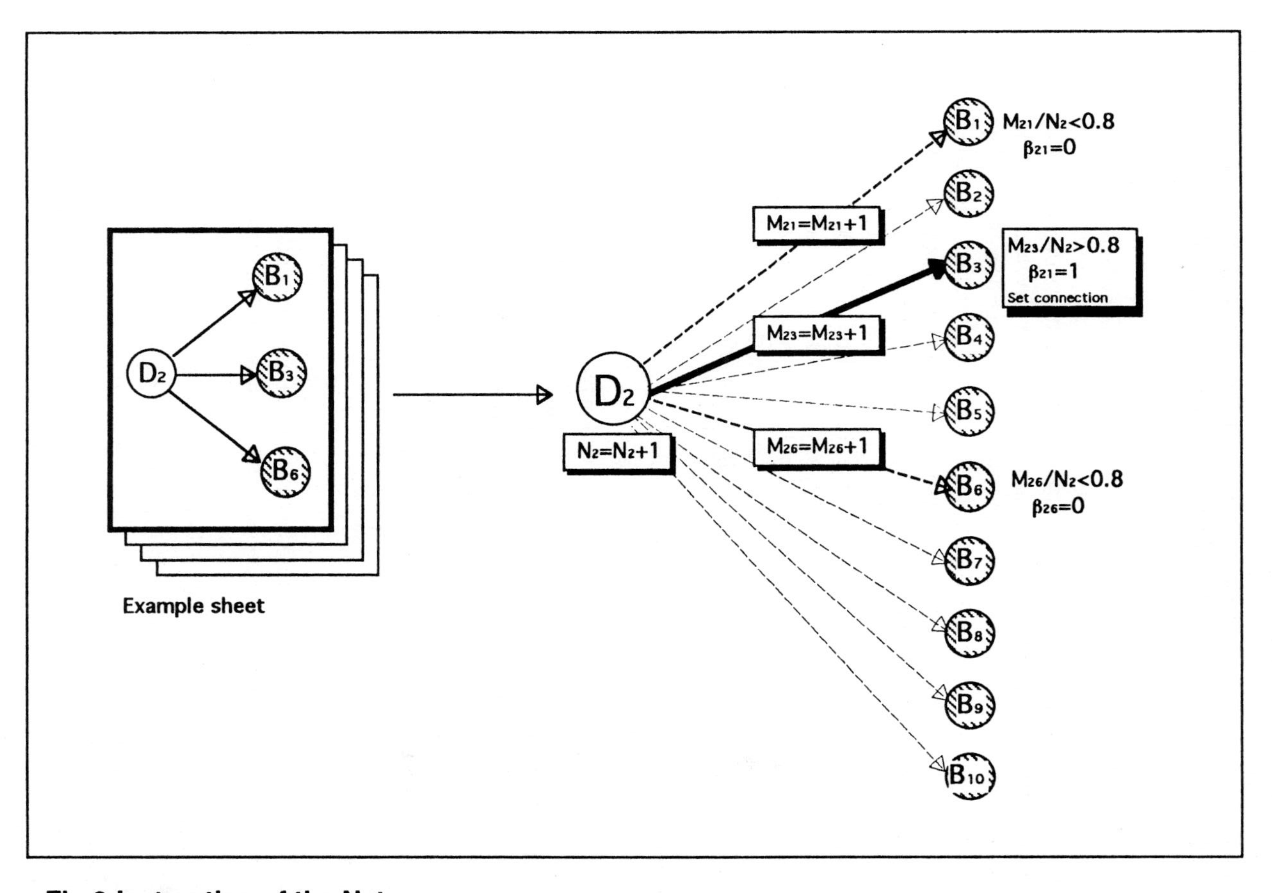

Fig.2-Instruction of the Net

defines the degree of importance of the rule. The association found for a single example could be the result of misjudgements by the designer; it is however unlikely that 80% of designers who have chosen a given type have made the same mistake. The kind of knowledge which can be acquired in this way is therefore based on past experience, on the assumption that most (but not all) considered designers were competent. It is not in this way that innovative designs can be elaborated; however in structural design past experience has always be, also for safety reasons, the main source of information. In the past centuries, before the development of methods of structural analysis this approach was the only one that was possible to adopt. Even in the cases where innovative solutions are required the solutions suggested by these procedures could be useful for reference: the "degree of innovation" could be evaluated, with the connected risks.

Summing it up, the expert system is not the "boss" of the designer. Its only purpose is to transmit to him in the most rational way the past experience. In addition, this procedure, although mechanical, is very similar in nature to the empirical knowledge acquired by the structural engineer with a difference: the number of data that is elaborated is much greater and the knowledge is not affected by subjective opinions.

4 Use of the "Instructed" Net

At this point, the "instructed net" could be used in two ways:

- Directly, as a decision making neural network.

In this case the layer containing performances is used as input layer for the required behaviours. The matrix [b] becomes the matrix of weights. The units of the output layer process the signals coming from the connections as sums of the weights of the connections with units representing the required behaviours(fig.3) .The threshold value is represented in this case by the number of required performances. When the threshold is reached in a given unit, it "fires" that is the design choice corresponding to that unit is made.

- Indirectly, that is the net is used as a source of models or rules in explicit form to be processed in a inference engine as the one described in references, [6],[10] and [12].

As an example, in the simplified net of fig.4 the way the choice of a bridge shape can be performed is illustrated

5 Individuation of the Design Levels

When a designer performs his task he is firstly concerned with the most general "strategic" choices and then proceeds gradually to more specific choices until even the smallest details have been defined. This simple procedure is not without drawbacks. However in our view, if an expert system is going to be accepted by the practicing engineer it must basically proceed in a similar way.

In references [6,10,12] a procedure of this kind has been described in the case of bridges and tall buildings and called "top down refinement with constraint

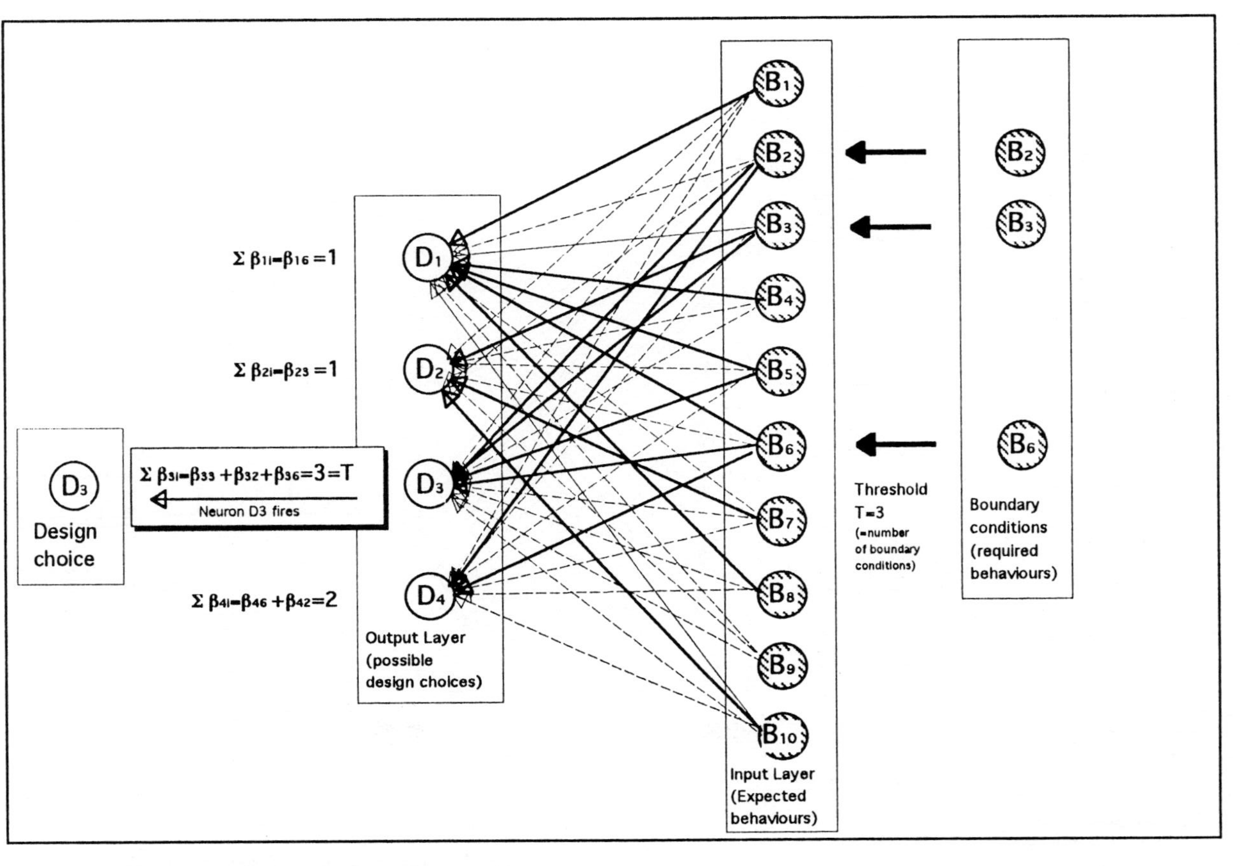

Fig.3-Use of the net for the design choice

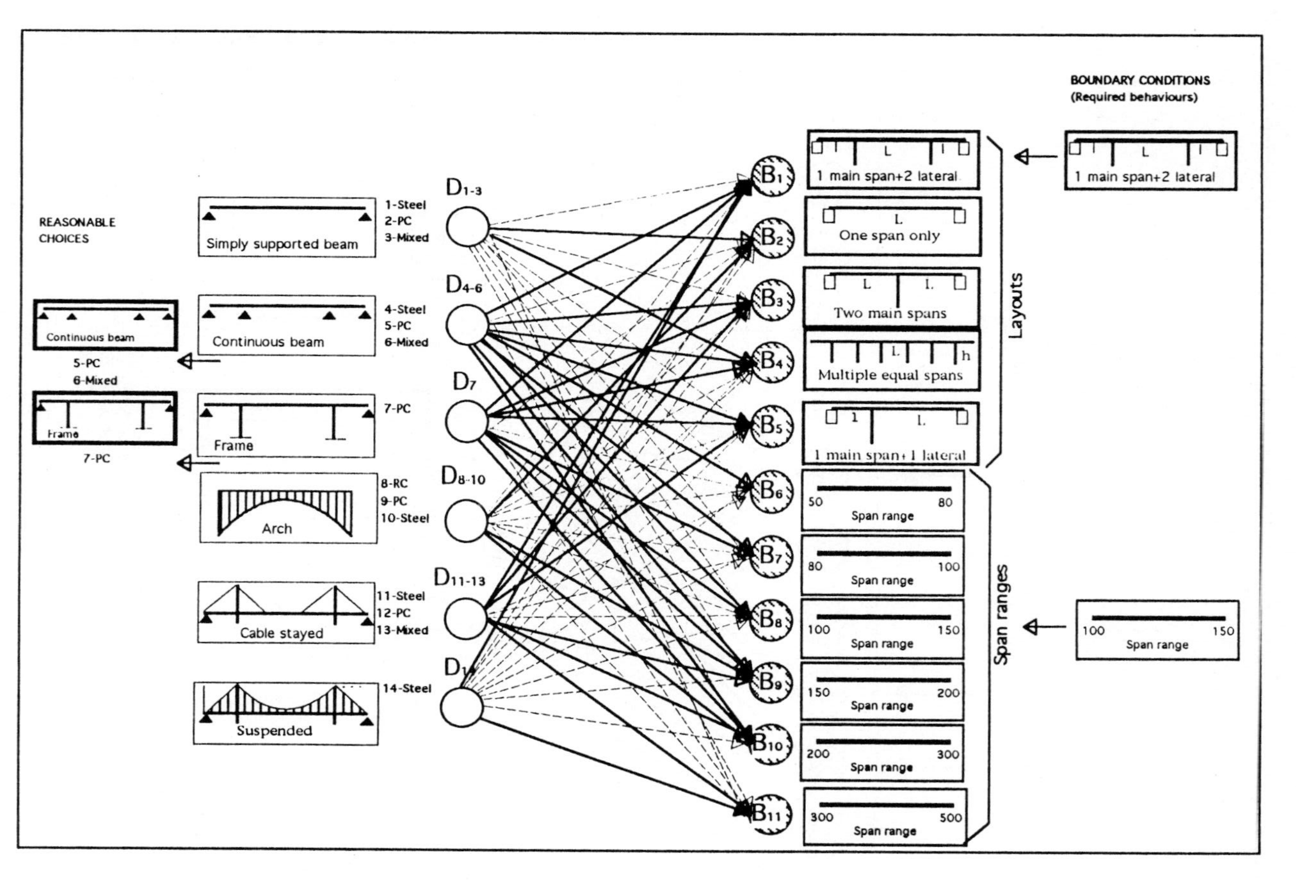

Fig.4-Choice of a bridge shape

propagation" according to reference [9]. With this method a neural network is instructed and used for every design level. Every performed choice at the upper levels introduces different required behaviours at the lower levels: in this way, at every design level two kind of behaviours are stored in the second layer of the neural network: those which are a consequence of the previously performed choices and those which are peculiar of the level under consideration.

6 Dimensioning and Prototype Assemblage

At this point the general shape of the structure and of its main components are clearly defined. To evaluate the obtained design and, in case, to perform preliminary structural analyses on it, it is needed to identify the dimensions which permit to define univocally each structural component. The sizes needed to uniquely determine the project can be divided into four categories:

- Sizes which are defined from the beginning or as a consequence of the basic layout choice(the length of the spans, the height of columns)
- Sizes which can be defined statistically by comparison with designs of about the same shapes, building heights and span lengths.
- Sizes which can be determined using empirical "rule of thumb" rules in function of the already determined shapes. For example the depth of the web in a beam can be established in function of its height.
- Sizes which are established to insure compatibility between connected parameterized parts of the design.

If the system can access to a graphic file where parametric prototype components are stored, once the dimensions of these components are defined they can be assembled in a "prototype" on which the design evaluation can be performed. To perform these tasks the system must be connected (or operate within) a CAD System. A design procedure of this kind is now being prepared and tested in the University of Pavia using the "state of the art" programming language Visual C++.

7 Macro Flow-Chart of System

The self explanatory macro flow chart of the system is represented on fig.5 It is interesting to notice that, at each design level the use of the inference engine is not mandatory, but the design choice can be made directly in those cases where the knowledge base is not considered sufficient, reliable or it lacks completely; it is thus possible to use the procedure even when the collection of knowledge is far from complete.

8 Procedure for the Implementation of the Expert System

The procedure for the implementation of an expert system of this kind can be described by the flow chart of fig.6. It is extremely important that basic data concerning a given type of structure be collected in a database that is organized according to a suitable classification of structural types [5,11]. This classification can be more or less easy to perform according to the structural types under consideration;

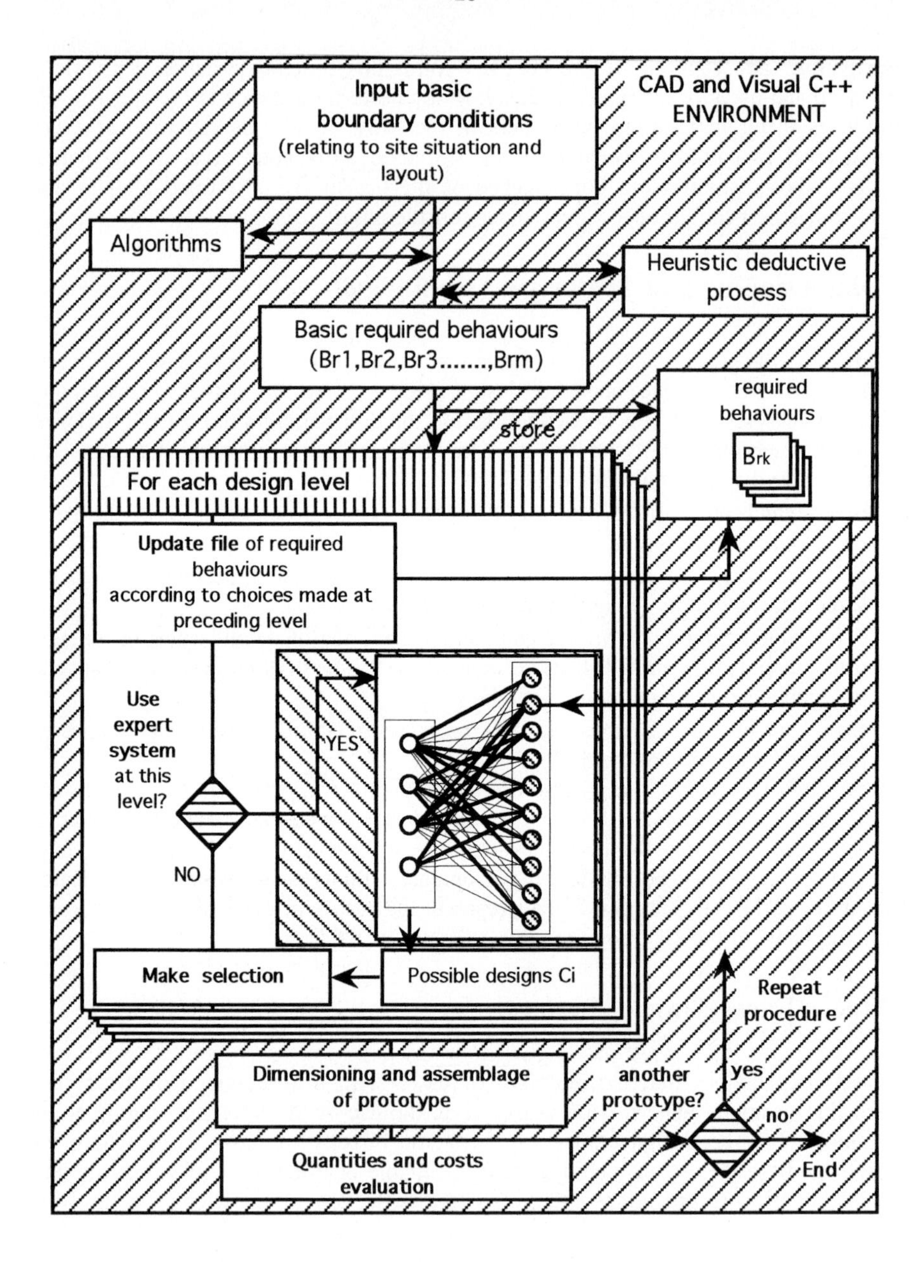

Fig.5-Basic flow chart of the Expert System

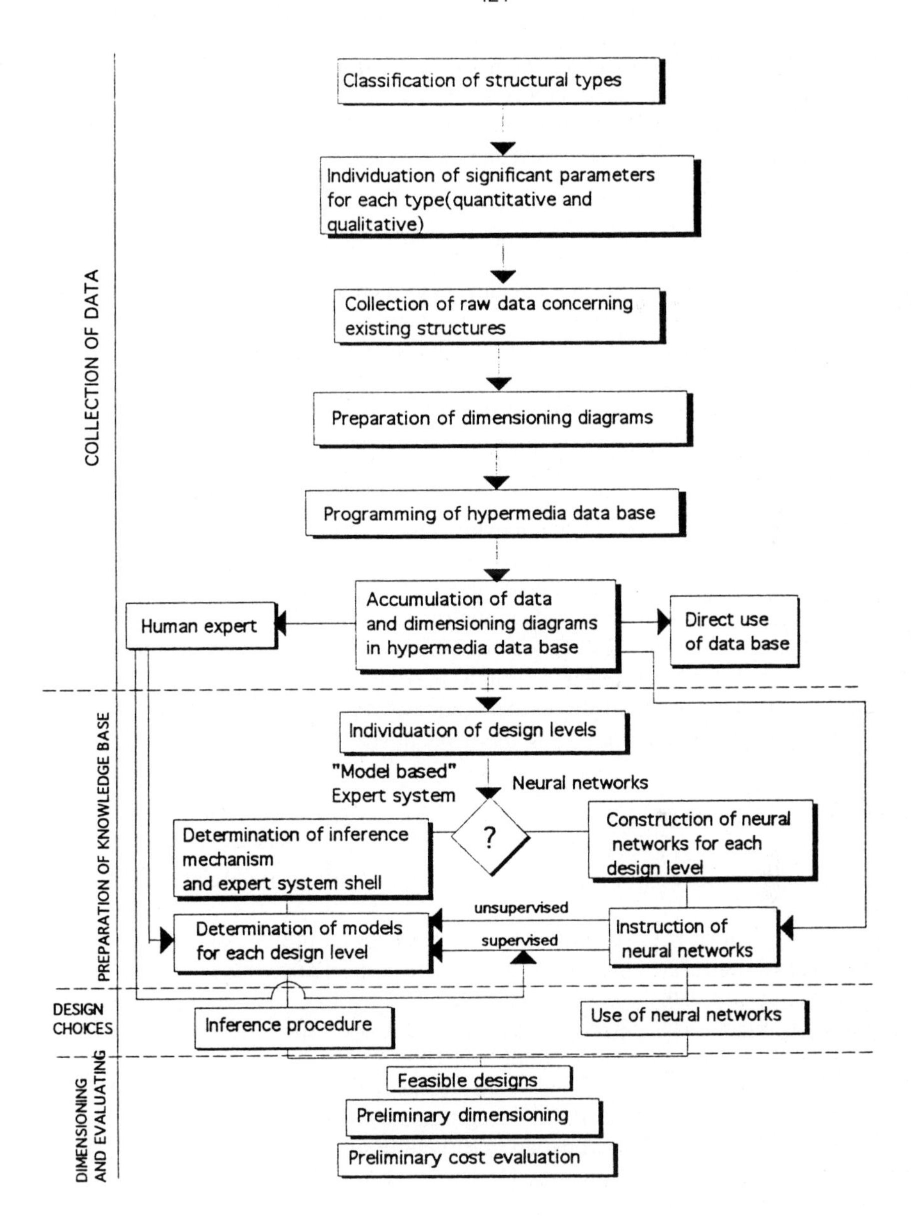

Fig.6-adopted procedure for construction of expert systems for design

it is relatively easy for bridges (and, to some extent for tall buildings) where the number of structural types is limited and well defined. It is also important to note that the advice of a human expert can be necessary to examine critically the collected knowledge rather than building the knowledge himself.

9 Conclusions

It is interesting to note that a simple two layered, "Perceptron type" neural net can usefully assist in building, departing from generic technical information concerning realized designs, the needed knowledge base to be used in an expert system devoted to preliminary structural design. The same network, duly instructed, can also be used as an inference engine to perform the needed choices. This procedure represent in some way the "missing link" in the preparation of an expert system for structural design by performing automatically the crucial task of preparing the knowledge base.

In fact while a "shell" has already been prepared and tested [6,10], and a collection of data to be stored in an hypermedia data base is now being prepared for bridges [5] and long span floors [11], the illustrated method will permit to convert these data bases in knowledge bases which will be stored implicitly in neural networks or, if needed, explicitly in "models" derived by the instructed neural network. The procedure, as required by the practical goals of this research, is straightforward, simple and best of all, follows the procedures which are normally adopted by the practising engineer with the additional advantages of memory and inference power of the computer.

References

1. Dayhoff J., Neural Network Architectures, Van Nostrand Reinhold,New York, 1990
2. Maher M.L., Fenves S.J., HI-RISE: an expert system for preliminary design of High Rise buildings, Technical Report, Dept of Civil Engineering, Carnegie Mellon University, 1984
3. Fenves S J , Maher M.L., Sriram D., Knowledge based expert systems in Civil Engineering", IABSE Periodica, Number 4, 1985
4. Fenves S J, Expert systems, Expectations versus realities, Proceedings of IABSE Colloquium " Expert Systems in Civil Engineering", Bergamo 1989.
5. Cauvin A, Stagnitto G., Classification of Structures and organization of hypermedia data bases for Structural Design, First workshop of EG-SEA-AI,EPFL, Lausanne 1994.
6. Cauvin A, Stagnitto G., A "Model Oriented" Expert system for the preliminary design of bridge structures, First workshop of EG-SEA-AI, EPFL, Lausanne 1994.
7. Tong C, Sriram,Artificial Intelligence in Engineering Design, Vol.1, chapt.1, introduction, Academic Press Inc., 1992
8. Faltings B, Reasoning Strategies for Engineering Problems, Proceedings of IABSE Colloquium "Knowledge Based Systems in Civil Engineering", Bejing 1993.

9. Steinberg L, Design as Top Down Refinement Plus Constraint Propagation, from Chapter 8 of book: Artificial Intelligence in Engineering Design, Vol.1, Academic Press Inc., 1992
10. Cauvin A., Stagnitto G., Framework of a general purpose expert system for preliminary structural design.Proceedings of IABSE COLLOQUIUM, Bergamo, 1995, Session 1-Knowledge support for Structural Design and Construction
11. Cauvin A., Passera R., Stagnitto G., Collection of relevant data and definition of a knowledge base for structural design using expert systems. Proceedings of IABSE COLLOQUIUM, Bergamo, 1995
12. Cauvin A, Stagnitto G. ,A "Top down" procedure for preliminary design of Tall Building Structures using expert systems. Proceedings of the World Conference on Tall Buildings and Urban Habitat, Amsterdam 1995.
13. Cauvin A, Use of Neural Networks for Preliminary Structural Design ,Proceedings of the 6th International Conference on Computing in Civil and Building Engineering, Berlin, Germany. Published by Balkema 1995.

Teaching Knowledge Engineering: Experiences

Tom Andersen & Susanne C. Hartvig

The Technical University of Denmark (DTU), Department of Planning,
Building 115, DK2800 Lyngby, Denmark
{ta, sh}@ifa.dtu.dk

Abstract. A knowledge engineering course is presented, and the paper sums up the experiences gained. It is concluded that a knowledge classification system and narrow scope domains are good starting points for teaching applied AI.

1. Introduction

This short paper outlines experiences gained from a course in DTU's graduate program called: 'Advanced Planning: Knowledge Engineering'.

The course is the only one in its category offered in the sector for Civil Engineering at DTU (approx. 120 candidates per year), and is as such crucial in creating a certain AI understanding with the new generations of civil engineers.

The form of the course is intensive, as we in only 3 weeks attempt to mimic a realistic process which usually takes months. Due to its intensive character the course is currently limited to 10 students.

2. Scope and Objective

The course seeks to provide the students with an understanding of the process of knowledge engineering. The scope is the development of a decision support system in domains related to the students future job situations. The domains are selected as to reflect typical and actual situations, i.e. domains are practical oriented.

Even so, that the end product of the course is a prototype system, the major objective of teaching knowledge engineering is to enable the students to cope with a 'new' (to them unknown) professional area in an effective and structured matter. This objective is motivated by the observation that working with AI-approaches results, if nothing else, in a profound insight in the domain under consideration [1]. This insight is a very important by-product of Knowledge engineering, and in fact KE can be regarded more broadly; KE is Engineering of Knowledge, and not only a process ultimately leading to an IT-solution.

2.1 Time-Frame and Structure

The course goes though several stages:

- *Theory*: Basic, and pragmatic theory in the areas: knowledge classification, knowledge acquisition, representation and problem solving (1.5day)
- *Process*: Practical guidance to the different stages of knowledge engineering: idea, acquisition, structuring, concept, prototype, test and implementation (0.5day)
- *Tutorial*: Exercises in knowledge structuring and in programming, KAPPA PC (1day)
- *Development*: Developing a decision support system including interviewing real world experts. Duration from idea to program 2.5 week, or 12 working days

Before the students prepare their reports, we stress to them that when evaluating (grading) the problem solutions immediately after end of course, we will focus on the following qualities: a) the knowledge structure (see section below) must be evident and clearly understood and b) the concept model (model of passive knowledge, e.g. small product model) must be clear and well defined.

2.2 Pragmatic Theory Presented (in brief)

We classify knowledge into 3 distinct main types [2]:

- Passive Knowledge: knowledge of static elements like beam, pipe etc.
- Dynamic Knowledge: knowledge about events and actions
- Meta-Knowledge: Knowledge about knowledge

Moreover, the concept of attributes is used as a sub-category to passive knowledge; attributes refine our passive knowledge: a water pipe holds a pressure - pressure is an attribute to the passive knowledge element: water pipe.

A 'Recipe' for Structuring Knowledge

We teach the students a basic approach to knowledge acquisition: Search the written sources for passive knowledge, and locate attributes. The outcome is a list of 'key elements' and their potential behavior i.e. the attributes. Then, use the attributes as a guide to acquiring dynamic knowledge - ask yourself (and eventually the expert): if a water pipe has an important attribute: Pressure, what kind of events could that lead to?

Dynamic knowledge is primarily elicited from human sources by interviewing professionals from practice. These interviews are possibly supported by an IT-demo; an approach we have found very useful.

For knowledge representation we teach the students that passive knowledge can be held in frame-like structures and that dynamic knowledge is usually represented in rules.

It should be noted, that theory is presented in a pragmatic way. We are aware that we cut corners, but the outcome of the students work clearly indicates that our practical and pragmatic approach works well; as a starting point for the students, and as a tool to get a proper insight in a domain, and subsequently establishing a knowledge-based system.

3. Domains Covered - Experience Gained

In the last two years 8 prototypes have been developed by 16 students in teams of two. The following domains have been used in the course:

- Safety planning of construction sites
- Blasting and demolition
- Relining of sewers
- Restoration of facade walls
- Selection of crane type
- Selecting floor-types

As can be seen, all domains (expect 'safety plan') are rather narrow in scope, which may be one reason why the students have been able to provide, in our view, surprisingly good systems - some of those could, with a reasonable effort, be turned into industrial use. The claim that 'expert systems' work best in narrow scope domains [3] seems to be true - and we basically think that our approach is 'healthy' for the students, as they experience that they can come up with something workable in 3 weeks - starting from scratch!.

However, pursuing this pragmatic, very practical approach poses a potential danger, i.e. students thinks that AI is limited to just narrow scope, stand-alone type of problems. To overcome this problem we emphasize that this is *not* the case through a presentation in class of more general and complex issues.

The domains have one more thing in common - they all, more or less, present reductive problems. The problem in 5 of 6 areas are that of selecting/disregarding methods/equipment/components etc. We do not think that selecting rather simple reductive problems hamper the quality of the course, simply because a majority of engineering tasks are reductive. There is of course many tasks which are more complex e.g. bridge design, but the fact is that when problems are reduced adequately into manageable sub-problems, they can be handled as straight forward generate and test problems.

4. Course Evaluation

Evaluation consists of two parts: evaluation of the students' work, and evaluation of the course.

Each group of students must present their work in writing and orally. Their results are discussed with the teachers and at least one professional from practice participates in the evaluation.

Each student fills a form where he/she evaluates the course. This material and communication with the external contributors from practice, provides valuable guidelines on how we can improve the course. As a result of this evaluation, we have included more theory on problem-solving and widened the scope of relevant domains.

5. Conclusive Remarks

Teaching knowledge engineering in the form just described have given the following experiences:

- Classification of knowledge is important, and it is a useful tool in the daily routine of an engineer - beyond knowledge-based system development
- Narrow scope problems are suitable for students with limited abilities in knowledge engineering - and help to spread the news, i.e. that knowledge-based systems do work in reality
- The heavy use of people from industry is a great asset to the course. The students, and we, achieve valuable information on actual problems of the real world, and not least, we get a good understanding of the need for, and attitude towards, decision support in civil engineering practice
- The topic covered by this course has become very popular with civil engineering students at DTU, and it is clear to us that we have to expand the number of available spaces

In a time where the AAI (applied AI) community constantly discusses what should be done to promote AI in industry, we think that introducing courses like the one described at universities, eventually will contribute positively to a more widespread use of AI in practice.

6. References

1. Smithers T et al: Design as intelligent behavior: An AI in industry research program. In: Gero J S (ed): Artificial Intelligence in design, Springer-Verlag, (1995)
2. Andersen T: An introduction to knowledge engineering, lecture notes (in Danish), Dept. of Planning, DTU (1995)
3. Maher M L: Expert Systems in Civil Engineering, Technology and Application, ASCE (1989)

Design Support for Viaducts

J. Corte-Real Andrade* and João Bento†

*Military Academy,
Paço da Rainha, 29, 1198 Lisboa CODEX, Portugal
e-mail: jorge.corte-real@ip.pt

†Instituto Superior Técnico (IST), Civil Engineering Department
Av. Rovisco Pais, 1096 Lisboa CODEX, Portugal
e-mail: joao@civil.ist.utl.pt

Abstract. The paper deals with the development of a system for supporting the design of highway bridges. The adopted approach is considered to innovate in relation to previous approaches of design systems for viaducts, for making a pragmatic combined use of a number of AI techniques such as CBR, neural networks and notions of agency. An interesting feature of the proposal is related to the fact that, instead endorsing a complex, and usually rigid design model for the adaptation phase of CBR-based design, the authors provide a conceptual (and partly implemented) framework for designers to interactively *adapt* the retrieved solutions. This adaptation takes place by using a *constrained* graphical/drafting environment, i.e. a CAD environment where design entities may carry physical and functional constraints relating to them in isolation, or to themselves in relation to the surrounding ones. Such constraints may be posted prior to the adaptation phase or throughout that process.

Introduction

In this work, design is understood as an evolutionary process of exploration of previously unknown states in order to reach a set of feasible solutions (artefacts or actions); these are adapted to the surrounding situation in order to accomplish the established performance requirements. A so-called decision process is then used to designate the choice of the solution with the highest expected utility, weighted over the universe of significant possible solutions.

It is assumed that designers start this process with a set of core ideas. Often, these ideas emerge relating the current situation biased by the nearest specific previous experience. This memory approach avoids starting from scratch and provides proofed solutions or tested processes, although, normally, adapted to other situations (Zhang and Maher, 1993).

Reasoning based on cases, a well-known paradigm in AI, is founded on the premise that it is useful to infer using past experiences. Intuition is also involved in this process as it results from experiences originated by different concepts. Designers adapt potential solutions to the current situation establishing a set of assumptions and involving induction, deduction, abduction, intuition and experience. Creative design should also cope with personalised acts, such as preferences, styles, cultural behaviours and beliefs. Moreover, creative design is an activity involving vague, fluid, ambiguous and parallel thought processes. In its corresponding design space, the variables, processes and behaviours involved are not entirely known in advance. In this context, design systems should enable the expanding, reducing or changing of the confined space in which a set of variables and processes are valid.

On the other hand, having in mind that intelligence is the result of a collaborative participation, design modelling through an agents' approach and adequate user interaction will provide active support to the above high level characteristics of creative design (Bento and Feijó, 1997).

In the present model, each final solution emerges from the internal interaction of active entities, among themselves, and from their interactions with the external user. Also the choice of the solution with the highest expected utility will be the result of such a collaborative process.

Conceptual Methodology and System Architecture

The system under development was conceived to address analogical design processes where, as well known, deciding amongst a number of alternatives is an important feature. It includes a Case-Base Manager, a Project Manager, a Validation Manager and a Decision Manager supporting design activities via a CADrafting environment (ACAD14, in the present version). Artificial Neural Networks, Induction and Rule based inference techniques complement the blend of tools involved in the development of the system.

The basic "routine" of system usage consists of the more or less sequential performing of 1) retrieval, 2) adaptation, 3) validation and 4) decision tasks, as briefly described in Figure 1, outlining the system architecture.

2.1 Case-Based Manager

The Case-base Manager is formed by two main components: the Database Module and the Retrieval Module. In the former, the description of previously designed/built viaducts, the context and contents, is included in a simple, although lengthy, database containing features and feature value types similar to the ones found in real projects. The database also stores (1) the adaptation history of every retrieved case, (2) the interoperability files for inter-application communication, (3) learned rules, and constraints and (4) built attributes.

As for retrieval, the authors contend that searching, matching, selecting and recovering past cases has to be more than a simple comparison between equivalent features; otherwise such processes would be of lesser use for complex and subjective domains.

The system lower level of retrieval, either in a global or a local context, has been implemented by a traditional numerically based matching and ranking procedure (Kolodner, 1993). However, the use of induction techniques over past cases also enables the computing of averaged weights of similarity assessment, in order to improve, whether desired, the stated user preferences or to replace his choice.

Moreover, using a neural network trained in advance to recognise key feature typologies may extend the efficiency of the system in terms of retrieval. For instance, when applied to 3-spanned viaducts, the trained NN will provide a means of recovering cases only amongst the relevant structural typologies, thus reducing the search space and refining the matching process (Quah et al., 1994).

Also, the availability of rules defined by user at runtime has been considered as a means to improve the retrieval of cases. This is intended to impact not only on the referred metric similarity assessment, but also on the overall matching process, allowing deep and more oriented retrievals.

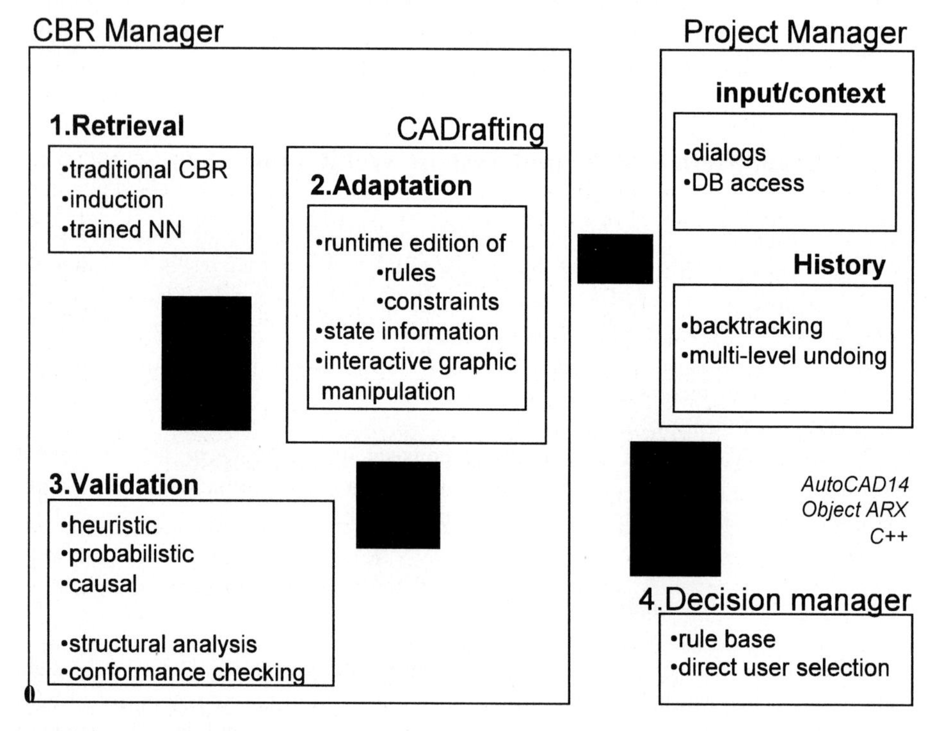

Fig. 1. Scheme of system architecture

2.2 Project Manager

One of the issues that deserve more attention in the work is that of providing an adequate level of user interface. Accordingly, whenever some of the nearest-neighbour cases are retrieved, their features are used to rebuild the original visual version of the artefact in a CAD (drafting) environment. Each entity – with a specific behaviour within the whole artefact – is turned into a 3D CAD object. Commitments, capabilities and choices (recovered or added by the user) are appended to each object in terms of attributes and functions. Relations between the entire community are re-established and the objects' initial states become defined. In the end, the initial "mental state" of each object and of the whole community is complete.

The user can modify these initial "mental states", at runtime, changing attribute values either by visual manipulation of entities or by written constraints (e.g. "*Face1:Column1=Face5:Beam2*" would produce the coincidence and equality of those parts of the two entities). An interpreted language was developed in order to deal with this and to support any required information about the community or component features. The language also provides the creation of new attributes based upon the existing ones (e.g. "*Area1:Column3=Dim1:Col1*Dim7:col1*").

The retrieved cases represent the initial "core ideas" that need further development and refining to adapt to the current situation. A structuring assumption of this work is that the artefacts are to be interactively adapted in a drafting environment based on CADrafting tools enriched by controlled and automated processes. These should support trivially known high level characteristics of design processes, such as pro-activity, reactivity, design biases and the fact that design may be an event-driven process (Feijó et al., 1998).

Modelling one entity or changing its position will induce in the surrounding environment, and also on possible clone entities, a potential to change. The automatic reaction by neighbours will produce modifications in order to re-establish bounds and the effect will be propagated throughout the entire community. Clones (e.g. columns of the same viaduct cross section) will automatically imitate the initiator.

Using constraints and rules can control the automatic propagation of modifications, as the result of entity pro-activity and reactivity. The latter may act immediately after the definition by the user or they can be learned and integrated, remaining active until desired. In the present context, constraints and design biases can be applied to entity attributes and/or to relationships among entities (e.g. entity columns cannot be positioned upon entity roads; all columns should have variable sections).

Constraints and design biases settled by the user will restrain the response of the affected entities and the propagation to others. Whenever mutations are dependent on multiple or impossible constraint resolutions the user will be prompted to interactively guide the choice or to input information that could not be modelled and thereby resolve such difficulties.

The provision of continuous state information has been identified as of utmost importance for bridge designers, as a permanent means of evaluation of the solution. Therefore, the following items are to be updated throughout the design process: contents of existing or created features, relations between existing or created value type features, constraints, and limits (global and local) of foregoing cases.

On the other hand, every state of the evolutionary process is stored in such a way as to enable the later retrieval and reuse of the adaptation history for each recovered case. The user can preview all these stages, go back to a previous one and resume adaptation from there. It will also be possible, at each design stage, to project the complete artefact to its final stage using costing and performance heuristics provided by bridge-design experts. Information from the nearest-neighbours matching cases is to complement possible insufficiencies of the descriptions provided by the user.

2.3 Validation Manager

The free-style iterative and evolutionary process enabled by the system has to be validated throughout the whole design process. If not before, such validation has to occur after a solution has been achieved. In the case of viaducts, this involves the satisfaction of design performance requirements stated in structural codes. The system is, thus, assisted by a structural analysis application whenever the user needs to evaluate a given performance. The required level of interoperability is achieved by loosely coupling a structural analysis package to the remaining modules through file-exchange (providing and interpreting the appropriate files). The output system file includes a finite element mesh (solid type) generated by splitting entities according to their dimensions, their shape and with the interaction among other entities embedded (e.g. voids, steel reinforcement). The computed results may be assessed and manipulated, at runtime, by making use of the developed interpreted language.

Validation based on heuristics will also be performed by user-defined rules. Once validated, solutions may be directly appended to the Case-Base Manager database, expanding as appropriate.

2.4 Decision Manager

Every achieved solution has its advantages and disadvantages with varying degrees of importance depending on the accomplishment of design performance requirements and user beliefs. Choosing among validated solutions implies the comparison of their expected utilities. In the present system, the definition of the benefits and drawbacks to compare, their scale and significance, are established either directly by the user or by making use of a rule base containing selection heuristics acquired from bridge design experts. The system will support user decision presenting the solution with the highest expected utility and the corresponding justification.

Conclusions

Repetitive domains of design, as clearly is the case of viaducts, are extremely well suited for CBR approaches. As in many other cases, the adaptation of retrieved cases is the hardest of the problems at hand. The present paper suggests the use of an environment in which the user is given the possibility of adapting by means of trivial drafting operations in a clear event-driven design environment. A reactive agents approach, by which knowledge and constraints are embedded in the design entities, is put forward with apparent success. Although implementation is well advanced in some directions, the work is still at an intermediate stage, reason why the drawing of performance conclusions, although apparently favourable, should be left uncommented.

References

Bento, J.; Feijó, B., (1997): "An Agent-Based Paradigm for Building Intelligent CAD Systems", Artificial Intelligence in Engineering, 11(3), 231-244, Elsevier Applied Science

Feijó, B., Rodarki, P.; Bento, J.P.; Scheer, S., Cerqueira, R., (1998): "Distributed agents supporting event-driven design processes", Artificial Intelligence in Design '98, ed.s John Gero and Fay Sudweeks, 557-577, Kluwer Academic Publishers.

Jennings, Nick R. (1993): "Commitments and conventions: The foundation of coordination in multi-agent systems", in The Knowledge Engineering Review, Vol. 8:3, pp 233-250, Cambridge University Press.

Kolodner, Janet (1993): "Case-Based Reasoning", Georgia Institute of Technology, Morgan Kaufmann Publishers Inc.

Quah; Tan; Raman (1994): "A Shell environment for developing connectionist Decision Support Systems", in Expert Systems, the International Journal of Knowledge Engineering and Neural Networks, Vol 11(4), 225-233.

Zhang; Maher (1993): "Using Case-Base Reasoning for the Synthesis of Structural Systems", Proceedings of the IABSE Colloquium.

Converting Function into Object

Adam Borkowski[1], and Ewa Grabska[2]

[1] Institute of Fundamental Technological Research,
Swietokrzyska 21, PL-00-049 Warsaw, Poland
abork@ippt.gov.pl
[2] Institute of Computer Science, Jagiellonian University,
Nawojki 11, PL-30-072 Cracow, Poland
uigrabsk@cyf-kr.edu.pl

Abstract. The aim of the present paper is to show how the multi-layered graph-based representation could stimulate creative thinking in design. The prototype software developed in this project allows the user to define functional requirements for the designed object and to transform them into the object itself. It is done in several phases and the representation scheme encourages the designer to look for alternatives, analogies and novel solutions at each level of abstraction. The proposed methodology is illustrated by simple example of designing the teapot but the same principles apply for each artefact.

1 Introduction

Design is a goal-directed activity. The objectives and the essence of a design task crystallise along with the development of the solution itself. Thus, the sequential model is unable to reflect the process of designing even a simple artefact. The trial-and-error nature of this process can be imagined as an exploration of a maze or a spiral climb in the search space .

Detailed design is very much domain dependent — designing a piece of furniture requires different skills than designing an aeroplane. However, at the conceptual level certain rules seem to be applicable irrespective of the domain. We believe that one of them is the priority of function: an object turns out to be well designed if it serves well the assumed purpose. This rule applies even for "pure form": the only purpose of a good sculpture or painting is to influence emotionally the spectator.

Following that rule the design process in any area of engineering begins with establishing functional requirements for the considered artefact. Architects speak with investors about the program for the future building whereas software engineers develop the specification of the future code. Answering the question "*What do we want to achieve*?" is very important since changes in the required functionality raise the costs of the design and diminish its quality.

The purpose of the present paper is to demonstrate how functional specification is integrated into the graph-oriented Composite Representation (CR) proposed earlier in

[1]. The CR-representation is composed of 3 type of knowledge: the syntactic and semantic knowledge about the designed object as well as the control knowledge about the design process. An editor of functional graphs, a graph grammar, a control diagram, a set of admissible primitives and a rendering program are tools for defining the CR-representation.

In order to comply with the restricted volume of the paper we illustrate our approach by a simple example of the teapot (similar problem of the coffee maker was discussed in [2]).

2 Specifying Function

A function of simple object can be described in few words, e.g. a pen should write on paper or an umbrella should protect against rain. On the other hand, a specification of complicated system can take hundreds of pages. The German textbook on design in Civil Engineering [3] provides a list of 110 questions that allow the architect to obtain from the client a complete functional description of the future building.

Functional requirements can be stored in different formats. One of them is the annotated and possibly hierarchical *list of goals* bearing the name of object. Such a list for a teapot could be:

teapot	=	*contains tea*	*(a)*
		stands firmly on table	*(b)*
		can be lifted by hand	*(c)*
		can be filled with water	*(d)*
		allows to pour tea into cup	*(e)*

Following [4] we distinguish between *goals* and *constraints*. Typical explicit constraints set by the client are the cost, weight, material or colour of the designed object. Many constraints are implicitly given by goals: for example, the goal „*a car must be safe*" implies a lot of subgoals and constraints.

In the prototype software that we are developing at present functional goals and constraints are entered by the user in textual form. Each goal is assigned a colour which helps the user to distribute goals between units in to which the object is decomposed (see Sec. 3).

3 Developing Composition Graph

After the list of functional requirements (goals and constraints) has been agreed-upon, the designer considers the decomposition of the object into smaller units. In general, one can distinguish the decomposition into *functional units, physical units* and *visual units*. For simple objects these classes of units may coincide and the obvious initial guess is the one-to-one mapping when each unit takes care of a single goal. For our illustrative example of the teapot such a *list of units* could be:

teapot	=	*container*	*(a)*
		base	*(b)*
		handle	*(c)*
		filler	*(d)*
		outlet	*(e)*

Establishing relations between particular units we obtain the first level of the CR, namely, the *Composition Graph* (CP-graph). The editor of CP-graphs allows the user to generate and manipulate nodes and edges.

Directed edges of the CP-graph represent relations between units. These edges bear labels describing the names of relations. An example of the CP-graph for the teapot is shown in Fig. 1. We distinguish here 2 relations: the relation *supports* deals with the stability of the teapot whereas the relation *connected* reflects merely the interaction of parts. The purpose of CR is to encourage the designer to pay enough attention to higher levels of abstraction before plunging into details. Thus, the CP-graph of teapot reflects only the mapping of functional requirements into the units building the object as well as the relations between those units. Decisions regarding the shape, the dimensions, the material, etc. are postponed until the designer comes to the lower level of CR.

Good representation should motivate the designer to look for alternative solutions

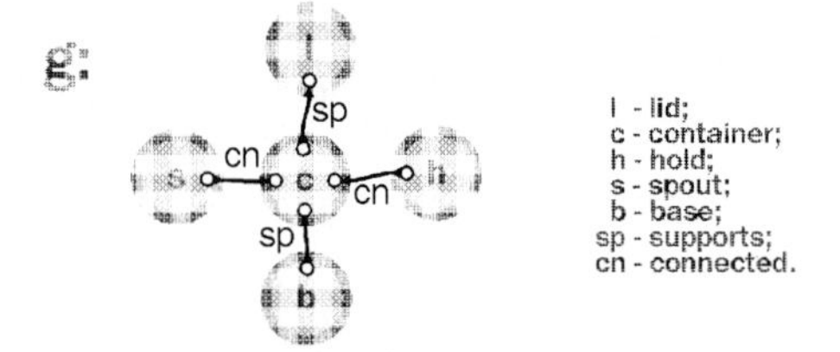

Fig. 1. CP-graph for teapot

whenever possible. Let us evaluate the CR from this point of view. The first opportunity of search occurs when the list of units is chosen: the satisfaction of goals can be distributed in many ways between units, the granularity of decomposition is also subject to choice by the designer. However, in most practical situations changes in the decomposition of designed object are taken on the higher loop of the spiral. The reason for that is that the visual perception of the CP-graph may often inspire either assembling certain modulae into larger units or, on the contrary, splitting certain units into smaller and more specialised functionally parts.

Even our simple example of teapot gives the opportunity for alternative solutions already at the decomposition level. Namely, the base and the container can be merged into a single unit: a container stable enough to stand firmly on the table. Merging the nodes labelled *c* and *b* of the CP-graph *g* into a single node labelled *c-b* can be described by means of the graph production p_1 shown in Fig. 2. Further, instead of attaching a spout to the container one could form the upper part of the latter in such a way that tea can be safely poured out (see production p_2). The next production p_3 says that the handle can be incorporated into the container as well.

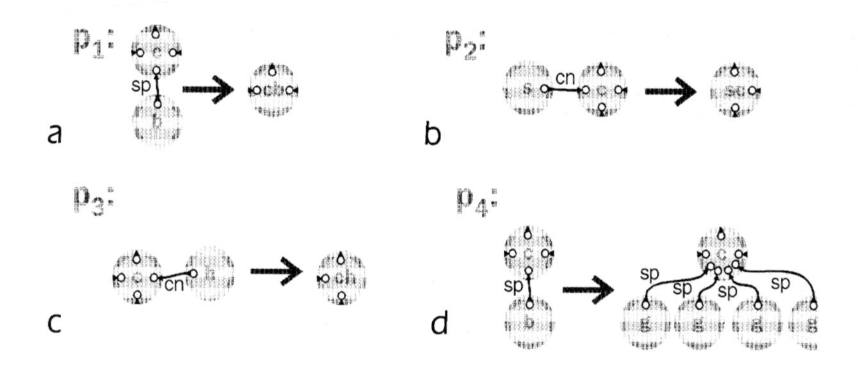

Fig. 2. Graph grammar for teapot

It is worth noticing that considering functional purpose of the designed object may stimulate the search for candidate analogues to the entity that is being designed. In our approach CP-graphs can represent abstractions of analogues. Transferring a CP-graph of a source structure into a target domain, however incomplete, enables the organization of units of the CP-graph determining the target structure in a coherent way. Let us illustrate it on the example of teapot. One of candidate analogues for the teapot appears to be a flask used in chemical laboratories. Such a flask equipped with an additional stand can suggest replacing teapot's base by 4 legs attached to the container. On the CP-graph level such a change is described by a node splitting operation which adds an internal structure (compare production p_4).

Given the CP-graph and the set of productions which can be applied to the CP-graph we want to generate those CP-graphs that are syntactic descriptions of plausible solutions for the considered object. Thus, we use graph grammars [5], [6] as a tool for generating CP-graphs. In the CR we represent control knowledge by a *Control Diagram* which is a connected directed labeled graph. With exception of the initial node *I* and the final node *F*, all other nodes of such a diagram are labeled by the names of productions. Applying them according to the order stated in the control diagram, we start with a production which corresponds to the label of the direct successor of the initial node. The derivation process stops when the final node is reached.
Fig. 3 presents an example of a very simple control diagram allowing us to introduce

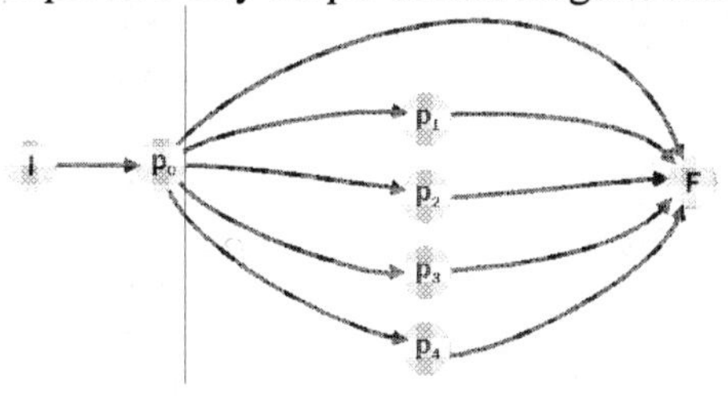

Fig. 3. Control diagram for teapot

at least one modification into the CP-graph g. The production p_0 is composed of the left-hand side defined as one node labelled by S without bonds and the right-hand side being the CP-graph g. This diagram can be transversed along one of the following paths:

path I	–	I, p_0, F	path IV	–	I, p_0, p_3, F
path II	–	I, p_0, p_1, F	path V	–	I, p_0, p_4, F
path III	–	I, p_0, p_2, F			

To introduce a sequence of modifications into the CP-graph g we have to define additional productions in which, for instance merging nodes being right-hand sides of p_1, p_2, p_3 should appear on their left-hand sides. Thus, the search for alternatives is natural and easy in our representation.

4 Creating Final Object

The bottom layer of the CR called the *Realisation Scheme* deals with a particular instance of the designed object. Therefore, at this level the designer makes up his mind regarding shape, geometry, material, colour and possibly other properties. Particular implementations of the CR, e.g. FLOOR [7] supporting the layout design for single-family houses or HDS [8] dealing with the design of heating for such houses, include the domain dependant libraries of primitives. Despite certain differences all implementations use similar interface. Fig. 4 shows how such an interface is used for finding the final shape of the teapot.

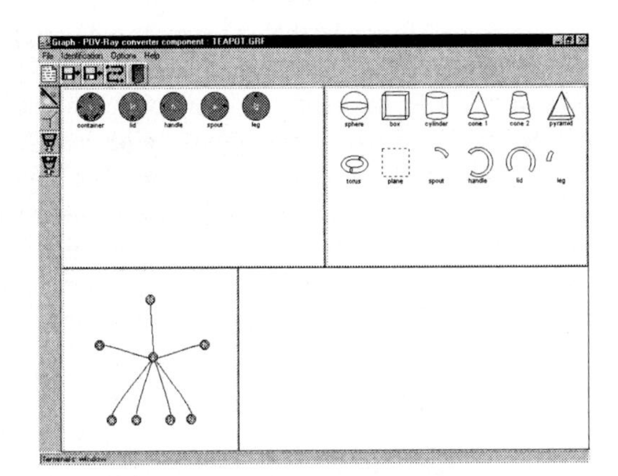

Fig. 4. User interface for realisation scheme

In the upper left corner of the window we see the set of terminal nodes. The CP-graph formed of these nodes is depicted in the bottom left corner. The set of admissible primitives is visualised in the upper right corner. Finally, the bottom right corner is occupied by the connection to the rendering program POV-RAY. The latter allows the designer to obtain and evaluate realistic image of the designed object. All geometric data as well as relations translated into proper transformations are automatically transferred to POV-RAY. Adjusting light sources and reflective properties of the surface the user can seek for the best artistic effect. Usually harmony between function and form is perceived as beauty of the artefact.

5 Conclusion

It would be naive to imagine that the specification of functional properties of the designed object could be automatically translated into its final description. This mapping is obviously non-unique and many crucial choices should be left for the designer. The aim of the proposed methodology is only to assist the user in taking such decisions. We believe that by allowing the designer to structure the knowledge into the layers of the Composite Representation and letting him move freely between those layers we stimulate creative thinking and search for innovative solutions.

Up to now we tested our methodology on small to medium scale problems (e.g., floor layouts [7] required 101 productions). However, it is known that especially for large scale problems graph grammars show their generative power. A good example is the application of graph grammar for the visualisation of iterative problems. This allows us to believe that the CR-paradigm will prove its efficiency in engineering practice.

References

3. Grabska E.: Graphs and designing. In: H. J. Schneider and H. Ehrig (eds.): Graph Transformations in Computer Science, Lecture Notes in Computer Science, Vol. 776, Springer-Verlag, Berlin (1994) 188—203.
4. Agarwal, M., and Cagan, J.: Shape Grammars and their Languages — A Methodology for Product Design and Product Representation. In: Proc. of the 1997 ASME Design Engineering Technical Conferences and Computers in Engineering Conference: Design Theory and Methodology Conference. DETC97/DTM-3867, Sacramento, CA, September 14—17 (1997)
5. Neufert. E.: Bauentwurfslehre. 33. Auflage, Vieweg & Sohn, Braunschweig-Wiesbaden (1992)
8. Pavlidis, T.: Linear and Context-Free Graph Grammars. Journal of the Association for Computing Machinery, Vol. 19, No. 1 (1972) 11—22
9. Fitzhorn, P.: Formal Graph Languages of Shape. AIEDAM, Vol. 4, No. 3 (1990) 151—163
10. Grabska E. and Borkowski. A.: Generating floor layouts by means of composite representation. In: Computers in the Practice of Building and Civil Engineering, Proc. of Worldwide ECCE Symposium, Lahti (1997) 154—158
11. Cichocki P., Gil M. and Pokojski. J.: Integrated system for heating system design. In: Applications of AI n Structural Engineering - IV, Proc. of the 4th EG-SEA-AI Workshop, Lahti (1997) 27—32

Software Agent Techniques in Design

Susanne C. Hartvig

Department of Planning, Technical University of Denmark, DK2800 Lyngby, Denmark
sh@ifa.dtu.dk

Abstract. This paper briefly presents studies of software agent techniques and outline aspects of these which can be applied in design agents in integrated civil engineering design environments.

1. Introduction

The research presented is part of an ongoing project which strives to develop guidelines for harmonizing civil engineering design applications to obtain integration. The project has been inspired by early research in integrated design environments such as ICADS [1] and IBDE [2], and by new advancing technologies e.g. software agents. This paper discusses software agent techniques and their possible use in development of design agents in an integrated environment.

2. Agent Theories

Three approaches have been studied: the Agent-oriented Programming paradigm (AOP), developed by Shoham at Stanford University [3], [4], the Knowledge Query Manipulation Language (KQML), developed by the Knowledge Sharing Effort (KSE) [5], [6], and finally the Knowledgeable Agent-oriented Systems (KAoS) approach, developed at Seattle University in co-operation with Boeing employees [7]. Three aspects are relevant for integrating civil engineering applications: Agent control, agent communication and system architecture.

Agent Control. In integrated agent environments attention has to be given to how agents are controlled. AOP and KAoS implement their agents with a mental state, which contains commitments or intentions. This state and its updating routines enable the agent to be autonomous - it acts on its own according to its internal state and the initial rules of commitment and intentions it has been given. KQML is not specific about the internal representation of an agent, but they propose the use of behavioral policies, to control the 'behavior' (activation/ execution) of agents.

The use of internal structures resembling a mental state and behavioral policies as control mechanisms are very common in autonomous agents systems, they enable the agents to be autonomous. In other environments, which are not as autonomous, other control mechanisms can be applied such as black board mechanisms, petri-nets or other mechanisms based on task decomposition [8].

Agent Communication. Communication is what integration is mainly about, thus the techniques for enabling communication between agents are of interests to anyone developing integrated environments.

In most agent systems, communication between agents are obtained by passing messages between them, much as in object oriented programming. What is different from the message passing between objects and between agents is the meaning of the messages. In object oriented programming a message to an object initiates a method, which is a function inside the object. The method is initiated using its specific name and predefined parameters. In agent communication, agents share a vocabulary of message types, they all know a subset of the same message set. Hence, an agent only needs to 'speak' the language to make itself understandable, it does not need to 'speak' n languages to 'talk' to n agents.

The communication languages used by KQML and KAoS are based on speech act theory. They define a number of primitives (verbs/ performatives), which are part of specific messages, and these primitives enable the agent to recognize the message types, and act according to them. The lists of primitives defined by KAoS and KQML are very similar, and include primitives such as: inform (tell) and query (ask). These primitives enable agents to communicate - they can request information from each other, they can answer questions, and so fourth, but to make the whole system work these simple ask-answer primitives are not enough. Hence, both KQML and KAoS suggest a number of primitives that are used to support system facilities, such as matchmaking (i.e. matching agents to requests) and system management (request (subscribe), offer (advertise), register, and unregister).

System Architecture. In agent systems, agents have to know ways of locating agents capable of answering their request. This can be done in two ways: 1) agents in the system know of all other agents in the system - they know exactly who to ask for some specific piece of information, 2) agents themselves are not aware of all other agents in the system, they have to 'lookup' the agent which provides the information they are requesting.

AOP is based on the first alternative - agents themselves should know who to contact. KAoS and KQML base their approaches on the second alternative. They have system components which supply the agents with a place to lookup other agents.

System facilities are necessary in larger agent systems, with many diverse agents, because every agent can not know of every other agent. This applies to non-autonomous agent systems as well, though in this situation it is not the agents that need to know which agents are attach, it is the user.

3. Software Agent Aspects in Design Agent Integration

Transportation. The KSE and KAoS suggestions for transportation of messages between agents is based on an object request broker and springs from hardware or platform considerations more than from domain considerations. Hence, as this approach is domain independent, it is applicable to any domain, including the civil engineering domain.

Agent Control. The AOP and KAoS approaches assume that agents work somewhat autonomously. This appears to be very appropriate for personal assistance agents, business applications, automated design applications, information sorting and collecting agents, but in civil engineering design this is a troublesome approach. Civil engineering design environments which perform automated design is not an immediate goal; human designers are not prepared to be substituted by design agents and within the research community there is some agreement that 'decision support' for the designer is more appropriate at the current stage [9], [10], [11] and [12]. Hence, as design agents are not expected to be autonomous, implementing them with a mental state will only complicate the development. Instead another control mechanism should be developed which aid the designer in his selection of tools. But this is a topic for further investigation.

Agent Communication. A communication language for civil engineering design agents and environments can be based on the approach used by KAoS and KQML, but some adaptation is required to support the specific characteristics of civil engineering design.

The content in a message can be formatted using different languages. KQML suggests a logic-based language called KIF [13]. In civil engineering design, agents will exchange information about artifacts which are of a general type and characteristics (class - generalization) and which are put together from smaller parts which themselves have their own characteristics (aggregation of instances). Hence, the possible use of an object oriented language should be investigated.

The primitives describe the kind of actions one can have with agents, and some of the primitives suggested by KSE and KAoS are applicable in civil engineering design agents. It is important to acknowledge though that the primitives to some extent depend on the domain of the applications as well as on the chosen control mechanism and the system architecture.

System Architecture. Both the KAoS and KSE architectures were developed to support an autonomous system, and as such they provide the agents with necessary system facilities.

The use of autonomous agents is not the immediate solution to integration problems in civil engineering design, instead we, as researchers, should first try to achieve user-controlled design-support systems. In such an environment, system facilities are expected to support the user in picking the 'right' agent for a particular task; thus facilities suggested by KAoS and KQML (e.g. register/ unregister) should be implemented in a civil engineering design environment, as it will enable the system to 'know' which agent is capable of what.

4. Final remarks

Software agent techniques promise to be useful when developing design agents, in particular the introduction of a common communication language based on speech act theory should be investigated. And furthermore the system facilities suggested are appropriate in any modular environment, as both agents and designers require some place to 'lookup' agents which can aid their assignments. Having said that, it is important to acknowledge that in civil engineering design environments control mechanisms and system facilities may deviate from those intended by the KQML and KAoS approaches, because of the special characteristics of civil engineering design (e.g. designer control, fragmentation, a large amount of human decision making).

5. References

1. Pohl, J., Myers, L.: A Distributed Cooperative Model for Architectural Design. Automation in Construction 3 (1994) 177-185
2. Fenves, S.J., Flemming, U., Hendrickson, C., Maher, M.L., Quadrel, R., Terk, M., Woodbury, R.: Concurrent Computer Integrated Building Design. Prentice Hall, Englewood Cliffs, N. J. (1994)
3. Shoham, Y.: An Overview of Agent-Oriented Programming. In: J.M. Bradshaw (ed.): Software Agents. AAAI PRESS, Cal. USA (1997)
4. Shoham, Y.: Agent-oriented programming. Artificial Intelligence, 60 (1993) 51-92
5. Finin, T., Labrou, Y., Mayfield, J. : KQML as an Agent Communication Language. In: J.M. Bradshaw (ed.): Software Agents. AAAI PRESS, Cal. USA (1997)
6. KQML specification document, DRAFT version, available at http://www.cs.umbc.edu/kqml.
7. Bradshaw, J.M., Dutfield, S., Benoit, P., Woolley, J.D.: KAoS: Towards An Industrial-Strength Open Agent Architecture. In: J.M. Bradshaw (ed.): Software Agents. AAAI PRESS, Cal. USA (1997)
8. Hartvig, S.C..: Issues Concerning Integrated Environments in Building Design. Ph.D. proficiency test report, Department of planning, The Technical University of Denmark (1997)
9. Smith, I.F.C.: Designers Like Designing. In: D. R. Rehak. (ed.) Bridging the Generations, Carnegie Mellon University, Pittsburgh (1994) 159-163
10. Galle, P.: Towards Integrated, "Intelligent", and Compliant Computer Modeling of Buildings. Automation in Construction 3 (1995) 189-211
11. Hartvig, S.C., Andersen T: Integrated Design Environments: Fundamental Specifications, submitted to International Journal of Computer-Integrated Design and Construction, Oct 1997.
12. Richens, P.: Does Knowledge Really Help ? CAD Research at the Martin Centre. Automation in Construction, 2 (1994) 219-227
13. Genesereth, M.R: An Agent-based Framework for Interoperability. In: J.M. Bradshaw (ed.): Software Agents. AAAI PRESS, Cal. USA (1997)

Case-Based Design Process Facilitating Collaboration and Information Evolution

Peter Johansson[1] and Stoylmina Popova[2]

[1] Division of Steel and Timber Structures, Chalmers University of Technology, S-412 96 Göteborg, Sweden
Peter.Johansson@ste.chalmers.se
[2] Division of Design Computing, Chalmers University of Technology, S-412 96 Göteborg, Sweden
Mina@arch.chalmers.se

Abstract. This paper describes a cased-based design process facilitating collaboration and information evolution from the point of view of architects and structural designers. It is discussed how case-based techniques together with 3D visualization techniques can contribute to a closer and more effective cooperation between these two actors and the client which will result in a product of higher quality and lower cost. It is also discussed how techniques such as real-time manipulations (variational design) and derivational analogy facilitate collaboration and speed up the design process. The paper presents a process model, which takes advantage of these new techniques, and discusses some of the advantages compared to the sequential work methods traditionally used in construction. We describe a vision of how the collaboration between designers and clients will affect their position among the other actors in the building industry. The process is being currently implemented at Chalmers University of Technology.

1 Background

Before the 19th century, the architect and the structural engineer was one person. Along with the industrialization during the 20th century, these two roles began to separate successively [9]. Today, we have a rather distinct picture of both professions and it is even pointed out that their languages differ [3, 7]. A parallel process to and a consequence of this separation is that both architects and structural engineers have lost influence in design and construction projects. The architect is only hired as a consultant to do his/her work during a limited part of the project time, after which the professional has very little insight, influence or control over the project. This segmented process makes it hard for architects to receive appropriate feedback from the construction site. As a result, some of the knowledge architects gain when designing is lost and cannot be reused later in the project nor in future projects. [10, 7]. Structural designers are in a similar position, the difference being, though, that they often enter into a project later than an architect does. This situation forces the architect to deal with problems consisting the very foundation of the structural designer's profession and which the engineer is more capable of solving. The research community has pointed out that an architect and a structural engineer need each other's competence and advice from the very start in order to propose a satisfactory solution [7]. It has also been pointed out that the conflict between the architects and the engineers is traditionally one of the greatest problems in building design [2].

The process model described here is focused on facilitating collaboration and information reuse/evolution from the point of view of architects and structural designers by using information technology. It is our opinion that to control design information and reuse, evaluate, and evolve it would increase the designers' influence and control of the projects and in this way eliminate the problems described above.

2 Process Model

The design process model described in this paper can be divided into three parts:
1. Conceptual design
2. Detailed design
3. Feedback and evaluation.

2.1 Conceptual Design

The process model presented in this paper starts with a meeting between a client and an architect, a discussion about needs, wishes, and aims with the design of a particular artifact takes place. Afterwards, the architect works on a project brief which is the outcome of the first stage. The aim is to specify the objectives with the project.

In the next step, one or more preliminary geometrical prototypes of the project/building are created and the primary method used here is case-based reasoning. That is, parts from old projects are retrieved using the project brief as input. These parts are then preliminary adapted to the new project using relatively simple methods of parametric adaptation. It is in this step the collaboration between the architect and the structural designer starts. Because they work with geometry 3D-models from old projects, and not sketches as in the traditional process, the designers can directly start discussing the advantages and disadvantages of both the space layout and the bearing system, conflicts can be noticed and resolved promptly which will decrease the cost of failure.

> "What we need is a combination of the integrity of the engineer with the vision, sense of beauty and human understanding of the architect.... Only a few can do the whole thing alone these days, unless it is a very simple job" [1].

Allowing the structural designer's participation this early in the project will not force the architect to deal with problems which the structural engineer is more capable of solving.

If there is some part in the new project that cannot be solved by reusing parts from old projects the designers, of course, have to work on a new solution.

In the third step, the prototypes are presented to the client. It is very important that the client gets this opportunity as quickly as possible in a project so that he can actively participate in the design process and in that way prevent misunderstandings. Because he is shown full geometrical models and not only sketches, he has greater possibilities to understand and discuss alternative solutions by commenting and criticizing the prototypes. Using modern techniques, such as surround-screen display systems and variational design, it is also possible to achieve very realistic visualization and even manipulate the prototypes in real time while discussing.

The process described here does not need full adaptation in the conceptual stage. Nevertheless, the gaining described will occur even with a preliminary adaptation using some relatively simple methods of parametric adaptation. The changes made in this preliminary adaptation will be recorded for use in the detailed design stage.

2.2 Detailed Design

In conceptual design, the major outlines of the artifact are decided. The designers concentrate on elaborating the details in the detailed design stage. Here the major adaptation is performed, reusing parts of old design cases being an adaptation process

Because of the fact that the information about the process is, in general, more adaptable than the information about the solution [11, 2, 12], the primary method in this process is derivation replay. That is, the design processes of the old parts reused together with the changes made to these parts in the conceptual design stage.

For the structural designer the information in the design calculation documents can be used in the form of derivational replay which means primarily reusing the dependency structures of the documents [6]. In the same way as old solutions have to be adapted, these dependency structures do. This can be done by using transformation of parameters but also by substituting some part of or a whole dependency structure. In the case of substitution, analogous reasoning can be applied to find a suitable substitute [6].

This will not be a fully automatic process. Even though the changes collected from the preliminary adaptation process in the conceptual stage make it possible for automatic adaptation of the design calculations (design process) to a certain extent, this will probably not be the case for all changes necessary. Still, it is our opinion that the detailed design process described here is going to be quicker than the one in the traditional design process.

Meetings with the client interrupt regularly the detailed design process in order to visualize the artifact to the client. It is done in the same way as in step three above, the difference being that the artifact is more elaborated. The client is thus given a chance to participate actively, understand, and discuss the solutions. This is very important since he usually has a somewhat vague idea of the outcome before the project is completed. The understanding evolves gradually as the design process proceeds [13]. In the design process a number of options are explored in the search of the optimal one.

The changes emerging during these meetings can be collected, in the some way as during the conceptual design stage, and together with the information about the current design process (e.g. design calculation documents) it can facilitate the process needed because of these changes.

2.3 Feedback and Evaluation

Feedback has been a problem in the traditional design process. Most agree that it is very important but there has been no practical way to collect and store the feedback so that it can be used in future projects. Using case-based reasoning for storing feedback is rather natural [8]. The information from the feedback is stored and linked directly to the part that is criticized e.g. if a frame is hard to erect on the constructions site this information and the reasoning why can be keyed in and linked to the frame. When reusing the part, this information is easily retrieved and maybe the problem can be avoided by changing the solution.

2.4 Information Evolution

If a designer starts to reuse design information, he also has the opportunity to make this information more usable. When the designer becomes aware of that he reuses a design solution repeatedly, he will probably also be more eager to eliminate the solution's disadvantages, mentioned in the feedback information, in order to make it more usable. In the same way, he will try to make the design process linked to a particular design solution more reusable and more general [5]. The same will count for the geometrical model of the solution, which can be generalized using different techniques of variational design. Reusing design information, it is our belief, will make information evolution a natural part of the design process. And again, having access to a case-base with evolved design information will help the designers obtain more influence and control over the projects which, in our opinion, would eliminate some of the problems of the traditional design process.

3 Summary

The case-based design process described here offers the necessary means to facilitate collaboration between architects, structural designers, and clients. The use of case-based reasoning makes it possible to create preliminary prototypes much earlier than in the traditional process. These prototypes can be discussed and criticised and misunderstandings can be prevented early in the process. In order to receive the advantages described it is not necessary to make full adaptation at the conceptual design stage. It is here suggested that the major part of the adaptation is accomplished at the detailed design stage. By using the design processes of the parts reused and the changes of these design parts it is possible to speed up the detailed design process. In addition, the process described here supports information evolution through design reuse of tested solutions. Having access to evolved design information will also help the designers to gain more influence and control in the building process.

References

1. Arup, O.: The engineer looks back. The Architectural Review, Volume CLXVII Number 993 (1979)
2. Carbonell, J.G.: Derivational Analogy: A Theory of Reconstructive Problem Solving and Expertise Acquisition. In Mikalski, Carbonell and Mitchell, Eds. Machine Learning, An Artificial Intelligence Approach, Volume II, Morgan Kaufmann, Boston (1986)
3. Fruchter, R: Conceptual Collaborative Building Design Through Shared Graphics. in IEEE Expert, Volume 11, Number 3, Los Alamitos (1996)
4. Holgate, A.: The art in structural design. Oxford University Press (1986)
5. Johansson, P.: Case-Based Structural Design –reusing design calculation document information. Chalmers University of Technology, Division of Steel and Timber Structures, Publ. S 96:3, Göteborg (1996)
6. Johansson, P., Edlund, B.: Analogical Reasoning using Design Calculation Documents. EG-SEA-AI Workshop "Applications of Artificial Intelligence in Structural Engineering-IV" Lahti, Finland (1997) 181-188
7. Kalay, Y. E.: P3: An Integrated Environment to Support Design Collaboration. In Representation and Design, ACADIA, Cincinnati, Ohio (1997)
8. Kolodner, J.: Case-Based Reasoning. Morgan Kaufmann, San Mateo (1993)
9. Larsson, U.: Brobyggaren (In Swedish). Carlsson Bokförlag, (1997)
10. Lindgren, H.: A Framework for Computer-aided Reuse of Architectural Design Solutions. CIB W78 workshop on Computer-Integrated Construction, 12-14th of May, Montreal, Canada (1992)
11. Mostow, J.: Design by derivational analogy: Issues in the Automatic replay of design plans. Artificial Intelligence Vol. 40, No. 1-3, Elsevier (1989) 119-184
12. Raphael, B., Kumar, B., McLeod, I. A. M.: Case based design based on methods. EG-SEA-AI Workshop "Application of Artificial Intelligence in Structural Engineering", Lausanne, Switzerland (1994) 305-317
13. Wikforss, Ö.: Information Technology Straight Across Swedish Architecture and Construction. Division of Computer Aided Design and Visualisation, Chalmers University of Technology, CHT-A CADLAB 1994:3, Gothenburg (1994)

Shared Experiences: Management of Experiential Knowledge in the Building Industry

Erik Leijten, Ger Maas, Eric Vastert

Eindhoven University of Technology, Faculty of Building and Architecture
P/O Box 513, 5600 MB, Eindhoven, the Netherlands
e.leijten@bwk.tue.nl

Abstract. Organizations in the building industry have major difficulties with identifying, saving and distributing knowledge gained during the production process. In this article the authors describe this matter and focus on knowledge exchange between planners and experienced site and project managers. After intensive field study to the nature of planning an approach based on design methodology is suggested to better structure planning (both work and output). Experiences of site and project managers can be taken back into the organization with a tool based on case based reasoning. When designing the construction process the planner is then supported by knowledge gained at earlier planning and construction processes. A conceptual model of the architecture of this tool and a few examples of cases are presented here. The authors finish this article by bringing forward rising questions that give direction to further research.

1 Introduction

In the building industry the planning process has always been considered a crucial part of the construction process. During planning the underlying conditions are formed to acquire an optimal production. Time, money and efforts invested in work preparation will always pay back later on in the process. At least this is what we always hear form managers and other experts in the industry. In practice the available time for planning is always too short and managers are afraid of increasing costs of overhead. The overburdened property developers division or the highly respected project manager often regard a planner to be their personal assistant. Besides this a planner has to participate in various projects under great time pressure.

Work preparation is often defined in terms of 'directing' and 'checking'. During this research the planner appeared to be much more the spider in the web of projects who is despite his crucial position not held in great esteem. This not only counts for his colleagues but for planners themselves as well. This is one of the reasons why planning as a profession is mostly conceived to be just a career starter and planners are soon eager to move on. The consequence of this is that planners in general are more or less inexperienced and knowledge and experiences from the more experienced employees is not passed on into the planning division of the organization.

2 Knowledge management

Construction planning is a complex task. Projects interfere, tasks are ill defined and constantly extended, a lot of different factors have to be adjusted to each other and there is substantial time pressure. Earlier research indicated that planning is not quit the type of process as was thought before. Planning was described as a problem solving design process, as a iterative process in which analyzing, synthesizing and evaluating phases can be recognized.

In Eindhoven University of Technology we developed a concept of the construction plan. With this we have a tool to structure the labor and the output of the planner. The concept basically consists of a matrix structure with plans and scripts.

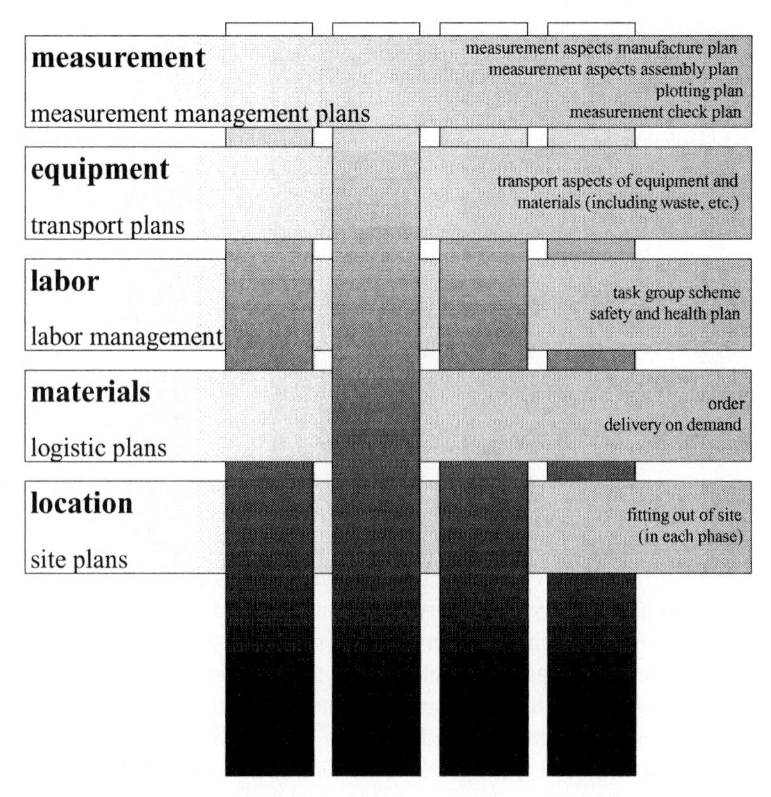

Fig. 1. Construction plan

A plan contains data concerning a specific production aspect like setting out, safety or equipment needed. All data concerning the production of a single part of the project or a building element are collected in a script [1].

Also upcoming theories like design & build or concurrent engineering, which are becoming very influential in the building industry nowadays, can better be implemented in a companies processes if the structure of these processes is more clear.

Another relevant issue is important to consider besides organizing data. By various causes a planner is not always up to date with experiences learned by site personal or project manager with the output of the planning process. Next to this it is most of the time hard to become acquainted with plans made by other planners in the same organization.

Because of this the planner lacks the possibility to learn from failures in the past or from changing conditions although the knowledge is already there, is already inside the organization. The knowledge is just not available to the right person on the right moment.

3 Case-based reasoning

We discovered the knowledge is there, it is in the organization, but wrongly spread. In this research we are searching for a technique with which it is possible to store experience in making plans and experience in using these plans and to bring back this knowledge into the organization again.

Case-based reasoning (CBR) is a technique within Artificial Intelligence that uses earlier encountered problems and earlier developed solutions. A new problem is compared with stored problems and accompanying solutions. If possible one or more is selected from the database and presented to the user. If a solution is not immediately available the most satisfying ones are adapted (if possible). The modified solution can then be used by the user and is in the same time added to the expanding database.

As is said by others often before, CBR has large advances compared to systems based solitarily on rules. Advances are for example the easier data collection, the fact that a system is learning as new cases are stored, companies consider the development of a system to be of low risk and experts are more comfortable in sharing their knowledge as it more or less speaks their own language [2] [3].

The link between design and CBR is appealing. Much of design knowledge comes through the experience of more or less unique previous design situations. A CBR system uses specific design knowledge to generate a new design [4]. Closely watching at old solutions is very common to every designer.

Considering planners to be designers makes the step to support planning by CBR rather small. Case based planning as described in [5] describes planning as actually remembering and recollecting old tasks.

To feedback experiences from production to the organization CBR seems to be a proper tool.

4 Postulate

Planning output and planning itself can be better structured by approaching work preparation as a design problem. Besides an improved and more structured output it becomes easier to keep a good overview over the various planning aspects. When a planner is able to use knowledge gained by experiences from earlier projects the organization is enabled to learn from earlier designed solutions and from failures as well. A tool based on CBR might be capable of supporting this knowledge management.

5 Research

This research program started early 1996. During the first year an introductory study after the phenomenon work preparation in the building industry was carried out in cooperation with TNO building research. Starting from 1997 and ongoing till the end of this year the research is continued with a new partner. This large construction firm called Hurks contractors is also based in Eindhoven. By working together with Hurks we develop an inside view over a building company and it's planning processes and we collect data for our theoretical research. We think to succeed in this by working very closely together with Hurks and by participating in daily practice.

Together with Hurks we search for the absent knowledge and we investigate if and where this knowledge is present. Besides tracing knowledge we try to identify the errors caused by these lacks of knowledge. To do this we need to get a complete overview how a plan, or part of a plan, can be characterized and recognized. The research is mainly carried out as a number of sequential case studies.

6 Preliminary conclusions

The result of this research will consist of three parts.

- An elaborated model of the planning process based on a design approach. Starting from the model of the construction plan we describe a detailed model of the planning process. The various plans summed up in figure one will become better defined and a method for composing scripts will be developed. Also we will define how to select the most complicated or risky product parts for which a script has be composed.
- Within this process model we develop an intelligent, supporting system. From this system we will finally present the architecture and a set of specifications. Up till now we have developed a conceptual architecture of this supporting system (figure two). It shows our first, basic ideas about planners retrieving and using knowledge gained by experiences from the past. It is based on common CBR theory and on our experiences with the nature of planning.

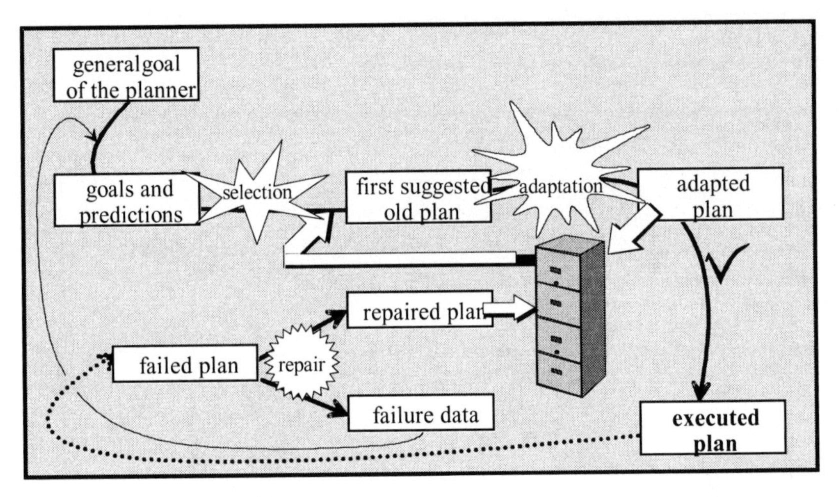

Fig. 2. Conceptual system architecture

- A part of this system will be further developed during this research and is currently under construction. With this we will describe the specifications of a tool which supports the planner when designing a plan for setting out prefabricated concrete elements (e.g. walls, columns or beams). We will develop this part of the system up till the level of specifications and system analysis.

The tool will contain a set of descriptive cases of past setting out experiences. These cases will be indexed by indices which are already familiar to the planner. Momentarily three independent indices are defined; production aspect, management aspect and building part. Production aspects refer to the already described model of the construction plan. The management aspects part covers themes like cost, time, quality, information and communication control. This index is being further explored by using the Baan IV software for the building industry, a major tool for modeling organizational processes. For the last defined index so far we use the so called 'Stabu' classification. This is the major method to describe a building and it's materialization in the Dutch construction industry. Because it is so widely used it is very useable and practical for indexing and retrieving. The use of these indices is illustrated by the example shown in figure three (still in a very early stage of development). Figure three shows a case that contains a description of a plan for the optimal techniques for setting out prefabricated concrete columns in both the X and the Y direction on a prefabricated floor using a total-station. The plan is represented through a detail of the situation with relevant setting out points, the location of the origin for setting out these points and the specific column. The indices through which the case selected are given in the first rows in this illustration.

Up till now about ten cases were collected. It is our goal to select about 20 more cases and to describe them in a more detailed way as can be presented at this moment. Also we want to test our so far selected indices in daily practice and if necessary extend or improve the set of indices.

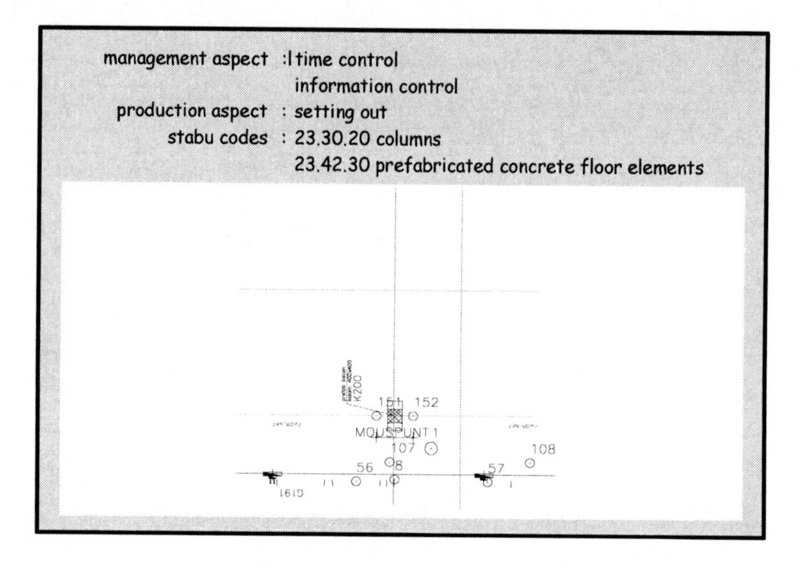

Fig. 3. Example of a case for setting out prefabricated concrete columns

As written earlier in this article we are momentarily collecting and describing cases and developing proper indices.

The hereby rising questions we have to deal with as how to prepare data for proper saving in a case and how far we have to go in defining and specifying indices to have them be optimal for use in planning departments will hopefully be answered in the forthcoming months.

References

1. Leijten, E., Maas, G., Vastert E.: Knowledge aspects of construction planning redesign. In: Drogemuller, R. (ed.): CIB W78 - Information technology in construction, proceedings. CIB/James Cook University, Cairns, 1997.
2. Kolodner, J.L.: Case-based reasoning. Morgan Kaufmann, San Francisco, 1993.
3. Leake, D.B.: CBR in context: the present and the future. in: Leake, D.B. (ed.): Case-based reasoning: experiences, lessons and future directions. MIT Press, Menlo Park California, 1996.
4. Maher, M.L., Balachandran, M.B., Zhang, D.M.: Case-based reasoning in design. Lawrence Erlbaum Ass., Mahwah, New Jersey, 1995.
5. Hammond, K.J.: Case-based planning: viewing planning as a memory task. Academic Press, San Diego, California, 1989.

Dam Safety: Improving Management

Eliane Alves Portela [*] and João Bento [†]

[*] Laboratório Nacional de Engenharia Civil (LNEC), Dams Department
Av. do Brasil 101, 1799 Lisboa CODEX, Portugal
e-mail: eliane@lnec.pt

[†] Instituto Superior Técnico (IST), Civil Engineering Department
Av. Rovisco Pais, 1096 Lisboa CODEX, Portugal
e-mail: joao@civil2.ist.utl.pt

Abstract. Dam engineering emerged as a structured branch of engineering in the earlier parts of this century. Because many large dams are now over 50 years old, special attention has been devoted to the issue of their efficient management. The act of managing large dams is very complex and requires the effort of multi-disciplinary teams involving highly specialised knowledge. In the case of dams, such knowledge is accumulated through a process of learning and understanding how each structure responds to the current and exceptional actions. Material properties may change during the course of their life, which means the structural responses to certain actions may change with time. Through the surveillance activity the changes in structural behaviour can be monitored and any anomalous situation should be identified by a thoroughly analysis. As old structures tend to deteriorate and request special attention from managers and as more structures are built every day, managers are faced with large amounts of information which need to be evaluated in order to highlight any anomalous situation which may endanger the structure serviceability or its safety, and trigger possible remedial actions. The present work has been dedicated to investigate how knowledge based system technology could help managers in the everyday duty of the dam safety control activity. The system is meant to support dam managers in the decision-making process but not to act as a replacement of sound human and engineering judgements.

1. Introduction

Safety and serviceability requirements of existing dams must endorse all relevant aspects of its history, namely: design, construction and operation. Such task is very comprehensive, demanding and often very puzzling. To ensure that a dam remains in good health, surveillance must be continuos and performed at a professional level. To assure timely perceptive analysis, those who process surveillance data must be selective. They must be able to sort out what may be important and study it quickly. Otherwise, there may be a tendency to bog down in the voluminous detail that can be generated by a comprehensive system of observation. After initial basic reviews, those chunks of data that may indicate questionable trends should be examined in greater depth [1].

AI concepts and techniques, namely knowledge based systems, are tools with a recognisable potential to assist the engineering activities related to monitoring, interpretation and diagnosis.

A few interesting associations between expert system technology and dam safety control activities have been reported with success in the literature [2, 3].

The work reported in the present paper concerns the development of one such knowledge based system aimed at identifying abnormal structural dam behaviour through the analysis of data originated both at the automatic monitoring system and as a result of regular visual inspections. The system is intended to encapsulate different types of information and knowledge about dam safety. It relies mainly in the results of both quantitative and qualitative monitoring data, design parameters, specific tests, operation history, site characterisation, expert knowledge on dam safety and on the historical behaviour of the specific structure being monitored.

2. Safety Assessment

The lifecycle of monitoring data starts with its acquisition on site, either by manual or automatic acquisition means. The original raw data suffers a first level check against basic parameters, such as the instrumentation range, and the sensor readings are translated to engineering quantities. On a subsequent second level such "raw" engineering quantities are validated against pre-defined mathematical models. Once the data is considered "reliable", the actual safety assessment may take place. Traditionally, most of the safety assessment tasks have been carried out by very experienced engineers, who have accumulated a great amount of experience about dam engineering and most specifically about each particular dam being assessed.

One of the main goals of the present research is to elicit engineering reasoning processes out of such a few dam experts, so that a model of that expert reasoning may be developed. By doing so, one expects to support less experienced staff and relieve experienced engineers from the everyday duty of routine safety assessment, allowing the expert more time for more complex problems.

Basically, three different levels of knowledge are identified: i) general background; ii) specific knowledge and iii) expert knowledge. The first type of knowledge corresponds to general scientific knowledge, such as structural mechanics, hydraulics and geotechnical engineering. Such general background knowledge has a very broad base of application, even though it is applied here within the relatively specialised context of dam engineering. The second form a separate chunk since each particular dam is a unique system and must be treated as such; its particularities must be identified to build the specific knowledge base on the dam. Finally, expert knowledge forms a fundamental part of the system, combining general background knowledge and specific knowledge about the dam. It enables reasoning processes leading to an overall evaluation of the whole system.

It is therefore obvious that the efficiency of the system depends deeply on the experience and adherence of the experts involved.

3. System Development

The system under development is being applied to a specific case of a double curvature arch dam in Portugal during its operation. It attempts to monitor a number of specific data values and spot abnormal ones or trends suggesting remedy measures to solve or mitigate foreseen problems. Monitored quantities are *water level*, *temperatures*, *displacements*, *strains*, *stresses*, *joint movements*, *seepage* and *uplift*. Whenever one of these quantities is detected to fall out of the expected range, the system should trigger an inference process leading to both a diagnosis and proposal for action [4].

Following the data collection, the monitoring data must be immediately processed and analysed with the results being checked in order to ascertain whether they conform to the pre-established behaviour models; this comparison may give rise to a reformulation of the models and a reassessment of the safety conditions of the works.

To explain an abnormal value found in the monitoring data, a dam engineering expert searches for a cause of the abnormality using his own expertise and general knowledge of the structure. Trying to mimic such reasoning processes, the system uses a set of complex and very large causal networks to identify possible scenarios associated to the detected anomalies. The causal networks and the relevant associated scenarios have been produced as a result of a thorough knowledge acquisition process.

Three groups of scenarios were defined:

- General scenarios - related mainly with the unusual or exceptional loads and deficiencies of design and construction;
- Foundation scenarios - related mainly to the dam foundation, such as weakness of the foundation under permanent or repeated loading, deterioration of grout curtains, erosion and solution, sealing of fractures, clogging of drainage system, plastic displacement and sliding of foundation, settlement and swelling of foundation material, and
- Dam body scenarios - related mainly to problems on the dam body, such as contraction, creep, swelling of concrete, loss of structural continuity, loss of structural strength under permanent or repeated actions and degradation due to chemical reactions of materials with the environment.

For each of the above scenarios a set of symptoms is established to confirm the relation between the abnormal measurement and the identified scenario, which will lead to a diagnosis. The set of conditions established for each scenario is based on the available cause-effect relationship, which may also affect the properties of the materials; and may be reflected in the monitored quantities in a specific way.

For each of the conditions it is defined a general descriptor, referring to its state: an example being *a slow decrease* or *a rapid increase*. The description by qualifiers is, by definition, subjective and a formal means of propagating subjective information is required so that consistent results can be obtained. Nevertheless, it is not an easy task to find “indicators” univocally connected with the safety state of the structure, and consequently one has to rely on rather indirect and tortuous procedures of “inference”, on the basis of easily readable “indicators”.

Each scenario is also linked to more specific symptoms: geological and geotechnical conditions, operation schemes, visual inspection data, water quality data, concrete properties, design and construction data, etc.

The correct identification of a scenario depends on the description of the appropriate symptoms. Part of the ability of the system to focus the search for possible scenarios on the ones that might afflict the structure, results from an automated identification of which symptoms are of primary importance and which are of secondary importance for the diagnosis of abnormal behaviours.

Primary symptoms are sure signs that a particular scenario has indeed taken or not taken place and, therefore, present strong evidence for the diagnosis of a particular type of scenario. Secondary symptoms provide some information about possible scenarios and represent weaker evidence for such particular scenario. Secondary symptoms become relevant when a fine-tuning of the diagnosis is required or if insufficient information is initially available to make a diagnosis.

Once a scenario is identified by the system through association of symptoms, the system compares its findings with a knowledge source containing a series of problems with possible recommendations. If a match occurs, the system displays the result.

Recommendations may consist of remedial measures, methods leading to further investigation to confirm a diagnosis, a preventive measure, or even a mere (justified) statement that no action is required.

4. Conclusions

The only way to prevent deterioration of structures and its natural consequences – danger of loss of serviceability and safety– is by enforcing a proper, thorough and timely assessment of the structure.

A knowledge-based system which identifies abnormal dam behaviours through the analysis of data coming from a monitoring system and visual inspection was presented. The system is intended to help managers overcoming the difficulties related with this activity and to support the full understanding of a dam's behaviour, while promptly identifying, at early stages, a path to a possible incident. Indeed, an immense set of domain knowledge needs to be invoked, including that of previous cases, in order to spot a problem, diagnosing it, and enabling the proposal of therapeutic action. No such task could be easily and effectively performed without the aid of AI.

Acknowledgements

The present work has been funded by the PRAXIS XXI Program, through research grants no. BD/278/94 and 3/3.1/CEG/2547/95.

References

1. Jansen, R. B. (1983) *Dams and Public Safety*. A Water Resources Technical Publication. U.S. Department of the Interior, Bureau of Reclamation, 332 p..
2. Salvaneschi, P.; Cadei, M.; Lazzari, M. (1997) "A Causal Modelling Framework for the Simulation and Explanation of the Behaviour of Structures" in *Artificial Intelligence in Engineering*, 11,205-216, Elsevier.
3. Zaozhan, S.; Zhongru, Wu (1994) "Cause Analysis in Dam Safety Assessment", in Dam Engineering, V-1, 31-41.
4. Portela, E.A.; Bento, J.; Ramos, J.M.; Silva, H.S. (1998) "The Safety Control of Dams: Improving Management Through Expert System Technology", Proceedings of the Fourth World Congress on Computational Mechanics, Buenos Aires, 29/June-2/July/1998 (to appear).

Integrating Virtual Reality and Telepresence to Remotely Monitor Construction Sites: A ViRTUE Project

Arkady Retik[1], Gordon Mair[2], Richard Fryer[3], and Douglas McGregor[3]

[1]Department of Civil Engineering, University of Strathclyde, Glasgow G4 0NG a.retik@strath.ac.uk
[2]Department of Design, Manufacture and Engineering Management, University of Strathclyde, Glasgow G4 0NG g.m.mair@strath.ac.uk
[3]Department of Computer Science, University of Strathclyde, Glasgow G4 0NG {rjf, g.mcgregor}@cs.strath.ac.uk

Abstract. The paper presents ViRTUE, a funded research project, which aims to integrate the Virtual Reality (VR), Telepresence (TP) and mobile video telecommunications technologies. A mobile, real-time, 3D-hybrid VR/TP system is being built at Strathclyde University, Glasgow. A system prototype has been completed and is being tested. The system will permit the user to integrate telepresence images with computer generated virtual environments superimposed over the remote real worldview. This integrated system incorporates emerging mobile telecommunications technologies to give rapid and easy access to the real and virtual construction sites from arbitrary locations. This system allows remote surveillance of the construction site, and integration of real world images of the site with virtual reality representations, derived from planning models, for progress monitoring.

1 Background

The complexity of the operations required to construct complex buildings or civil engineering structures makes efficient monitoring and control the key management issues. The gathering of site-specific information and its integration with current and planned activity is, therefore, vital.

However, there are still many problems of efficient monitoring and control as a consequence of the scale of operations undertaken. Moreover, new procurement techniques (such as *design-build*) which have been introduced to speed up delivery of projects, demand closer interaction with sites by parties involved, such as designers, clients, etc. Efficient communication with the sites from remote locations is, therefore, a crucial issue for the project success.

The project addresses these problems by providing remote (fixed or mobile) access to a 3D visual presentation of the project status (presenting actual vs. planned) and thus enabling a correct understanding of both current progress and work to be done. Reviews of research work in this area are presented in [1] and [2].

2 Project Approach

This project integrates the Virtual Reality (VR), Telepresence (TP), and mobile video telecommunications technologies. In the project context, VR involves the impression of participating in a computer generated synthetic environment and TP the impression of being present at a remote construction site. A mobile, real-time, 3D-hybrid VR/TP system is currently being built. Once fully completed and tested, it will permit the user to integrate telepresence images with computer generated virtual environments superimposed over the remote real worldview. This integrated system incorporates emerging mobile telecommunications technologies to give rapid and easy access to the real and virtual construction sites from arbitrary locations. It also allows remote surveillance of the construction site, and integration of real world images of the site with virtual reality representations, derived from planning models, for progress monitoring.

The key elements of the approach are:

a) *Integration of real and synthetic images* (also called augmented reality): The intent is to allow real images to be integrated into the 3D world. Such an ability will be crucial for design and planning of construction projects, which are always an integral part of a project environment.

b*) Access to remote sites*: Both the project progress monitoring and site control can be performed remotely and examined interactively providing more efficient use of the manager's time. Comparison using intelligent superimposition of 'actual vs. planned' situations will create possibilities which are not available today, i.e. remote site inspection and progress verification by the client.

3 System Description

The system consists of three main parts:

1) VR module for planning and scheduling of construction projects as well as a platform deployment;
2) Telepresence platforms for acquiring in real time real world data on the remote site;
3) Communications module for linking a construction site with remote users.

The conceptual scheme of the system presented in Figure 1.

VR Module. This module allows creation of a 'virtual construction project' from a schedule and subsequent visual monitoring of, and interaction with, the progress of the simulated project. It is based on the prototype developed at Strathclyde's Virtual Construction Simulation Research Group (VCSRG) and described in detail in [3].

Camera Platforms. In this project the intention is to construct two camera platforms. The first will be operated from a fixed position and have a simple pan and tilt monocular camera system with zoom capability. The compact camera, drive, and control system may be powered either from a mains or battery supply.

The second platform will be mounted on a remotely controlled tracked vehicle, which is provided by the vehicle manufacturer (Alvis Logistics Ltd.). This platform has a stereo camera pair capable of pan, tilt, and roll movements and it will be powered from the vehicle batteries. The vehicle and platforms at the remote site will be controlled from a home site.

The camera platforms and the associated control system are being developed from the "Mark I" Strathclyde Telepresence System developed by the University's Transparent Telepresence Research Group [4]. An illustration of the original system is shown below

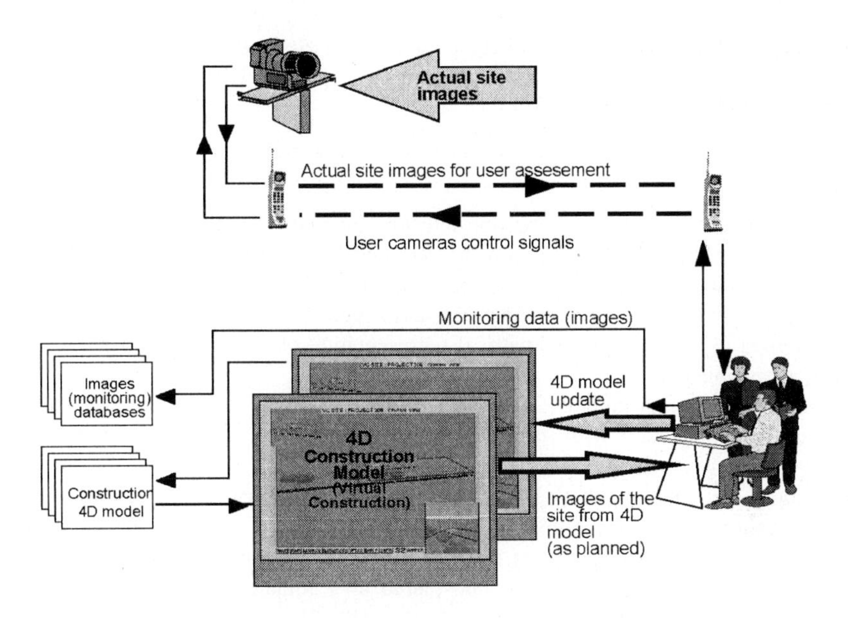

Fig. 1. Interactive visual monitoring of a remote site – a system concept

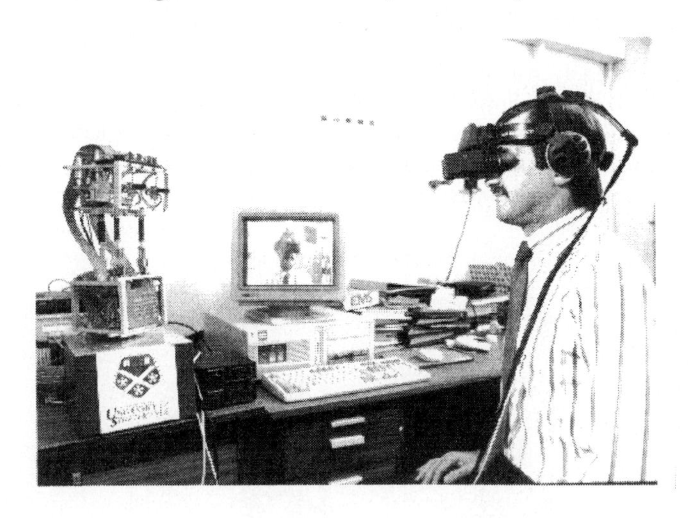

Fig. 2. "Mark I" telepresence system

(see Fig.2). This system has already been operated over long range using ISDN2 (128 kb/s) on a number of occasions. However the new system will have unlimited range and be able to operate from battery power thus making it truly mobile.

Communications Module. The user of the system must be linked to the remote site via a communications channel. A wireless system has been selected using domestic mobile telephones. In order to send video images over the mobile telephones, video compression technology is required. This is based on software developed in the Computer

Science Department, University of Strathclyde, and optimised for the low bandwidths of mobile telephones [4]. Both monoscopic and stereoscopic pictures will be transmitted for display on a VR Headset or user's screen.

4 Discussion

Various problems specific to the construction industry have been identified which should be resolved by using the output of this project. Visualisation of progress of construction requires representation of changes in the geometry of a building or structure. Such changes prevent efficient use of existing 3D architectural and structural models for planning purposes [5]. Moreover, geometrical representation of construction activities is not always 'compatible with'
or even presented by design models. Therefore, integration of real images of a site with simulation of, and interaction with, the construction process will require the ability to model dynamic changes to the site geometry.

Integration of real and synthetic worlds is commonly known as Hybrid VR or Augmented Reality (AR). Azuma [6] defines an AR system as a system which: (a) combines real and virtual worlds; (b) is interactive in real time and (c) is registered in three dimensions. AR, therefore, can be considered as a 'middle ground' between VE (completely synthetic) and telepresence (completely real) [7], or as a virtual world that enriches rather than replaces the real world [8].

Creation of AR worlds can be done by 'blending' real and synthetic images in several ways. These will mainly depend on the technology used, which can be either optical or video based. An optical technology uses an optical see-through Head Mounted Display (HMD), which allows the user to view the merging of real word with the virtual world on an optical combiner in front of the user's eyes. Video based systems employ a video compositor to merge images, which are displayed either using a HMD (immersive mode) or monitor (non-immersive mode). This approach is a basis for the system.

Fig. 3. 'Actual *vs* planned' project monitoring

There are many ways in which the video composition can be carried out. Chroma-keying, overlaying, superimposing, combining, annotating are among the techniques already being reported [6,7,8]. In this case, one of the possible ways of superimposing actual (real images) versus planned (virtual images) construction work progress is demonstrated in Fig.3. At this stage these images are calibrated and superimposed manually. However, time based simulation of the construction process and complexity of construction sites create difficulties even for simple superimposition and overlapping cases. Issues such as organising and the management of video information, hierarchical representation of the project, separate presentation of critical activities, image based retrieval, 3D matching and viewpoint calibration may well require the use of advanced AI techniques in order to produce efficient solutions.

5 Summary and Conclusions

The project aims to research, develop, build, test and critically analyse the performance of a prototype mobile hybrid VR/TP system for use on construction sites. The system has being deployed for progress monitoring and the site control of the real project. The performance and effectiveness of the system and approach will thus be assessed and examined in the real situation. Whatever the results of the experiment as far as the effectiveness of remote access and monitoring concerned, it seems that the means of visual information gathering and integration of real and virtual worlds will create possibilities which are not available today.

Acknowledgements

This work is supported by the EPSRC MNA programme (Grant GR/L 06164) in collaboration with ALVIS Logistics, Orange (Nokia) & Babtie Engineering. The contribution of N.Clark, R. Hardiman, N.Retik and K Revie is acknowledged.

References

1. Mair, G.M., Fryer, R., Heng, J., Clark, N., Sheat, D.: Long Range Telepresence Using ISDN, 'Mechatronics'96 Conference, Portugal (1996)
2. Retik, A.: VR System Prototype for Visual Simulation of the Construction Process, EPSRC Conference, Salford (1995) 90-93
3. Retik, A., Clark, N., Fryer, R., Hardiman, R., Mair, G., McGregor, D., Retik, N., Revie, K.: Mobile Hybrid Virtual Reality and Telepresence for Planning and Monitoring Engineering Projects, Proc. 4th UK VRSIG Conference, London (1997) 80 – 89
4. Lambert, R. B., Fryer, R. J., Cockshott, W. P., McGregor, D. R.: Low Bandwidth Video Compression with Variable Dimension Vector Quantisation, Proc. ADViCE'96, Cambridge, UK (1996)
5. Retik, A.: Geometric-Based Reasoning System for Project Planning, Discussion Paper, ASCE J. Computing in Civil Eng-g, Oct (1995) 293-294
6. Azuma, R.T.: A Survey of Augmented Reality, Presence, Vol6, No4 (1997) 355-380
7. Milgram, P., Rastogi, A., Grodski, J.: Telerobotic Control Using Augmented Reality. IEEE Intern Workshop on Robot and Human Communication (1995) 211-29
8. Feiner, MacIntyre, Sellgmann.: Knowledge Based Augmented Reality. Communications of ACM, Vol36, No7 (1993) 53-62

Proposal for 4.5 Dimensional Design via Product Models and Expert System

Mika Salonen, Student, Tampere University of Technology, Tampere, Finland

Juha Rautakorpi, Student, Tampere University of Technology, Tampere, Finland

Markku Heinisuo, Tech. Dir., KPM-Engineering, Tampere, Finland

Introduction

Four dimensional design including space and time has been introduced recently [1] in civil engineering projects. The main interest of that paper is visualization following the stages of construction and scheduling. The costs involved in the project have been proposed in new industrial foundation classes (IFC) proposed by the IAI organization [2]. These classes will be used in the future when applying product models in building projects. The present paper deals with a design system where it is possible to combine the space and time (4d) with costs. One can variate space, time and costs in the beginning of the project. Later, during the project, these coordinates are non-orthogonal. Moreover, the norm in five dimensional space is not clear. So called 4.5 dimensional design is proposed in the paper. The tool to perform this is to apply profound neutral product models [3, 4] which have been formulated using parts 11 [5] and 21 [6] of the standard ISO 10303 (STEP). The control of the models is made by applying a knowledge based expert system, called FST-EXPERT [7]. The application field is the design of steel structures.

Applied Design Process

In the following the reference is made to the design of steel structures, but many proposals hold when considering other structures, too. In many papers [8, 9] it has been proposed, that the starting point of the design process is an analysis model of the structure. After which the geometrical model of the structure is made. It has been shown in practical examples [10, 11] that real working design systems can be made only by starting from the geometrical model. The same conclusion can be made considering the simple k-joint of a truss: how it is possible to design cost effective joint without difficult cuttings without starting from the geometrical model?

The design process must somehow follow the process presented in Fig. 1 in order to make reasonable designs.

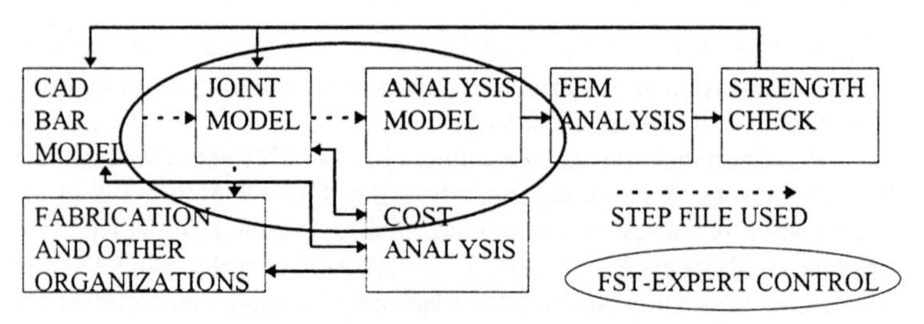

Fig. 1. Proposed design process [12]

A prototype of this type of design process has been carried out using the knowledge based expert system FST-EXPERT. The main features of the product models applied in the study and by taking into account the demands of the 4.5 dimensional design are explained in the following, briefly. The cost estimation process and control of the models using FST-EXPERT will also be described.

Product Models

No hardware or software will ever reach the monopoly status. Therefore, the use of neutral formats when transferring data between models and organizations is recommended. The use of the enormous STEP as a whole is still far away in the future (at least in the field of civil engineering). Of course, this situation is constantly changing due to increasing capacity of software, hardware and know how. The two parts, 11 and 21, are widely used as stand alone tools in practice (e.g. in IFC and in this study). This means that we are working in a hardware and software independent environment. There are some good and some bad features when using these type's of systems [13].

The product models used in this study have been made applying the concept of primitives and careful combinations of them. The primitives for steel structures and the main ideas to combine these are given in [14, 9, 15]. Similar primitives for wooden structures are given in [8]. Two different product models are used: geometrical and analysis. The union of these models is called a FST-model. The geometrical model and the analysis model (structural analysis, FEM) are kept as separate models because there are not many software that can handle both geometrical and analysis modelling. The automatic generation of the analysis model from the geometrical model is made using FST-EXPERT, almost without extra work by the designer. There does exist other systems to produce the analysis model from the geometrical model [16, 17], but they do not use neutral data files.

An atomic part, a primitive, of the building (in this case the steel skeleton) is starting point in the product model (bottom up). From the point of view of the designer the primitives can be profile HEA200, steel grade S355J2K3, bolt M20 or similar. From the point of view of some other organization (fabricator, architect, end user etc) the primitives may be something else. When making a model of the final product the next stage is to define some composition mechanism to the primitives. The designer may need to combine e.g. the profile HEA200 to the steel grade S355J2K3 in order to make a more final product. The goal is to produce the final product from the scope of the designer, in this case the steel skeleton.

It must be noted that the composition of the primitives and some parts at the higher levels in the hierarchy (the same composition mechamism can be used in the FST-model not only to compose primitives, but other entities also) always includes some logistics transportation and costs are involved in the compositions. Composition of the artifacts can be done by simulating the fabrication process. The designer is not an expert on the fabrication process, but he must make some proposals for the combinations of the primitives and higher level parts in order to define the final product. The FST-model is made especially by taking into account of the need to simulate the fabrication and erection processes of steel skeletons as accurately as possible. The expert knowledge dealing with fabrication and erection processes have been used when defining the FST-model. The designer can propose some order to make the final assembly (beam, column, truss or similar) and the fabricator can change the order of combinations so that it fits well to his fabrication process. Moreover, the final skeleton on site can be composed many ways from the assemblies made in the workshop. Here the expertise from the erection group is also needed and the product model must be flexible in order to take into account all needs.

The product model including the geometry of the building must include all the space information of the entities, the way to compose the building starting from the primitives in order to simulate the fabrication and the erection processes. By these means it is possible to put the time scale to the model. The costs are not put into the model in this study by defining the unit costs to all the entities as in the IFCs. In this study a special cost aggregation form [18] is used when estimating the costs effects of the different combinations of the primitives. In this study this means the estimation of the effect for different joint layouts to the costs of the steel skeleton. An Australian cost information [19] is used for units costs. The visualization of the entities is outside the scope of this study and so, the time scale is not demostrated here.

The product model of the analysis model is made by composing the primitives of the finite element method such as nodes and finite elements [20]. The analysis model is made automatically from the geometrical model by the control of the FST-EXPERT. The control system is explained in [7] and it includes such information as the element types available in the FEM application, number of elements used for mid bars (bars between joints), the local joint models (beam models or plate models) etc.

FST-EXPERT

As explained above, the product model of the geometry must include spatial information and much more in order to be used to simulate the real construction process. The time scale and cost information can be included into the composition entities. The FST-model used in this study apply the composition entities for cost estimation. Control of the model is made using the FST-EXPERT.

The time scale is not present (yet), but the costs are estimated using the cost aggregation form of the steel skeleton. The use of a very detailed cost aggregation form is extremely difficult due to lack of proper information in the free literature. The Australian information was the only one found in this study. In distinct phases of the fabrication and erection can be used rather specific data (e.g. flame cutting or welding cost). The use of specific data and the information derived from that is called deep knowledge.

Due to lack of the information of the proper costs and other fabrication and erection data, so called shallow knowledge can also be used within FST-EXPERT. FST-EXPERT can be used to take into account e.g. the fabricator's capacity to produce the steel skeleton using the joints especially suitable to the fabricator's production system. The shallow knowledge must be used also, when evaluating e.g. the structural behavior of joints especially in three dimensions. There is very little information available in the literature dealing with the latter. Special weighted tables with weighted properties with "standard" suitability factors of expert systems are applied in FST-EXPERT to assist in the decision making process of the structural designer. The expertise knowledge in the tables has been collected from Finnish experts in the field.

The coding of FST-EXPERT is done using C++ and Visual Basic for the graphical user interface. The model check is done by applying ECCO-toolkit [21]. In examples the geometrical models have been made using the AutoCAD™ program and the finite element analysis have been run using the ABAQUS™ and the FINNSAP™ programs. The deep knowledge of the behaviour of some structural steel joints have been derived using the QSE Structural Office™ program.

Conclusions

A 4.5 dimensional design system is proposed in this paper. The application field is the structural design of steel structures. The four dimensions are space and time. The costs of the final structure is an additional, non-orthogonal coordinate. The fabrication and the erection of the structure can be simulated if the product model includes composition classes which are used to compose the primitives and other entities of the case. The cost distribution of the steel skeleton is simulated using the separate cost aggregation form, combined to the product model using the composition classes, not using cost attributes for the classes of the geometrical model. Visualization is not included in the system (yet). The knowledge based expert system is used to control the models and the KBES is used as a graphical user interface. Not only deep, but also shallow knowledge can be used in the system. This is an important possibility, when dealing with incomplete information, such as costs and structural behavior of joints in three dimensions in some cases.

The preliminary tests have shown, that the system gives quite reasonable results both for preliminary and for final design tasks. The costs effectivity and the "fabrication led design" or "fertigungsorientierte Produktmodellierung" must be the goal in the future in order to improve the competition ability for different structural systems.

Acknowledgements

The financial support of TEKES, Kvaerner Pulping Oy, Rautaruukki Oy, A-Insinöörit Oy and PR-Steel Oy are gratefully acknowledged. The study is part of the FINNSTEEL technology program.

References

1. http://gaudi.stanford.edu/4D-CAD/INTRO-4DCAD.HTML
2. International Alliance of Interoperability, http://www.interoperability.com.
3. Heinisuo M., Product model for the analysis of skeletal structures, Applications of Artificial Intelligence in Structural Engineering - III, Ed. I. MacLeod, Proceedings of the 3rd Workshop of the European Group for Structural Engineering Applicationd of Atrificial Intelligence (EG-SEA-AI), Ross Priory, University of Strathclyde, UK, August 12 - 13, 1996, pp. 55 - 56.
4. Hyvärinen J., FST - Finnish product model of steel skeleton, Applications of Artificial Intelligence in Structural Engineering - III, Ed. I. MacLeod, Proceedings of the 3rd Workshop of the European Group for Structural Engineering Applicationd of Atrificial Intelligence (EG-SEA-AI), Ross Priory, University of Strathclyde, UK, August 12 - 13, 1996, pp. 41 - 46.
5. ISO 10303-11: Industrial automation systems and integration - Product data representation and exchange - Part 11: Overview and fundamental principles. International Organisation for Standardisation, Geneva, 1994.
6. ISO 10303-21: Industrial automation system and integration - Product data representation and exchange - Part 21: Implementation Methods: Clear text encoring of the exchange structure. International Organisation for Standardisation, Geneva, 1994.
7. Heinisuo M., Hyvärinen J., Knowledge based approach to the design of structural steel joints, Information Processing in Civil and Structural Engineering Design, B. Kumar (Editor), CIVIL-COMP Ltd, Edinburgh, Scotland, 1996, pp. 75-83.
8. Osterrieder P., Haller H.-W., Saal H., Bauspezifische, fertigungsorientierte Produktmodellierung im Stahlbau und Holzbau, Bauenginieur 72, 1997, pp. 489-496.
9. CIMSteel: Integration Standards (Release One), University of Leeds, Department of Civil Engineering, 1995.
10. Mikkola M., Computer Aided Design of Wood Trusses using Nail Plates, Proceedings of CIVIL-COMP 85, Volume 1, CIVIL-COMP PRESS, Edinburgh, 1985, pp. 149 - 154.
11. Heinisuo M., Möttönen A., Paloniemi T., Nevalainen P., Automatic design of steel frames in a CAD-system, Proceedings of the 4th Finnish Mechanics Days, Ed. Niemi E., Research Papers 17, Lappeenranta University of Technology, Lappeenranta, 1991, pp. 197 - 204.
12. Hyvärinen J., Expert systems for design of steel structures, Licenciate thesis, Tampere University of Technology, 1998, (in preparation).
13. Heinisuo M., Product Models for the Data Transferring in Building Projects, Application Steel Structures, Proceedings of the 7th International Conference on Computing in Civil and Building Engineering (ICCCBE-VII), 19 - 21 August, 1997, Seoul, Korea.
14. Heinisuo M., Hyvärinen J., Hierarhical Aggregation Form of Steel Skeleton, Product and Process Modelling in the Building industry, Ed. R. J. Scherer, A. A. Balkema, Rotterdam, 1995, pp. 147 - 154.
15. Haller H. - W., Ein Produktmodell fur den Stahlbau, Dissertation, Bauingenieur- und Vermesseungswesen der Universität Friedericiana zu Karlsruhe (TH), Karlsruhe, 1994.
16. Turkiyyah G. M., Fenves S. J., Knowledge-Based Assistance for Finite-element Modeling, IEEE EXPERT INTELLIGENT SYSTEM & THEIR APPLICATIONS, Vol. 11, N. 3, 1996, pp. 23 - 32.
17. Remondini L., Leon J. C., Trompette P., Towards an integrated architecture for the structural analysis of mechanical structures, Information Processing in Civil and Structural Engineering Design, B. Kumar (Editor), CIVIL-COMP Ltd, Edinburgh, Scotland, 1996, pp. 65 - 73.
18. Heinisuo M., Jenu M., Knowledge Based Tool of Steel Structures Including Economical and Structural Aspects, Computers in the Practice of Building and Civil Engineering, Wolrdwide ECCE Symposium, Fellmanni Institute, Lahti, Finland, September 3 - 5, 1997, pp. 136 - 140.

19. Watson K. B., Dallas S., Van Der Kreek N., Main T., Costing of steelwork from feasibility through to completion, Steel construction, Vol 30, Number 2, June 1996, pp. 2 - 49.
20. Hededal O., Object Oriented Structuring of Finite Elements, Ph. D. - Thesis, Department of Building Technology and Structural Engineering, Aalborg Universitet, AUC, Aalborg, Danmark, Engineering Mechanics, Special Report No 1, 1994.
21. ECCO Tool Kit, An Environment for the Evaluation of EXPRESS Models and the Development of STE based IT Applications, http://www-rpk.mach.uni-karlsruhe.de/kompetenzen/Ecco.

A Product Information System Based on Dynamic Classification

Steffen Scheler;
Technische Universität Dresden
Lehrstuhl für Computeranwendung im Bauwesen;
Germany; Steffen.Scheler@cib.bau.tu-dresden.de

Introduction

Product data provided by product information servers are a valuable property. However, without a flexible classification structure the data is not much worth, because the client is not able to find easily the information he needs. The classification structure of the information system is of upmost importance to support the client in his navigation process and retrieve the right data for the right problem at the right time.

A pre-given classification structure, as it is good practice in current product information systems, means a predefined and fixed ordering of the data and therefore a fixed representation of the domain knowledge. It represents a particular predefined search strategy and not - as it would be desirable - a generic search strategy. What we would like to have is a highly flexible, adaptive classification structure, adaptive to the requirements of the particular searches, and flexible to the intention and objectives of the user. Therefore catalogues and the reasoning context of the engineering task have to be dynamically linked, and not only simply merged [1].

Concept

Product information systems provide information on products on the basis of attribute/value pairs. To be useful for a given design task this information has to be interpreted in specific engineering contexts. The gap between the pure data of catalogues and the reasoning context in the engineering task has to be bridged, and not only simply merged.

The context in which a product solution is evaluated and classified is dynamic and varies with the intention and objectives of the user. Two levels of product representations exist:

- *The catalogue data level*, that describes the product properties based on attributes and associated values.

- The *task level*, that classifies the product in categories associated with the actual design task.

Consequently two different representations have to exist. The representation on the catalogue level data can be a data structure enumerating the product properties in fixed form. This is an abstraction of arbitrary implementations of persistence of product data.

In general the relational model of conventional databases may be chosen for the implementation of this level because of its efficiency. In the context of product information that is traditionally document-oriented, it might also consist of documents structured via pre-defined tags with associated values in a philosophy corresponding to the Standard Generalised Markup Language (SGML) that structures information in documents for easy navigation and retrieval.

In any case we can characterise the catalogue data level as a flat and fixed data structure. An implicit hierarchical organisation of product items in categories linked with similarity relations, specialisation and generalisation is not considered appropriate, since these taxonomies have to correspond to categories that are context-dependent and not only supplier-dependent.

This context related to project, designer, enterprise and discipline is dynamically captured on the task level. The representation on the task level has to group products, in possibly differing supplier-dependent specifications, in categories relevant to the actual design task.

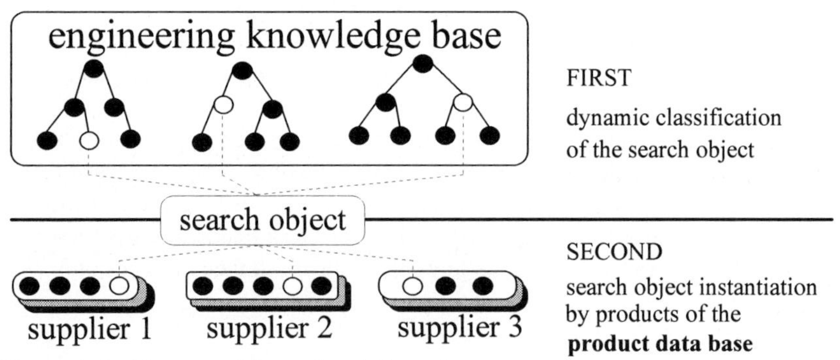

Figure 1: Architecture of the product information system

Our approach is a strict separation between supplier product information and domain knowledge representation and an explicit bridging of the two when needed. This bridging is to be achieved with a special object called *search object.* The task of the search object is twofold [Figure 1]. First, the search object can be matched on an abstract level to the knowledge taxonomies, and knowledge properties can be inherited from the knowledge taxonomies to the search object. With this step, we are able to fulfil the first user requirements of being advised by the system, i.e. the system will support the user in figuring out his search problem by providing domain knowledge. In the second step, the enriched search object is linked to the supplied product data by instantiation. Only those product objects are linked, which fulfil the search requirements. This architecture allows to distinguish clearly between supplier-independent advice about products and supplier-dependent providing of product data. Besides the user-friendly advisory service, this architecture allows also distributed product data storage and a liberal, because supplier-oriented, storage of the product data. The efficiency of the search process depends on the classification structure, the matching technique and the inheritance power offered by the Product Information System. Above all, the architecture gives the possibility of competition on a supplier independent domain knowledge level between different Product Information System providers.

A successful approach to the representation of the associated knowledge in such highly structured domains, are the frame-based respectively object-oriented formalisms. In these approaches the states of objects represent their properties, methods allow to represent the behaviour of objects, inheritance can be used to abstract from unimportant features, encapsulation hides inner mechanisms and referencing expresses the relations and dependencies between objects.

The object-centred approach, which is described in the next chapter, allows not only the dynamic association of search requirements to a knowledge taxonomy and a stepwise refinement of the search requirements by inference of knowledge and user intention, but also provides such a high flexibility that the distributed knowledge taxonomies are dynamically or adaptively configured for each particular search problem.

Dynamic Classification of Supplied Objects

The essential goal of our research is the development of a method for the dynamic linking of the object-oriented representation of engineering knowledge available in a product data model with the usually relational product data representation, available on the servers of product suppliers. Such servers are simplified in our study by one relational, flat database. The process of classification is based on the object-centred approach as implemented in LOOM [2] and applied in [3] for civil engineering problems. The essence independent of the object-centred approach is the definition of class structures from the objects which are instantiated at run-time.

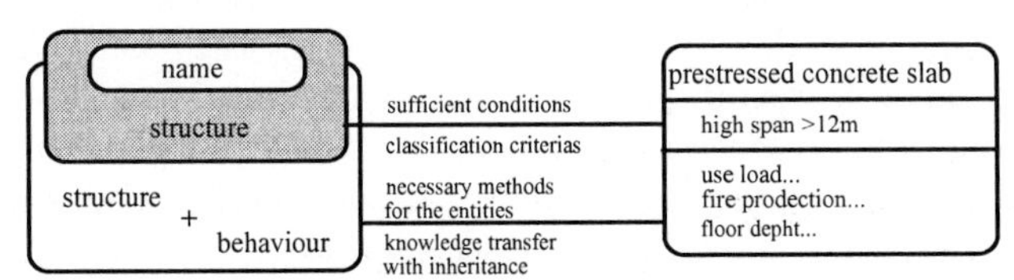

Figure 2: Definition of an instance in the approach

In the object-centred approach, a class defines (a) sufficient conditions, which are the criteria for class membership, and (b) necessary conditions, which represent declarative and procedural knowledge applicable to each member instance of the class. An object which fulfils all sufficient conditions is classified to be a member of this class. With this approach it is possible to implement active knowledge transfer, because each object fulfilling the sufficient conditions can be attached to the class and inherits all necessary conditions [4] [Figure 2]. Furthermore, in the object-centred approach it is possible that objects are instances of different classes at the same time and, more than that, due to the dynamic change of the object properties, an evolution of the class membership is possible [5].

From our point of view the application of the object-centred approach provides the following important advantages for a flexible representation of products in an object-oriented model:

- Dynamic classification based on the sufficient conditions represented in the class structure.
- The multifunctional behaviour of an object in different aspects (engineering sub-domains), which are not necessarily pre-linked. The advantage of the object-centred approach here is that an object can be an instance of two mutually independent classes belonging to different aspects.
- The properties of the search object can be changed during the design process on the one side and during the product search due to the refinement of search requirements on the other side.

Establishing of Knowledge Base and the Search Process

First the engineering **knowledge base** has to be established, because only a generic data structure and the basic subsumption algorithms are made available by LOOM. The basic domain knowledge has to be programmed with the help of LOOM routines [6]. Based on this initialisation program combined with the deductive support provided by LOOM, a knowledge base is generated which consists of definitions, rules, facts, defaults and alternatives. This semantic network, enriched with declarative knowledge, is established with the aid of the above mentioned classifier of LOOM. The structure of the knowledge base is comparable with a generic product model similar to STEP (ISO 10303). It consists of classes respectively objects, relationships and methods. In this network of the classes and their relationships the attributes which characterise each class are determined. All necessary conditions for further specification of classes and relations automatically result from the inherent logical connections.

The initiation of the **search process** is based on the requirements specified by the user according to his specific design task. Based on this requirements a *search object* is instantiated and then dynamically associated to the existing knowledge classes. The dedicated steps of the navigation process are:

1. Instantiation of the search object with arbitrary start-up requirements, produced by the user.
2. Classification of the search object.
3. Inheritance of all necessary conditions of the knowledge objects by the search object. These are used in the matching process with the product data.
4. If no classification of the search object can be carried out, LOOM assumes that the search object represents completely new knowledge and generates a new single instance without classification.

The above mentioned process is repeated if the specification of the search object is altered by the user for the matter of refinement of the requirements of the search. As a consequence, the search object absorbs gradually more and more refined specifications, which are then used, for the query of the relational data base containing the products provided by the product suppliers. Finally, all appropriate products found in the product data base will be instantiated to the search object and on the appropriately visualized to the user and/or embedded on a particular application (e.g. a CAD-system).

Summary

The quintessence of the performed product data selection is an interactive and step-by-step creation of a search object. At each step, the search object is dynamically classified by LOOM according to the knowledge base and inherits all the necessary attributes of the matched classes. This cycle of user-altered requirements and automatically inherited attributes is one of the powerful aspects of the navigation by dynamic classification. Another important result of our research is that the matching of the product data with the search object is carried out by the dynamic classification of LOOM. This is instead of using a data base query with possibly insufficient attributes of the products, i.e. attributes which are not provided by the product supplier are deduced automatically from the knowledge base, as a side effect of the dynamic classification process.

References

[1] SCHERER, R.: *A Product Information System with an adaptive Classification Structure*, GEN'97 Symposium on Global Design Engineering Networking; Antwerpen, Belgien, 1997.

[2] MACGREGOR, R.: *Using a Description Classifier to Enhance Deductive Inference*, Proceedings. of the 7th IEEE, Conference on AI Applications, 1991.

[3] HAKIM, M.: *Modelling evolving information about engineering design products*, PhD thesis, Department of Civil engineering if Carnegie Mellon University, 1993.

[4] BRACHA, G.; COOK, W.: *Mixin-based inheritance*, Proceedings OOPSLA/ECOOP'90, Band 25 der ACM SIGPLAN Notices, pp. 303-311, 1990.

[5] DELCAMBRE, L. M. L.; LIM, B. B. L.; URBAN, S. D.: *Object-centered constraints*, Proceedings IEEE International Conference on Data Engineering, Kobe, Japan, p. 368ff, 1991.

[6] BRILL, D.: *LOOM - Reference Manual*, University of Southern California-School of Engineering, Information Sciences Institute, 1993.

Structural Monitoring: Decision-Support through Multiple Data Interpretations

Ruth Stalker and Ian Smith

Institute of Structural Engineering and Mechanics (ISS-IMAC)
EPFL-Federal Institute of Technology
CH-1015 Lausanne
Switzerland

Abstract. Decision-support for monitoring is performed using models in order to provide multiple interpretations of the same data set. This results in a space of possible interpretations of structural behaviour. Incremental addition of information for each interpretation modifies this space and helps engineers converge on realistic behaviours. Such experimentation and exploration of data interpretation leads to more rational decision-making in structural maintenance and life-cycle economies.

1 Introduction

An important stage in structural monitoring is the determination of structural behaviour through data analysis. Appropriate decision-making maintenance only leads to reductions in life-cycle costs and cost-effective repair strategies when data interpretation is performed accurately. However, in [1] it has been observed that, large quantities of data can be so overwhelming, crude methods of analysis are employed for data reduction. Hence, more sophisticated methods have been investigated. Neural nets have been used in a pattern-recognition approach in order to identify damage in structures [2]. Other applications use them only for data retrieval in order to support other data analysis methods [3]. Traditional knowledge-based systems (KBS) have been used and shown to be too inflexible for large structures [4]. Model-based reasoning (MBR) has been coupled with KBS as discussed in [5] in order to increase KBS flexibility, (see overviews [6]). Models are normally used to represent devices or structures [7, 8] where underlying assumptions can be changed. However, structural analysis performed through modelling alone can be unsatisfactory because, in the absence of measurements, theoretical calculations, due to conservation assumptions, result in stress values that are several times greater than measured values [9]. Until now, analyses of measurement data most often lead to only single interpretations. This is unjustified since an interpretation is only as accurate as the model used, and a model is only an approximation. Therefore, in this paper, an MBR approach for reducing measurement data using several models is presented. The following section illustrates multiple data interpretations and their advantages. Section 3 presents a decision-support framework for multiple data interpretations and Section 4 draws the conclusion that multiple data interpretations are imperative for more rational decision-making.

2 An Example of Multiple Data Interpretations

Figure 1 is an illustration of multiple data interpretations of bending deformation measurements. These measurements are taken from SOFO, a monitoring system based on fibre optic sensors [10]. SOFO measures local deformations, and with sufficient sensors, information is extrapolated for the whole structure. Each bending deformation measurement can be treated as a data point which is plotted against a curve representing the expected deformation [11]. In other words the data is reduced using a model. These curves, however, are only approximations. MBR is only as good as the model itself [6]. In the figure, the left hand side shows two beams, one with variable moment of inertia and a crack, the other with constant moment of inertia and a crack, both under non-uniform loading. The middle figures assume stiff supports whereas the figures on the right hand side have supports that may settle and rotate. It can be seen that these assumptions lead to very different deflection curves. The data itself may indicate which curve (model) is the correct one to choose. Thus, it is better to multiply interpret the data in order to converge on the best model for better maintenance decisions.

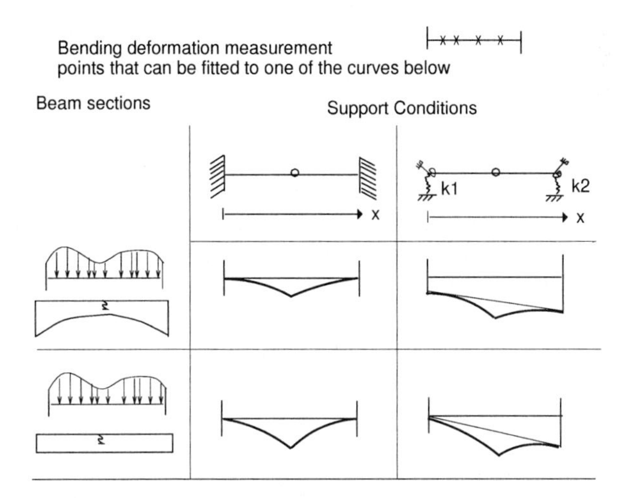

Fig. 1. *Deflected shapes for a range of beam sections and support conditions. Depending on these assumptions, data from the beam is interpreted in many ways. Hence, multiple data interpretations may lead to unanticipated discoveries about the structural behaviour of these beams*

3 Decision-Support for Structural Monitoring

Through multiple data interpretations, information about the size of the space of data interpretations is shown prior to decision making. This is known as active decision-support [12]. Active support requires human-computer interaction to allow an engineer to interact with a system in order to add information which enhances a given model. In Figure 2 it can be seen how data is interpreted and re-interpreted until a suitable interpretation is decided upon. This interpretation can be used for better maintenance decision-making for improved life-cycle economies. Until now, decision-support in engineering has been limited almost exclusively to critiquing [13].

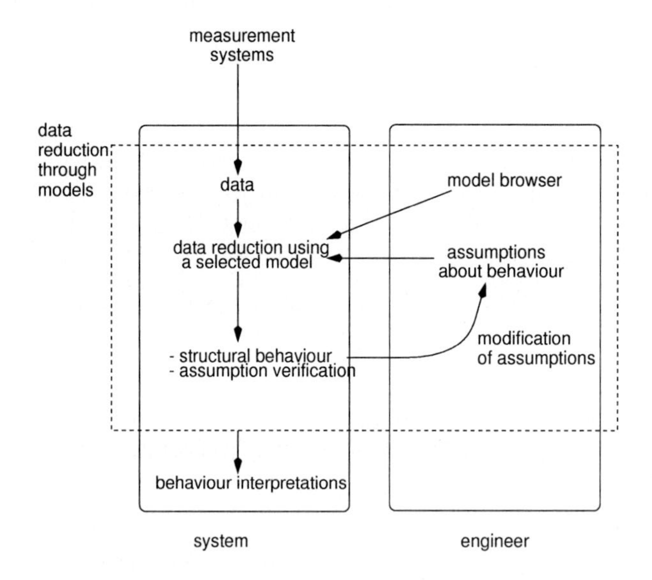

Fig. 2. *Multiple data interpretations are performed iteratively through choosing relevant models. Each model represents different assumptions. Engineers choose and revise assumptions in order to achieve better interpretations*

4 Conclusions and Further Work

Multiple interpretations of data through the application of several models, which represent assumptions, may lead to discoveries about structural behaviour which otherwise may not have been made. This can only be done through systems offering the exploration and exploitation of data for decision-support. One crucial factor in the successful applications of these systems is human-computer interaction with advanced visualisation techniques. This is in order to encourage engineers to use such systems. Consequently, work in the area of visualisation has already been undertaken and will continue in order to support such decision making (see [14]). Results will be implemented in a decision-support framework in order to provide justifications for life-cycle economies such as structural life extension and more cost-effective repair and retrofit strategies, and more appropriate functional modifications.

Acknowledgements

The authors would like to thank the Swiss Commission for Technology and Innovation (Project CTI-3290.1) , Logitech SA, Silicon Graphics Incorporated for supporting this work, Palais de Beaulieu for their collaboration, and Denis Clement, Branko Glisic, Benny Raphael, Kristi Shea, and Samuel Vurpillot for discussions.

References

1. J.P. Davis and A.M. Vann.: "Monitoring Instrumentation Fault Diagnosis and Data Interpretation" Artificial Intelligence in Engineering (1995) 923–938
2. P.Kirkegaard and A Rytter.: "Vibration Based Damage Assessment of Civil Engineering Structures using Neural Networks" Proceedings of the 1st Workshop of the European Group for Structural Engineering Application of Artificial Intelligence(EG-SEA-AI) (1994) 90–103
3. Luisito Brembilla, Marco Lazzari, and Paolo Salvaneschi.: "Neural Associative Memories for Detecting Analogies: an Application to Structural Safety Management" Proceedings of the 2nd Workshop of the European Group for Structural Engineering Application of Artificial Intelligence(EG-SEA-AI) (1995) 93–100
4. A. Goodier and S. Matthews.: "Knowledge Based System applied to Real-Time Structural Monitoring" Information Processing in Civil and Structural Engineering Design Civil-Comp Press (1996) 263–270
5. Boi Faltings.: "Reasoning Strategies for Engineering Problems" Knowledge-Based Systems in Civil Engineering, IABSE Colloquium Beijing (1993) 79–87
6. Randall Davis and Walter Hamscher.: "Exploring Artificial Intelligence: Survey Talks from the National Conferences on Artificial Intelligence" Chapter 8: Model-based Reasoning: Troubleshooting Ed.Howard E. Schrobe (1988) 297–346
7. George M. Turkiyyah and Steven J. Fenves.: "Knowledge-Based Assistance for Finite-Element Modelling" AI in Civil and Structural Engineering (1996) 23–32
8. Paolo Salvaneschi, Mauro Cadei, and Maroc Lazzari.: "Applying AI to Structural Safety Monitoring and Evaluation",AI in Civil and Structural Engineering IEEE Expert, Intelligent Systems and their Applications (1996) 24–34
9. R.A.P. Sweeney, G. Oommen, and L. Hoat.: "Impact of Site Measurements on the Evaluation of Steel Railway Bridges" IABSE Workshop (1997) 139–147
10. Daniele Inaudi.: "PhD Thesis Fiber Optic Sensor Network for the Monitoring of Civil Engineering Structures" (1997)
11. S.Vurpillot, D. Inaudi, and A. Scano.: "Mathematical model for the determination of the vertical displacement from internal horizontal measurements of a bridge" Smart Structures and Materials SPIE Vol. 2719-05 (1996)
12. Ian F.C. Smith.: "Abductive Engineering Support" ICCCBE VII, Korea (1997)

13. Markus Stolze.: "Visual critiquing in domain oriented design environments; showing the right thing at the right place" Artificial Intelligence in Design '94 Kluwer Academic Publishers (1994) 467-482
14. Ruth Stalker and Ian Smith.: "Structural Monitoring Support through Augmented Reality" International Meeting on Artificial Intelligence in Structural Engineering Centro Stefano Franscini, Monte Verita Ascona (1998)

Augmented Reality Applications to Structural Monitoring

Ruth Stalker and Ian Smith

Institute of Structural Engineering and Mechanics (ISS-IMAC)
EPFL-Federal Institute of Technology
CH-1015 Lausanne
Switzerland

Abstract. Support for complex monitoring tasks can be provided through the application of Augmented Reality(AR) techniques. This is done through on-site applications as well as subsequent image enhancement, which are used to visualise structural monitoring systems and measured structural behaviour. This paper discusses appropriate applications that have much potential for improving the performance of maintenance engineers.

1 Introduction

Fig. 1. *Visualisation of structural behaviour, predicted or actual, is a powerful complement to mathematical and graphical representations*

Structural monitoring research currently focuses on the interpretation of measurement data [1, 2]. Visualisation and presentation of the results remains a secondary consideration. Figure 1 illustrates how an image is a powerful complement to other representations, particularly if it is sufficiently accurate. Augmented Reality (AR) produces such images. It combines real representations (pictures) with virtual representations (CAD). It enhances a users perception of, and interaction with, the real world by displaying information that cannot be directly detected with a users own senses. AR has been applied to domains such as telemedicine training, devices for the disabled, robot path planning and military applications (see [3] for an AR overview). Research programs are underway in architectural anatomy and design intent [4] where an "x-ray vision" approach enables the visualisation of reinforcing bars in columns. Thus, the anatomy (reinforcing bar) of a structure (column) is shown. Work by Klinker [5] in the domain of building site management has identified various alignment issues between the real and the virtual. However, no application of AR to monitoring of existing structures has been identified.

This paper, therefore, discusses the potential for applying augmented reality during on-site structural monitoring and subsequent image enhancement in order to facilitate structural monitoring tasks.

2 Augmented Reality for Structural Monitoring on-site

Mann in [6] describes a head-mounted display (HMD) which enables the visualisation of information to aid the memory, by the insertion of virtual information into the visual stream. A HMD has the potential to be useful for on-site structural monitoring because correlating plans with full-scale structures is not easy. Plans and sections can be digitised, indexed and super-imposed on the HMD as well as prestressing cables and measurement systems. Through combination with a knowledge-based system as suggested by Feiner *et al* in [7], diagnosis and repair annotations can be displayed thus allowing for more productive on-site inspection. Relevant plans and behaviours, for example deformations that are measured and used, in order to calculate the deflection of a structure as in [2] can be requested by an inspector and new observations relayed back to the KBS. This visualisation of real behaviours is important as it is often very different from theoretical behaviour [8].

Open HMDs - HMDs designed in particular for AR - can be grouped into two categories: optical see-through and video see-through. However, video see-through HMDs are potentially dangerous, for a power failure leaves the wearer blind. Optical see-through HMDs need a head tracker for positioning [9] or the wearer must perform several orientation tasks in order to establish position [10]. Thus, tracking an AR system outdoors in real time with the required accuracy remains an open problem [3]. Consequently, much work has been undertaken for image enhancement but again these efforts are not directly applicable to the domain of structural monitoring.

3 Image Enhancement for Structural Monitoring Support Visualisation

Image enhancement is performed using two approaches. Firstly, a video of a structure is taken and then combined with CAD images. The second approach involves mounting cameras on the structure and combining "live" images with

CAD images [3]. Such image enhancement has the potential to support remote monitoring which is performed via the web [13]. Photogrammetry techniques accurately calibrate the 2D (virtual) and 3D (real). This process has been semi-automated and added to a feature extraction process in order to build up detailed models of structures from crude CAD models [11]. Hence, models of structures can be quickly constructed and extrapolated before applying them to enhance images. These models can contain positions of measurement systems.

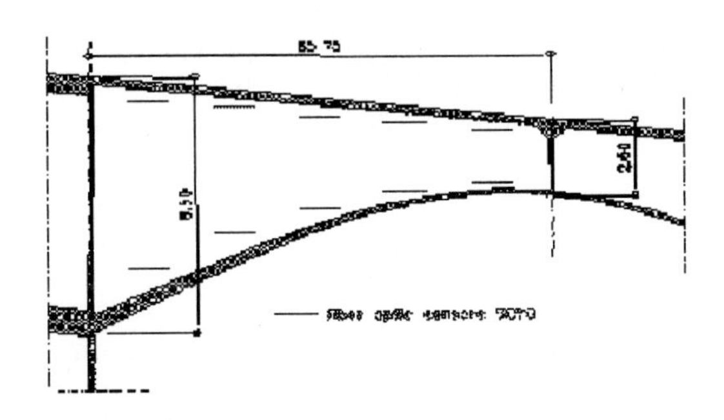

Fig. 2. *The top figure show typical bridge drawings of captors that have been placed in it for deformation measurement. The bottom figure shows an AR images where the captors and the drawing have been superimposed*

SOFO [12] is a monitoring system based on fibre optic sensors which measures local deformations, and with sufficient sensors, information is extrapolated for the entire structure. SOFO captors are often embedded in concrete. Current diagrammatic representations such as those shown in Figure 2 are enhanced in order to aid comprehension of their precise location. This approach becomes much more useful when these visualisations are linked up to a knowledge-based system as discussed in the previous section. In this instance, data and the behaviour extrapolated from captor data can be retrieved by using the AR image as an interface and information retrieval is done by clicking on a captor. This is much simpler than querying a database. Figure 2 shows captors on images of a structure from differing views. When a view of the structure is shown, the relevant captors are superimposed. When the footage is taken from mounted cameras, extra information concerning the environment which affects structural behaviour and, which cannot be ascertained from looking at plans, is also given. Thus, it can be seen that, image enhancement is useful for proposals related to repairs, maintenance and further inspection. Timely reaction to the potential problems identified reduces future maintenance efforts and thus lowers life-cycle costs.

4 Conclusions

Careful employment of AR techniques will increase effectiveness, and reduce complexity of, structural monitoring tasks in order to perform effective maintenance decision making, thereby lowering life-cycle costs.

Acknowledgements

The authors would like to thank the Swiss Commission for Technology and Innovation (Project CTI-3290.1), Logitech SA, and Silicon Graphics Incorporated for supporting this work and Samuel Vurpillot, Daniele Inaudi for discussion.

References

1. Paolo Salvaneschi, Mauro Cadei, and Maroc Lazzari.: "Applying AI to Structural Safety Monitoring and Evaluation",AI in Civil and Structural Engineering IEEE Expert, Intelligent Systems and their Applications (1996) 24–34
2. S.Vurpillot, D. Inaudi, and A. Scano.: "Mathematical model for the determination of the vertical displacement from internal horizontal measurements of a bridge" Smart Structures and Materials SPIE Vol. 2719-05 (1996)
3. Ronald T. Azuma.: " A Survey of Augmented Reality" Presence: Teleoperators and Virtual Environments 6,4 August (1997)
4. Anthony Webster, Steven Feiner, Blair MacIntyre, William Massie, and Theodore Krueger.: " Augmented reality in architectural construction, inspection and renovation" Proc. ASCE Third Congress on Computing in Civil Engineering 913–919
5. Gudrun Klinker.: "Augmented Reality" http://www.ecrc.de/staff/gudrun/ar/ar.html (1997)
6. Steve Mann.: "Wearable Computing : A First Step towards Personal Imaging" IEEE Computer (1997) 25–31
7. Steven Feiner, Blair MacIntyre, and Dore Seligmann.: "Knowledge Based Augmented Reality" Communications of the ACM 36, 7 53–62

8. R.A.P. Sweeney, G. Oommen, and L. Hoat.: "Impact of Site Measurements on the Evaluation of Steel Railway Bridges" IABSE Workshop (1997) 139–147
9. Andrei State, Gentaro Hirota, David T. Chen, William F. Garrett, and Mark A. Livingston.: "Superior Augmented Reality Registration by Integrating Landmark Tracking and Magnetic Tracking" COMPUTER GRAPHICS Proceedings, Annual Conference Series (1996) 429–438
10. Ronald Azuma.: " Improving Static and Dynamic Registration in a See-Through HMD" Proceedings of SIGGRAPH'94, Computer Graphics Annual Conference Series (1994) 197–204
11. Streilin95 and U Hirschberg .: "Integration of digital Photogrammetry and CAAD: Constraint-based Modelling and Semi-automatic Measurement" CAAD Futures'95 International Conference (1995)
12. Daniele Inaudi.: "PhD Thesis Fiber Optic Sensor Network for the Monitoring of Civil Engineering Structures" (1997)
13. S.S Chen and C.M.Ballard.: "Retrofit Design Verification Via In-Service Monitoring over the Internet: A Case Study" Information Processing in Civil and Structural Engineering Design (1996) 183–187

Analysis and Design of the As-Built Model

W. Tizani

Department of Civil Engineering, University of Nottingham, Nottingham, UK
walid.tizani@nottingham.ac.uk

Abstract. This paper describes an integrated system that enables designers to work with the virtual model of the steel structural frame. It provides methods for modelling the connections therefore ensuring fidelity with the expected behaviour of the frame in practice. The behaviour of the joints is based on the strength, stiffness and ductility of connections rather than simply assumed to be 'pinned' or 'rigid'. The paper outlines a research programme that has the above as one of its objectives and gives an example of its application.

1 Introduction

In the structural engineering of buildings, conceptual design, frame analysis and member sizing, taking full account of the consequences that the decisions taken at these stages have on all the processes that follow, hold the key to a more streamlined production process. However, the above stages are not adequately supported considering what can now be delivered using today's information technology techniques.

The analysis and design tools, available to designers, are still too closely based on stick models and do not support the creation of analysis and design assumptions that are compatible with the physical model nor the creative conceptual design process which involves experimentation with various layouts and frame behaviour systems, linked to economic appraisal. The member/connection interface is a much-ignored aspect of the analysis and design - despite its substantial influence on cost. There are many benefits in modelling more accurately such an interface, including the opportunity to use more sophisticated design approaches, leading to greater economy and improved safety in matching the design assumptions with actual frame behaviour. Thus, for the current way of working and when using the available tools, attempting to generate and appraise a variety of feasible solutions where economy and practicality of production are taken into account is not really possible. The consequence is that designers are forced to work with a limited subset of the options that might be available to them and do not make the most of the many advances in information technology, structural engineering understanding and manufacturing techniques that are available.

2 A Research Programme for the Design of Virtual Structures

A research programme which is aimed at providing a more intuitive way of working that closely matches reality and offers measurable benefits with implications on the total production process is underway. One of the objectives of this project is:

- To conduct in an integrated environment the processes of analysis, design and detailing using models that are compatible with the as-built building prototype.

The project has the wider aim of supporting designers in carrying out an effective and creative process of conceptual design with the ability to quickly appraise alternative schemes for economy and workability while ensuring such schemes fulfil the required functions. Other objectives include:

- **Economic appraisal of design options:** The provision of an evaluation module and a parametric cost model for supply, fabrication, transport and erection. This will allow for the economic appraisal of design options using relative data and a means for project cost estimating by modelling different production practices.
- **Advice on workable options:** The provision of distributed knowledge bases, dealing with subjects such as frame analysis and design options, connection detailing and standardisation, rationalisation of the global options for the structure.
- **Visualisation:** The option to view the as-built prototype with walk-through.
- **Decision support for integration, concurrency and simultaneity:** Integrate the above in a system that manages inter-dependability and controls the integrity of the overall design. The seamless integration of the above will ensure that all options can be effectively assessed by designers at the conceptual and detailed design processes in an environment facilitating the conduct of what-if and what-consequences scenarios. The system can be driven in such a way as to provide the required decision support at various levels or stages of the design process.

This paper however concentrates on one of its major aspects, that is ensuring the fidelity of the analysis and design models to the as-built product.

3 Current State of Analysis and Design Tools

Traditionally, the processes of analysis and design have been treated as separate. More recently, the two processes can be carried out within a 'linked' environment. This has advanced the process in that data used in the analysis can be used for the design. However, essentially, the two processes are kept separate.

It has long been identified in the research community that the conduct of design and in a wider context the whole construction process should be carried out in an integrated environment. Not simply to share data between the various processes but to ensure consistency between them and facilitate the sharing of information and minimising re-work.

However, from the structural engineer's point of view and apart from the integration aspect, it is essential to ensure compatibility between the analysis and the design models and between the analysis and the structure as it is going to be built. Designers should ensure that the chosen elements can behave within the completed structure in a manner compatible with the assumptions made at the analysis stage.

One major issue of compatibility is that of modelling connections. Often connections are assumed to behave as pins or as rigid. The design of the connections is carried out after the beam and column elements have been chosen. It is not unknown that in many cases, especially in steelwork, fitting a connection behaviour onto prescribed member sizes can be problematic often leading to re-design. For example, depending on the selected member sizes and the level of loading a connection may not be able to resist the level of moment applied on it for it to be classified as rigid.

More recently, the concept for semi-continuous or semi-rigid construction has started to gather acceptance, as this will offer more economy [1]. The economy stems from accounting for the capacity of these connections to resist some of the moment that are normally distributed to the beams. However, using this semi-continuous or semi-rigid approach would require more sophisticated analysis methods. The modelling of the behaviour of these connections in frame design remains a difficult task [2]. A number of models have been suggested, e.g. [3]. The suggested models tends to rely on the designer assuming the behaviour and then impose it onto the frame analysis model.

4 The IFD System

The research programme tackles the above issues by providing a tool to model the properties and behaviour of the virtual structural steelwork elements. An Integrated Frame Design system (IFD) has been implemented using the Object-Orientated Programming (OOP) methodology and includes an OOP inference engine. The inference engine is a general purpose object that has been developed in order to integrate knowledge-based technology within an environment (namely C++) capable of demanding computations.

The system 'fully' integrates the processes of analysis of the frame and the design of members and connections by using a single data model. It accounts for the behaviour of the virtual connections for the purpose of the analysis. The connection behaviour is modelled using an element having stiffness relative to that of the connecting structural members. The stiffness of the connection is determined using a set of rules that considers the ductility of the connection and its strength relative to that of the connecting members. The moment capacity (strength) of each connection is calculated to identify whether it should be modelled as a pin (moment capacity equal to zero) or as moment-resisting (moment capacity greater than zero). A moment-resisting connection can be modelled as a plastic hinge with a known capacity (if it exhibits the required ductility and stiffness and allows for the plastic distribution of forces between its components), as a semi-rigid joint (if it exhibit some stiffness but cannot develop the full plastic moment capacity), or as a rigid joint if it exhibits infinite stiffness. The stiffness matrix of the whole frame including the connection elements is assembled and the stiffness matrix method is used for the analysis. In the above, KBS techniques have been used in judging what relative stiffness value should be used for the connection and OOP techniques have been essential in the adequate representation of this relatively complex problem.

KBS techniques have further been used in the generation of 'best-practice' member sizes and connection detailing. In using the system the user specifies total structural frame data (members, connections and supports). Complete multi-storey frames can be generated with initial sizes for beams and columns determined based on rules of thumb taking into account level of loading and span lengths [4]. The system will also generate 'best practice' standard connections that ensure adequate detailing fulfilling both structural and fabrication requirements [5,6].

The extent of the contribution of this work to the structural analysis and design of steel structures will be better evaluated once the on-going validation process has been completed. It is clear, however, that integration of the design processes and, from the outset, accounting for actual detailing in modelling connection behaviour in a 'practical' environment is a step change from attempting to force-fit behaviour and details onto prescribed assumptions.

The Next section gives an illustrative example describing the approach used in the system for the design of a 'portalised' steelwork frame.

5 An illustrative example

A simplified example is provided here to illustrate the aspect of the IFD system that deals with modelling analysis and design for compatibility with the as-built frame.

Figure 1 shows the layout of a two storeys portal frame. Typically, connections at the eaves and the apex (Nodes 7, 8 and 9) will be assumed to be moment-resisting connections. The remaining of the connections will be assumed to be shear-only resisting connections.

If the 'conventional' methods of analysis and design were used, Nodes 7, 8 and 9 will be assumed rigid and the remaining pinned. A cycle of analysis and design will then be carried out leading to sizing of the columns and beams (or rafters) members based on the calculated forces as a result of these assumptions. Next, the connections are designed to resist the forces at the nodes. In many countries (including the U.K.) the step of designing the connections is carried out by a different organisation than that that carried out the sizing of members. The connection detailing is restricted by having to utilise the sizes selected at the frame analysis and design stage. In many instances, the assumptions made at the nodes (i.e. fully rigid and pinned connections) cannot be accommodated.

An alternative process for the complete design (members and connections) can be carried out using the IFD System. The approach used in the system is that the designer works with the as-built structural components rather than idealised elements and assumed behaviour. A default connection type (i.e. simple, e.g. flexible end plate, or moment, e.g. flush end-plate and extended end-plate) can be selected to be applied at all joints. Variations on the default type can then be specified for specific joints. The properties of the nodes at the joints are taken to be that of the detailed connections (unless the designer chooses to over-ride this).

At this stage of the process, the designer is dealing with the virtual structural frame and no assumptions as to the behaviour of the nodes or supports have been made. The designer can analyse the 'as-built' frame. As described earlier this is carried out by modelling the connections using dummy elements having stiffnesses compatible with the properties of the connections. Figure 1 shows the bending moment diagram obtained from applying this analysis option. Noting the bending moment at the eaves and apex, where a moment-resisting connection have been used, and at the first floor level, where shear-only resisting connections have been used, this Figure shows how this diagram takes account of the behaviours of the connections.

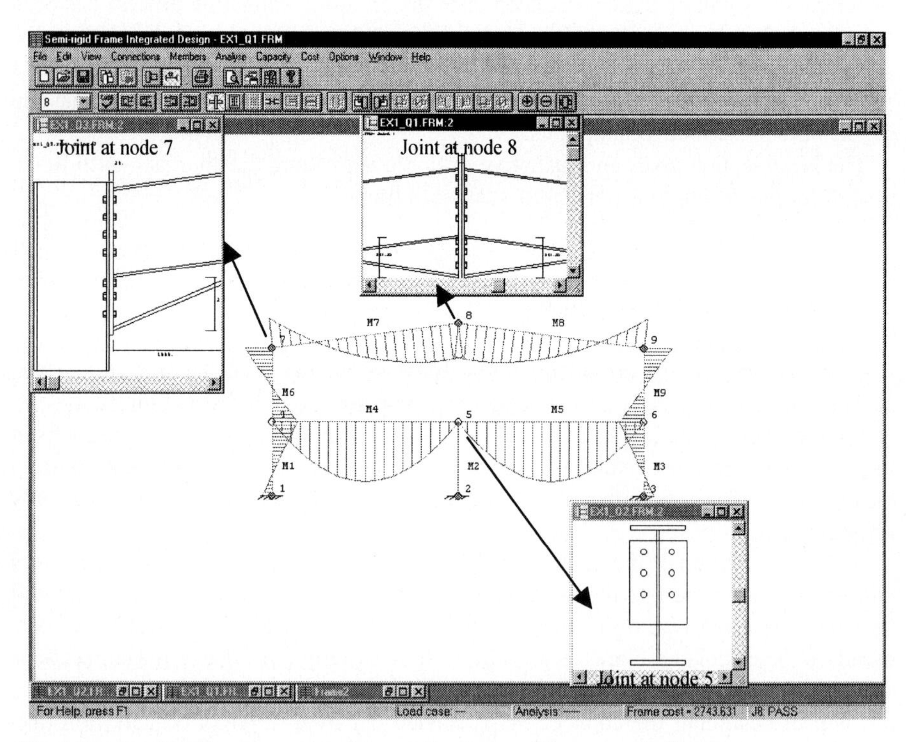

Fig. 1. Layout, bending moment and connection types of the example-frame

6 Future enhancement

The project has the further aim of allowing for the modelling of three-dimensional frames that take into account the global connectivity and stability options and allow for the local focus on sub-assemblies such as 2D frames. During the assessment of the potential benefits of the system it was realised that a number of limitations exists and that the 'total' three-dimensional frame should be considered. This is because many of the 2D elements interacts with others in the third dimension and assumptions has to be made as to their effects on the 2-D behaviour of these (mainly dealing with restraints and members that are connected-to from both directions). Other reasons for this effort is to allow for the cost appraisal of the 'total' building cost.

7 Conclusions

This paper described an integrated system for the analysis and design of two-dimensional steel frames using a method that allows for compatibility of the assumptions made at these stages with the as-built frame. This is done by incorporating the detailing and design of connections within the frame and by modelling the behaviour of these connections in a manner that is compatible with their strength, stiffness and ductility.

The system enables the designer to work with the virtual model of the structural frame and therefore ensuring fidelity with the actual behaviour of the frame in practice. This will also allow the designer to exploit the strength available in the structure with implication on economy.

In addition to the above aspect, the system integrates an economic appraisal model and a knowledge-based system for the advice on remedial actions for element failure. However, these aspects are outside the scope of this paper.

The development of the system is part of a continuing project that has the further aim of modelling three-dimensional frames, that take into account the global connectivity and stability options, and allows for the local focus on sub-assemblies such as 2D frames.

References

1. Nethercot, D. A. and Zandonini, R.: Beam-to-columns connections in steel frames: Discussion. Canadian Journal of Civil Engineering. 15 (1987) 282-284
2. Morris, G.A., Huang, J and Scerbo, M.: Accounting for Connection Behavior in Steel Frame Design. Canadian Journal of Civil Engineering. **22** (1995) 955-969
3. CEN.: Eurocode 3: Design of Steel Structures. European Community for Standardisation (1992)
4. Boys, B.W.J.: Initial Sizing and Design of Steel Members in Buildings. British Steel Structural Advisory Service. British Steel, UK (1991)
5. BCSA and SCI.: Joints in Simple Construction - Volume 1: Design Methods. BCSA/SCI, Publication 205. The Steel Construction Institute, UK (1993)
6. BCSA and SCI.: Joints in Steel Construction - Moment Connections. BCSA/SCI, Publication 207/95. The Steel Construction Institute, UK (1995)

On Theoretical Backgrounds of CAD

Ziga Turk

Asst.Prof., Faculty of Civil and Geodetic Engineering, University of Ljubljana, Slovenia.
IKPIR-FGG, Jamova 2, Ljubljana, Slovenia; email: ziga.turk@fagg.uni-lj.si

Abstract. In the past, some information technologies (IT) have quickly been adopted by the eng i-neering practice while the implementation of others has been slower. In the paper, the author looks for a solid theoretical background that could be used to explain this difference and warn about po s-sible obstacles in applying some technologies to civil and structural engineering. The prevailing theoretical background for the contemporary development in computer aided design is the ratio n-alistic philosophical tradition, which claims that intelligent human behavior is based on mental manipulation of symbolic representations of the real world; that problem solving is a search in a space of potential solutions and that communication is exchange of information. This tradition has been challenged by a theory referred to as "hermeneutic constructivism", that claims that the r a-tionalistic premises are wrong and suggests an alternative philosophical background for the use of computers. The author confronts the two theories and discusses their impact on the use of inform a-tion technology in construction. He finds that the new theory can clarify some of the problems we face in current CAD research, as well justify some optimistic expectations of the approaches that avoid complex representations of data, activities and knowledge.

1 Introduction

The use of computers in structural engineering has largely depended on the developments in the field of information technology (IT). Technology-push and client-pull are forming a spiral where market demand and technological development reinforce improvements in each other [1]. In the past, a few breakthrough works and underlying technologies were identified [2,3] that have opened up new directions in the structural IT research. Since the mid 1990s, the communication, collaboration and coordination technologies, invented in the 1960s, became broadly available, most evidently in the form of the Internet and the World Wide Web. The research community embraced the new technology and started to explore the possibilities it might offer to in the domain of construction and structural engineering. There has been a lot of enthusiasm and relatively rapid penetration into the practice. In comparison, the influence of the expert systems has been small and standard product models developed in STEP and IAI take longer then expected to be developed and used. Working in both of the mentioned areas of IT in structural engineering, the author examines what causes this difference.

1.1 Philosophical Background

Since the mid 1950s, the cognitive science has been trying to explain human, animal and machine cognition and intelligence (overview in [4]). Researchers in this field include pioneer computer scientists (Turing, Von Neumann), leading AI figures (Wizenbaum, Misky, Simon) and linguists (Chomsky). Cognitive science had a profound impact on the way we think about computer programs. It suggests that intelligent systems (humans, animals and computers) achieve their intelligence by manipulating symbols of real world items; that symbol-manipulating processes are similar in all such systems. Cognitive science postulates that because computers are symbol-manipulating systems too, they can achieve intelligence as well, and moreover, computerized simulation is believed to be an excellent tool for the insight into human cognitive processes. Philosophically, cognitive science is based on the rationalistic tradition that can be traced back to the ancient Greece.

The traditional understanding of cognition, language and intelligence has been challenged by Martin Heidegger and other philosophers (Gadamer), linguists (Austin, Searle), and neurobiologists

(Maturana). Winograd and Flores [5] summarized the effects that these non-traditional ideas should have on our understanding of the role of computers. They suggested a new approach to the design of software that stresses the aspects of human-to-human and human-to-machine interactions. Winograd [6] calls it "hermeneutic constructivism".

1.2 About the paper

In the paper we present their analysis, particularly the problems of representation and problem solving that are a central issue both in the product and process modeling and to AI in structural engineering. Discussion highlights some implications that hermeneutic constructivism has on some current directions of construction IT and outlines some sample applications that are based on technologies that build on the theories of hermeneutic constructivism.

2 Representation

Rationalistic tradition believes that our ability to intelligently act in the world around us is due to the mental images or representations of the real world that we have in our brains. It claims that we think in terms of objects and their properties that are (more or less faithful) models of objective reality.

Example: Lets consider a statement, "*this beam is cracked*". The terms "beam" and attribute "cracked" apply to some objectively present beam that objectively is cracked. Knowledge is storage of such representations. When we need to think about a problem, we retrieve a relevant representation (or frame of representations) and adapt it to the current situation. This tradition can be traced back to Aristotle, who introduced a three-way distinction between *words*, "*experiences in the psyche*", and *things*.

Early in this century a similar schema was proposed by Ogden and Richards [7] in the form of the Ogden meaning triangle that connects words, concepts and things (Fig. 1). The *object* is any entity from some real or imagined world about which an idea is held, for example the beam in the Nada Ward Museum of Modern Art in Kobe damaged by an earthquake [8]. The *concept* is the idea or thought of the object as held in the mind of a person, for example a structural concept of a beam as illustrated. Another likely illustration of the "beam" concept would have been an EXPRESS diagram of the data required to describe a beam). The *symbol* is an auditory, visual, or other form of utterance that is taken to stand for the object when communicated as part of a language.

Such understanding of our thought processes had a profound impact on computer programming. Writing computer programs means teaching the computer about the concepts we have about the world. The symbols used in the computer (zeros and ones) will be different from what humans use to symbol-

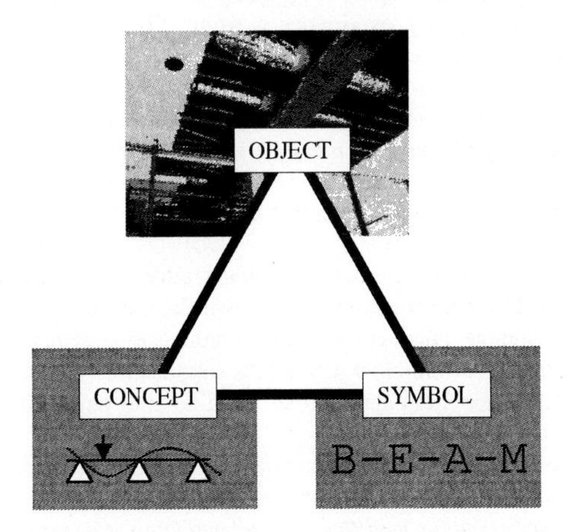

Fig. 1 Ogden's meaning triangle.

ize the concepts (sounds, words, graphical notation) but, so the theory claims, this is not important as long as the basic mechanisms are preserved. Using programs is mapping real world objects into computer's symbols and the interpretation of the results reverses this process. Such approach is also the foundation for contemporary database design and standards such as ISO-STEP[9] and ISO-IRDS[10]. It is the foundation for the product modeling approach to computer integrated construction.

The essence of Heidegger's and Maturana's philosophy is that knowledge and understanding do not result from formal operations on mental representations of an objectively existing world. Things do not have properties independent of the interpretation. *"Heidegger insists that it is meaningless to talk about the existence of objects and their properties in the absence of concernfull activity with the potential of breaking down"* ([6] pg. 37).

In their view, a statement "*this beam is cracked*" does not stand for some objective reality, but gets its meaning only if and when we know the context or background of when this statement was voiced. The speaker knew that the audience expects a technical evaluation of a nearby structural element. On a finest level every beam has some microscopic cracks so an objective statement that a beam has cracks is pointless. Statements like this make sense if the speaker can assume how the audience will interpret the term "cracked". The main ideas of hermeneutic constructivism related to the representation and "information exchange" are:

- Language is not an exchange of information. It is not a medium for description but one of action. The statement about cracks in the beam in fact engages the audience to do something about the cracks. It creates a commitment network, between the speaker and the hearer, to do something about the cracks.
- Language is a form of social behavior that creates a "mutual orientation". By talking to each other we establish common views and ideas. A consensual view on the world around us is not based on what that world objectively is, but what the speaker and the audience, collectively, consider it to be. Structural engineering audience does not expect the word beam to mean that of light.
- Meaning of a statement is created at the receiving end and is not independent of it. Meaning is created within commitment networks.
- Mutual orientation - understanding of a message - can only be achieved if the speaker and the audience share mutual concern, and a common background. The unspoken is at least as important as the spoken part of a message.

2.1 Product Models are Representations

Conceptual modeling approach to computer integrated construction, whether product model or process model based, depends heavily on the rationalistic approach. In fact it assumes that conceptual models for the building industry should be built and often it looks at the objective reality as a reference for the definition of these models. Many developers of building product models experienced that models are difficult to agree upon. Author's recent experience is from ESPRIT-ToCEE project [11]. This can be explained by the fact that we, humans, do not know about objective reality and the developers of the models do not share the same orientation and concern. Conceptual models are difficult to prove correct, because we do not have an objective reference (like reality) to measure them against. Models are even more difficult to prove wrong - because the meaning is not within the model but is given to it by the receiver -a flexible and intelligent human being. This is quite different to mechanical or mathematical models, where, for example, a theoretical model of concrete-mix cementation can be verified against the empiric tests in the laboratory.

Not all kinds of models are affected equally. Product models that are being developed without a concrete application in mind (outside a commitment network and out of context) are more likely to be affected. On the other hand, models that are limited to synthetic or well confined worlds, like structural mechanics, or to data exchange between few pre-selected applications and not directly to the real world do not suffer as much. First successful applications of IT to civil engineering were designed in such domains.

Winograd and Flores claim that *"the most successful* (software) *designs are not those that try to full model the domain in which they operate, but hose that are 'in-alignment', with the fundamental structure of that domain, and that allow for modification and evolution to generate new structural coupling"* ([5], pg. 53).

3 Problem solving

Traditional problem solving would claim that where there is a problem, there is a set of possible solutions. Problem solving is the process that finds one, possibly the best of all solutions. Graphically this has been presented in the format of the Venn diagrams which would show (as in Fig. 2) the set of potential solutions and subsets of feasible, candidate, or constraint satisfying solutions [12]. Similar solution spaces for architectural have also been proposed by Coyne [13], and consist of generative and interpretive design space. Problem solving is a search process (arrow) that can be more or less intelligent in terms of how smart the search arrow, from initial state to satisfying the goal, behaves.

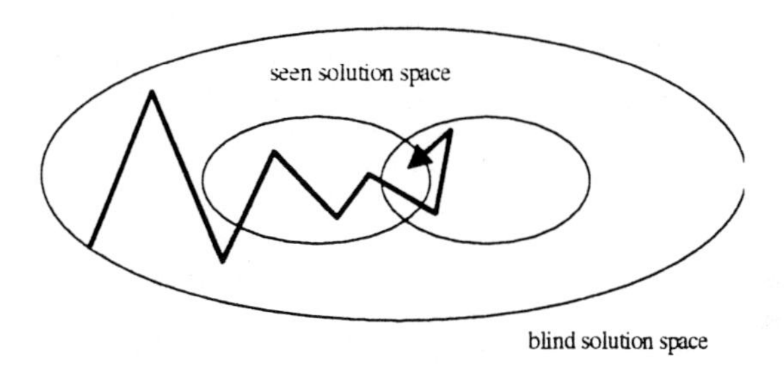

Fig. 2 Problem solving as a search (arrow) in the solution space.

Typical examples of such problem solving in architecture is the design of floor layouts where certain rules govern the topology of the rooms, e.g. you do not enter the sleeping room through the kitchen. There are perhaps thousands of the possible combinations of the floor layout, but using rules or grammars one can extract the required solutions. Similarly, given the shape of the valley, "intelligent" design systems can generate several different concepts for the bridge, and, using rules, select the best solution.

Programs that do problem solving are created so that a model of the task environment is created, then it is formally represented as a database and populated with the data of a specific problem. A search procedure is implemented that, for example using generate and test strategy, finds plausible solutions.

The critique of this understanding of problem solving is based on Heidegger's belief that the basis for our everyday action is the ability to act pre-reflectively when "thrown" into a situation. He claims, contrary to the tradition that would consider reflective analysis of a detached observer as the basic intelligent behavior, that reflective thought about objects and properties is derived from pre-conscious experience of them as "ready-to-hand". The essence of intelligence is "throwness", not reflection. To hammer a nail, he argues, we do not require conscious reflective knowledge about the physical properties of a hammer and the physics of hammering. The tool is *ready-to-hand* and we just hammer the nail into the wall. Similar pre-reflective actions can also explain the so-called "intuition", "insight" or plain "common sense", that are sometimes used by the designers or engineers to explain their creative process. A good idea just pops out and is not a result of a systematic search in the set of potential solution. "*The essence of intelligence is to act appropriately when there is no simple pre-definition of the problem or the space of states in which to search for a solution*" ([5], pg. 98).

The very instant we begin to think reflectively about a situation and analyze it in terms of object and properties, we disconnect ourselves from the *being-in-the-world*, we try not to be *thrown* into a situation. We limit our view of the problem to that, which can be expressed by the objects and properties we have adopted, and create *blindness* for all other possible solutions. For example in the expert system to select a bridge type, only the structural designs built into the expert system can be "designed". Blindness is best demonstrated by parametric design where the set of potential solution is the number of permutations of different parameter values. This creates blindness for all other kinds of possible designs. Predefined conceptual product models create blindness for designs that cannot be expressed using the objects and relationships of the conceptual product model. The fact is a bit more obscure, but nevertheless still present, when designs are generated by combinations or through structure evolution.

However, computers again can be efficient in closed, formally defined domains. Problems suitable for AI application in engineering are such where there is narrow domain of expertise, limited language to express facts and relations, limiting assumptions about problem and solution methods, little knowledge of own scope and limitations solving problems [14]. The game of chess is such systematic, formally defined space; detailed engineering design can be similar to such closed domains. On the other hand, conceptual design and particularly architectural design, is quite from it.

4 Conclusions and Discussion

Hermeneutic constructivism would explain that the relatively slower adoption of expert system, product modeling and other "representation centered" approaches is caused by (1) some inherent philosophical problems of representing the real world in the computer and (2) the blindness that any representation causes. The theory suggests to concentrate on software, that is not necessarily intelligent, but is "ready-to-hand" and on software that supports the commitment networks among humans. It explains the limitations of software that depends heavily on representations of complex domains (like that of civil engineering) and of domains that lack a natural or synthetic reality (like IT in civil and structural engineering). In this section we discuss some positive and some negative examples.

Islands of automation. In the history of construction information technology we have witnessed the application of several bottom up approaches to the use of computing technology to help engineers in what they do. These approaches were aimed at assisting in finding a solution in fairly formalized, limited and discrete areas. Most such problems appear in the detailed design where the so-called "islands of automation" appeared. The application of information technology has been successful. Application of IT to structural analysis, for example, has relied on mathematical modeling of physical behavior of structures. Representation could be verified against the "nature" itself. Another successful application of IT have been the computer-aided drafting packages, such as the AutoCAD. Hermeutic constructivism would explain the popularity of generic drafting packages that deal with geometric entities such as lines with their generality, simple underlying models and therefore smaller space of "blindness" when compared with design packages that deal with engineering entities, such as walls.

Product model based CIC. The research of computer integrated construction (CIC) has been looking for a top-down solution that would connect the islands of automation. It seemed that building product models are a suitable glue technology for the task. But as argued in section 2, engineers do not think in terms of objective reality that we would like to model in product models.

The problems that we (e.g. [11]) have with product modeling include (1) the difficulties in which the conceptual product models are being formalized by the experts in the field and (2) the incompleteness of the models, particularly in the areas of unconventional and creative design.

Expert systems. Similarly, the main difficulties with the expert systems are believed to be (1) the difficulty to extract expert knowledge for the experts and (2) the problem of blindness that is created by excluding large portions of the background knowledge.

One of the pitfalls of expert system development has been the apparent gap between the success of the research prototypes and the acceptance in the practice. Expert systems performed well in small isolated domains. So do product models. It was believed that for the real world applications one just adds some more rules, a richer representation and more background knowledge. And it is believed that for practical use on should just build more complete product models. However, in neither case this solves the fundamental problem – that of blindness that is created when we define representations of any level of sophistication.

Neural networks. In contrast to expert systems, the neural networks are almost representation-less. Should a further study prove that such systems been more successful in engineering, hermeneutic constructivism explains why.

Document management systems. The difficulties with the product model based solutions to CIC have been acknowledged by the search for shorter-term solutions such as document management. We have found out, that in combination with some rather simple agent-oriented technology quite useful solutions emerge [15]. Documentary models are an order of magnitude (or more) simpler than the product models. They are also much more generic and not specific to civil and structural engineering.

Process and activity based CIC. One of the reasons to consider product models as the core of computer integrated construction was a belief that construction processes and practices are too different to be standardized. This approach is being revisited [16, 17] because processes appear simpler than products and more intimately involve humans and their interrelationships.

Collaboration, coordination and workflow. This technology even personally descends from the inventors of hermeneutic constructivism. It has been applied successfully in other businesses. Its application in the area of civil and structural engineering has started in the mid 1990s [18,19,20]. It focuses in facilitating the human to human and human to machine interactions and streamlines the flow of work between the actors involved in the design and construction processes.

The above list is not exhaustive neither is the argumentation backed by quantifiable empirical studies. It has been argued that CAD related research is quite hard to evaluate objectively [21]. Comparisons of different approaches could prove even more difficult. Informally, the theory can be backed by the success of weak AI techniques, minimal generic product models and the implementation of coordination and collaboration tools in construction practice.

Although hermeneutic constructivism can be understood as if it questions theoretical foundations of the mainstream of building CAD research, the results achieved prove the validity of those approaches as well. A fresh perspective and a solid theoretical background, however, can help us to understand the limitations and, by combining a thesis and anti-thesis into a synthesis, tackle some of the problems in a way that combines the best of both worlds.

References

1 Brandon, P., Betts. M.: Veni, Vidi, Vici: in Brandon P and Betts M (eds), The Armathwaite Initiative, ISBN 1-900491-03-6, University of Salford, UK. (1997), 76-79

2 Fenves, S.J.: Information technologies in construction: a personal journey, in Z. Turk (ed), Construction on the information highway, ISBN 961-6167-11-1, CIB Publication 198, University of Ljubljana, Slovenia, (1996) 13-19

3 Gierson, D.E.: Information Technology in Civil and Structural Engineering Design: Taking Stock to 1996, in B. Kumar, (ed), Information Processing in Civil and Structural Engineering Design, Civil-Comp Press, (1996) 7-16.

4 Gardner, H.E. (1987). The Mind's New Science, Basic Books, USA, ISBN 0-465-04635-5.

5 Winograd, T. and Flores, F.: Understanding Computers and Cognition, Addison Wesley (first published by Ablex Corporation in 1986). (1997)

6 Winograd, T.: Thinking Machines: Can There Be? Are We?, Informatica, Vol. 19 No.4, November (1995) 443-460

7 Ogden, C. K., Richards, I. A.: The Meaning of Meaning, Harcourt Brace Jovanovich, New York, (1989) (First published 1923).

8 Fischiner, M., Cerovsek, T., Turk, Z.: http://www.ikpir.fgg.uni-lj.si/easy/

9 ISO-STEP. ISO 10303, Industrial automation systems and integration -- Product data representation and e x-change (1994)

10 ISO-IRDS: ISO/IEC 10728, Information technology -- Information Resource Dictionary System (IRDS) Ser v-ices Interface (1993)

11 Turk, Z., Wasserfuhr R., Katranuschkov P., Amor R., Hannus, M., Scherer, R.J. Conceptual Modelling of a Concurrent Engineering Environment, in C.J. Anumba, N.F.O. Evbuomwan (eds), Concurrent Engineering in Construction, Institution of Civil Engineers, London, UK, ISBN 1 874266 35 2, (1997) 195-205

12 Galle, P.: Computer Methods in Architectural Problem Solving: Critique and Proposals, Journal of Arcitectural Planning and Research, Vol. 6., No. 1 (1989)

13 Coyne, R.: Logic Models of Design, Pitman Publishing, London, UK (1988)

14 Buchanan, B.: New Research on Expert Systems, in Hayes, J.E., Michie, D., Pao, Y., Machine Intelligence 10, Ellis Horwood, (1982), 269-299

15 Turk, Z.: Software Agents and Robots in Construction: An Outlook, in R. Drogemuller (ed), Information Tec h-nology Support for Construction Process Reengineering, CIB proceedings, Publication 208, James Cook Un i-versity, Queensland, Australia, (1997), 375-388

16 Luiten G., Froese T., Björk B-C., Cooper G., Junge R., Karstila K., Oxman R. (1993). *An information reference model for architecture, engineering and construction*, in Mathur K.S., Betts M.P., Tham K.W. (eds) Manag e-ment of information technology for construction, World Scientific Publishing, Singapore

17 Turk, Z., Björk, B-C, Karstilla, K.: Towards a Generic Process Model for AEC, Extended Abstract, Intern a-tional Congress on Computing in Civil Engineering, Boston, MA, USA, Oct. 18-21 (1998)
18 Hussein, K., Feniosky, Pena-Mora, Sriram, R.D.: Cairo: A System for Facilitating Communication in a Distri b-uted Collaborative Engineering, Proceedings of the Fourth Workshop on Enabling Technologies: Infrastructure for Collaborative Enterprises, IEEE Computer Society, April 20-22, (1995)
19 Frucher, R..: Conceptual, Collaborative Building Design Through Shared Graphics, IEE, June, (1996) 33-41
20 Maher, M.L., Saad, M.: The experience of virtual design studios at The University of Sydney, ANNZAScA Conference, University of Canberra (1995).
21 Clayton M. J., Fischer M. A., Teicholz P., Kunz J. C.: The Charrette Testing Method for CAD Research, A p-plied Research in Architecture and Planning Vol. 2, ed. Robert Hershberger and Mary Kihl. Tucson, AZ, He r-berger Center for Design Excellence. (1997)

Author Index

Lecture Notes in Artificial Intelligence (LNAI)

Vol. 1314: S. Muggleton (Ed.), Inductive Logic Programming. Proceedings, 1996. VIII, 397 pages. 1997.

Vol. 1316: M.Li, A. Maruoka (Eds.), Algorithmic Learning Theory. Proceedings, 1997. XI, 461 pages. 1997.

Vol. 1317: M. Leman (Ed.), Music, Gestalt, and Computing. IX, 524 pages. 1997.

Vol. 1319: E. Plaza, R. Benjamins (Eds.), Knowledge Acquisition, Modelling and Management. Proceedings, 1997. XI, 389 pages. 1997.

Vol. 1321: M. Lenzerini (Ed.), AI*IA 97: Advances in Artificial Intelligence. Proceedings, 1997. XII, 459 pages. 1997.

Vol. 1323: E. Costa, A. Cardoso (Eds.), Progress in Artificial Intelligence. Proceedings, 1997. XIV, 393 pages. 1997.

Vol. 1325: Z.W. Raś, A. Skowron (Eds.), Foundations of Intelligent Systems. Proceedings, 1997. XI, 630 pages. 1997.

Vol. 1328: C. Retoré (Ed.), Logical Aspects of Computational Linguistics. Proceedings, 1996. VIII, 435 pages. 1997.

Vol. 1342: A. Sattar (Ed.), Advanced Topics in Artificial Intelligence. Proceedings, 1997. XVIII, 516 pages. 1997.

Vol. 1348: S. Steel, R. Alami (Eds.), Recent Advances in AI Planning. Proceedings, 1997, IX, 454 pages. 1997.

Vol. 1359: G. Antoniou, A.K. Ghose, M. Truszczyński (Eds.), Learning and Reasoning with Complex Representations. Proceedings, 1996. X, 283 pages. 1998.

Vol. 1360: D. Wang (Ed.), Automated Deduction in Geometry. Proceedings, 1996. VII, 235 pages. 1998.

Vol. 1365: M.P. Singh, A. Rao, M.J. Wooldridge (Eds.), Intelligent Agents IV. Proceedings, 1997. XII, 351 pages. 1998.

Vol. 1371: I. Wachsmuth, M. Fröhlich (Eds.), Gesture and Sign Language in Human-Computer Interaction. Proceedings, 1997. XI, 309 pages. 1998.

Vol. 1374: H. Bunt, R.-J. Beun, T. Borghuis (Eds.), Multimodal Human-Computer Communication. VIII, 345 pages. 1998.

Vol. 1387: C. Lee Giles, M. Gori (Eds.), Adaptive Processing of Sequences and Data Structures. Proceedings, 1997. XII, 434 pages. 1998.

Vol. 1394: X. Wu, R. Kotagiri, K.B. Korb (Eds.), Research and Development in Knowledge Discovery and Data Mining. Proceedings, 1998. XVI, 424 pages. 1998.

Vol. 1395: H. Kitano (Ed.), RoboCup-97: Robot Soccer World Cup I. XIV, 520 pages. 1998.

Vol. 1397: H. de Swart (Ed.), Automated Reasoning with Analytic Tableaux and Related Methods. Proceedings, 1998. X, 325 pages. 1998.

Vol. 1398: C. Nédellec, C. Rouveirol (Eds.), Machine Learning: ECML-98. Proceedings, 1998. XII, 420 pages. 1998.

Vol. 1400: M. Lenz, B. Bartsch-Spörl, H.-D. Burkhard, S. Wess (Eds.), Case-Based Reasoning Technology. XVIII, 405 pages. 1998.

Vol. 1404: C. Freksa, C. Habel. K.F. Wender (Eds.), Spatial Cognition. VIII, 491 pages. 1998.

Vol. 1409: T. Schaub, The Automation of Reasoning with Incomplete Information. XI, 159 pages. 1998.

Vol. 1415: J. Mira, A.P. del Pobil, M. Ali (Eds.), Methodology and Tools in Knowledge-Based Systems. Vol. I. Proceedings, 1998. XXIV, 887 pages. 1998.

Vol. 1416: A.P. del Pobil, J. Mira, M. Ali (Eds.), Tasks and Methods in Applied Artificial Intelligence. Vol. II. Proceedings, 1998. XXIII, 943 pages. 1998.

Vol. 1418: R. Mercer, E. Neufeld (Eds.), Advances in Artificial Intelligence. Proceedings, 1998. XII, 467 pages. 1998.

Vol. 1421: C. Kirchner, H. Kirchner (Eds.), Automated Deduction – CADE-15. Proceedings, 1998. XIV, 443 pages. 1998.

Vol. 1424: L. Polkowski, A. Skowron (Eds.), Rough Sets and Current Trends in Computing. Proceedings, 1998. XIII, 626 pages. 1998.

Vol. 1433: V. Honavar, G. Slutzki (Eds.), Grammatical Inference. Proceedings, 1998. X, 271 pages. 1998.

Vol. 1434: J.-C. Heudin (Ed.), Virtual Worlds. Proceedings, 1998. XII, 412 pages. 1998.

Vol. 1435: M. Klusch, G. Weiß (Eds.), Cooperative Information Agents II. Proceedings, 1998. IX, 307 pages. 1998.

Vol. 1437: S. Albayrak, F.J. Garijo (Eds.), Intelligent Agents for Telecommunication Applications. Proceedings, 1998. XII, 251 pages. 1998.

Vol. 1441: W. Wobcke, M. Pagnucco, C. Zhang (Eds.), Agents and Multi-Agent Systems. Proceedings, 1997. XII, 241 pages. 1998.

Vol. 1446: D. Page (Ed.), Inductive Logic Programming. Proceedings, 1998. VIII, 301 pages. 1998.

Vol. 1453: M.-L. Mugnier, M. Chein (Eds.), Conceptual Structures: Theory, Tools and Applications. Proceedings, 1998. XIII, 439 pages. 1998.

Vol. 1454: I. Smith (Ed.), Artificial Intelligence in Structural Engineering.XI, 497 pages. 1998.

Vol. 1456: A. Drogoul, M. Tambe, T. Fukuda (Eds.), Collective Robotics. Proceedings, 1998. VII, 161 pages. 1998.

Vol. 1458: V.O. Mittal, H.A. Yanco, J. Aronis, R. Simpson (Eds.), Assistive Technology in Artificial Intelligence. X, 273 pages. 1998.

Lecture Notes in Computer Science